Introduction to Actuarial and Financial Mathematical Methods

Companion website: http://booksite.elsevier.com/9780128037379/

Introduction to Actuarial and Financial Mathematical Methods

S. J. GARRETT

AMSTERDAM • BOSTON • HEIDELBERG • LONDON
NEW YORK • OXFORD • PARIS • SAN DIEGO
SAN FRANCISCO • SINGAPORE • SYDNEY • TOKYO
Academic Press is an imprint of Elsevier

Academic Press is an imprint of Elsevier
125 London Wall, London, EC2Y 5AS, UK
525 B Street, Suite 1800, San Diego, CA 92101-4495, USA
225 Wyman Street, Waltham, MA 02451, USA
The Boulevard, Langford Lane, Kidlington, Oxford OX5 1GB, UK

Library of Congress Cataloging-in-Publication Data
A catalog record for this book is available from the Library of Congress

British Library Cataloguing in Publication Data
A catalogue record for this book is available from the British Library

For information on all Academic Press publications
visit our website at http://store.elsevier.com/

Printed and bound in the United States

ISBN: 978-0-12-800156-1

Publisher: Nikki Levy
Acquisition Editor: Scott Bentley
Editorial Project Manager: Susan Ikeda
Production Project Manager: Jason Mitchell
Designer: Mark Rogers

To Yvette, for everything past, present, and future.

CONTENTS

PREFACE

Mathematics is a huge subject of fundamental importance to our individual lives and the collective progress we make in shaping the modern world. The importance of mathematics is recognized in formal education systems around the world and indeed how we choose to teach our own children prior to their formal schooling. For example, from a very early age children are taught their native language in parallel to the fundamentals of working with numbers: children learn the names of objects *and* also how to count these objects.

Despite claims to the contrary, most adults do have considerable mathematical knowledge and an intuitive understanding of numbers. Irrespective of their educational choices and natural ability, most people can count and understand simple arithmetical operations. For example, most know how to check their receipt at the supermarket; that is, they understand the fundamental concepts of addition and subtraction, even if they prefer to use a calculator to perform the actual arithmetic. As a further example, given a distance to travel, most people would intuitively know how to calculate an approximate time of arrival from an estimate of their average speed.

Given that mathematics is so engrained in our childhood and used in our everyday adult lives, any book on the practical use of mathematics must begin by drawing a line that separates the material that is assumed as prerequisite and that which the book wishes to develop. The correct place to draw this line is difficult to determine and must, of course, depend on the intended audience of the book. This particular book, as the title suggests, is intended for people who ultimately wish to study and apply mathematics in the highly technical areas of actuarial science and finance. It is therefore assumed that the reader has a prior interest in mathematics that has manifest in some kind of formal mathematical study to, say, the high-school level at least. It is at the level of high-school mathematics that the line is drawn for this book.

It may be that high school was a long time ago and the mathematics learned there has since left you. For this reason I begin by softening what could be a sharp line. The first chapter on preliminary concepts summarizes some relevant mathematical terminology that you are likely to have seen before. Chapters 2–6 then proceed to discuss mathematical concepts and methods that you may also have been familiar with at some point, possibly at high school or maybe during the early months of an undergraduate program in a numerate subject. The material in Chapters 1–7 forms Part I which is intended to give you the foundation for the more technical Part II.

A number of chapters close with a brief section on an example use of the ideas developed in that chapter within actuarial science. While this book does not aim to cover topics in actuarial science in any detail, these sections are included where a taste of a topic can be given without also providing a significant amount of background material. Not all chapters include such a section and this reflects either that the application would be too esoteric or that sufficient "real-world" examples have already been included in the main material of the chapter.

As you may be aware, mathematics can be an extremely formal and rigorous subject. While such rigor is essential for the development of new mathematics and its application to novel areas, there are many instances where formal rigor is a hindrance and a distraction from the real application and purpose of using the mathematics. Rather than over complicating the descriptions with excessive technical considerations, the aim of this book is to present a concise account of the application of mathematical methods that may be required when studying for actuarial examinations under the *Institute and Faculty of Actuaries* (IFoA) or the *Society of Actuaries* (SoA) in the UK and the USA, respectively. The book should also be of use for those studying under the *CFA Institute*, for example, and many other professional bodies related to finance professionals. The book does not give any formal proofs of the concepts used, although some attempt will be made to justify many of the ideas. After studying this book the reader should expect to possess a well-stocked tool box of mathematical concepts, a practical understanding of when and how to use each tool, and an intuitive understanding of why the tools work.

The scope of the material discussed in this book has been heavily influenced by the statements of prerequisite knowledge for commencing studies with the IFoA and SoA. Certainly the book should be considered as covering all prerequisite material required for beginning studies with the IFoA and SoA. However, I have gone further and included some additional topics that, in my experience, students from diverse academic backgrounds have found useful to refresh during their early studies of actuarial science and financial mathematics at the postgraduate level.

I am grateful for the many discussions regarding the content of this book with numerous students on the various actuarial and financial mathematics programs at the University of Leicester, particularly Marco De Virgillis. I would also like to thank Dr Jacqueline Butter who provided the additional perspective of someone who has gone through an actuarial program from a background in physics and entered industry on the other side.

Writing a book is a lonely task and I would like thank my two sons, Adam and Matthew, and my wife, Yvette, for giving me the time and space to impose this loneliness on myself. This book is dedicated to Yvette who is an unfailing supporter of everything I do.

Professor Stephen Garrett
Leicestershire, UK
Spring 2015

PART I

Fundamental Mathematics

CHAPTER 1

Mathematical Language

Contents

Prerequisite knowledge

- "School" mathematics
  - use of a calculator
  - algebraic manipulation
  - analytical solution of simple polynomial expressions
- Familiarity with basic use of Excel

Learning objectives

- Define, recognize, and use
  - number systems
  - mathematical notation including set notation
  - bracket notation
  - quantifiers
  - equations, identities, and inequalities

In this chapter, we state and illustrate the use of common mathematical notation that will be used without further comment throughout this book. It is assumed that much of this section will have been familiar to you at some point of your education and is included as an *aide-mémoire*. Of course, given that the book will explore many areas of the application of mathematics, the material presented here may well prove to be incomplete. It should therefore be considered as an illustration of the level of mathematics that will be assumed as prerequisite, rather than a definitive list.

1.1 COMMON MATHEMATICAL NOTATION

1.1.1 Number systems

We begin by summarizing the types of numbers that exist. As this book in concerned with the practical application of mathematics, it should be unsurprising that the set of

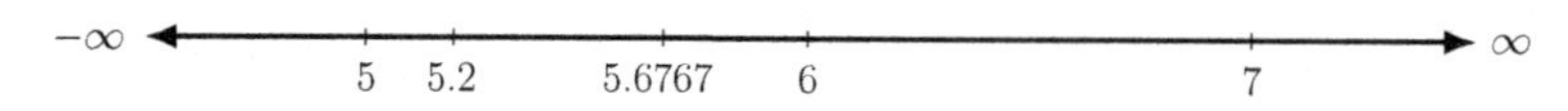

Figure 1.1 The real number line.

real numbers forms the building blocks of most (but not quite all, see Chapter 8) of what we will study.

A real number is a value that represents a position along a continuous number line. For example, numbers 5 and 6 have clear positions on the number line in Figure 1.1 and so are real numbers. The number 5.2 also has a position on the number line, a fifth between 5 and 6. Going further we see that 5.6767 is also on the line. In fact, we can keep going and, with a sharp enough pencil, mark a number with any number of decimal places on the number line. With this intuitive understanding, it should be clear that the set of all real numbers includes numbers to any number of decimal places and that we can also freely expand the number line without limit. As illustrated in Figure 1.1, real numbers can be positive or negative. The set of all real numbers, denoted $\mathbb{R}$, is therefore seen as the fundamental collection of numbers that we might want to work with in real-world applications.

As we can in principle define a real number with an infinite number of decimal places, there is in some sense an "infinity of infinities" of real numbers. It should then be of no surprise that the set $\mathbb{R}$ has many subsets, each with an infinite number of members. Such subsets include

- positive real numbers, $\mathbb{R}^+$
- negative real numbers, $\mathbb{R}^-$
- integers, $\mathbb{Z}$
- natural numbers, $\mathbb{N}$
- rational numbers, $\mathbb{Q}$
- irrational numbers, $\mathbb{J}$

The meaning of the terms *positive real numbers* and *negative real numbers* should be clear, although note that 0 is technically neither. You may however need to be reminded that the *integers* are the subset of real numbers that are "whole." For example, 0, -10, and 34 are integers, but -10.1 and 34.8 are not.

The *natural numbers* are easily understood as the positive integers and zero.[1] For example, 57 and -6 are both integers, but only 57 is a natural number. Natural numbers

[1] Note that there is some disagreement as to whether zero is a natural number. Some authors claim that it does not belong to the set of natural numbers, instead is a member of an additional set called the *whole numbers* which are the positive integers and zero.

are useful for counting and are the first number system we work with as children. It will prove useful to define $\mathbb{N}^+$ as the nonzero natural numbers.

In addition to the sets of whole and natural numbers, a *rational number* is any real number that can be expressed as the fraction of two integers. It should be clear that the set of integers are also rational numbers, for example, $32 = 32/1$ and $-7 = -7/1$, but so are numbers like $45/2$ and $-98,736/345,298$.

In contrast, *irrational numbers* are those which cannot be represented as a fraction of two integers. Irrational numbers are numbers which have an infinite number of decimal places, for example, π, e, and $\sqrt{2}$. Irrational numbers cannot therefore be integers or natural numbers.

The relationship between the different sets of real numbers is summarized in Figure 1.2. From this it is clear that the "sum" of the sets of rational and irrational numbers form the broader set of real numbers. The set of rational numbers can be further subdivided into integers and nonintegers; the set of integers contains the natural numbers.

EXAMPLE 1.1

Where would 0 appear in the Venn diagram of Figure 1.2?

Solution

According to the definitions given here, zero is a real number, a rational number, an integer, and a natural number. It will then sit inside of the circle indicated by $\mathbb{N}$. However, other authors claim that it is not a natural number and so sits inside of the circle indicated by $\mathbb{Z}$ but outside of $\mathbb{N}$.

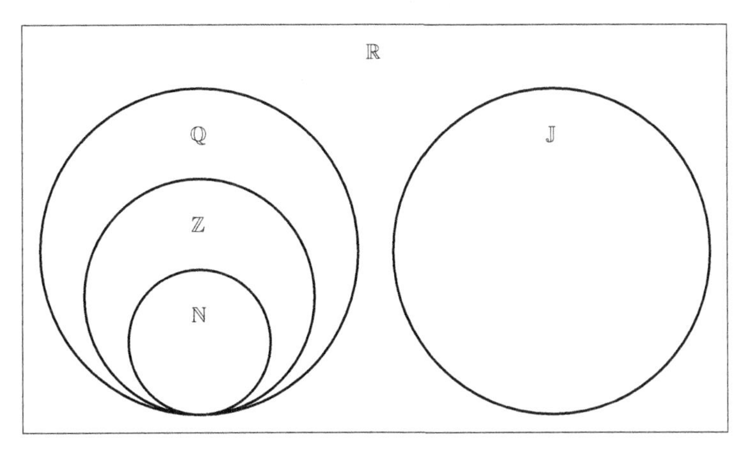

Figure 1.2 Venn diagram of the real number systems.

EXAMPLE 1.2

Give three examples for each of the following number systems.

a. $\mathbb{R}^+$
b. $\mathbb{Z}$
c. $\mathbb{N}$
d. $\mathbb{Q}$
e. $\mathbb{J}$

Solution

a. $\mathbb{R}^+$ is the set of *positive real numbers*. Examples could be 0.0001, 3.2, and 100.
b. $\mathbb{Z}$ is the set of *integers*. Examples could be -10, 0, and 35.
c. $\mathbb{N}$ is the set of *natural numbers*. Examples could be 1, 7, and 92.
d. $\mathbb{Q}$ is the set of *rational numbers*. Examples could be -2, 5/6, and 9,883/3.
e. $\mathbb{J}$ is the set of *irrational numbers*. Examples could be $\sqrt{3}$, $\sqrt{100} - \pi$, and $^3\sqrt{4}$.

1.1.2 Mathematical symbols

In addition to the symbols used to denote the different number systems, mathematics is full of notation. At this stage, it is useful to list the most common items of notation that you should be able to identify, this done in Table 1.1.

Note that the list of basic notation in Table 1.1 identifies two symbols for "is approximately equal to," $\approx$, and $\simeq$, and this prompts discussion of our first mathematical subtly: The symbol $\approx$ is commonly used to reflect that in practical situations we are often forced to report approximate values of exact values. For example, the mathematical constant e is an irrational number and so has a numerical value with an infinite number of decimal places

Table 1.1 Basic mathematical notation

$=$	is equal to	$\equiv$	is identical to
$\approx$	is approximately equal to	$\simeq$	is approximately equal to
$>$	is greater than	$<$	is less than
$\geq$	is greater than or equal to	$\leq$	is less than or equal to
$\Rightarrow$	implies that	$\Leftarrow$	is implied by
$\Leftrightarrow$	implies and is implied by	$\rightarrow$	goes to
: or |	such that	$\ldots$	and continues
!	factorial	$\|x\|$	the modulus of x, i.e., $\|x\| = \begin{cases} = x \text{ for } x \geq 0 \\ = -x \text{ for } x < 0 \end{cases}$

Note that a strike through indicates its negation, e.g., $\neq$ denotes "is not equal to."

$$e = 2.71828182845904523536028747135266249775724709369995\ldots$$

The practical use of e therefore requires one to truncate this to a manageable number of decimal places, say three or four. This truncation is an approximation of the actual value and we write

$$e \approx 2.7183$$

Similarly, the mathematical constant π is an irrational number with value

$$\pi = 3.14159265358979323846264338327950288419716939937510\ldots$$

In practice one might use

$$\pi \approx 3.1416$$

In contrast, the symbol $\simeq$ is used when a mathematical method is used explicitly to generate an approximation. For example, as we shall see in Chapter 5, it is often possible to develop a method to approximate the value of an equation. In this case, one would acknowledge that the method is not intended to deliver the *exact* numerical value by using the symbol $\simeq$. The practical use of this symbol can be seen in Chapters 5 and 13, and in particular Eq. (5.9), for example.

The other items of notation in Table 1.1 are assumed to be self explanatory.

EXAMPLE 1.3

Interpret the following mathematical statements in words and give two examples in each case. You should work in the set of real numbers, $\mathbb{R}$.

a. $y > 5.4$
b. $z \leq 10$
c. $x + 2 > 4$
d. $y = x$ and $y = z \Leftrightarrow x = z$
e. $x \approx \frac{1}{2}$
f. $|q| = 7$

Solution

a. y is greater than 5.4. For example, $y = 5.41$ or $y = 6$.
b. z is less than or equal to 10. For example, $z = 10$ or $z = 9.6$.
c. $x + 2$ is greater than 4. For example, $x = 2.1$ or $x = 3$.
d. $y = x$ and $y = z$ implies and is implied by $x = z$. Any identical values of x, y, and z are examples of this.
e. x has approximate value $\frac{1}{2}$. For example, $x = 0.503866281774$ or $x = 0.4986827$.
f. The modulus of q is 7. For example, $q = 7$ or $q = -7$, that is $q = \pm 7$.

1.2 MORE ADVANCED NOTATION

1.2.1 Set notation

Table 1.2 lists the basic items of set notation. We have loosely used the term *set* when discussing the number systems, for example, we have discussed "subsets of the set of real numbers," without a proper definition of what a set actually is.

For all intents and purposes in this book, a set is simply understood as a collection of *distinct* objects. The set of real numbers, $\mathbb{R}$, is interpreted as the collection of all possible real numbers. Any particular real number is *a member* of the set of real numbers, denoted by $\in$. For example, $4.56 \in \mathbb{R}$ is read as "4.56 is a member of the set of real numbers." Any set formed from a collection of particular real numbers is considered to be a *subset*, denoted $\subset$, of real numbers. For example, $\{\pi, 4.56, 456/23\} \subset \mathbb{R}$. From Figure 1.2 it should be clear that

$$\begin{array}{lll} \mathbb{R}^+ \subset \mathbb{R} & \mathbb{R}^- \subset \mathbb{R} & \mathbb{Q} \subset \mathbb{R} \\ \mathbb{J} \subset \mathbb{R} & \mathbb{Z} \subset \mathbb{R} & \mathbb{N} \subset \mathbb{R} \\ \mathbb{Z} \subset \mathbb{Q} & \mathbb{N} \subset \mathbb{Q} & \mathbb{N} \subset \mathbb{Z} \end{array}$$

Consider the sets

$$A = \{-1, 1, 2\} \quad B = \{0, 2, 3\} \quad \text{and} \quad C = \{-3, -2, 4\}. \tag{1.1}$$

It is clear that A and C are subsets of the sets of real numbers, rational numbers, and integers, and B is a subset of real numbers, rational numbers, integers, and natural numbers. None is a subset of the irrational numbers. The mathematical shorthand for these statements would be

$$\begin{array}{lllll} A \subset \mathbb{R} & A \subset \mathbb{Q} & A \subset \mathbb{Z} & A \not\subset \mathbb{N} & A \not\subset \mathbb{J} \\ B \subset \mathbb{R} & B \subset \mathbb{Q} & B \subset \mathbb{Z} & B \subset \mathbb{N} & B \not\subset \mathbb{J} \\ C \subset \mathbb{R} & C \subset \mathbb{Q} & C \subset \mathbb{Z} & C \not\subset \mathbb{N} & C \not\subset \mathbb{J} \end{array}$$

Of course, set theory is not limited to discussing number systems and we can work with sets of any objects. Where appropriate in this book we will use capital letters to refer to sets and lower case letters to refer to a particular *member*, that is *element*, of a set.

Table 1.2 Basic set notation

$\{\}$	a set	$\in$	is a member of
$\emptyset$	empty set	$\subset$	is a subset of
$\cap$	set intersection	$\cup$	set union
$\setminus$	relative complement		

EXAMPLE 1.4

Give all possible values of x that would satisfy the following statements concerning the sets in Eq. (1.1).

a. $x \in A$
b. $x \in A$ and $x \in B$
c. $x \in B$ and $-x \in C$

Solution

a. $x \in A \Rightarrow x = -1, 1,$ or 2.
b. $x \in A$ and $x \in B \Rightarrow x - 2$
c. $x \in B$ and $-x \in C \Rightarrow x = 2$ and 3.

It is possible to form larger sets by "adding" two sets using the *union* operation, ∪. For example,

$$A \cup B = \{-1, 1, 2\} \cup \{0, 2, 3\} = \{-1, 0, 1, 2, 3\}$$

The union operation forms a new set that consists of all members of the original two sets. Note that $2 \in A$ and $2 \in B$ but it is only listed once in the resulting union of A and B. This is because a set is a list of *distinct* elements. The idea of a union can be extended to three or more sets in the obvious way.

$$\begin{aligned} A \cup B \cup C &= \{-1, 1, 2\} \cup \{0, 2, 3\} \cup \{-3, -2, 4\} \\ &= \{-1, 0, 1, 2, 3\} \cup \{-3, -2, 4\} = \{-3, -2, -1, 0, 1, 2, 3, 4\} \end{aligned}$$

Furthermore, we can form the set that consists of the common elements of two sets using the *intersection* notation, ∩. For example,

$$A \cap B = \{-1, 1, 2\} \cap \{0, 2, 3\} = \{2\}$$

A set with no elements is called an *empty set* for obvious reasons, and is denoted by $\emptyset$. For example, since A and C have no common elements

$$A \cap C = \{-1, 1, 2\} \cap \{-3, -2, 4\} = \emptyset$$

The intersection of three or more sets, for example, $A \cap B \cap C$, has an obvious meaning.

In terms of the number system, we can write the following statements with the union and intersection notation

$$\mathbb{Q} \cup \mathbb{J} = \mathbb{R} \quad \mathbb{Q} \cap \mathbb{J} = \emptyset$$

The *complement* of a set can be understood in broad terms as the set of items outside of the set. However, in order to define the items outside of a set, we need to define the space of items that the set exists in. For example, the complement of the set of irrational numbers is the set of all items that are not irrational numbers; without somehow specifying that we actually meant "the complement of the irrational numbers within the set of real numbers," there is nothing stopping us listing cats, dogs, and apples alongside the set of rational numbers as members of the complement! For this reason it is useful to define the *absolute complement* of a set within some broad space of all possible elements Ω, and the *relative complement* of two sets that are both within Ω.

If it is clear that we are concerned only with real numbers, then the space of all possible elements is limited to the set of real numbers and $\Omega = \mathbb{R}$. Now that Ω is defined, we can consider the absolute complement of subsets of Ω. The absolute complement of set A is denoted by either $\bar{A}$ or A^c. For example, in the space of real numbers, $\mathbb{J}^c = \mathbb{Q}$.

In contrast, the relative complement of two sets provides a means of subtracting one set from another, assuming that both sets exist in Ω. In general, for $A, B \subset \Omega$, we define the relative complement of A in B as

$$B \setminus A = \{x \in B : x \notin A\}$$

Using Tables 1.1 and 1.2, we can translate this to words as "the relative complement of A in B are those things, x, in B such that are not in A." Even more simply, it is what remains of set B after having removed those items also in A. The analogue to the subtraction $B - A$ should be clear.

EXAMPLE 1.5

Using the sets in Eq. (1.1), determine the relative complement of A in B.

Solution

The relative complement of $A = \{-1, 1, 2\}$ in $B = \{0, 2, 3\}$ is all the elements in B that are not in A. Therefore,

$$B \setminus A = \{0, 2, 3\} \setminus \{-1, 1, 2\} = \{0, 3\}$$

Back to our motivating example of number systems, we can broaden our space Ω to include both the real and imaginary number systems (see Chapter 8), and define the relative complement of the irrational numbers in the real numbers,

$$\mathbb{R} \setminus \mathbb{J} = \mathbb{Q}$$

EXAMPLE 1.6

Interpret the following mathematical statements in words and give an example in each case. You assume that $\Omega = \mathbb{R}$.

a. $x \in \mathbb{R}$
b. $y \in \mathbb{Z}$
c. $z \in \{0, 1, 2, 3\} \cup \{5, 6\}$
d. $y \in \mathbb{Z} \cap \mathbb{R}^+$
e. $\mathbb{R}^+ \cup \{0\} \cap \mathbb{Z} = \mathbb{N}$
f. $B \setminus B = \emptyset$

Solution

a. x is a real number. For example, $x = 1.53$.
b. y is an integer. For example, $y = 9$.
c. z is a member of set formed from the union of the two sets $\{0, 1, 2, 3\}$ and $\{5, 6\}$, i.e., z is from $\{0, 1, 2, 3, 5, 6\}$. For example, $z = 2$.
d. y is a member of the set formed from the intersection (i.e., overlap) of the integers and positive real numbers. For example, $y = 892$.
e. The intersection of the set of the union positive real numbers and zero with the set of integers is the set of natural numbers. For example, 3 is a positive real number (one can label it on the positive half of the number line), it is an integer, and is also a natural number.
f. The complement of set B within itself is the empty set. That is, there are no elements outside of B than are simultaneously also in B.

The basic set operations discussed here are summarized visually in Figure 1.3.

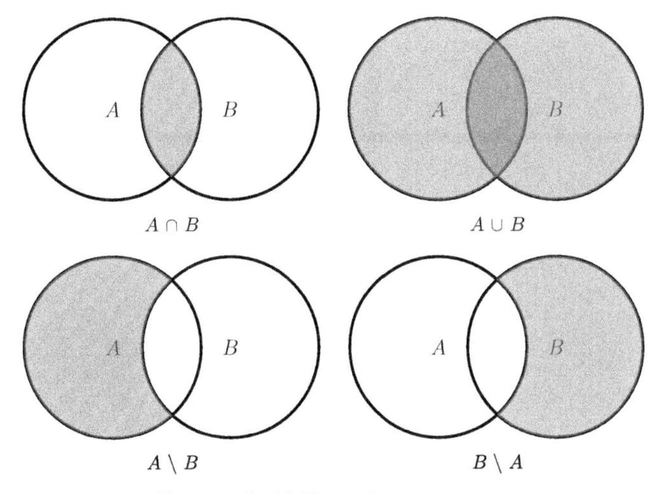

Figure 1.3 Basic set operations illustrated with Venn diagrams.

We now leave aside explicit mention of set theory for the while and return to this in Chapter 9 on probability theory. Unless otherwise stated, you should assume that all mathematical quantities represent *real* numbers in all that follows.

1.2.2 Interval notation

Throughout this book we will make extensive use of *interval bracket notation*. In particular, we will use the following bracket notation

- $[a, b]$ denotes the interval $\{x : a \leq x \leq b\}$
- $[a, b)$ denotes the interval $\{x : a \leq x < b\}$
- $(a, b]$ denotes the interval $\{x : a < x \leq b\}$
- (a, b) denotes the interval $\{x : a < x < b\}$

where the term *interval* can be interpreted as subset of the real number line. Using Tables 1.1 and 1.2 to translate these statements into words, it should be clear that the interval $[a, b]$ is read as "the set of numbers x such that x is between and including a and b." In contrast, the interval (a, b) is read as "the set of numbers x such that x is between but not including a and b." The interpretation of the intervals $[a, b)$ and $(a, b]$ follows in a similar manner. The key point, of course, is that a square bracket denotes an inclusive endpoint of the interval, and a rounded bracket does not.

We refer to an interval that does not include its endpoints as an *open interval*. For example, $(1, 5)$ consists of all numbers x such that $1 < x < 5$ and is open. A *closed interval*, however, does include its endpoints. For example, $[10, 102]$ consists of all numbers x such that $10 \leq x \leq 102$ and is closed.

When working with an endpoint at infinity, a closed interval is meaningless and $\pm\infty$ should appear only next to a rounded bracket: $(-\infty, \infty)$, $(-\infty, b]$, $[a, \infty)$. The interpretation of this is that ∞ is not a number that one can draw on a number line, rather it represents that we can keep on using more and more of the number line without imposing any bound.

EXAMPLE 1.7

Interpret the following mathematical statements in words and give two examples in each case.

a. $x \in [100, \infty)$
b. $y \in (0, 10]$
c. $p \in [0, 1]$
d. $z \in (-9.9, -9.8)$

Solution

a. x is such that $100 \leq x < \infty$. For example, $x = 100$ or $x = 564.3$.
b. y is such that $0 < y \leq 10$. For example, $y = 0.1$ or $y = 10$.
c. p is such that $0 \leq p \leq 1$. For example, $p = 0$ or $p = 1$.
d. z is such that $-9.9 < z < -9.8$. For example, $z = -9.87$ or $z = -9.82$.

1.2.3 Quantifiers and statements

There are two mathematical quantifiers which, when combined with the notation described previously, form a powerful means of writing a wide variety of mathematical statements in a concise way. These are

- $\forall$, read as "for all"
- $\exists$, read as "there exists"

The quantifier $\forall$ is often referred to as the *universal quantifier*, and $\exists$ as the *existential quantifier*. The meaning of both should be immediately apparent, although their power might not be. To hint at the power of the two quantifiers in simplifying statements, we begin with an example:

EXAMPLE 1.8

Demonstrate the intuitive fact that it is possible to find a rational number that approximates the value of π to any finite level of accuracy. Use concise mathematical notation to express that this is true for all real numbers.

Solution

We list the approximations to the value of the irrational number π to an increasing number of decimal places, expressed as a rational number:

$$\begin{aligned}
\pi &\approx 3.14 &&= 157/50 \\
\pi &\approx 3.142 &&= 1571/500 \\
\pi &\approx 3.1416 &&= 3927/1250 \\
&\cdots \\
\pi &\approx 3.1415926536 &&= 3,926,990,817/1,250,000,000 \\
&\cdots
\end{aligned}$$

That this is true for all real numbers (not just the irrational π) is expressed by

$$\forall x \in \mathbb{R}, \quad \forall \epsilon \in \mathbb{R}^+ \quad \exists r \in \mathbb{Q} : |x - r| < \epsilon$$

The mathematical statement given in the solution to Example 1.8 is translated to words as

> *for all x in the set of real numbers and for all ϵ in the set of positive real numbers, there exists r in the set of rational numbers such that the absolute value of the difference between x and r is smaller than the value of ϵ*

Some thought should convince you that this statement is a reflection of our process for approximating π. However, aside from that this is an interesting mathematical fact, the benefits of using the concise mathematical statement formed from the two quantifiers should be immediately apparent.

EXAMPLE 1.9

Translate the following mathematical statements into words. Give a numerical example in each case.

a. $\forall x, y \in \mathbb{R}, x \times y \in \mathbb{R}$

b. $\forall p, q \in \mathbb{R}^-, p \times q \in \mathbb{R}^+$

c. $\exists z \in \mathbb{Z} : z < 7$ and z is odd.

d. $\forall p \in \mathbb{Q}, \exists q \in \mathbb{Q} : 3p = q$

Solution

a. For all x and y in the set of real numbers, the product of x and y is in the set of real numbers. For example, $2.1, 3.2 \in \mathbb{R}$, and $6.72 \in \mathbb{R}$.

b. For all p and q in the set of negative real numbers, the product of p and q is in the set of positive real numbers. For example, $-10.6, -5.3 \in \mathbb{R}^-$, and $56.18 \in \mathbb{R}^+$.

c. There exists z in the set of integers such that z is less then 7 and is odd. For example, $z = 5$.

d. For all p in the set of rational numbers, there exists q in the set of rational numbers such that $3p = q$. For example, $p = 4/3$ and $q = 4$.

1.3 ALGEBRAIC EXPRESSIONS

As we shall see throughout this book, mathematical methods require the manipulation of mathematical expressions. At this stage, it is important to define what we mean by the distinct types of mathematical expressions: *equations*, *identities*, *inequalities*, and *functions*. The distinction between these terms is the topic of this section, and functions will be considered in detail in Chapter 2.

1.3.1 Equations and identities

The key distinction between an equation and an identity is the number of values of the independent variable (x in Eqs. 1.2 and 1.3) for which the expression is true. An identity is true for all values of the independent variable, but this is not true of an equation.

Put another way, for an identity, it is possible to show that the expression on the left-hand side (LHS) of the equal sign is algebraically equal to that on the right-hand side (RHS). This is not true of an equation and one might be required to find the particular values of the independent variable for which the equality between the LHS and RHS holds. In general, an equation could have a finite or infinite number of values of the independent variable for which the equality holds; typically we might refer to these values of the independent variable as the *solutions* of the equation.

Consider the following expression:

$$(x + 4)^2 = x + 10 \tag{1.2}$$

It should be immediately clear that Eq. (1.2) is *not* an identity. In particular, we might note that the LHS is a quadratic expression, that is the highest power of x is 2, and the

RHS is linear, that is the highest power of x is 1. For this reason the behavior of the LHS and the RHS will be very different as x takes different values. We return to a discussion of polynomials in Chapter 2.

It is natural to enquire which values of x satisfy Eq. (1.2); that is, for which values of x does LHS = RHS? It is assumed that you will be familiar with the algebraic manipulations required to solve such equations, however, for completeness, we detail the process below.

$$\begin{aligned}(x+4)^2 &= x+10\\ x^2+8x+16 &= x+10\\ x^2+7x+6 &= 0\end{aligned}$$

and so, by the standard quadratic formula,

$$x = \frac{-7 \pm \sqrt{7^2 - 4 \times 1 \times 6}}{2 \times 1}$$

Equation (1.2) is therefore shown to be true for $x = -6$ and $x = -1$ only. The reader is invited to confirm that this is true.

The solution of general equations is not always as simple in practice and you are likely to be familiar with some analytical approaches to finding the solution to polynomial expressions, for example, the quadratic formula or factorization. However, in many practical instances it may not be possible to find an analytical solution. With this in mind, various numerical approaches to solving equations are discussed in Chapter 13.

Consider now the expression

$$(x+4)^2 = x^2 + 8x + 16 \tag{1.3}$$

In this case, it should be immediately clear that the RHS is algebraically identical to the LHS, and the mathematical statement is true for all values of x. Expression (1.3) is therefore an example of an *identity*, and, using the notation of Table 1.1, it would be correct to write

$$(x+4)^2 \equiv x^2 + 8x + 16$$

It is not always possible to confirm an identity by simple observation and we should consider alternative methods. The most obvious approach is the algebraic manipulation of both the LHS and RHS to confirm that they are indeed identical.

EXAMPLE 1.10

Using an algebraic approach, classify the following expression as an equation or an identity:

$$\frac{y^2+1}{y(2y^2-1)(y-1)} = \frac{1}{y} - \frac{6y+3}{2y^2-1} + \frac{2}{y-1} \tag{1.4}$$

Solution

The answer is not immediately obvious. One should attempt to cast the LHS in terms of a *partial fraction* of the following form:

$$\frac{A}{y} + \frac{By + C}{2y^2 - 1} + \frac{D}{y - 1}$$

with A, B, C, and D unknown constants to be determined. Some manipulation leads to $A = 1$, $B = -6$, $C = -3$, and $D = 2$ which confirms the original expression as an *identity*.

One might consider the algebraic manipulations of complicated expressions to be time consuming and prone to error. In practical situations, an alternative to the algebraic approach could be to plot both sides of an expression over some particular interval of the independent variable that we deem appropriate. In the case of Eq. (1.4), it might be quicker to plot the RHS and the LHS for a reasonable interval in y and compare the result, and this is particularly true if you have access to a computer. The result of such a graphical approach is given in Figure 1.4 and it does appear to be an identity, at least over the values of y considered. We should of course also explore the comparison over different intervals of y to convince ourselves that it is actually an identity, that is, it is true for all y. However, the graphical approach can never be as rigorous as an algebraic approach and Example 1.11 is given as a warning against its use without proper

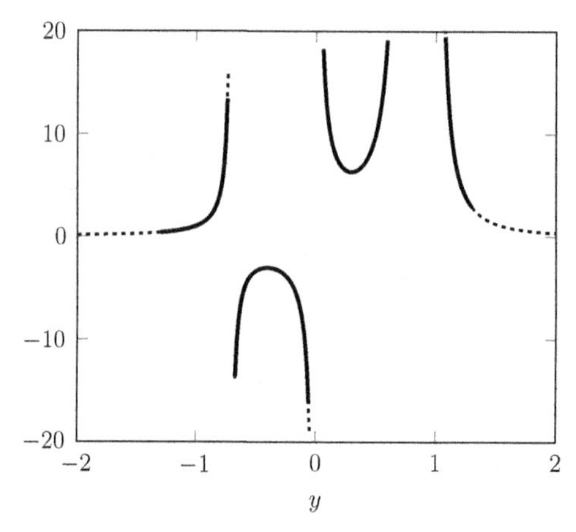

Figure 1.4 Computational plot LHS and RHS of Eq. (1.4). LHS (–) over $y \in [-1.3, 1.3]$ and RHS $(\cdots)$ over $y \in [-2, 2]$.

consideration of the ranges used. Before considering that example, let us first discuss how it is possible to use computers in our studies.

1.3.2 An introduction to mathematics on your computer

Computers are extremely useful tools in mathematics, as will be reflected throughout this book. The advent of relatively cheap and powerful computers has led to the parallel development of programming languages and specialist software that can be used to perform highly complicated mathematical operations at great speed and accuracy. Indeed many commercially available packages are at the very center of the modern financial and scientific industries. However, rather than learning how to "do" the mathematics on a computer, in this book we will use computers to complement our understanding of the mathematical techniques developed. Specialist software packages are therefore over and above our needs.

A particularly powerful and free online tool is Wolfram Alpha, available at the url http://www.wolframalpha.com. Wolfram Alpha is an extremely useful computational engine that can be used for checking or exploring much of the mathematics we will study in this book and we will make repeated mention of it. The great advantage of Wolfram Alpha is that it does not require knowledge of a specialist computational language, indeed it is usually possible to write the mathematical request in plain English.

For example, mathematical expressions can be very quickly plotted on Wolfram Alpha. If, say, we would like to plot the expression $x^2 - 4x + 4$ over $x \in [-5, 5]$ we would navigate to the Web site and simply enter the instruction

```
plot x^2-4x+4 between x=-5 and 5
```

The engine will then report the required plot. Two or more expressions can be plotted simultaneously using, for example, the input

```
plot x^2-4x+4, (x-2)^2, x^2-2x+2 between x=-5 and 5
```

The online tool is therefore a useful means for visually comparing and exploring mathematical expressions.

Wolfram Alpha can also be used for algebraic manipulations. For example, the expression in Example 1.10 can very quickly be confirmed as an identity using the instruction

```
express (y^2+1)/(y(2y^2-1)(y-1)) as a partial fraction
```

It is likely that you have access to Excel on your computer. While Excel is not as easy to use as Wolfram Alpha and is not aimed at "doing mathematics," it is an extremely powerful numerical tool that is used very widely in business. You should therefore spend some time familiarizing yourself with both Excel and Wolfram Alpha if you are not already familiar with them.

Producing plots in Excel is slightly more cumbersome than with Wolfram Alpha. In particular, one would need to specify the values of x, calculate the associated value of the expression at each of these values, and produce a "scatter plot" from these data points. It is assumed that you are familiar enough with Excel to do this.

We will make further reference to both Wolfram Alpha and Excel throughout this book, however, the focus will always be as a complement to the concepts under discussion. It is extremely important to realize that computers should not be used a replacement for mathematical knowledge. While much of the mathematics presented in this book can be performed using Wolfram Alpha and similar tools, it is still crucial that you understand the operations and mathematical concepts behind the results. This is the aim of this book.

EXAMPLE 1.11

Use Wolfram Alpha to investigate whether

$$(x-2)^4 = x^4 - 8x^3 + 24x^2 - 32x + 15$$

is an equation or an identity.

Solution

This example is given as a warning against the graphical approach for confirming identities. The output from the following command demonstrates the reason for this warning,

```
plot (x-2)^4 and x^4-8x^3+24x^2-32x+15
```

Without specifying the range for x, the engine chooses two ranges, a narrow range, say, $x \in [0.5, 3.5]$ and a broad range, say, $x \in [-20, 20]$. The plot over the narrow range clearly shows that the LHS and RHS are not equal. However, had we seen only the plot over the broader range, we might have falsely concluded the curves were identical without further experimentation.

A much better approach would be to base the decision on the algebraic expansion of $(x-2)^4$. This could be done using the command

```
expand (x-2)^4
```

which yields $x^4 - 8x^3 + 24x^2 - 32x + 16$. The RHS and LHS are therefore seen to differ and the expression is correctly classified as an equation.

1.3.3 Inequalities

To complete this section on the types of mathematical expressions, consider

$$x - 4 > 2 \tag{1.5}$$

The use of $>$ makes it clear that this expression is neither an equation nor an identity; it is in fact an example of an *inequality*. In general, an inequality will be true within particular intervals of the independent variable on the real line, or possibly not true over any interval. An inequality is therefore distinct from identities and more akin to equations. As with equations, the appearance of an inequality will usually trigger the need to find the particular ranges of the independent variable for which it is true. It is assumed that the reader is familiar with handling inequalities algebraically and the straightforward solution to Eq. (1.5) is detailed for completeness only.

$$
\begin{aligned}
x - 4 &> 2 \\
x &> 2 + 4 \\
x &> 6
\end{aligned}
$$

That is, Eq. (1.5) is satisfied for $x \in (6, \infty)$. Inequalities that arise in practice are unlikely to be so simple and one must be prepared to resort to more involved algebraic manipulations or graphical techniques to solve them. We will return to techniques for solving equations and inequalities at many points throughout this book, including Section 2.1.2 in the next chapter. However the following examples are given to illustrate the process that we will be required to follow.

EXAMPLE 1.12

Find the values of q such that the following inequality is true:

$$q^3 - 3q^2 - 4q + 12 \geq 0$$

Solution

The answer is not immediately obvious. One's first instinct might be to find the values of q which set the LHS to 0 and proceed from there. Taking this approach, we note that $q = \pm 2$ and 3 are such of the LHS, which can be obtained, for example, from factoring the LHS expression as $(x - 2)(x + 2)(x - 3)$. The cubic term will dominate for large values of q and is such that the LHS is negative as $q \to -\infty$ and positive as $q \to \infty$. With the information that the LHS must remain negative for large negative q, positive for large positive q and cross zero at the points identified, a little thought will conclude that the inequality is true for $q \in [-2, 2]$ and $q \in [3, \infty)$. Note that square bracket notation is used as the end points are included in the solution to the inequality.

Alternately, if one had access to a computer, it would be possible to plot the LHS and quickly determine the same ranges for q without the need to explicitly determine the values that zero the expression. Such a plot is shown in Figure 1.5.

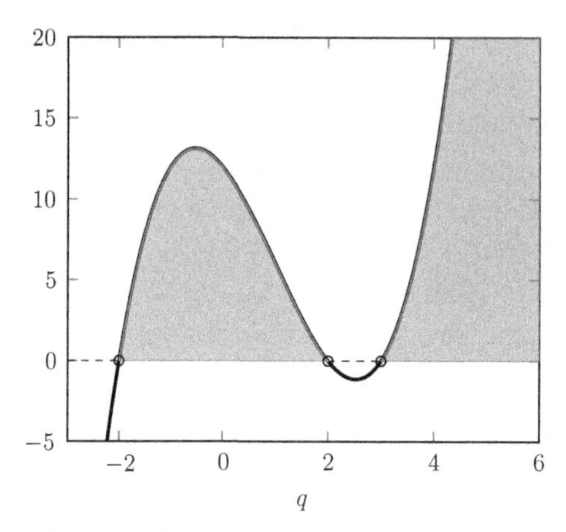

Figure 1.5 Plot showing the inequality in Example 1.12.

EXAMPLE 1.13

Find the values of q such that the following inequality is true:

$$q^3 - 3q^2 - 4q + 12 > 0$$

Solution

The solution follows that presented for Example 1.12. However, in this case, the solution ranges do not include the roots of the LHS and open intervals are required, that is $q \in (-2, 2)$ and $q \in (3, \infty)$.

In contrast to Eqs. (1.2)–(1.5), Eq. (1.6) provides a means of returning an output to any input value of x without imposing any equality or inequality constraints on that value.

$$(x + 4)^2 \tag{1.6}$$

For example, $x = 1$ returns $(1 + 4)^2 = 25$ and $x = 2.2$ returns $(2.2 + 4)^2 = 38.44$. It is clear from the previous descriptions that Eq. (1.6) is neither an identity, an equation, nor an inequality. In fact, the expression motivates the use of the term *function* which is a fundamental concept in mathematical methods that will be discussed in Chapter 2.

1.4 QUESTIONS

The following questions are intended to test your knowledge of the preliminary material discussed in this chapter. Full solutions are available in Chapter 1 Solutions of Part III. You should use an algebraic approach unless otherwise stated.

Question 1.1. Identify the real number systems that the following belong to.

a. 5
b. 6.48763
c. π^2
d. $\frac{43}{7}$
e. -6

Question 1.2. Translate the following mathematical statements into words.

a. $\forall x \in (-\infty, 0), x^2 \in \mathbb{R}^+ \cup \{0\}$
b. $\forall p \in \mathbb{Z}, q \in \mathbb{Z} \setminus \{0\}, \frac{p}{q} \in \mathbb{Q}$
c. $\exists y \in \mathbb{Z} : y < 3$ and y is odd
d. $\forall z \in \mathbb{R}^+, z > 0$
e. $\forall a \in \mathbb{R}, \exists b \in \mathbb{R} : a \times b = 2$

Question 1.3. In the particular case that $A = \{0, 1, 2, 3, 4, 5\}$, $B = \{-2, -1, 1, 2\}$, and $C = \{2, 3, 4, 5, 6\}$, demonstrate that each of the following identities are true.

a. $A \cup (B \cap C) \equiv (A \cup B) \cap (A \cup C)$
b. $A \cap (B \cup C) \equiv (A \cap B) \cup (A \cap C)$
c. $C \setminus (A \cap B) \equiv (C \setminus A) \cup (C \setminus B)$
d. $B \setminus \emptyset \equiv B$
e. $(B \setminus A) \cap C \equiv (B \cap C) \setminus A$

Question 1.4. State whether each of the following expressions are identities or equations. Where appropriate, use any method to identify the values of y such that the equations hold.

a. $y^2 - 16 = (y - 4)(y + 4)$
b. $y^4 + 9 = 10y^2$
c. $y^3 + 25ay = (25 + a)y^2$ where $a \in \mathbb{R}$ is a constant
d. $-y^4 + 5y^3 - 9y^2 + 7y = 2$

Question 1.5. Use an algebraic method to find intervals for z that solve the following inequalities.

a. $z - 1 > 0$
b. $2z + 1 \leq -2$
c. $4z < 3z + 2$
d. $z - 4 \leq 2z + 1$

CHAPTER 2

Exploring Functions

Contents

Prerequisite knowledge	**Intented learning outcomes**

• Chapter 1
• Inverse trigonometric operations
	• Identify and work with examples of
	  ◦ mappings
	  ◦ functions
	  ◦ composite functions
	  ◦ inverse functions
	• Recall properties of the standard classes of functions (polynomial, rational, exponential, logarithmic, and circular/trigonometric), including
	  ◦ continuity and smoothness
	  ◦ limits
	  ◦ asymptotes
	  ◦ singularities
	• Ability to explore properties of more complicated functions using algebraic and visual methods

In this chapter, we give a detailed discussion of functions of a single independent variable. Functions are a hugely important concept in mathematics and will be used extensively in all that follows in this book. We begin with a general discussion of functions in the broad sense, and then proceed to discuss particular fundamental classes of functions: polynomial, rational, exponential, logarithmic, and circular (trigonometric) functions. These fundamental classes of functions will form the "building blocks" for the expressions that arise in the actuarial and financial context.

A crucial aim of this chapter is to encourage you to think of functions as mathematical objects and to be able to explore their properties. You should assume that we are working with the *real numbers* in all that follows.

2.1 GENERAL PROPERTIES AND METHODS

2.1.1 Mappings

Chapter 1 ended with a discussion of the expression

$$(x+4)^2 \tag{2.1}$$

After studying the previous chapter, it should be clear that Eq. (2.1) is not an example of an *equation* or an *identity*; that is, it does not have an equal sign separating a LHS and RHS. Nor is it an inequality. In fact, the expression is an example of a *function*. As we shall see, functions have a precise meaning and not all expressions are necessarily functions. We now build toward an understanding of what is meant by the term "function."

For the moment, you should think of Eq. (2.1) as a "rule" that returns an *output* for a given value of x as an *input*. For example, particular input and output values are stated in Table 2.1.

With the results of Table 2.1 in mind, it makes sense to think of Eq. (2.1) as a *mapping*. For example, in this particular case, −4 has been *mapped* to 0, 1 has been mapped to

Table 2.1 Example input and associated output values of $(x+4)^2$

x	$\longrightarrow$	$(x+4)^2$
−5	$\longrightarrow$	1
−4	$\longrightarrow$	0
−3	$\longrightarrow$	1
−2	$\longrightarrow$	4
−1	$\longrightarrow$	9
0	$\longrightarrow$	16
1	$\longrightarrow$	25
2	$\longrightarrow$	36
3	$\longrightarrow$	49
4	$\longrightarrow$	64
5	$\longrightarrow$	81

25, and 4 has been mapped to 64. Alternatively, using the "goes to" notation listed in Table 1.1, we can write these particular mappings as

$$-4 \xrightarrow{(x+4)^2} 0, \quad 1 \xrightarrow{(x+4)^2} 25, \quad 4 \xrightarrow{(x+4)^2} 64$$

Rather than always writing the full expression that defines the mapping, it is convenient to label it with a single letter. For example, we could denote Eq. (2.1) by the letter f and write

$$f : x \to (x + 4)^2$$

In fact, the mapping is fully defined by the statement

$$f : x \to (x + 4)^2, \quad \forall x \in \mathbb{R} \tag{2.2}$$

which is read as

f is such that x goes to $(x + 4)^2$ for all x in the set of real numbers.

It is often convenient to write this simply as

$$f(x) = (x + 4)^2$$

which states the mapping's label, f, and the symbol used to define the independent variable, x. Of course, the shorthand notation $f(x)$ gives no information about the properties of the independent variable, for example, that this f is defined for all real numbers. However, properties of the input will typically be known from the context of the problem.

There is nothing special about using the letter f to denote a mapping and x to denote the independent variable. For example, the following statement defines exactly the same mapping as in Eq. (2.2)

$$g : y \to (y + 4)^2, \quad \forall y \in \mathbb{R}$$

and can be summarized as $g(y) = (y + 4)^2$.

The set of possible values of the independent variable on which the mapping is defined is called the *domain*. For example, the domain of mapping (2.2) is the set of real numbers. However, the mapping

$$h : z \to (z + 2)^3, \quad \forall z \in [-2, 2] \tag{2.3}$$

has a domain formed from the closed interval of real numbers between -2 and 2 (including the end points).

If we understand that the domain defines the extent of the input of a mapping, the *range* defines the extent of the output of the domain under that mapping. For example, the range of mapping (2.2) is $\mathbb{R}^+ \cup \{0\}$, equivalently $[0, \infty)$, and the range of function (2.3) is $h(z) \in [0, 64]$. We can interpret this as it being impossible to find an

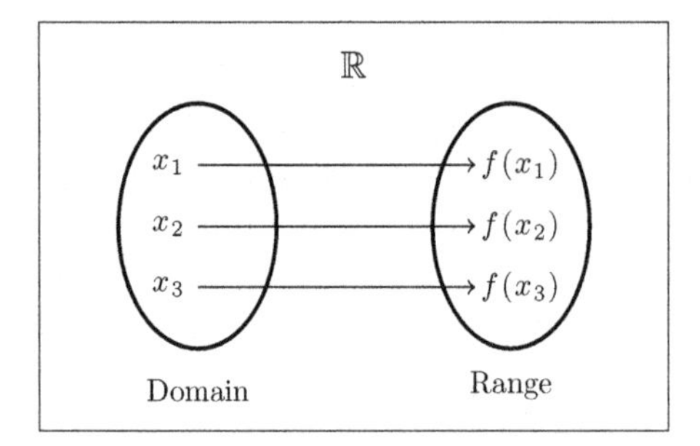

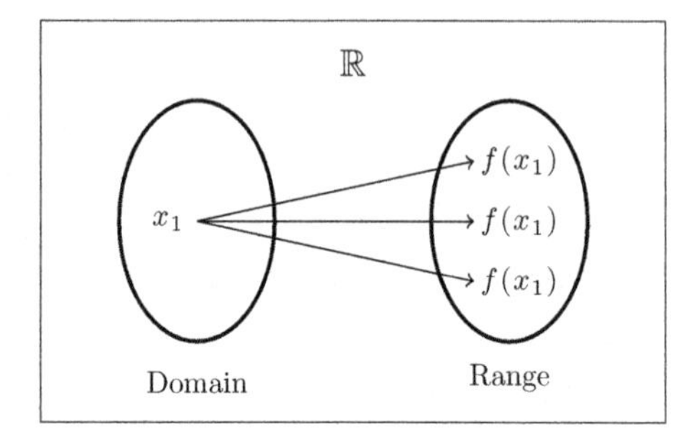

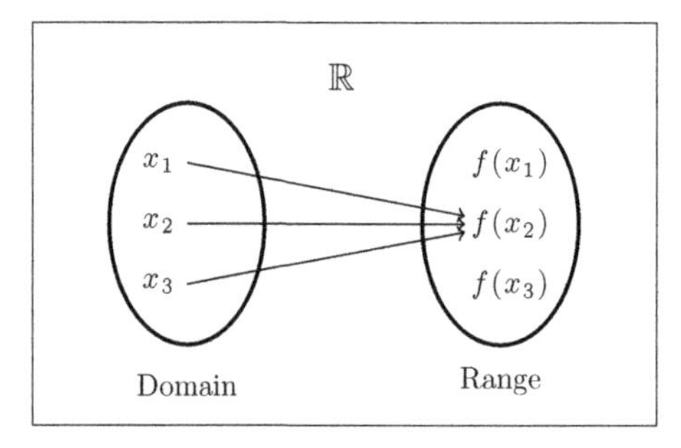

Figure 2.1 An illustration of classifications of mappings between real numbers. (a) One-to-one mapping, (b) one-to-many mapping, and (c) many-to-one mapping.

input within the domain that gives an output outside of the range. We return to question of how to determine the mapping's range later on.

2.1.2 Functions

Functions vs. mappings

Mappings can be classified as being either *one-to-one*, *many-to-one*, or *one-to-many*, as illustrated in Figure 2.1. As the name suggests, a one-to-one mapping maps each unique input in the domain to a unique output in the range. A many-to-one mapping maps more than one input to a particular output, and a one-to-many mapping maps a single input to more than one output. One-to-one and many-to-one mappings are very useful and are also said to define *functions*. The potential ambiguity regarding the output for a given input means that one-to-many mappings are less useful; they are not said to define functions.

EXAMPLE 2.1

State whether the following mappings are one-to-one, many-to-one, or one-to-many. Also state whether each is also a function.

a. $g(x) = x + 10$
b. $f(z) = (z - 2)^2$
c. $h(p) = \sqrt{p}$

Solution

a. Each input is mapped to a unique output, 10 further along the real number line. The mapping $g(x)$ is therefore one-to-one and also defines a function.

b. We note that, for example, $f(4) = 2$ and $f(0) = 2$. The mapping $f(z)$ is therefore many-to-one mapping and also defines a function.

c. The square root operation returns a positive and a negative value. For example, $h(9) = \pm 3$. For this reason, the mapping $h(p)$ is a one-to-many mapping does not define a function.

Despite the technical definition of a function, a one-to-many mapping might be perfectly usable as a function if some restrictions are imposed. As demonstrated in Example 2.1(c), for example, the square root operation is an example of a one-to-many operation on the real domain; mappings that include this operation without clarification of the value to be taken cannot therefore be functions. However, in this case, we could insist that only the positive value is to be taken and reduce the one-to-many mapping to a one-to-one mapping. The mapping $h(p) = \sqrt[+]{p}$ is then a function.

Odd and even functions

The terms *odd* and *even* refer to particular properties of symmetry that functions may possess. Although a particular function must be either a one-to-one or a many-to-one mapping, it does not necessarily have to be either odd or even.

By definition, an even function is one that has the property

$$f(-x) = f(x) \tag{2.4}$$

Similarly, an odd function is one that has the property

$$f(-x) = -f(x) \tag{2.5}$$

From a practical perspective, an even function can be thought of as a mapping that is symmetrical about $x = 0$. In contrast, an odd function is a mapping that produces a negative image either side of $x = 0$. It should then be clear why the odd-even property is referred to a "symmetry" property. The odd or evenness of a function is sometimes referred to as its *parity*.

Odd and even functions have particular properties that are of interest in both the abstract and practical use of mathematics, as is our interest here. Such properties will be flagged as necessary as we proceed through the book. For the moment, it is sufficient to understand the terms odd and even and be able to identify whether any particular function possesses either property.

EXAMPLE 2.2

Determine the parity of the following functions.

a. $f(q) = q^2$

b. $g(y) = y^3$

c. $h(z) = |z|$

d. $l(x) = x - 1$

Solution

a. We note that, for example, $f(2) = f(-2)$ and this is true of every $q, -q$ pair. The function $f(q)$ is therefore even.

b. We note that, for example, $f(-2) = -f(2)$ and this is true for every $y, -y$ pair. The function $g(y)$ is therefore odd.

c. Recall from Table 1.1 that

$$|z| = \begin{cases} z, & \text{for } z \geq 0, \\ -z, & \text{for } z < 0. \end{cases}$$

We see that, for example, $h(2) = h(-2)$ and this is true for every $z, -z$ pair. The function $h(z)$ is therefore even.

d. We note that, function $l(x)$ possess no symmetry about $x = 0$, that is, $l(-x) \neq \pm l(x)$. Function $l(x)$ is therefore neither odd nor even.

We return to the visual classification of odd and even functions in Section 2.1.4, and plots of the functions in Example 2.2 can be seen in Figure 2.4.

Roots

Another extremely useful concept is that of the *roots* of a function. A function $f(x)$ is said to have a root at $x = c$ iff $f(c) = 0$. A function may have one root, many roots, or no roots in the domain of real numbers. You may find this phrasing rather awkward, but, as we shall see in Chapter 8, although a function may not have any real roots, it may have *complex roots*; that is, roots in the particular number system called *complex numbers*, $\mathbb{C}$. This is stated without explanation here and we will continue to ignore complex roots until Chapter 8. Until that point, we will assume that everything is firmly based in terms of real numbers and the term "roots" should be understood as meaning "real roots."

As we can immediately see from the definition, the roots of a function are the values within the domain that solve the *equation* $f(x) = 0$. This concept provides a clear link between functions and the material of Section 1.3. Strategies that are used to find the roots of functions are identical to the strategies used to solve equations.

EXAMPLE 2.3

Use an algebraic approach to find the roots of the function

$$g(y) = y^2 - 4y + 3$$

Solution

The roots are the values of y such that

$$y^2 - 4y + 3 = 0$$

Two strategies might immediately come to mind. The first is to use the standard quadratic formula. The second is to note that the LHS factorizes as $(y - 3)(y - 1)$. Both strategies lead to the correct result that $g(y)$ has roots at $y = 1$ and $y = 3$.

EXAMPLE 2.4

If

$$g(x) = (x+1)(x^2 - 4x + 3)$$

find the values of x such that $g(x) > 0$.

Solution

The first stage in solving an inequality is to find the roots of the function. In this case, we note that the expression is already part factorized and so we can immediately see that $x = -1$ is a root. Other roots can be determined by noting that, in this case, the quadratic term is identical to that in Example 2.3. We then identify $x = 1$ and $x = 3$ as the remaining roots. We now need to determine the broader behavior of the function in order to solve the inequality. The function expands to

$$f(x) = x^3 - 3x^2 - x + 3$$

and we see its value at large values of x is dominated by x^3. The function is therefore negative for large negative values of x and positive for large positive values of x. With this information and knowledge that it crosses the horizontal axis three times at $x = \pm 1$ and $x = 3$, it should be clear that $g(x) > 0$ for $x \in (-1, 1)$ and $x \in (3, \infty)$. We might opt to write these intervals as $x \in (-1, 1) \cup (3, \infty)$, but this is not essential to communicate the answer correctly.

At this stage in the book, it is assumed that your mathematical armory is not yet sufficient to attack root-finding problems in anything but simple polynomial functions. Where necessary, you should revert to graphical techniques, as discussed below, or software tools such as Wolfram Alpha or Excel's *Goal Seek* function. More advanced methods will be developed in later chapters.

2.1.3 Root finding on your computer

Consider the function $f(x) = x^3 - 2x^2 - x + 2$. Wolfram Alpha can be used to very quickly determine the roots of $f(x)$ with the intuitive command

```
roots x^3-2x^2-x+2
```

Alternatively, we could understand the problem as needing to find the solution to the equation $x^3 - 2x^2 - x + 2 = 0$ and use the command

```
solution x^3-2x^2-x+2=0
```

Both approaches lead to the result that $f(x)$ has roots at $x = \pm 1$ and 2. Indeed, you may have noticed that $f(x)$ can be factorized as $(x-1)(x+1)(x-2)$ and so immediately identify the same roots.

Consider now the function $h(x) = x^3 - 4x^2 - x + 2$. This expression is not easily factorized and so Wolfram Alpha is particularly useful. Indeed, the command

```
roots x^3-4x^2-x+2
```

leads to three roots at $x \approx -0.7616, 0.6367$, and 4.1249, it also gives a plot of the function where it identifies the three roots. Note that Wolfram Alpha will return all roots, both real and complex.

Excel's *Goal Seek* function can be used to find real roots of functions. To implement Goal Seek, we formulate a spreadsheet that evaluates the function in one cell, the output, at an input value stated in another cell. Goal Seek then determines the input value such that the output is zero. Goal Seek can be accessed via the Tools menu. Of course, Goal Seek can only find one root at a time and the particular root it finds is determined by the starting value of the input.

For example, we can use Goal Seek to find the roots of $h(x)$. Using an initial input starting value of $x = 5$, Goal Seek converges to root $x \approx 4.1249$. Using a starting value of $x = -5$, it obtains $x \approx -0.7616$. Furthermore, using a starting value of $x = 0$, it obtains $x \approx 0.6367$. While Goal Seek has correctly obtained the three real roots in this case, some familiarity with the function and the approximation location of the roots is required. We return to these issues later in this chapter.

Both Wolfram Alpha and Excel use numerical methods to obtain the roots of functions. Numerical methods for root finding and other mathematical operations are discussed in Chapter 13.

2.1.4 Plotting functions

Consider the function

$$f(x) = (x - 2)^2 + 1 \tag{2.6}$$

As previously discussed, it is possible to determine the output of a mapping for any given input in its domain. Furthermore, as we are now concerned with functions, each input will result in a single output, as is the case here. With this in mind, it is not difficult to envisage a way of obtaining a graphical representation of a function $f(x)$ over any finite interval of the domain. Specifically, one could specify a number of values of x within a particular interval, compute the associated value of $f(x)$, and plot this either by hand or using software.

For example, let us plot function (2.6) over the interval $x \in [-5, 10]$. The associated values of the function at integer values of x within this interval are given in Table 2.2.[1] We can then plot the function $f(x)$ using the values obtained; this is shown in Figure 2.2 with the symbols $\circ$. The resulting plot is then a graphical representation of function (2.6). With a higher resolution, or using the default interpolation package in Excel, for example, or indeed with a simple free-hand sketching method between the data points, we could fill in the space between the discrete points $\circ$ and show the complete curve for the function over the interval, as indicated by a solid line.

[1] Of course, there is no reason why we could not obtain values of the function at a higher resolution than shown here; for example, we could use steps of 0.1 rather 1.

Table 2.2 Values of $f(x) = (x-2)^2 + 1$ at various points on the interval $[-5, 10]$

x	$f(x)$	x	$f(x)$
−5	50	3	2
−4	37	4	5
−3	26	5	10
−2	17	6	17
−1	10	7	26
0	5	8	37
1	2	9	50
2	1	10	65

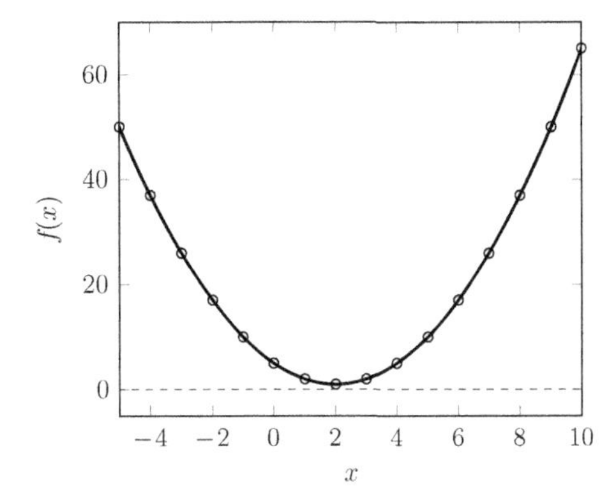

Figure 2.2 Plot of $f(x) = (x-2)^2 + 1$.

The benefit of having a graphical representation of a function, whether it be an accurate computational plot or merely a paper sketch, is that we can immediately deduce its salient properties. In the case of Figure 2.2, we immediately see the following properties of function (2.6).

1. It takes its lowest value of 1 at $x = 2$ and appears to increase in the positive direction without bound outside of the interval of x considered. From this we can determine that the *range* of the function is $f(x) \geq 1$.
2. Although the function is symmetrical in the even sense, it is not symmetrical about $x = 0$ and so is not an even function. Rather it is symmetrical about $x = 2$, such that $f(2+a) = f(2-a)$. There is clear evidence for this in Table 2.2.
3. The graph does not cross the horizontal axis (where $f(x) = 0$) and so does not have any roots on the real axis.
4. Any horizontal line above $f(x) = 1$ crosses the curve at two points; $f(x)$ therefore is a two-to-one mapping.

Of course, the above properties may have been evident to you from the functional form of Eq. (2.6). In particular, all the properties stem from the symmetry and range properties of the quadratic component "$(x-2)^2$." More specifically, $(x-2)^2 \geq 0$ and $([2+a]-2)^2 = a^2 = ([2-a]-2)^2$. However, in general, you might find it quicker and easier to determine the properties of a function using a graphical approach rather than algebraic reasoning.

EXAMPLE 2.5

Plot the following function using your preferred method, for example, using either Excel or the `plot` command in Wolfram Alpha. Use this plot to discuss the properties of the function.

$$h(y) = y^5 + y^4 - 24y^3 - 4y^2 + 80y$$

Solution

A plot of $h(y)$ over $y \in [-6, 5]$ is given in Figure 2.3. From this we can immediately deduce the following properties.

1. The range for $h(y)$ is the set of all real numbers, $\mathbb{R}$.
2. It has no symmetry.
3. It has five roots at $y = -5, -2, 0, 2$, and 4.
4. Any horizontal line will, at most, cross the curve at five places and any vertical line will cross the curve at one place. The function is therefore a many-to-one mapping.

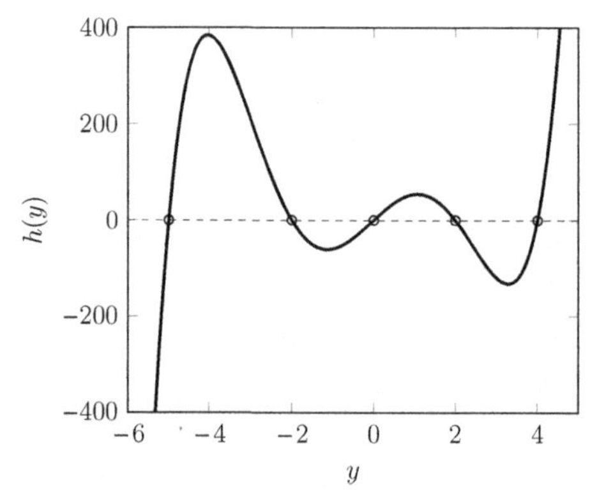

Figure 2.3 Plot of $h(y) = y^5 + y^4 - 24y^3 - 4y^2 + 80y$.

EXAMPLE 2.6

Determine the values of y such that $h(y) \leq 0$, where $h(y)$ is stated in Example 2.5.

Solution

Figure 2.3 shows that the inequality is solved for $y \in [-\infty, -5] \cup [-2, 0] \cup [2, 4]$.

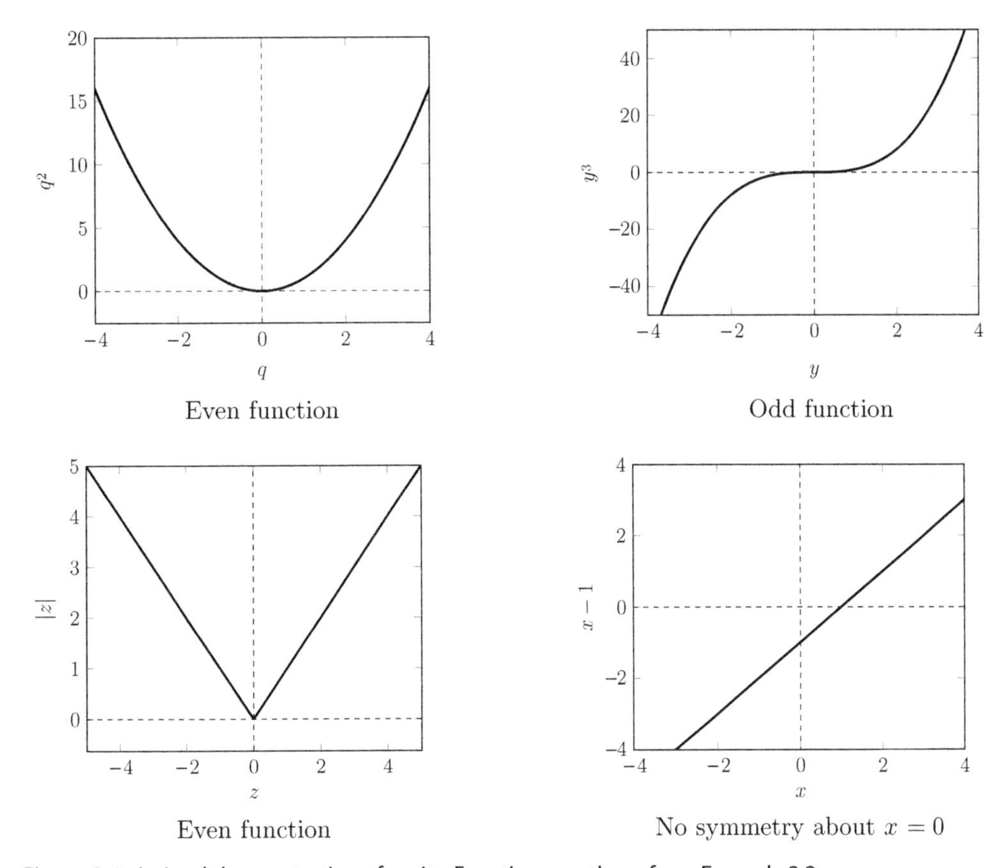

Figure 2.4 A visual demonstration of parity. Functions are those from Example 2.2.

The concept of symmetry within odd and even functions is easily seen from plotting functions; this is shown in Figure 2.4.

2.2 COMBINING FUNCTIONS

2.2.1 Simple combinations

It is possible to define new functions from combinations of more simpler functions and this is the topic of this section. For example, the function $x^2 + 2x$ can be considered as being the sum of the single-term functions x^2 and $2x$. Indeed, the most obvious way to form new functions is by adding other functions. More specifically, if $f(x)$ and $g(x)$ are functions and a and b are constants, it is clear that

$$h(x) = af(x) + bg(x)$$

defines another function $h(x)$. This is true for the sum of any number of basic functions.

EXAMPLE 2.7

If $f(x) = x^2 + 2$ and $g(x) = x^3 - 2x^2 - 3$, derive expressions for the functions formed by

a. $f(x) + g(x)$
b. $-2f(x) + 4g(x)$

Solution

a. $h(x) = (x^2 + 2) + (x^3 - 2x^2 - 3) = x^3 - x^2 - 1$
b. $l(x) = -2(x^2 + 2) + 4(x^3 - 2x^2 - 3) = 4x^3 - 10x^2 - 16$

An alternative way of defining new functions is by multiplying powers of the basic functions. More specifically, it is clear that

$$h(x) = (f(x))^a \times (g(x))^b$$

defines another function $h(x)$. Again, this is true for the product of any number of terms formed by integer powers of other functions.

EXAMPLE 2.8

If $f(x)$ and $g(x)$ are as defined in Example 2.7, derive expressions for the functions formed by

a. $f(x) \times g(x)$
b. $(f(x))^2 \times (g(x))^{-1} = \frac{f^2(x)}{g(x)}$

Solution

a. $h(x) = (x^2 + 2) \times (x^3 - 2x^2 - 3) = x^5 - 2x^4 + 2x^3 - 7x^2 - 6$
b. $l(x) = \frac{(x^2+2)^2}{x^3-2x^2-3}$

The properties of functions resulting from some combination of basic functions can sometimes be related back to the properties of the basic functions. For example, the following properties of sums of odd and even functions should be clear.

i. The sum of two or more even functions is necessarily an even function.
ii. The sum of two or more odd functions is necessarily an odd function.
iii. The sum of odd functions and even functions is not necessarily odd or even.

EXAMPLE 2.9

If $f(x) = x^2$, $g(x) = x^4$, $h(x) = x$, and $l(x) = x^3$, determine the parity of the following functions.

a. $f(x) + g(x)$
b. $h(x) + l(x)$
c. $f(x) + l(x)$

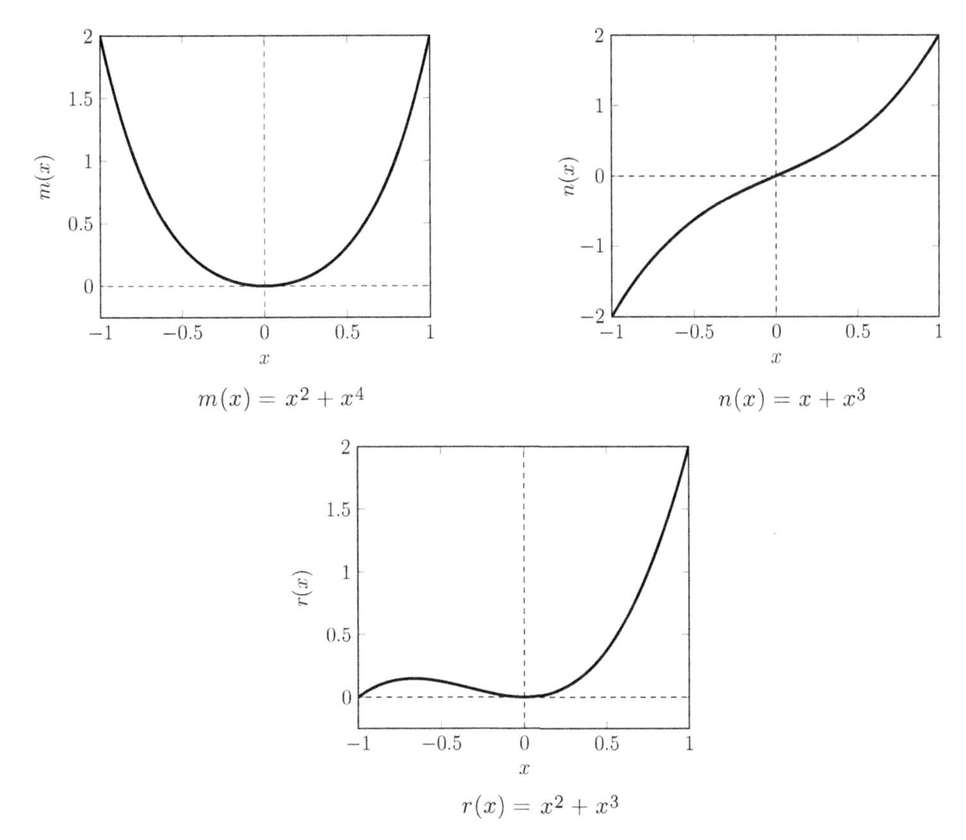

Figure 2.5 Plots of functions resulting in Example 2.9.

Solution

We begin by noting that $f(x)$ and $g(x)$ are even functions, and $h(x)$ and $l(x)$ are odd functions.

a. We define $m(x) = x^2 + x^4$. Note that $m(3) = m(-3)$ and that this is true for all $x, -x$ pairs; $m(x)$ is therefore an even function. This is consistent with property i.

b. We define $n(x) = x + x^3$. Note that $n(3) = -n(-3)$ and that this is true for all $x, -x$ pairs; $n(x)$ is therefore an odd function. This is consistent with property ii.

c. We define $r(x) = x^2 + x^3$. Note that $r(x) \neq \pm r(x)$ if $x \neq 0$; $r(x)$ is therefore neither an even nor an odd function. This is consistent with property iii.

The parity of functions $m(x)$, $n(x)$, and $r(x)$ are illustrated in Figure 2.5.

Similar properties of the products of odd and even functions can be determined. This is left as an exercise in Question 2.9.

2.2.2 Composite functions

If we consider a function as a method of associating an output to a particular input, it is possible to visualize a string of functions such that the output of one provides the input for the next. In the simplest case, we can have the string $f \rightarrow g$ formed from two functions $f(x)$ and $g(x)$. An input value, $x = a$, would then pass through the string as

$$a \xrightarrow{f} f(a) \xrightarrow{g} g(f(a))$$

That is, the input a passes through the string leading to the output $g(f(a))$, i.e., the function g evaluated at $x = f(a)$.

EXAMPLE 2.10

If $f(x) = 4x^2 + x + 2$ and $g(x) = 2x^2 + 2x - 1$ calculate the output of the string $f \rightarrow g$ for input values

a. $x = 0$
b. $x = 1$
c. $x = 4$

Solution

a. $0 \xrightarrow{f} -2 \xrightarrow{g} 11$
b. $1 \xrightarrow{f} 7 \xrightarrow{g} 111$
c. $4 \xrightarrow{f} 70 \xrightarrow{g} 9939$

In a sense, all we are doing is defining a new function, $c(x)$, say, from the string $f \rightarrow g$. The new function, $c(x)$, is said to be the *composite* of f and g. It is possible to determine an explicit expression for a composite function with some algebraic manipulation. For example, if $f(x) = 2x$ and $g(x) = x + 10$ then

$$c(x) = g(f(x)) = (2x) + 10 = 2x + 10$$

As a slightly more complicated example, let $f(x) = x^2$ and $g(x) = x^3 + 10x$ then

$$c(x) = g(f(x)) = (x^2)^3 + 10(x^2) = x^6 + 10x^2$$

We can go further and define a composite function of three other functions $f \rightarrow g \rightarrow h$. An input value a would pass through the string

$$a \xrightarrow{f} f(a) \xrightarrow{g} g(f(a)) \xrightarrow{h} h(g(f(a)))$$

The composite would then be calculated as $c(x) = h(g(f(x)))$. Following this process, we can form a composite function of any number of constituent functions. However, it should be clear that an explicit expression for a composite could become extremely complicated. With this in mind, we introduce some standard notation for composite functions.

$$(g \circ f)(x) = g(f(x)) \tag{2.7}$$

$$(h \circ g \circ f)(x) = h(g(f(x))) \tag{2.8}$$

The order that the functions are written in a composition is crucial. For example, if $f(x) = 2x$ and $g(x) = x + 10$, then

$$(g \circ f)(x) = g(f(x)) = 2x + 10$$
$$(f \circ g)(x) = f(g(x)) = 2(x + 10) = 2x + 20$$

It is true that, in general,

$$(f \circ g)(x) \not\equiv (g \circ f)(x)$$

EXAMPLE 2.11

If $f(x) = 4x^2 + x + 2$ and $g(x) = 2x^2 + 2x - 1$, determine an explicit expression for the following composite functions.

a. $(g \circ f)(x)$
b. $(f \circ g)(x)$

Solution

a. $(g \circ f)(x) = g(f(x)) = 2(4x^2 + x + 2)^2 + 2(4x^2 + x + 2) - 1 = 32x^4 + 16x^3 + 42x^2 + 10x + 11$
b. $(f \circ g)(x) = f(g(x)) = 4(2x^2 + 2x - 1)^2 + (2x^2 + 2x - 1) + 2 = 16x^4 + 32x^3 + 2x^2 - 14x + 5$

Of course, it is possible to form composites of the same function. For example, if $g(x) = 2x$ then

$$(g \circ g)(x) = 2(2x) = 4x$$
$$(g \circ g \circ g \circ g)(x) = (2)^4 x = 16x$$

2.3 COMMON CLASSES OF FUNCTIONS

In Section 2.1, we discussed some fundamental properties of functions in general. We now turn to discussing the particular properties of the most common classes of functions. These basic functions can be considered the "building blocks" for more complicated functions that occur in practical applications.

2.3.1 Polynomial functions

Our discussions to this point have been limited to polynomial expressions exclusively. That is, expressions formed from the sum of different natural number powers of the independent variable. We briefly summarize the particular properties of the polynomial functions now.

A polynomial function has general the form

$$f(x) = a_n x^n + a_{n-1} x^{n-1} + \cdots + a_1 x^1 + a_0, \quad \text{with } n \in \mathbb{N},\ a_i \in \mathbb{R} \tag{2.9}$$

Table 2.3 The first six orders of polynomials

n	Order	Name	Example
0	Zeroth order	Constant	2
1	First order	Linear	$2x - 1$
2	Second order	Quadratic	$x^2 + 2.6x - 3$
3	Third order	Cubic	$2x^3 + x^2 - 4x - 2$
4	Fourth order	Quartic	$0.9x^4 + 2x^3 - 7x^2 - 8x + 12$
5	Fifth order	Quintic	$-x^5 + 5x^4 - 20x^2 + 16x$
6	Sixth order	Sextic	$x^6 - x^5 - 33x^4 + 13x^3 + 296x^2 - 36x - 720$

The largest power, n, that appears gives the *degree* of the polynomial. For example, $f(x) = 2x^2 + x + 1$ is a *second-order polynomial*, alternatively, we might call this a *quadratic*. Similarly, $g(x) = -x^5 + 2x^3 + 5$ is a *fifth-order polynomial*, or a *quintic*. Such terms are summarized to the sixth order in Table 2.3.

Note that for an nth-order polynomial, only the coefficient a_n needs be nonzero. For example, $h(x) = 2x^3$ is a third-order polynomial, as is $2x^3 + 2$.

You may well already be familiar with the "shapes" of plots of the basic polynomials x^n for different n from your previous studies. These are summarized to the sixth order in Figure 2.6. Note that plots of general polynomial functions of the form in Eq. (2.9) are very dependent on the particular form of the function and each should be investigated separately.

An important observation from plots of general nth-order polynomials is that they can have at most n roots; that is, the curve crosses the horizontal axis at most n times. For example, Figure 2.3 shows the five roots of a particular quintic expression. The plot of a polynomial does not necessarily have to cross the axis at all if n is even (that is, the polynomial does not have real roots), or it may cross the axis fewer than n times if n is odd (that is, has fewer than n real roots). Such properties are explored in Question 2.8. An alternative situation is that of *repeated roots*, where a particular root exists more than once. For example, it should be clear that the cubic function

$$h(x) = x^3 - 3x^2 + 4 = (x - 2)(x - 2)(x + 1)$$

has roots at $x = -1$ and $x = 2$, with this second root occurring twice. Our previous interpretation of a root was a point at which the curve crossed the horizontal axis, but how can a curve cross the axis twice at the same point? Of course, the answer is that it does not cross the axis. In fact, a repeated root can be identified graphically by the curve touching the axis at a *turning point*, that is, a point that is locally flat and tangential to the x-axis. We return to turning points in Chapter 4. As can be seen in Figure 2.6, each basic polynomial x^n possess n repeated roots at $x = 0$. All possible situations for general quadratic and cubic polynomials are illustrated in Figures 2.7 and 2.8, respectively.

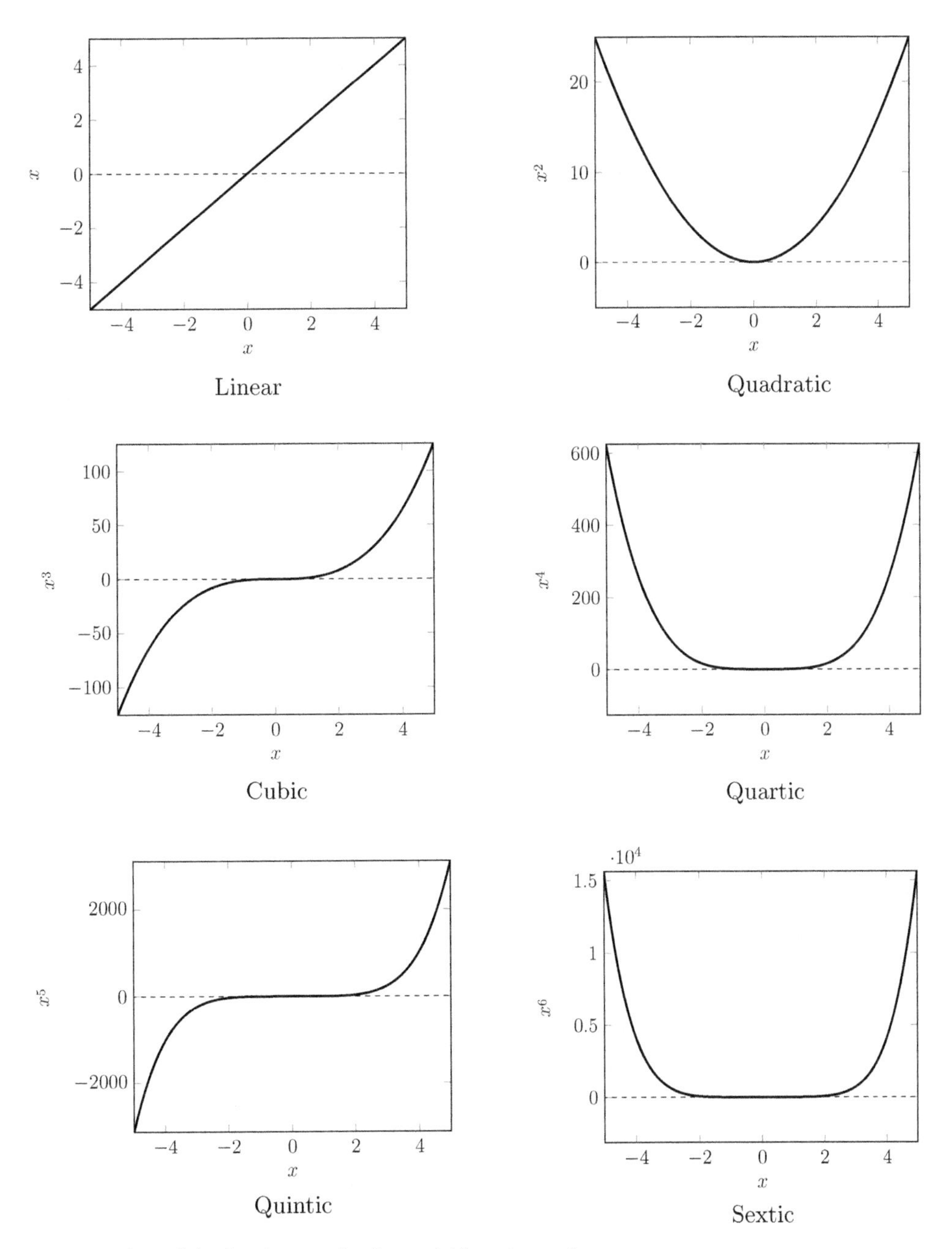

Figure 2.6 Plots of the fundamental polynomial functions, x^n.

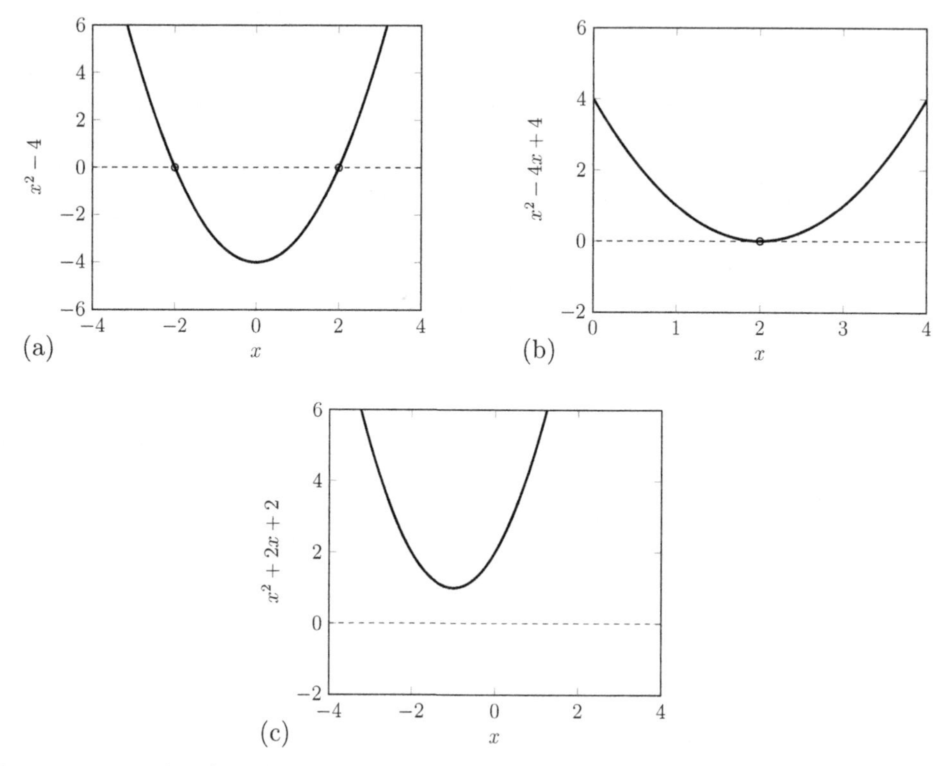

Figure 2.7 Examples of quadratic polynomials with two or fewer real roots. (a) Two real roots, (b) one repeat root, and (c) no real root.

In addition to the maximum number of roots, polynomial functions have other properties that we wish to emphasize at this point.

i. Polynomials are *smooth* and *continuous*, that is, the plots in Figure 2.6, for example, do not show any "corners" or "cusps" and it is possible to draw each figure with a single line.

ii. Polynomials have a domain equal the set of real numbers, $\mathbb{R}$, that is, there are no restrictions on the input to the function.

iii. Polynomials have a range equal to the entire set of real numbers, $\mathbb{R}$, when n is odd, and some subset of the real numbers when n is even.

Note that the definitions of smoothness and continuity will be revisited more formally in Chapter 3. A further property that we have made implicit use of in our prior discussions concerning roots is that

iv. Polynomials of order n are dominated by the term $a_n x^n$ as $x \to \pm\infty$.

This property is related to Property iii. In particular, for a polynomial of even order, the value of the function as $x \to \pm\infty$ depends only on the sign of a_n. For example,

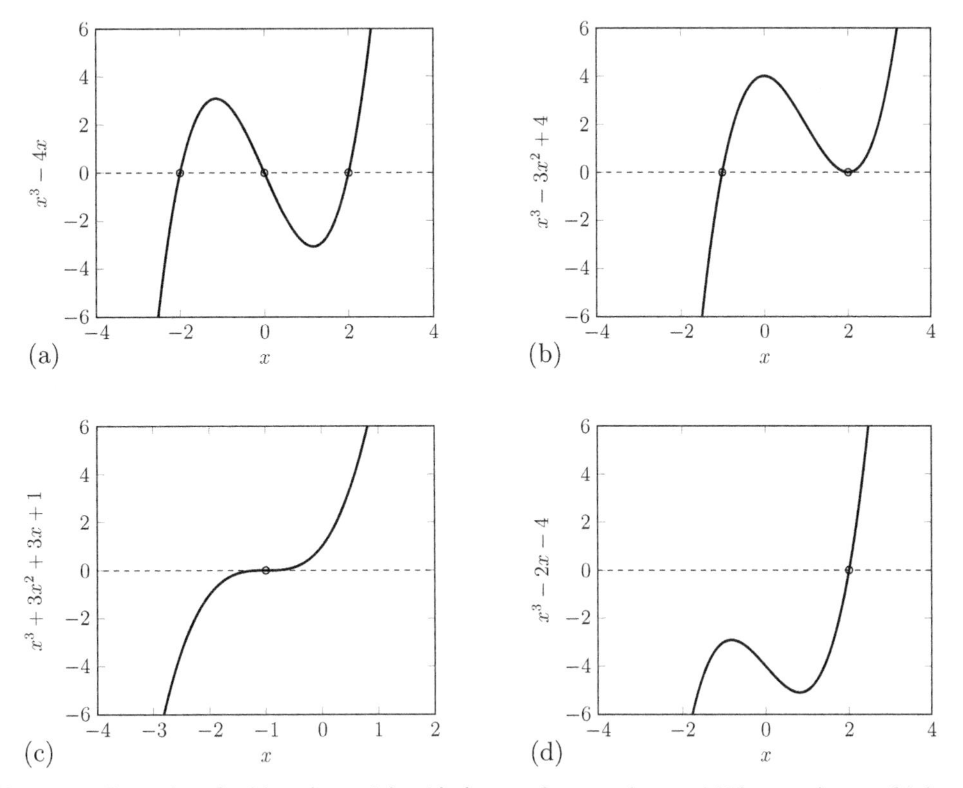

Figure 2.8 Examples of cubic polynomials with three or fewer real roots. (a) Three real roots, (b) three real roots with one repeated, (c) one repeat root, and (d) one real root.

$$x^2 + 4x - 2 \to \infty \quad \text{as } x \to \pm\infty$$
$$-x^2 + 4x - 2 \to -\infty \quad \text{as } x \to \pm\infty.$$

Whereas for a polynomial of odd order, the value of the function as $x \to \infty$ takes the same sign as a_n, and its value as $x \to -\infty$ takes the opposite sign as a_n. For example,

$$x^3 + 2x^2 - x - 2 \to \pm\infty \quad \text{as } x \to \pm\infty$$
$$-x^3 + 2x^2 - x - 2 \to \mp\infty \quad \text{as } x \to \pm\infty.$$

This property is exactly why a polynomial of even order might have no real roots: it does not necessarily have to cross the horizontal axis. A odd-order polynomial must, however, cross the horizontal axis and so has at least one root. Furthermore, an even-order polynomial has a range restricted to be between some minimum value and ∞ when a_n is positive, and some maximum value and $-\infty$ when a_n is negative. These properties should be evident from many of the previous figures.

A further property of polynomial functions is that the sum and products of any number of polynomial functions are also polynomial functions. Furthermore, the composite function formed from any string of polynomial functions is also a polynomial function. Consider the functions $f(x)$, an nth-order polynomial, and $g(x)$, an mth-order polynomial, with $m > n$ and a, b, c real constants. The following yet further properties of polynomials should then be clear,

v. $af(x) + bg(x)$ is, at most, an mth-order polynomial
vi. $c \times f(x) \times g(x)$ is an $(n \times m)$-order polynomial
vii. $(f \circ g)(x)$ and $(g \circ f)(x)$ are both $(m \times n)$-order polynomials.

These properties are demonstrated in the following examples.

EXAMPLE 2.12

If $f(x) = 2x^2 + 2$ and $g(x) = x^3 + 2x + 1$ demonstrate that functions resulting from the following are also polynomial. State the order of the resulting polynomial function in each case.

a. $f(x) + g(x)$
b. $f(x) \times g(x)$
c. $(f \circ g)(x)$
d. $(g \circ f)(x)$

Solution

a. $h(x) = (2x^2 + 2) + (x^3 + 2x + 1) = x^3 + 2x^2 + 2x + 3$. A third-order polynomial.
b. $k(x) = (2x^2 + 2) \times (x^3 + 2x + 1) = 2x^5 + 6x^3 + 2x^2 + 4x + 2$. A fifth-order polynomial.
c. $(f \circ g)(x) = 2x^6 + 8x^4 + 4x^3 + 8x^2 + 8x + 4$. A sixth-order polynomial.
d. $(g \circ f)(x) = 8x^6 + 24x^4 + 28x^2 + 13$. A sixth-order polynomial.

EXAMPLE 2.13

If $j(x) = 2x^5 - x^4 + 2x - 2$ and $k(x) = 4x^5 + 3x^3$, determine values of a and b such that $aj(x) + bk(x)$ is *not* fifth order.

Solution

Property v. is appropriate here. Note that

$$aj(x) + bk(x) = (2a + 4b)x^5 - ax^4 + 3bx^3 + 2ax - 2a$$

which will be fourth order if $2a + 4b = 0$, that is, $a = -2b$ for any $b \neq 0$. For example, if $b = 1$ then $a = -2$ and

$$aj(x) + bk(x) = 2x^4 + 3x^3 - 4x + 4$$

which is fourth order. In the other case that $b = 0$ and $a = 0$, $aj(x) + bk(x) \equiv 0$, which is zeroth order, or a constant.

2.3.2 Rational functions

It should be clear from our previous discussions that the functions

$$f(x) = x^{-1} \quad \text{and} \quad g(x) = \frac{2x+1}{3x^2+x-2},$$

are *not* examples of polynomial functions. In fact, these functions are examples of the broader class of *rational functions*. A rational function is any function that can be expressed as an algebraic fraction with numerator and denominator both given by polynomial expressions.

We can write the general form of a rational function as

$$f(x) = \frac{a_n x^n + a_{n-1}x^{n-1} + \cdots + a_1 x^1 + a_0}{b_m x^m + b_{m-1}x^{m-1} + \cdots + b_1 x^1 + b_0} \tag{2.10}$$

where $n, m \in \mathbb{N}$ and $a, b, \in \mathbb{R}$. It should be clear that all polynomial functions are also rational functions; that is, the polynomial function (2.9) is simply the special case of Eq. (2.10) with $m = 0$ and $b_m = 1$.

As with polynomial functions, it is possible to define the order of a rational function. However, given that the order of the numerator and denominator polynomials are, in general, distinct and independent, we define a rational function's order with two numbers. For example,

$$g(x) = \frac{x^3 + 6x^2 - x - 30}{x^2 + x + 1} \tag{2.11}$$

has a third-order numerator and a second-order denominator, and $g(x)$ is said to have "order 3/2."[2] Alternatively, we might say that $g(x)$ is a cubic/quadratic rational function. We see that Eq. (2.10) is of order n/m.

It should be clear that the functions resulting from the addition, multiplication, and composition of rational functions are also rational functions. However, the resulting functions will be extremely complicated to manipulate algebraically.

We now proceed to discuss the properties of rational functions. Some basic properties are as follows.

i. They are smooth and continuous on every interval of the real numbers that does not contain a root of the denominator function.

ii. They have domain consisting of the real numbers *excluding* roots of the denominator function.

[2] It is important to understand that by order 3/2 we do not mean 3 divided by 2 = order 1.5. The statement of the order is simply denoted as n/m and it is clear that a rational function of order 3/2 is quite distinct from a rational function of order 6/4.

These statements imply that, despite being similar in construction to polynomial functions, rational functions can have very different properties. These different properties relate to the existence of real roots of the denominator function.

To understand why the existence of roots of the denominator is so important, we plot Eq. (2.11), whose denominator has no real roots, and also

$$h(x) = \frac{x^3 + 6x^2 - x - 30}{x^2 + x - 12} \tag{2.12}$$

which has the same numerator but a denominator with roots at $x = 3$ and $x = -4$. These plots are shown in Figure 2.9(a) and (b), respectively. The figure shows that, despite having an unusual shape compared to a polynomial function, function $g(x)$ is smooth, continuous and has three roots at $x = -5$, $x = -3$, and $x = 2$, corresponding to the roots of the numerator function. Function $h(x)$ has these same roots arising from the same numerator, but also has *discontinuities* at $x = 3$ and $x = -4$ which correspond to the roots of the denominator. We might say that the rational function "blows up" at the roots of the denominator function and leads to a *singularity*.

To understand why rational functions behave in this way around the roots of their denominator, we consider the very simple function

$$l(x) = \frac{1}{x}$$

It should be clear that $l(x)$ is a 0/1-order rational function with a constant numerator function, 1, and a linear denominator function, x. The denominator has a single real root at $x = 0$ and so we might expect $l(x)$ to posses a discontinuity at this value. To see why this is so, we evaluate $l(x)$ at points successively close to $x = 0$ from above, this is starting from $x = 1$ and moving toward $x = 0$; numerical values are shown in the top portion of

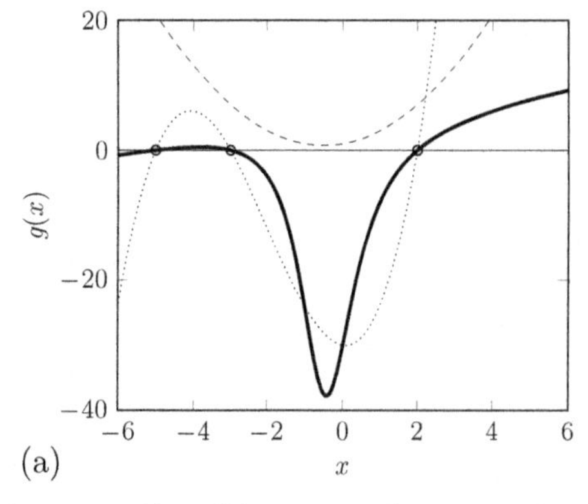

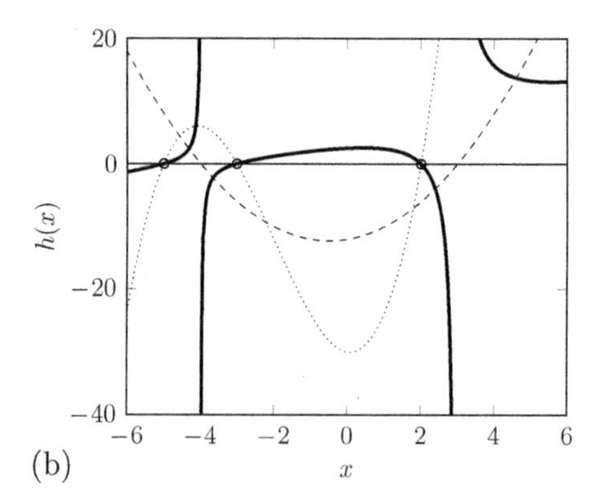

Figure 2.9 Plot of the rational functions (–), together with their numerator (· · ·) and denominator (– –). (a) $g(x)$ given by Eq. (2.11) and (b) $h(x)$ given by Eq. (2.12).

Table 2.4 Evaluations of the function $l(x) = 1/x$ close to $x = 0$

x	1	0.1	0.01	0.001	0.0001	0.00001	0.000001	0.0000001	...
$l(x)$	1	10	100	1000	10,000	100,000	1,000,000	10,000,000	...
x	−1	−0.1	−0.01	−0.001	−0.0001	−0.00001	−0.000001	−0.0000001	...
$l(x)$	−1	−10	−100	−1000	−10, 000	−100, 000	−1, 000, 000	−10, 000, 000	...

Table 2.4. If we extrapolate the patten, we see that as we approach $x = 0$, the function approaches a positive infinite value. Furthermore, if we approach $x = 0$ from below, that is starting at $x = -1$, the function approaches a negative infinite value. This jump from infinitely negative to infinitely positive values as x moves across the singularity at 0 is the reason for the discontinuity.

A very similar argument can be used close to the roots of the denominator in any rational function. Such discontinuities at these values of x account for the constraints on the continuity and smoothness properties stated above. The rational function is said to be *undefined* at the roots of the denominator and we must remove such points from its domain.

We note that this discussion necessarily assumes that the rational function under consideration is in its simplest form; that is, any common factors of the numerator and denominator have been canceled. In the process of deriving rational functions in practical situations, one may inadvertently work with functions under the false assumption that they are in their simplest form, and this needs care. For example, consider the function

$$m(x) = \frac{x^3 - 3x^2 - 4x + 12}{x + 2} \tag{2.13}$$

At first glance, this is a rational function of order 3/2, defined on the entire real line apart from $x = -2$ (the single root of the denominator). However, a plot of $m(x)$ shown in Figure 2.10 appears in contrast to this. In particular, we note that the function does not possess a discontinuity at $x = -2$ and takes a shape consistent with a quadric polynomial (a rational function of order 2/1). The reason for this becomes clear after a little exploration of the numerator and denominator functions: *both* functions have a root at $x = -2$, which can also be seen from Figure 2.10. The implication of this is that the numerator can be factorized with a $(x + 2)$ term that cancels with the denominator. Some algebraic manipulation of Eq. (2.13) then demonstrates that

$$m(x) = \frac{x^3 - 3x^2 - 4x + 12}{x + 2} = \frac{(x + 2)(x^2 - 5x + 6)}{x + 2} = x^2 - 5x + 6$$

The range of a rational function is clearly determined by the presence of singularities and also the behavior of the function as $x \to \pm\infty$. This *asymptotic* behavior is determined by the relative size of n and m and the particular possibilities are summarized by the following properties.

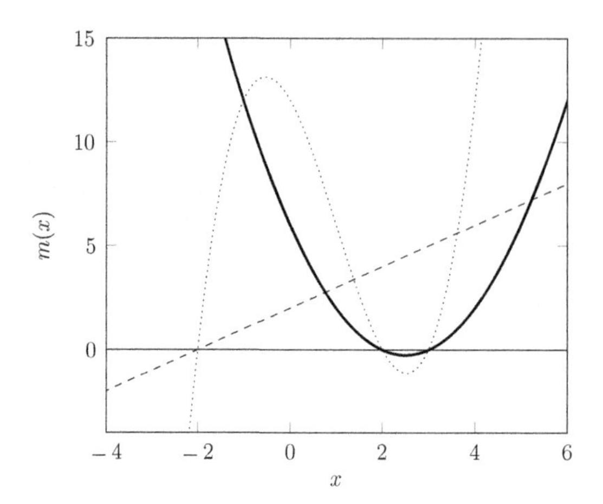

Figure 2.10 Plot of Eq. (2.13) (–), together with its numerator (· · ·) and denominator (– –).

iii. If $n = m$, a rational function has a horizontal asymptote equal to a_n/b_m.
iv. If $n < m$, a rational function has a horizontal asymptote equal to 0.
v. If $n > m$, there is no horizontal asymptote.

The term *horizontal asymptote* may be unfamiliar to you. From a practical viewpoint, if a function has a horizontal asymptote equal to c it simply means that the function tends to value c as the independent variable increases to either $\pm\infty$. We would write $f(x) \to c$ as $x \to \pm\infty$. The particular case that $f(x) \to c$ as $x \to \infty$ is referred to as a *right-hand asymptote*, and $f(x) \to c$ as $x \to -\infty$ is referred to as a *left-hand asymptote*. While the left- and right-hand asymptotes always occur together for rational functions, other classes of function can have one without the other, or indeed different valued asymptotes in either direction.

The reasons behind these asymptotic properties are easily understand if we recall that polynomials are dominated by the term with the largest order as $x \to \pm\infty$. With this in mind, it should be clear from Eq. (2.10) that, as $x \to \pm\infty$, a rational function is dominated by

$$\frac{a_n x^n}{b_m x^m} = \frac{a_n}{b_m} x^{n-m}$$

Therefore,

- If $n = m$, the rational function tends to be $\frac{a_n}{b_m} x^0 = \frac{a_n}{b_m}$ as $x \to \pm\infty$, that is, Property iii.
- If $n < m$, then $d = m - n > 0$ and the rational function is dominated by $\frac{a_n}{b_m}\frac{1}{x^d} \to 0$ as $x \to \pm\infty$, that is, Property iv.
- If $n > m$, then $d = n - m > 0$ and the rational function is dominated by $\frac{a_n}{b_m} x^d$ which is unbounded as $x \to \pm\infty$ and so behaves like a polynomial of order d, that is, Property v.

Examples of these three asymptotic properties are shown in Figure 2.11(a)-(c), respectively.

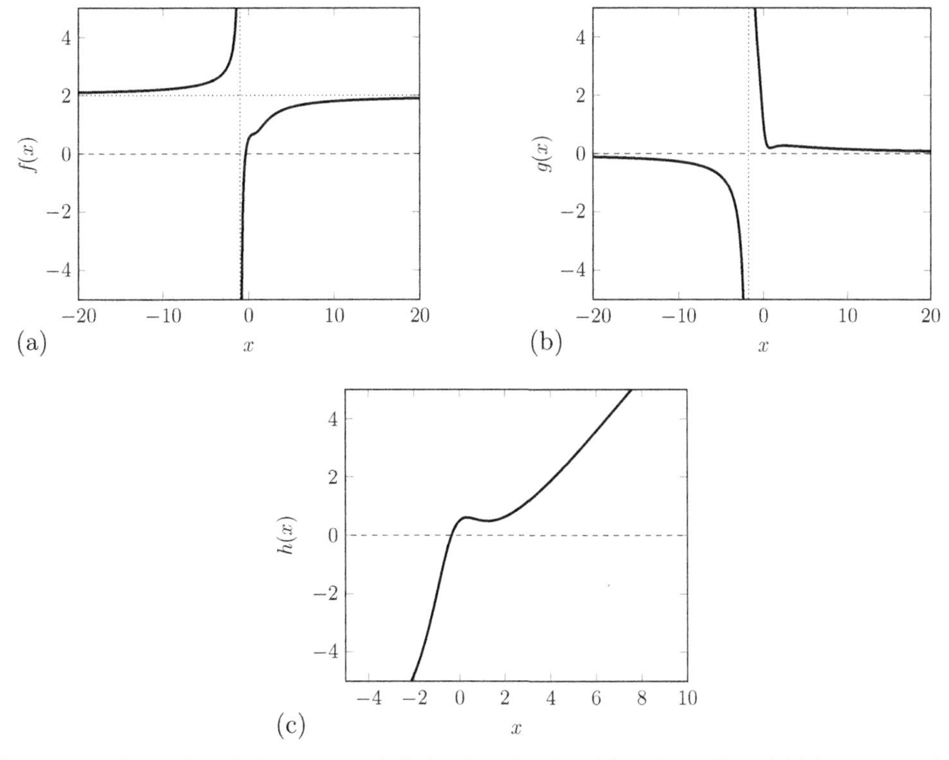

Figure 2.11 Examples of the asymptotic behavior of rational functions. Plots (a)-(c) correspond to Properties iii.-v., respectively. (a) $f(x) = \frac{2x^3-2x^2+2x+1}{x^3+x+2} \sim 2$ as $x \to \infty$, (b) $g(x) = \frac{2x^2-2x+1}{x^3+2x^2+x+1} \sim \frac{2}{x}$ as $x \to \infty$, and (c) $h(x) = \frac{x^3-2x^2+2x+1}{x^2+x+2} \sim x$ as $x \to \infty$.

In the "bounded" cases, that is, if a rational function possesses Property iii. or iv., the range would be determined by the behavior of the function at finite x. In the case of Property v., that the rational function is unbounded means that the range will be determined by the function's behavior at large $\pm x$. Of course, if the denominator of any rational function has real roots, the associated singularities mean that the rational function has a large range within the real numbers. This is true for all relative values of n and m. These ideas are illustrated in the following examples.

EXAMPLE 2.14

Determine the horizontal asymptotes (if any) of the following functions. Use a computer to plot each function to check your answer and also to investigate the range.

a. $f(x) = \frac{3x^2-2x+2}{2x^2+x+1}$

b. $g(y) = \frac{5y^2}{y^4+8y^3+24y^2+32y+16}$

c. $h(z) = \frac{2z^5-5}{z^2+2z+5}$

d. $k(p) = \frac{2p^4-5}{p^2+2p+5}$

Solution

a. $f(x) \to 1.5$ as $x \to \pm\infty$. The range is determined by the behavior of the function at finite values of x. We note that the denominator has no real roots and so the range will be finite. This is seen in Figure 2.12(a).

b. $g(y) \to 0$ as $y \to \pm\infty$. The denominator has a single real root, leading to a singularity and a discontinuity in $g(y)$. The range is seen to be $\mathbb{R}^+$ in Figure 2.12(b).

c. $h(z) \to \pm\infty$ as $z \to \pm\infty$ and so has no horizontal asymptotes. The denominator has no real roots and the leading order behavior is an odd-order polynomial. The function can therefore take all positive and negative values and the range is $\mathbb{R}$, as can be seen in Figure 2.12(c).

d. $k(p) \to \pm\infty$ as $p \to \infty$ and so has no horizontal asymptotes. The denominator has no real roots and the leading order behavior is an even-order polynomial. The function can therefore take all values above its minimum value, as can be seen in Figure 2.12(d).

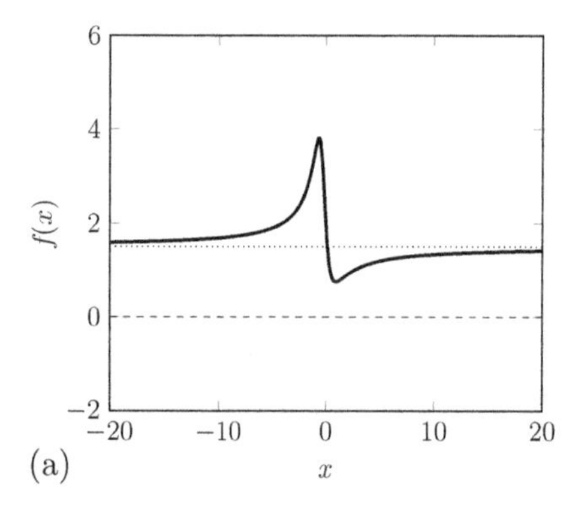

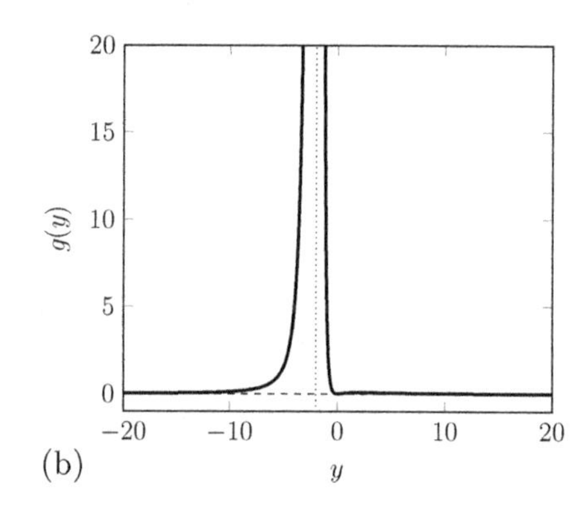

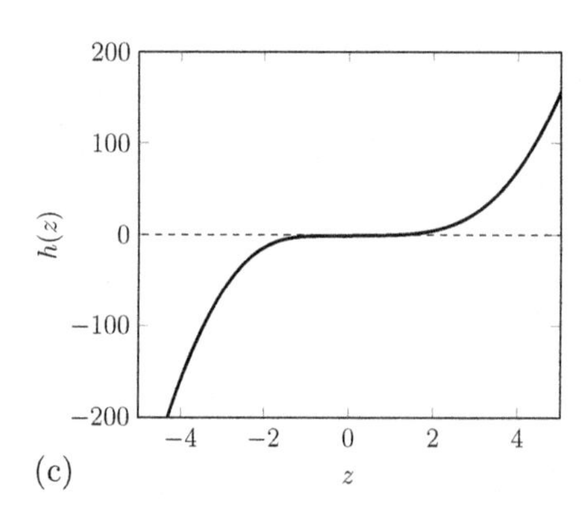

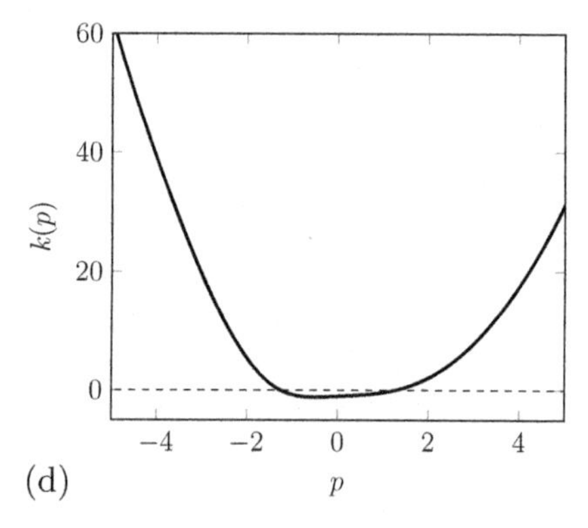

Figure 2.12 Plots for Example 2.14. (a) $f(x) = \frac{3x^2-2x+2}{2x^2+x+1}$, (b) $g(y) = \frac{5y^2}{y^4+8y^3+24y^2+32y+16}$, (c) $h(z) = \frac{2z^5-5}{z^2+2z+5}$, and (d) $k(p) = \frac{2p^4-5}{p^2+2p+5}$.

It is clear from Example 2.14(a) that, in the absence of any roots of the denominator, the range of a rational function with horizontal asymptotes is determined by the maximum and minimum values that the function can take for finite x. This is also true for even-order polynomials, as previously mentioned. We return to strategies for calculating the maximum and minimum values of functions, and therefore such ranges, in Chapter 4 on using differential calculus.

Now that we have introduced the concept of horizontal asymptotes, it is interesting to point out that the roots of the denominator in a rational function can be considered as examples of *vertical asymptotes*. These can be seen by the vertical lines placed at the roots of the denominator in Figures 2.11(a), (b) and 2.12(b), for example. The meaning of the terms *upward asymptote* and *downward asymptote* should then be self evident as the vertical analogue of left and right asymptotes. Figure 2.12(b), for example, demonstrates that a rational function can posses an upward-only asymptote and Figure 2.11(a) and (b) demonstrates that rational functions can also posses simultaneous upward and downward asymptotes.

We return briefly to rational functions that possess Property v., for example, the functions appearing in Figures 2.11(c) and 2.12(c), (d). As we now know, such functions will be dominated by the ratio the leading-order terms in the numerator and denominator as $x \to \pm\infty$. It is therefore possible to consider that these functions possess asymptotic behavior determined not by horizontal lines, but by polynomials. For example, we have that

$$h(x) = \frac{2x^5 - 5}{x^2 + 2x + 5} \sim 2x^3 \quad \text{as } x \to \pm\infty \tag{2.14}$$

which can be clearly seen in Figure 2.13(a). Furthermore, the asymptotic behavior of

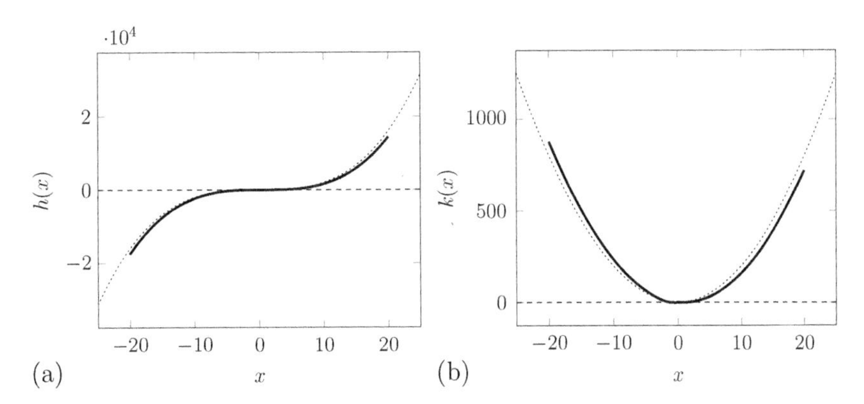

Figure 2.13 Asymptotic behavior of rational functions that possess Property v. (a) Asymptotic behavior of rational function (2.14) (–) and $2x^3$ ($\cdots$). (b) Asymptotic behavior of rational function (2.15) (–) and $2x^2$ ($\cdots$).

$$k(x) = \frac{2x^4 - 5}{x^2 + 2x + 5} \sim 2x^2 \quad \text{as } x \to \pm\infty \tag{2.15}$$

can be seen in Figure 2.13(b).

Note that Wolfram Alpha can be used to determine vertical and horizontal asymptotes. Again, the command is very intuitive and is simply `asymptote`.

EXAMPLE 2.15

Use Wolfram Alpha to confirm the horizontal and vertical asymptotes of

$$f(x) = \frac{2x^3 - 2x^2 + 2x + 1}{x^3 + x + 2} \quad \text{and} \quad g(y) = \frac{5y^2}{y^4 + 8y^3 + 24y^2 + 32y + 16}$$

as illustrated in Figures 2.11(a) and 2.12(b).

Solution

Using the command

```
asymptotes (2x^3-2x^2+2x+1)/(x^3+x+2)
```

we obtain the results

$$\text{Horizontal asymptote:} \quad \frac{2x^3 - 2x^2 + 2x + 1}{x^3 + x + 2} \to 2 \quad \text{as } x \to \pm\infty$$

and

$$\text{Vertical asymptote:} \quad \frac{2x^3 - 2x^2 + 2x + 1}{x^3 + x + 2} \to \pm\infty \quad \text{as } x \to -1$$

Similarly, the command

```
asymptotes (5y^2)/(y^4+8y^3+24y^2+32y+16)
```

results in

$$\text{Horizontal asymptote:} \quad \frac{5y^2}{y^4 + 8y^3 + 24y^2 + 32y + 16} \to 0 \quad \text{as } y \to \pm\infty$$

and

$$\text{Vertical asymptote:} \quad \frac{5y^2}{y^4 + 8y^3 + 24y^2 + 32y + 16} \to \infty \quad \text{as } y \to -2$$

Both results are consistent with the figures.

2.3.3 Exponential functions

Exponential functions are distinct from polynomial and rational functions in that the independent variable, x, say, appears in the *exponent*, that is the power. The following functions are examples of exponential functions.

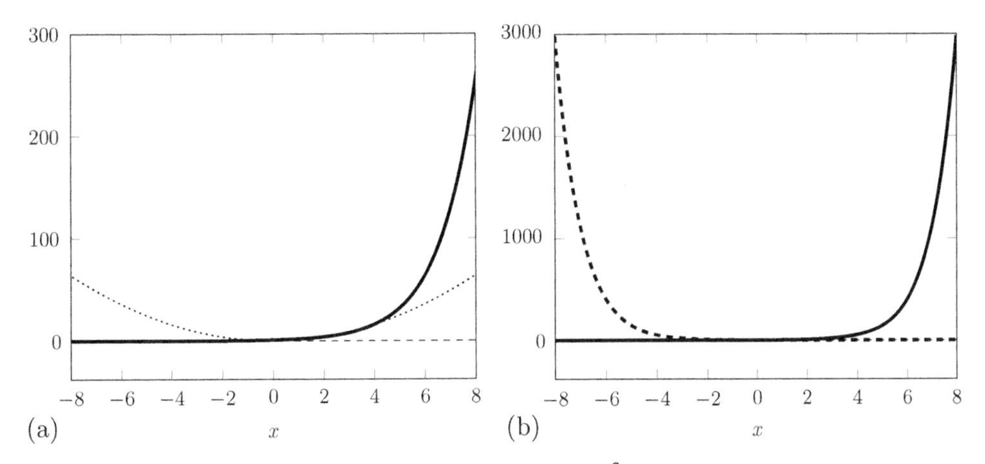

Figure 2.14 Introducing exponential functions. (a) 2^x (–) and x^2 (···). (b) e^x (–) and e^{-x} (···).

$$4^x, \quad 60^{-x}, \quad e^x, \quad e^{-2x}+1, \quad e^{-(x^2+x-2)}$$

Note that e is the irrational number discussed in Chapter 1.

This is the first time that we have seen functions that are not based on polynomials. Exponential functions are distinct from polynomials but can be considered as providing an additional means of giving an output to a particular input. The inclusion of exponential functions to our mathematical armory merely provides an additional set of operations that we can work with.

Since the independent variable appears in the exponent, exponential functions can grow very quickly as x increases. This property is demonstrated in Figure 2.14(a) with a comparison between the exponential function 2^x and the quadratic polynomial function x^2.

For reasons that will become apparent in later chapters, the exponential function

$$f(x) = e^x \tag{2.16}$$

is of particular importance in many physical and financial systems and will form the focus of our discussion of exponential functions here. Although we are restricting our attention to e^x, e is simply a particular real number and, in principle, there is no difference between the properties of e^x and a^x, where a is any real number greater than 1.

Note the alternative equivalent notation

$$\exp(x) \equiv e^x$$

which is often useful when we need to write expressions in the exponent. We shall use these two forms interchangeably throughout this book.

An important variation of the "growing" exponential e^x is

$$g(x) = \frac{1}{e^x} = e^{-x} \tag{2.17}$$

which we call the *decaying exponential* for reasons that are obvious from Figure 2.14(b) where e^x and e^{-x} are compared. It should be clear from the plots of $f(x)$ and $g(x)$ that the basic exponential functions have the following properties.

i. It is smooth and continuous
ii. It has the domain $\mathbb{R}$
iii. It has range $\mathbb{R}^+$.
iv. $e^{\pm x} \to 0$ as $x \to \mp\infty$.

It should also be clear that the basic exponential functions are one-to-one mappings. Property iii. arises from the fact that $f(x) = e^x$ left-hand asymptotes to zero and $g(x) = e^{-x}$ right-hand asymptotes zero, that is, Property iv. Each of these properties can be seen clearly in Figure 2.14(b).

EXAMPLE 2.16

Demonstrate the respective left- and right-hand asymptotes of $\exp(x)$ and $\exp(-x)$ using numerical values.

Solution

Using a calculator or Excel, we quickly see that

- $\exp(-1) \approx 0.37$
- $\exp(-10) \approx 5.4 \times 10^{-5}$
- $\exp(-100) \approx 3.7 \times 10^{-44}$

These values are consistent with the behavior shown in Figure 2.14(b) for both e^x as $x \to -\infty$ and e^{-x} as $x \to \infty$.

As it currently stands, the basic exponential functions do not look particularly interesting. However, their importance in every aspect of actuarial science and finance cannot be underestimated. One particularly useful aspect is the limited range and the related fact that any composite function that is formed from a string ending in either function $\exp(\pm x)$ is necessarily positive.

EXAMPLE 2.17

If $f(z) = \exp(z)$ and $g(z) = \exp(-z)$, use a computer to visually explore the range of each of the following composite functions.

a. $(f \circ h)(z)$, where $h(z) = z^2 - 5z + 6$.
b. $(g \circ h)(z)$, where $h(z)$ is as in a.
c. $(g \circ m)(z)$, where $m(z) = \frac{1}{z^2-4}$

Solution

The composite functions are plotted in Figure 2.15.

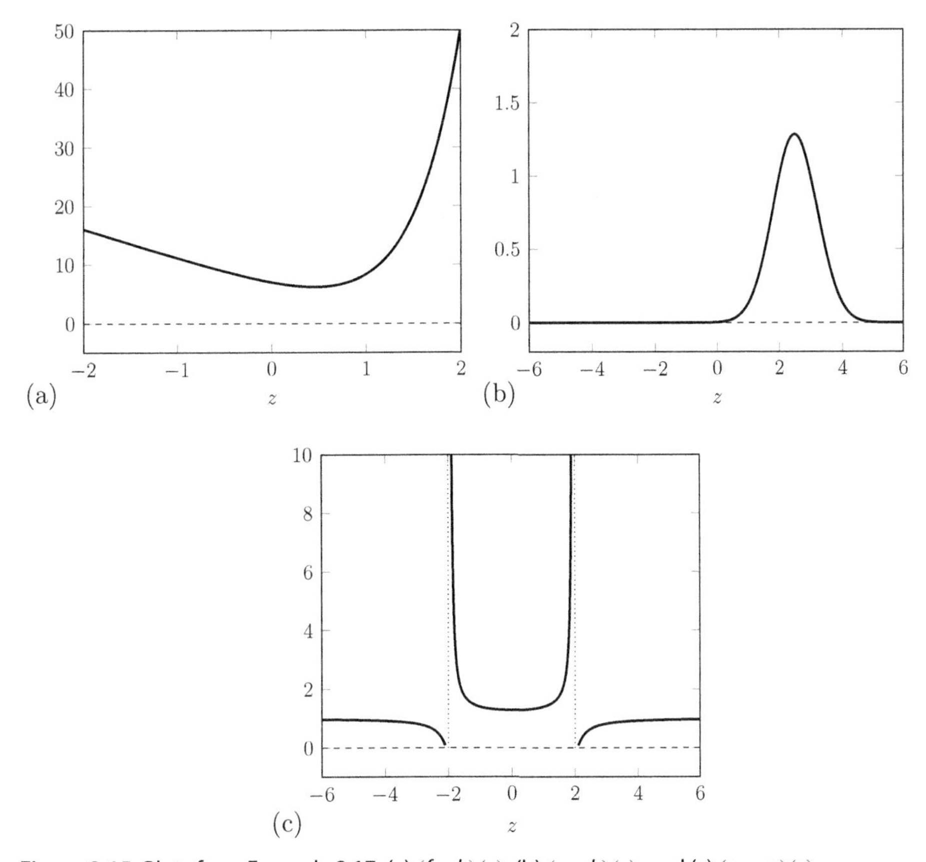

Figure 2.15 Plots from Example 2.17. (a) $(f \circ h)(z)$, (b) $(g \circ h)(z)$, and (c) $(g \circ m)(z)$.

a. The composite function in question is $\exp(z^2 - 5z + 6)$. The fact that it is exponential immediately restricts the range to positive real numbers. Furthermore, the quadratic exponent takes unbounded positive values as $z \to \pm\infty$ but has a minimum of -0.25 at $z = 2.5$; these translate to a positive minimum of the composite. The composite function therefore has a range equal to the set of positive real numbers greater than $e^{-1/4}$, more formally we might write that the range is $\{y \in \mathbb{R} : y \in [e^{-1/4}, \infty)\}$.

b. The composite function in question is $\exp(-(z^2 - 5z + 6))$. The range is again restricted to positive real numbers. However, the exponent now has a maximum value of 0.25 at $z = 2.5$ and this translates to a maximum value of $e^{1/4}$ for the composite function and a restricted range. The range is then $\{y \in \mathbb{R} : y \in (0, e^{1/4}]\}$.

c. The composite function in question is $\exp(-1/(z^2 - 4))$. We note that the exponent has singularities at $p = \pm 2$ and a turning point at $z = 0$. The exponent is positive and greater than 0.25 over the interval $(-2, 2)$ and is negative for $z \in \{(-\infty, -2) \cup (2, \infty)\}$. This translates to a range for the composite function of $\{z \in \mathbb{R} : z \in (0, 1) \cup [e^{1/4}, \infty)\}$. Note also that we must modify the domain of the composite function to remove $z \pm 2$; there are upward asymptotes at these points.

The composite notation used in Example 2.17 is rather cumbersome and, in practice, one would usually work directly with expressions such as $e^{-(z^2-4)}$, for example.

A further property of the basic exponential functions is that

v. They have no real roots.

The property follows immediately from Properties iii. and iv., that is, neither e^x nor e^{-x} takes the value 0. However, that is not to say that no exponential functions have real roots. To demonstrate this, we show plots of the following exponential functions in Figure 2.16.

$$e^{2x} - 2, \quad e^{3x} + e^{2x} - e^2, \quad e^{x^2-1} - 4$$

The figure clearly shows real roots for each function. The next question is, given an exponential function with real roots, how does one find these roots? We begin to answer this question in the following example.

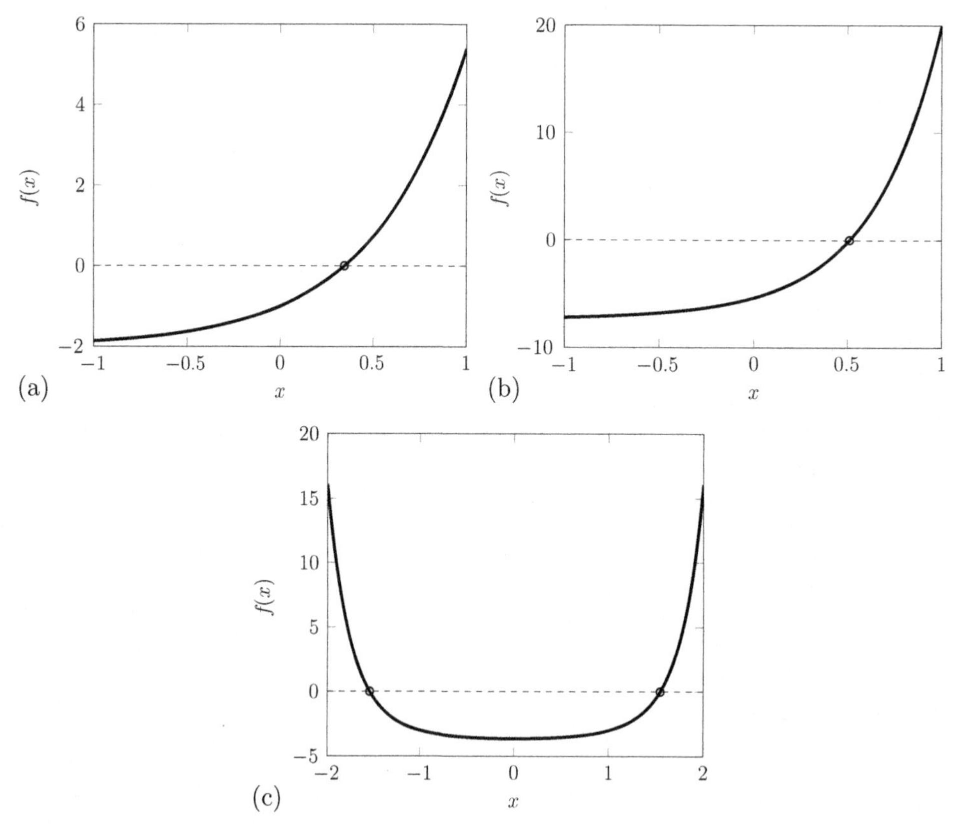

Figure 2.16 Roots of particular composite exponential functions. (a) $f(x) = e^{2x} - 2$, (b) $f(x) = e^{3x} + e^{2x} - e^2$, and (c) $f(x) = e^{x^2-1} - 4$.

EXAMPLE 2.18

Find the single real root of the function $g(x) = e^{2x} - 2$.

Solution

We require the single real value of x such that $g(x) = 0$. Note that we know that there is only a single root from Figure 2.16. The algebraic manipulations proceed as follows:

$$\begin{aligned} e^{2x} &= 2 \\ \Rightarrow (e^x)^2 &= 2 \\ \Rightarrow e^x &= \overset{+}{\sqrt{2}} \end{aligned}$$

Note that we know not to consider the negative square root of $\sqrt{2}$ since e^x has a strictly positive range. At this stage, we need a means of reversing the basic exponential function; that is, we need a method of calculating x such that $e^x = 2$. In the absence of such a method, trial and error can be used to give that $x \approx 0.34657$. This value is consistent with the position of the root evident in Figure 2.16 and can be confirmed using Wolfram Alpha or Excel's Goal Seek command.

Example 2.18 motivates *logarithmic* functions which we introduce in the next section as being the *inverse* of exponential functions. However, it should be clear that in more complicated examples of exponential functions, it can be extremely difficult to find roots. Often explicit expressions for the roots do not exist and one has to resort to numerical or computational methods.

EXAMPLE 2.19

It is clear from Figure 2.16 that the function $g(x) = e^{3x} + e^{2x} - e^2$ has a single root at $x \approx 0.5$. Find the value of this root, accurate to four decimal places.

Solution

Two approaches can be taken:

1. We use numerical methods to solve $g(x)$ directly. For example, Wolfram Alpha or Goal Seek leads to $x \approx 0.5099$.
2. We note that $g(x)$ is a cubic expression in e^x. That is, if we define $y = e^x$ we can write $g(y) = y^3 + y^2 - e^2$. This is again solved using Wolfram Alpha or Goal Seek to show that $y = e^x \approx 1.6651$ (note that it must be real and positive) and so $x \approx 0.5099$.

Although Method 2 used to find the root in Example 2.19 may seem like an unnecessary complication at this stage (particularly, as we are yet to discuss how to find the unknown power using logarithms), it does hint at an important strategy for solving some expressions. This is further elaborated in Example 2.20.

EXAMPLE 2.20

Find all real roots of the following functions, accurate to four decimal places.

a. $f(x) = e^{2x} - 5e^{x} + 6$

b. $g(x) = e^{4x} - 9e^{2x} + 20$

Solution

a. We note that $f(x)$ is a quadratic in e^x. If $y = e^x$ then $f(y) = y^2 - 5y + 6$ which has roots at $y = 2$ and $y = 3$. These correspond to $x \approx 0.6931$ and 1.0986.

b. We note that $g(x)$ is a quadratic in e^{2x}. If $z = e^{2x}$ then $g(x) = z^2 - 9z + 20$ which has roots at $z = 4$ and $z = 5$. These correspond to $x \approx 0.6931$ and $x \approx 0.8047$.

The use of such *substitution techniques* is further explored in Question 2.10

2.3.4 Logarithmic functions

As mentioned in the previous section, *logarithmic* functions can be considered as the inverse of exponential functions. More fundamentally, taking the *logarithm* is the inverse operation to raising to a power in the same way that subtraction is the inverse operation to addition, and division is the inverse operation to multiplication. Indeed, a "logarithm" is simply another word for a "power."

For example, $4^2 = 16$ can be interpreted as saying that 2 is the power to which 4 must be raised to equal 16. Put another way, 2 is the logarithm which, to base 4, equals 16. We would write this mathematically as

$$4^2 = 16 \Leftrightarrow \log_4 16 = 2$$

The inverse nature of the logarithm operation should be clear from this expression. The use of $\Leftrightarrow$ simply means that the logic of the statement works in both directions.

Rather than getting distracted by the term *base*, it is sufficient to understand that a base is actually defined by the above expression. More specifically, the base is simply the positive number that needs to be raised by a power to equal another number.

In principle, we can use any positive real number as a base in a logarithm calculation. However, in most practical applications within actuarial science and finance, we would be concerned with base 10 or base e, and these will be our focus from now on. It is convenient to distinguish between logarithms to base 10 and logarithms to base e by use of log and ln, respectively. That is,

$$\log(x) \equiv \log_{10}(x) \text{ and } \ln(x) = \log_e(x)$$

The notation ln is read as the "natural log."[3]

[3] Note that care is needed when reading around this subject as some authors use log for both bases, with the hope that the context makes it clear which base is actually meant.

EXAMPLE 2.21

Evaluate the following logarithms and confirm that they are correct using the inverse operation.

a. $\log(0.1)$
b. $\log(4)$
c. $\ln(e)$
d. $\ln(2)$
e. $\log(100)$
f. $\log(976)$
g. $\ln(e^{10})$
h. $\ln(22.3)$

Solution

a. $\log(0.1) = -1 \Leftrightarrow 10^{-1} = 0.1$
b. $\log(4) \approx 0.60206 \Leftrightarrow 10^{0.60206} \approx 4$
c. $\ln(e) = 1 \Leftrightarrow e^1 = e$
d. $\ln(2) \approx 0.69315 \Leftrightarrow e^{0.69315} \approx 2$
e. $\log(100) = 2 \Leftrightarrow 10^2 = 100$
f. $\log(976) \approx 2.98945 \Leftrightarrow 10^{2.98945} \approx 976$
g. $\ln(e^{10}) = 10 \Leftrightarrow e^{10} = e^{10}$
h. $\ln(22.3) \approx 3.10459 \Leftrightarrow e^{3.10459} \approx 22.3$

The following algebraic properties of logarithms stem from the properties of exponents and powers. These are stated without further justification and you are asked to show that they are true in Question 2.14.

$$\begin{aligned} \log\left(a^b\right) &\equiv b\log(a), & \ln\left(a^b\right) &\equiv b\ln(a) \\ \log(ab) &\equiv \log(a) + \log(b), & \ln(ab) &\equiv \ln(a) + \ln(b) \\ \log\left(\tfrac{a}{b}\right) &\equiv \log(a) - \log(b), & \ln\left(\tfrac{a}{b}\right) &\equiv \ln(a) - \ln(b) \end{aligned}$$

EXAMPLE 2.22

Verify the following identities using a calculator.

a. $\log(12) = \log(6) + \log(2) = \log(3) + \log(4)$
b. $\ln(10) = \ln(20) - \ln(2) = \ln(15) - \ln(1.5)$
c. $\ln(100) = 2\ln(10)$

Solution

a. Each gives 1.07918. This arises as $12 = 6 \times 2 = 3 \times 4$.
b. Each gives 2.30259. This arises as $10 = 20/2 = 15/1.5$.
c. Each gives 4.60517. This arises as $100 = 10^2$.

Example 2.22(c) illustrates the practical use of logarithms, that is, they can be used to "bring down the exponent" in algebraic manipulations. With this in mind, you are invited to revisit Examples 2.18–2.19. For example, in Example 2.18, we can now see that the root of $e^{2x} - 2$ is actually $x = \ln(\sqrt{2})$. This has approximate numerical value 0.34657, as we had already determined.

Now that we are familiar with logarithmic operations, it is possible to briefly discuss the class of *logarithmic* functions. These are functions that contain a logarithmic operator. For example,

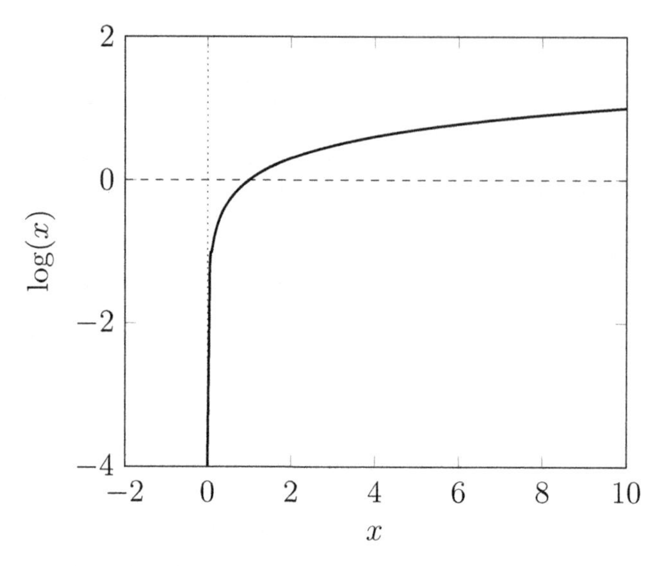

Figure 2.17 The logarithmic function, log(x).

$$\log(x), \quad \ln(3x+1), \quad \log_3(x-2)$$

are all logarithmic functions.

We proceed by explicitly discussing the properties of the basic logarithmic function

$$f(x) = \log(x)$$

and note that these properties extend to all functions $\log_a(x)$ with $a > 0$, including $\ln(x)$.

Some initial insight can be gained from Figure 2.17 which plots the function $f(x) = \log(x)$. In particular, we note that the basic logarithmic functions have the following obvious property.

i. They are smooth and continuous.

Since the values of the base are restricted to positive numbers, it is impossible to generate the logarithm of a negative number. This property arises from the fact that we unable to change the sign of a number by raising it to a particular power. Furthermore, the logarithm of a number in the interval $(0, 1)$ is negative and the logarithm of a number in the interval $(1, \infty)$ is positive. This leads us to the following further properties of basic logarithmic functions

ii. They have domain $\mathbb{R}^+$

iii. They have a range equal to $\mathbb{R}$

Furthermore, we see that the basic logarithmic functions are unbounded as $x \to \infty$. That is, they have no horizontal asymptotes. However, since log(0) is undefined, we could say that $x = 0$ leads to a vertical asymptote. Figure 2.17 shows that this asymptote is such that $\log(x) \to -\infty$ as $x \to 0$ and so is a downward asymptote.

The following property of the basic logarithmic functions should also be clear from Figure 2.17.

iv. They have a single root at $x = 1$

We note that the position of this root is independent of the particular base. This is true because any real number raised to power 0 is one. More specifically, to find the real root of the general expression $h(x) = \log_a(x)$, we would perform the following manipulations

$$\log_a(x) = 0 \Leftrightarrow x = a^0 = 1$$

Now that the fundamental properties of the basic logarithmic functions are established, we look at the properties of more complicated functions involving logarithms. However, we must be careful around the property that the basic logarithmic functions are undefined on $x \in (-\infty, 0]$. In particular, if we consider the composite function $h(x) = (f \circ g)(x)$ with $f(x) = \log(x)$ and $g(x)$ a rational function, the range of $g(x)$, the location of its roots and the existence of any singularities have implications for the behavior of the composite $h(x)$.

For example, consider

$$h(x) = \ln(x^2 - 4) \tag{2.18}$$

It should be immediately clear that the *argument* (the expression appearing inside the ln), $x^2 - 4$, takes positive values for $x \in (-\infty, -2)$ and $x \in (2, \infty)$, negative values for $x \in (-2, 2)$ and has roots at $x = \pm 2$. The function $h(x)$ is therefore undefined in the interval $x \in [-2, 2]$. Furthermore, despite $\ln(x)$ being undefined for negative x, $h(x)$ is defined on the interval $x \in (-\infty, -2)$ in addition to $x \in (2, \infty)$. The response of $h(x)$ to the roots of the argument are downward vertical asymptotes at $x = \pm 2$. Furthermore, $h(x) \to \infty$ as $x =\to \pm\infty$. The range of $h(x)$ is then $\mathbb{R}$. This behavior is shown in Figure 2.18(a).

Consider instead

$$l(x) = \ln\left(\frac{1}{x^2 - 4}\right) \tag{2.19}$$

The argument function has singularities points at $x = \pm 2$, takes negative values for $x \in (-2, 2)$, and positive values for $x \in (-\infty, -2)$ and $x \in (2, \infty)$; in addition, it has a horizontal asymptote to zero as $x \to \pm\infty$. The composite function $l(x)$ is therefore undefined for $x \in [-2, 2]$. In addition, it responds to the singular points by upward vertical asymptotes at $x = \pm 2$. Furthermore, the horizontal asymptote of the argument function causes $l(x)$ to decrease unbounded as $x \to \pm\infty$. The range of $l(x)$ is then $\mathbb{R}$. The behavior is shown in Figure 2.18(b). Note that $l(x) = -k(x)$.

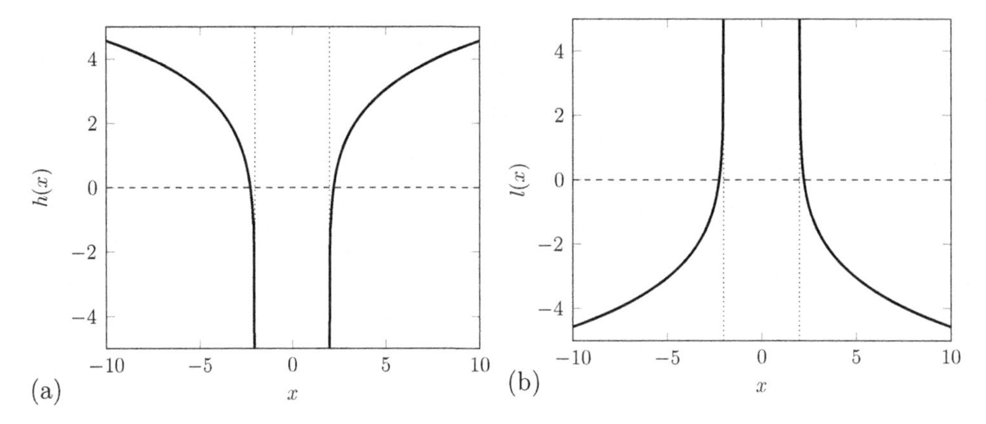

Figure 2.18 Plots of Eqs. (2.18) and (2.19). (a) $h(x) = \ln(x^2 - 4)$ and (b) $l(x) = \ln\left(\frac{1}{x^2-4}\right)$.

EXAMPLE 2.23

Demonstrate algebraically that $h(x) = -l(x)$ as defined in Eqs. (2.18) and (2.19).

Solution

We note that

$$\begin{aligned} l(x) &= \ln\left(\frac{1}{x^2-4}\right) \\ &= \ln\left((x^2-4)^{-1}\right) \\ &= -1 \times \ln\left(x^2-4\right) \\ &= -k(x) \end{aligned}$$

Note that we have used the ln operation to "bring down the power."

As with the composite exponential functions discussed in the previous section, finding roots of composite logarithmic functions is often difficult and one must typically resort to numerical approaches. However, there are of course some relatively simple cases which permit analytical solution.

EXAMPLE 2.24

Find the real roots of the logarithmic functions $h(x)$ and $l(x)$ defined in Eqs. (2.18) and (2.19).

Solution

These are examples where analytical progress can be made.

a.

$$\begin{aligned} &\ln(x^2-4) = 0 \\ \Rightarrow\ &e^{\ln(x^2-4)} = e^0 \end{aligned}$$

$$\Rightarrow x^2 - 4 = 1$$
$$\Rightarrow x^2 = 4 \Leftrightarrow x = \pm 2$$

b.

$$\ln\left(\frac{1}{x^2 - 4}\right) = 0$$
$$\Rightarrow -\ln(x^2 - 4) = 0$$
$$\vdots$$
$$\Rightarrow x^2 = 4$$
$$\Leftrightarrow x = \pm 2$$

Note that we could have noted that $l(x) = -k(x)$ and written down these roots immediately following part a.

These roots are confirmed by Figure 2.18.

2.3.5 Circular (trigonometric) functions

You will be familiar with the basic trigonometric operations: sin, cos, and tan, from your school days. We begin with a brief summary of what one might recall from school trigonometry.

Consider the right-angled triangle given in Figure 2.19(a). The angle θ can be related to the lengths of the opposite (O), adjacent (A), and hypotenuse (H) by the expressions

$$\sin\theta = \frac{O}{H}, \quad \cos\theta = \frac{A}{H}, \quad \tan\theta = \frac{O}{A} \tag{2.20}$$

For example, a right-angled triangle with $H = 5$ and angle $\theta = 30°$ must be such that $O = 5\sin(30°) = 2.5$ and $A = 5\cos(30°) \approx 4.3301$.

Of course, the terms "opposite" and "adjacent" are defined with respect to a particular angle. If we instead work in terms of angle $\beta = 90° - \theta$, the labels O and

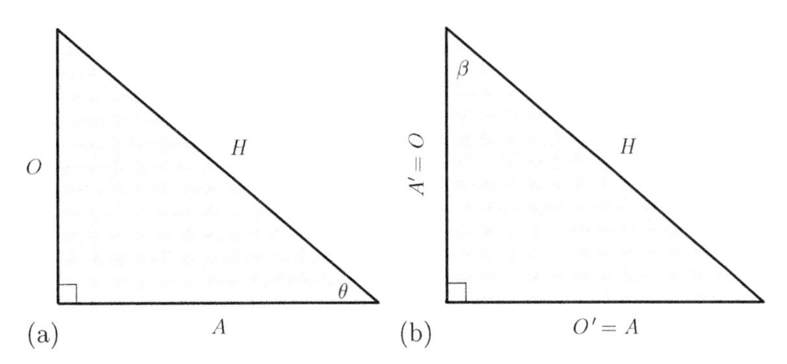

Figure 2.19 Right-angled triangles.

A are swapped over, as in Figure 2.19(b). The expressions in Eq. (2.20) therefore lead to the identities

$$\sin(\theta) \equiv \cos(90^\circ - \theta), \quad \cos(\theta) \equiv \sin(90^\circ - \theta), \quad \tan(\theta) \equiv \frac{1}{\tan(90^\circ - \theta)}$$

which of course assumes that $\tan(90^\circ - \theta) \neq 0$. In addition, expressions (2.20) are consistent with

$$\frac{\sin(\theta)}{\cos(\theta)} = \tan(\theta) \quad \sin^2(\theta) + \cos^2(\theta) = 1$$

The second identity follows immediately from Pythagoras' theorem, $H^2 = O^2 + A^2$.

The identities given above are only a small number of the very many trigonometric identities that are listed in Appendix A.

As you might imagine, our interest in trigonometric functions in this book is not motivated by triangles and simple geometry problems. Rather, trigonometric functions will prove very useful for their *cyclical* properties. The above descriptions hint at these properties. In particular, a circle consists of 360 degrees, denoted as 360°, and so any angle θ must, to all intents and purposes, be identical to the angle $\theta + 360^\circ n$, where $n \in \mathbb{Z}$. Going further, it should then be the case that

$$\begin{aligned} \sin(\theta) &\equiv \sin(\theta + 360^\circ n) \\ \cos(\theta) &\equiv \cos(\theta + 360^\circ n) \\ \tan(\theta) &\equiv \tan(\theta + 360^\circ n) \end{aligned} \tag{2.21}$$

where $n \in \mathbb{Z}$. These identities are indeed true and demonstrate the cyclical properties of the trigonometric functions. The cyclical nature of these trigonometric functions means that they are often referred to as *circular* functions. These two names will be used interchangeably in what follows.

In practice, it is typical to use *radians* rather than degrees when working with the circular functions outside of the study of simply geometry problems. A radian is the standard unit of angle measure in mathematics and is defined such that a circle has 2π radians. It should therefore be clear that an angle expressed in degrees, θ°, can be converted to radians and vice versa using the following expression which represents a simple scaling.

$$\frac{\theta^\circ}{360} \equiv \frac{\theta}{2\pi} \quad \text{and so} \quad \frac{\theta^\circ}{180} \equiv \frac{\theta}{\pi}$$

Wolfram Alpha, Excel, and other mathematical software use radians by default. You will, however, need to set your calculator to be in either degree "DEG" or radian "RAD" mode when evaluating the circular functions.

EXAMPLE 2.25

Convert the following angles to radians. Confirm the numerical value of $\cos(\theta)$ with your calculator in both degree and radian mode.

a. $30°$
b. $45°$
c. $60.24°$

Solution

a. $\theta = \frac{30}{180}\pi = \pi/6$
$\cos(30°) = \sqrt{3}/2 = \cos(\pi/6)$.

b. $\theta = \frac{45}{180}\pi = \pi/4$
$\cos(45°) = \sqrt{2}/2 = \cos(\pi/4)$.

c. $\theta = \frac{60.24}{180}\pi$
$\cos(60.24°) \approx 0.49637 \approx \cos\left(\frac{60.24}{180}\pi\right)$.

We proceed with a discussion of the properties of the basic circular functions; you should assume that radians are being used from this point on throughout this book. As we have seen, the operations sin, cos, and tan represent many-to-one mappings and so can be used to represent functions of real numbers

$$f(x) = \sin x, \quad g(x) = \cos x, \quad h(x) = \tan x$$

The cyclical nature of the functions is evident in Figure 2.20. In particular, we see that the sin and cos functions take the form of a repeating "wave." The "wavelength," that is the distance between equivalent points on consecutive cycles, is 2π. It is common to say that the basic sin and cos functions have *period* 2π. This property is stated mathematically as

$$\sin x \equiv \sin(x + 2\pi n)$$
$$\cos x \equiv \cos(x + 2\pi n)$$

where $n \in \mathbb{Z}$. These expressions are consistent with Eq. (2.21). Although the tan function does not take the form of a wave, it does repeat with period π. We therefore write

$$\tan x \equiv \tan(x + \pi n)$$

where $n \in \mathbb{Z}$. In addition, the tan function demonstrates singularities and discontinuities at $x = \pm\frac{\pi}{2}, \pm\frac{3\pi}{2}, \ldots$ with vertical asymptotes at these values. These arise from the identity

$$\tan x \equiv \frac{\sin x}{\cos x}$$

and correspond to roots of the cos function. How the tan function approaches the asymptote reflects whether the root of the cos function is approached from above or below as x moves forward or backward along the real line.

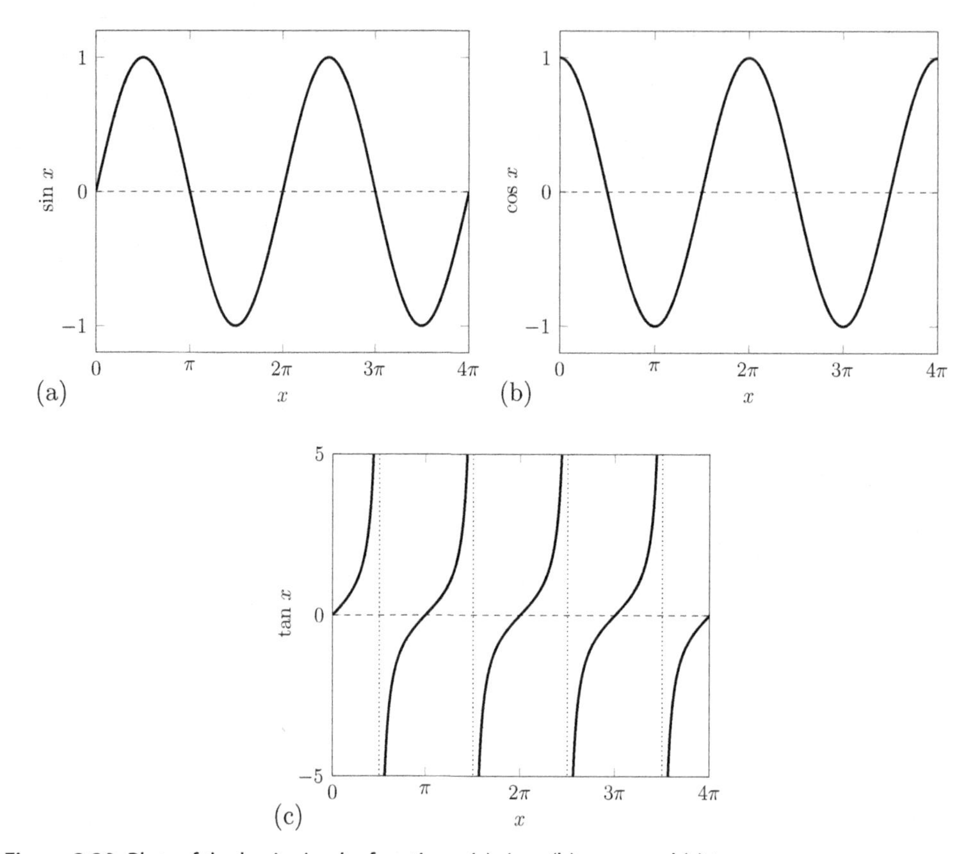

Figure 2.20 Plots of the basic circular functions. (a) $\sin x$, (b) $\cos x$, and (c) $\tan x$.

From Figure 2.20, we can immediately note the following properties of the circular functions:

i. The sin and cos functions are smooth and continuous. The tan function is smooth and continuous in intervals that do not contain $x = \pm\frac{\pi}{2}, \pm\frac{3\pi}{2}, \ldots$.

ii. The sin and cos functions have domain $\mathbb{R}$. The tan function is undefined at $x = \pm\frac{\pi}{2}, \pm\frac{3\pi}{2}, \ldots$ and these values are removed from $\mathbb{R}$ to form its domain.

iii. The sin and cos functions have range $\{y : y \in [-1, 1]\}$. The tan function has range $\mathbb{R}$.

iv. The sin, cos, and tan functions have an infinite number of roots:

- $\sin x$ has roots at $x = n\pi$ for $n \in \mathbb{Z}$
- $\cos x$ has roots at $x = \frac{(2n+1)\pi}{2}$ for $n \in \mathbb{Z}$
- $\tan x$ has roots at $n\pi$ for $n \in \mathbb{Z}$

It should also be evident from Figure 2.20 that the sin and cos functions are identical but for a translation in x, or, in angular terms, a *phase shift*. In particular, the cos function translated backward by $\pi/2$ is exactly the sin function. Mathematically, these statements can be generalized as

$$\sin(x) \equiv \cos\left(x - \frac{\pi}{2} + 2n\pi\right)$$
$$\cos(x) \equiv \sin\left(x + \frac{\pi}{2} + 2n\pi\right)$$

for $n \in \mathbb{Z}$.

As a closing comment on the basic circular functions, we note the further property that

v. $\sin(x)$ and $\tan(x)$ are odd functions and $\cos(x)$ is an even function.

The cyclical nature of the circular functions makes for interesting properties of their composite functions which we now turn to. We begin by considering the function

$$m(x) = \sin\left(x^2 - 4\right)$$

It is clear that the argument is a polynomial function with roots at $x = \pm 2$. Function $m(x)$ is shown graphically in Figure 2.21. The plot reveals a complicated sequence of roots that arise from the cyclical nature of the sin function and the quadratic nature of the argument.

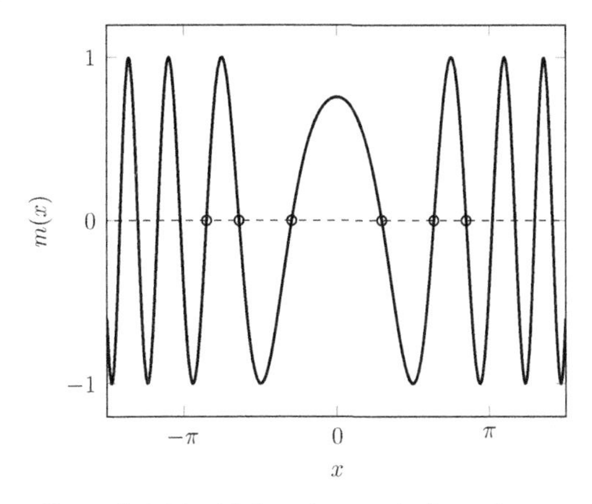

Figure 2.21 Plot of $m(x)$ in Example 2.26 with first six roots indicated.

EXAMPLE 2.26

Determine all real roots of the function $m(x) = \sin\left(x^2 - 4\right)$.

Solution

We begin by noting that the sin function has roots at πn for $n \in \mathbb{Z}$. It is therefore true that the real roots are given by the solution of the equation

$$x^2 - 4 = \pi n \Leftrightarrow x = \pm\sqrt{4 + k\pi} \quad \text{for } k \in \mathbb{N} \cup \{-1\}$$

The reason for the relatively complicated condition on the value of k is because we cannot have real square roots of negative values. We must therefore restrict the values of n to ensure that this does not happen. The reader is invited to check the numerical values of the first six roots, that is, when $k = -1, 0, 1$, against those roots indicated on Figure 2.21.

In order to calculate the range of composite functions of the form $\sin(f(x))$, we must first acknowledge that the basic $\sin(x)$ function takes its maximum value of 1 at $x = \pm\frac{\pi}{2}, \pm\frac{5\pi}{2}, \ldots$, and its minimum value of -1 at $x = \pm\frac{3\pi}{2}, \pm\frac{7\pi}{2}, \ldots$. In order for a composite function to have the range $[-1, 1]$, any function used as an argument in a sin function must have a range that spans at least one value from both of these sets. In the case of $m(x)$, the argument function has range $[-4, \infty]$ and so $m(x)$ clearly has the full range $[-1, 1]$.

Similar reasoning would apply to a composite function of the type used in Example 2.26 but formed with a cos function. However, note that in this case, the argument function would have to span at least one odd multiple of π and one even multiple of π to ensure that the composite had the full range of $[-1, 1]$.

It is also instructive to consider composite functions where circular functions form the arguments of other functions. For example,

$$p(x) = \cos^2(x) - 5\cos x + 6$$

is a "second-order polynomial in $\cos x$." That the range of $\cos x$ is $[-1, 1]$ has implications for range of $p(x)$. In particular, $p(x)$ can take a maximum value of $p(-1) = 12$ and a minimum value of $p(1) = 2$. These properties lead to $p(x)$ taking cyclical values within this range for all real x. Incidentally, the function $p(x)$ has no real roots since its range does not contain zero.

EXAMPLE 2.27

Demonstrate algebraically that the following statements are true for $f(x) = x^2 - 5x + 6$, $g(x) = \cos(x)$, and $h(x) = \tan(x)$.

a. **i.** The composite function $(f \circ g)(x)$ has no real roots.
ii. The composite function $(f \circ h)(x)$ has infinitely many real roots.

b. Find all real roots in part a(ii).

Solution

a. **i.** The composite function has the form $\cos^2(x) - 5\cos(x) + 6$. Using the substitution $y = \cos(x)$, we arrive back at $f(y) = y^2 - 5y + 6$ which has roots at $y = 2$ and $y = 3$. However, in order for the composite function to have real roots, we require values of x such that $\cos(x) = 2$ and $\cos(x) = 3$. The range for $\cos(x)$ is $[-1, 1]$ and so no real roots can exist.

ii. Following the method for a(i) but with the substitution $z = \tan(x)$, we arrive at the need to solve $\tan(x) = 2$ and $\tan(x) = 3$. The range of $\tan(x)$ is $\mathbb{R}$ and so real roots do exist in this case. Figure 2.20(c) shows that there are infinitely many roots.

b. We know that $\tan(x)$ has roots at $x = n\pi$ for $n \in \mathbb{Z}$, and so the two sets of roots of the composite are such that $x + n\pi = \tan^{-1}(2)$ and $x + n\pi = \tan^{-1}(3)$. The roots are therefore $x = -n\pi + \tan^{-1}(2)$ and $x = -n\pi + \tan^{-1}(3)$. Note that since $n \in Z$ and $-n \in \mathbb{Z}$, we can also write these as $x = n\pi + \tan^{-1}(2)$ and $x = n\pi + \tan^{-1}(3)$.

EXAMPLE 2.28

Determine all real roots of the function $q(x) = \cos^2(x) + \cos(x) - 2$.

Solution

Using the substitution $y = \cos(x)$, the function becomes $q(y) = y^2 + y - 2$ which has roots at $y = -2$ and $y = 1$. The root at $y = -2$ can be ignored as -2 is out of the range of $\cos(x)$. However, the root at $y = 1$ requires $\cos(x) = 1$ and so $x = 2n\pi$ for $n \in \mathbb{Z}$.

We used the notation $\tan^{-1}(x)$ in the solution of Example 2.27(b) to represent the "inverse tan" operation. It is assumed that the reader is comfortable with the inverse operations $\sin^{-1}$, $\cos^{-1}$, and $\tan^{-1}$ which are sometimes called arcsin, arccos, and arctan. We note that $\sin^{-1}(x)$ is distinct from $1/\sin(x)$, despite sometimes using the notation $\sin^n(x) \equiv (\sin(x))^n$, for $n \in \mathbb{N}$.

2.4 INVERSE FUNCTIONS

We have previously mentioned the concept of inverse operations. For example, the basic circular operations

$$a = \sin(b), \quad a = \cos(b), \quad a = \tan(b)$$

can be reversed as the operations

$$b = \sin^{-1}(a), \quad b = \cos^{-1}(a), \quad b = \tan^{-1}(a)$$

with appropriate restrictions to avoid the discontinuities of the tan operation and the limited range of the sin and cos operations. With inverse operations in mind, it is possible to understand the concept of *inverse functions*. For example, the inverse function of $f(x) = \sin x$ is $f^{-1}(x) = \sin^{-1}(x)$, which is commonly denoted as $\arcsin(x)$. Similarly, we have the $\arccos(x)$ and $\arctan(x)$ functions as inverses of the $\cos x$ and $\tan x$ functions, respectively.

With appropriate restrictions on the domain of the inverse function so that it corresponds to the range of the original function and other restrictions to avoid a many-to-one function leading to a one-to-many inverse mapping which, of course, cannot be a function in the technical sense, we can define an inverse function in general to be such that

$$(f^{-1} \circ f)(x) = x \quad \text{and} \quad (f \circ f^{-1})(x) = x$$

With the appropriate restrictions in mind, it is sometimes (but not always) possible to calculate explicit expressions for the inverse of a function by solving the equation

$$f(x) = y \Rightarrow x = f^{-1}(y)$$

and noting that the independent variable of the inverse function is merely a *dummy variable* that can be renamed at will to obtain $f^{-1}(x)$. This is illustrated in the following example.

EXAMPLE 2.29

Where possible, determine an explicit expression for the inverse of each of the following functions.

a. $f(x) = x + a$
b. $g(x) = x^2$
c. $h(x) = (4x - 1)^3$
d. $l(x) = \cos(x) + x$
e. $p(x) = \ln(x)$

Solution

a. $y = x + a \Rightarrow x = y - a$. Therefore, $f^{-1}(y) = y - a$ or $f^{-1}(x) = x - a$ for all $x \in \mathbb{R}$.
b. $y = x^2 \Rightarrow x = \sqrt[+]{y}$. Therefore, $f^{-1}(y) = \sqrt[+]{y}$ or $f^{-1}(x) = \sqrt[+]{x}$ for $x > 0 \in \mathbb{R}$.
c. $y = (4x - 1)^3 \Rightarrow x = (\sqrt[3]{y} + 1)/4$. Therefore, $f^{-1}(y) = (\sqrt[3]{y} + 1)/4$ or $f^{-1}(x) = (\sqrt[3]{x} + 1)/4$ for $x > 0 \in \mathbb{R}$.
d. $y = \cos(x) + x$. In this case, it is not possible to derive an explicit expression for $f^{-1}(x)$.
e. $y = \ln(x) \Rightarrow x = e^y$. Therefore, $f^{-1}(y) = e^y$ or $f^{-1}(x) = e^x$.

Some common examples of inverse functions that you may already be familiar with include

$$f(x) = \sin(x) \longleftrightarrow f^{-1}(x) = \arcsin(x)$$
$$g(x) = \cos(x) \longleftrightarrow g^{-1}(x) = \arccos(x)$$
$$h(x) = \tan(x) \longleftrightarrow h^{-1}(x) = \arctan(x)$$
$$l(x) = a^x \longleftrightarrow l^{-1}(x) = \log_a(x)$$

2.5 ACTUARIAL APPLICATION: THE TIME VALUE OF MONEY

The concept of the *time value of money* is at the core of actuarial science and indeed all aspects of financial mathematics. This concept therefore forms our starting point for illustrating where the mathematical ideas of this book are used within actuarial science. The time value of money has been written about extensively in the literature, including the author's other textbook (Garrett, 2013), however, we present a brief overview of the key concepts in this section while keeping an eye on the mathematics of functions, as is the mathematical focus of this chapter.

Most chapters in this book end with a brief discussion of a particular aspect of actuarial science in which the mathematical ideas developed there can be seen. This section is necessarily longer than those appearing in other chapters and reflects that a certain amount of groundwork is required in preparation for those later chapters.

You will be aware of the action of *compound interest* in your bank account. For example, let us suppose that you have a deposit account that pays interest at a fixed rate equal to $i = 3\%$ per annum and have deposited \$100 at some time we define to be $t = 0$. This investment will *accumulate* under the action of compound interest as time moves forward, such that the value of the initial \$100 after 1 year is

$$\$100(1.03) = \$103$$

If no other investments or withdrawals are made, the balance of your account after two and five and half years, say, will be

$$\$100(1.03)^2 \approx \$106.09 \quad \text{and} \quad \$100(1.03)^{5.5} \approx \$117.65,$$

respectively.

We can generalize this idea and note that the accumulation of a unit investment over t years under the action of a compound interest of annual rate i is given by

$$A = (1 + i)^t \tag{2.22}$$

In situations where the interest rate is fixed, we can consider this accumulation to be a function of time t. That is,

$$A(t) = (1 + i)^t \quad \text{for some fixed } i$$

and note that this is an *exponential function*. The use of this function is explored in the following examples.

EXAMPLE 2.30

A bank account offers an interest rate of 4% per annum. Calculate the accumulated value of the following investments made into this account.

a. \$1 made 10 years ago.
b. \$200 made 4 years 3 months ago.
c. \$3000 invested at $t = 4$ and withdrawn at $t = 6.5$.

Solution

a. The accumulated value of \$1 made 10 years ago is given by $A(4) = (1.04)^{10} = \$1.48$.
b. The accumulated value of \$200 made 4.25 years ago is given by $200A(4.25) = 200(1.04)^{4.25} = \$236.28$. We note that the function $A(4.25)$ applies to each \$1 and so acts as a multiplying factor.
c. At time $t = 6.5$, a \$3000 investment made at $t = 4$ will have accumulated to $3000A(2.5) = 3000(1.04)^{2.5} = \$3309.06$. Crucially, the investment has accumulated for 2.5 years and we apply the factor $A(2.5)$.

EXAMPLE 2.31

State the function $A(t)$ that gives the accumulated value after time t of a unit investment made in an account that pays compound interest at 7% per annum. State and interpret the domain and range of this function.

Solution

The accumulation function is given by

$$A(t) = (1.07)^t$$

Since we are looking forward, that is we are accumulating the investment, time is a nonnegative (real) number. The domain of $A(t)$ is therefore $t \in [0, \infty)$ and the range is clearly $\{A \in \mathbb{R} : A \geq 1\}$.

The case that $A = 1$ corresponds to $t = 0$ and is interpreted as that the investment has just been made and so no interest can be credited to the account. The case that A increases without bound corresponds to the investment being left untouched and so continues to earn interest as time moves forward without bound.

EXAMPLE 2.32

An investment of £200 was made sometime ago into an account that pays compound interest at a rate of 5.6% per annum. If the accumulated amount is now £352.09, determine how long ago the investment was made.

Solution

The expression for this accumulation is clearly

$$352.09 = 200(1.056)^t$$

where t is an unknown. It can however be determined by inverting the exponential function using logarithms. We proceed as follows:

$$\begin{aligned} 352.09 &= 200(1.056)^t \\ \Rightarrow \frac{352.09}{200} &= 1.056^t \\ \Rightarrow \ln\left(\frac{352.09}{200}\right) &= t\ln(1.056) \\ \Rightarrow t &= \frac{1}{\ln(1.056)} \ln\left(\frac{352.09}{200}\right) \approx 10.38 \end{aligned}$$

and determine that the investment was made approximately 10 years and 139 days ago. (Note that 139 days is approximately 38% of a year with 365 days.)

We might also consider the situation that the time over which a unit investment is made is fixed at some $t = n$. In this case, Eq. (2.22) is a function of the interest rate i and we have

$$A(i) = (1+i)^n \quad \text{for some fixed } n$$

This function is related to a polynomial function of order n. For example, if $n = 2$, $A(i) = i^2 + 2i + 1$ which is clearly a quadratic function. For the moment we are restricting n to be positive, that is we are looking forward in time, but do allow it to take any nonnegative (real) value; this is in contrast to standard polynomial functions where $n \in \mathbb{N}$.

EXAMPLE 2.33

An investor is looking to invest £2m for precisely 2 years. Calculate the accumulated value of his investment in the following situations.

a. The money is invested in a deposit account that pays a fixed interest rate of $i = 3\%$ per annum. As before, $A(i)$ applies to each £1 and so is used as a multiplying factor.

b. The money is invested in a deposit account that pays a fixed interest rate of $i = 7.4634\%$ per annum.

Solution

a. If the money is invested for 2 years at a fixed interest rate of $i = 3\%$ per annum, the accumulated value is given by $2(1.03)^2 =$ £2.1218m.

b. If the money is invested for 2 years at a fixed interest rate of $i = 7.4634\%$ per annum, the accumulated value is given by $2(1.074634)^2 \approx$ £2.3097m.

EXAMPLE 2.34

State a function that gives the accumulated value of \$5.6m after 5 years under a compounding rate of i% per annum. State and interpret the domain and range of this function.

Solution

In money units of \$m and time units of years, the function is given by

$$A(i) = 5.6(1+i)^5$$

Note that it is reasonable to assume that a rational investor would only invest in a deposit account that pays a nonnegative interest rate. If the interest rate were negative, the investor would be losing money and would be better simply keeping the money secure in a safe. Under this assumption, the domain of the function is $i \in [0, \infty)$ and the range is $\{A \in \mathbb{R} : A \geq 5.6\}$.

The case that $A(i) = 5.6$ corresponds to the investment being made a zero interest rate. A rational investor may do this to take advantage of the physical security that investing the cash in a bank may bring. The case that A is unbounded corresponds to the bank paying unbounded interest rates. Although this is unlikely, very large interest rates do occur in times of hyperinflation.

EXAMPLE 2.35

An investment of $\$100,000$ was made 2 years ago and expected to earn a fixed annual rate of interest. If the accumulated value is now $\$345,340$, calculate the interest rate.

Solution

In money units of $'000s and time units of years, the relevant expression is

$$345.34 = 100(1+i)^2$$

We are therefore required to invert a quadratic polynomial. In particular,

$$\begin{aligned} 345.34 &= 100(1+i)^2 \\ \Rightarrow i &= \sqrt[\pm]{3.4534} - 1 \end{aligned}$$

Since the investment is seen to grow, the interest rate must be positive and so we take only the positive square root. We therefore determine that $i \approx 85.83\%$. While this value looks extremely high and unfeasible in typical economic times, we note that the investment has more than tripled in just 2 years; the value is therefore correct.

As we have seen, *accumulations* deal with the growth of investments under positive interest rates as time moves *forward*. The investment return from interest payments is very useful in practice and will now lead to us the concept of *discounting*. Discounting essentially amounts to looking *backward* in time.

Consider a liability of, say, $100 due at some future time, say $t = 5$ years. Rather than ignoring this liability until it is due and worrying how to pay it then, it would be sensible to use compound interest to our favor. Let us also suppose that we have access to an account that pays a guaranteed interest rate of 4% per annum. Knowing that we will earn this return on an investment between now and when the liability falls due at $t = 5$ allows us to invest an amount smaller than $100. But how much should we invest to cover the debt?

If we denote the investment to be made at $t = 0$ by X, we know that X must be such that

$$100 = X(1.04)^4 \Rightarrow X = \frac{100}{1.04^5} = \$82.19$$

That is, we have computed that an investment of $82.19 made at $t = 0$ in an account that pays $i = 4\%$ per annum will be sufficient to meet the $100 liability at $t = 5$. More precisely, we might say

the present value of $100 due in five years under the action of $i = 4\%$ per annum is $89.19.

Generalizing this concept, we define the *present value* of a unit amount due at future time $t = n$ under the action of compound interest of i per annum as

$$V = \frac{1}{(1+i)^n} = (1+i)^{-n} \tag{2.23}$$

Note that, it is often convenient to understand the *present value* as existing now, that is, at $t = 0$, and the *accumulated* value to exist at some future $t = n > 0$. The process of determining the present value of a future cash flow is often called *discounting*. That is, we are discounting a future cash amount backward in time.

It should be clear that, for fixed n, $V(i)$ is a rational function of i of order $1/n$. Alternatively, for fixed i, $V(n)$ can be considered to be an exponential function of n.

EXAMPLE 2.36

Calculate the present value of the following liabilities under an interest rate of $i = 7\%$ per annum.

a. £1m due in 6.5 years.
b. £3.50 due in 4 months.
c. £10,000 due in 30 years.

Solution

a. We discount £1m backward by 6.5 years and obtain $(1.07)^{-6.5} \approx$ £0.644m.
b. We discount £3.50 backward by one third of year and obtain $3.50(1.07)^{-1/3} \approx$ £3.42.
c. We discount £10,000 backward by 30 years and obtain $10{,}000(1.07)^{-30} =$ £1313.67.

You will have noticed that the factor $(1 + i)^t$ is common to functions for both accumulating and discounting. Indeed, it should be clear that, assuming fixed i,

$$A(t) = (1 + i)^t = \frac{1}{V(t)}$$

With this in mind, we can allow the accumulation function $A(t)$ to be used for *discounting* by extending its domain to include *negative* t; that is, we extend the domain to $t \in \mathbb{R}$. The function over this domain is plotted in Figure 2.22 in the particular case that $i = 3\%$ per annum. The range of $A(t)$ for domain $t \in \mathbb{R}$ is $\{A \in \mathbb{R} : A > 0\}$. Furthermore, it is possible to show that A has a left asymptote to $A = 0$.

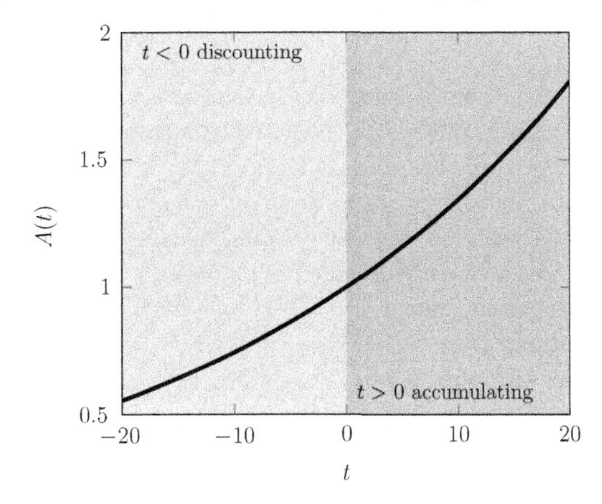

Figure 2.22 Plot of $A(t)$ for accumulating and discounting a unit investment at $i = 3\%$ per annum.

EXAMPLE 2.37

Use the accumulating function $A(t) = (1 + i)^t$ with domain $t \in \mathbb{R}$ to calculate the following. You should assume that $i = 5.5\%$ per annum.

a. The accumulated value of \$500 after 15.5 years.
b. The present value of \$6867 due in 4 years.
c. The present value of \$700m due in 35 years.

Solution

a. The accumulated value of \$500 after 15.5 years is given by $500A(15.5) = 500(1.055)^{15.5} \approx \$1146.52$.
b. The present value of \$6867 due in 4 years is given by $6687A(-4) = 6867(1.055)^{-4} \approx \$5397.86$.
c. The present value of \$700m due in 35 years is given by $700 \times (1.055)^{-35} \approx$ \$107.46m.

Real actuarial and business applications of these ideas are rarely restricted to single cash flows. Instead it is common to consider problems with many cash flows, some in flows and some out flows. We return to the mathematics of handling very many cash flows in Chapter 5. In principle though, there are no new ideas in handling multiple cash flows, as can be seen in the following example.

EXAMPLE 2.38

A project requires an initial outlay of £20,000 and generates cash flows of £5000 at each of $t = 1, 2, 3, 4$, and 5 years. If the outlay is funded by a loan at rate 7% per annum and the inflows are invested in a deposit account earning 7% per annum, calculate the following.

a. The *net present value* of the entire project.
b. The accumulated value of the entire project at $t = 5$.

Solution

a. The *net present value* is the present value, that is, value at $t = 0$, of all cash flows, whether in or out. By convention, inflows are treated as positive and outflows are treated as negative. A positive net present value is therefore required for a profitable project. In this example, we have an outflow at $t = 0$ and inflows at $t = 1, 2, 3, 4$, and 5. Working in units of £'000, we have cash flows of -20 and five lots of $+5$. The present value of each individual cash flow is easily obtained as follows:

- -20 at $t = 0$ has present value $-20(1.07)^0 = -20$
- 5 at $t = 1$ has present value $5(1.07)^{-1} \approx 4.6729$
- 5 at $t = 2$ has present value $5(1.07)^{-2} \approx 4.3672$
- 5 at $t = 3$ has present value $5(1.07)^{-3} \approx 4.0815$
- 5 at $t = 4$ has present value $5(1.07)^{-4} \approx 3.8145$
- 5 at $t = 5$ has present value $5(1.07)^{-5} \approx 3.5649$

This leads to a net present value equal to the sum of the six individual present values, that is, 0.5010 or £501.

b. The accumulated value at $t = 5$ is calculated in an entirely analogous way to the net present value, although we accumulate each cash flow to $t = 5$ rather than discounting to $t = 0$. In this case, the borrowing and investment rates are identical and so we can compute the accumulated values with the same i irrespective of if they are positive (deposit) or negative (loan). Individually, the accumulated values are given as

- -20 at $t = 0$ has accumulated value at $t = 5$ of $-20(1.07)^5 = -28.0510$
- 5 at $t = 1$ has accumulated value at $t = 5$ of $5(1.07)^4 = 6.5540$
- 5 at $t = 2$ has accumulated value at $t = 5$ of $5(1.07)^3 = 6.1252$
- 5 at $t = 3$ has accumulated value at $t = 5$ of $5(1.07)^2 = 5.7245$
- 5 at $t = 4$ has accumulated value at $t = 5$ of $5(1.07)^1 = 5.3500$
- 5 at $t = 5$ has accumulated value at $t = 5$ of $5(1.07)^0 = 5.0000$

This leads to an accumulated value of 0.7027 for the entire project. That is, £702.70. In this case, the net present value and accumulated values of the project are seen to be positive. This reflects that, despite taking out a loan and paying interest on it, the receipts and the interest earned on the deposit of those receipts are in excess of the cost of the loan; that is, the project is profitable at this interest rate.

EXAMPLE 2.39

A particular project requires a single investment of \$50,000 which is to be funded by a loan. The project is then expected to return \$20,000 and \$40,000 2 and 4 years later, respectively, which will be invested in a deposit account. Determine the *maximum* interest rate such that the project is profitable. You should assume that the same rate applies to both loans and deposits.

Solution

At first glance, two approaches can be taken to solve this problem: we may opt to work with an expression for the net present value at $t = 0$, or we might opt to work with an expression for the accumulated value at some point, say, $t = 4$. In both cases, we would determine i such that the expression is *zero*; that is, the project is neither profitable nor loss making. It should however be clear from the solution to Example 2.38, for example, that these expressions would be identical but for a factor of $(1 + i)^4$. This factor reflects that, despite there being many cash flows involved, the net present value is simply the present value of the accumulation value at $t = 5$. Since this factor is necessarily positive, a positive net present value necessarily leads to a positive accumulated value, and vice versa.

We opt to work with the net present value in units of \$10,000 and, for this project, it is given by the function,

$$f(i) = -5 + \frac{2}{(1+i)^2} + \frac{4}{(1+i)^4}$$

The required i is such that $f(i) = 0$. That is, i is a root of the function $f(i)$. At first glance, this may appear a difficult function to solve, however, the substitution $x = \frac{1}{(1+i)^2}$ leads the quadratic

$$-5 + 2x + 4x^2 = 0 \Rightarrow x = \frac{1}{4}\left(-1 \pm \sqrt{21}\right)$$

It should be clear that the solutions are

$$x = \frac{1}{(1+i)^2} = \frac{1}{4}\left(-1 + \sqrt{21}\right) \Rightarrow i \approx 0.0567 \text{ or } -2.0567$$

and the only relevant solution is $i \approx 5.67\%$.

The interpretation of this value is that, even though the total receipts exceed the initial loan, when we include the interest owed on the loan and that earned on the deposit of the receipts over the 4 years, the project is *not* profitable if $i > 5.67\%$ per annum. However, this is not the case for an interest rate less than 5.67%, and the project would be profitable.

These conclusions are confirmed from the values of the net present value at various i. In particular, $f(0.06) \approx -0.0516$ and $f(0.05) \approx 0.1049$.

Of course, business and investment projects such as those considered in Examples 2.38 and 2.39 are not always funded by a loan, nor are the receipts immediately invested in a fixed-interest deposit account. We might therefore wish to consider the more general situation that simply treats business ventures or financial investments as consisting of some outflows and inflows, irrespective of how the costs are funded or how the receipts are invested outside of the project.

To illustrate this, let us return to the venture in Example 2.39, but consider it simply as being an investment of \$50,000 that returns \$20,000 and \$40,000 2 and 4 years later, respectively. It is reasonable to then ask what is the "average" annual return generated from this investment over the 4 years? That is, what is the *internal rate of return* of this project expressed as an annualized figure?

To answer this question, we begin by forming the *equation of value* of the project. This is simply a mathematical statement of the value of all cash flows at some particular time as a function of interest rate i. Working in units of \$10,000, the equation of value at $t = 0$, for example, is obtained as

$$g(i) = -5 + \frac{2}{(1+i)^2} + \frac{4}{(1+i)^4}$$

We note that $g(i)$ is exactly the same as the net present value function $f(i)$ in Example 2.39. In some sense, this system is "closed" and so it is reasonable to have this equation of value equal to zero. That is, since there is no other source of income nor any other expenditures within the project, it must be that $g(i) = 0$ and so

$$5 = \frac{2}{(1+i)^2} + \frac{4}{(1+i)^4}$$

The value of i that solves this expression is interpreted as the annual return generated from the project, that is, the internal rate of return. It is obtained as the appropriate root of the equation of value function, $g(i)$.

The mathematical process for obtaining the internal rate of return in our example is identical to that taken in Example 2.39, and we would again obtain 5.67% per annum. Tying this back to Example 2.39, if we are required to service a loan that extracts any more than this rate from the project, the project would cease to be profitable; as was our conclusion in the example.

EXAMPLE 2.40

A business venture consists of an initial investment of £1m, requires a further investment of £0.5m after 2 years, and leads to an income of £1.25m after a further 2 years. Calculate the internal rate of return for this project and interpret your answer.

Solution

Working in money units of £m and time units of years, the equation of value of this project at $t = 0$ is

$$-1 - \frac{0.5}{(1+i)^2} + \frac{1.25}{(1+i)^4} = 0$$

This is a quadratic polynomial in $\frac{1}{(1+i)^2}$ and we find roots at

$$\frac{1}{(1+i)^2} = \frac{1}{5}\left(1+\sqrt{21}\right) \Rightarrow i \approx -0.0536 \text{ or } -1.95$$

The internal rate of return is therefore given by $i \approx -5.36\%$ per annum. We note that this is negative, which reflects that the project is unprofitable, as can be clearly seen from the fact that it costs £1.5m and only generates £1.25m over the 4 years.

Note that can form an equation of value at any time, say $t = n$. In this case, we would then be able to divide out a factor $(1+i)^n$ from each term and arrive back at the equation of value as at $t = 0$. It is therefore typical to work in terms of present values when forming an equation of value, although not essential.

2.6 QUESTIONS

The following questions are intended to test your knowledge of the material discussed in this chapter. Full solutions are available in Chapter 2 Solutions of Part III. You should use an algebraic approach unless otherwise stated.

Question 2.1. Determine whether the following mappings are one-to-one, many-to-one, or one-to-many mappings.

a. $f(x) = x^5 - 29x^3 + 100x$
b. $g(z) = z^4 - 6z^3 + 4z^2 + 24z - 32$
c. $h(y) = y \times (0.5 - \cos(y))$
d. $l(p) = (p^2 - 4) \times \cos(2p)$
e. $m(q) = \sqrt{q} - 2q\tan(q)$

Question 2.2. Determine which of the mappings in Question 2.1 are functions.

Question 2.3. Determine which of the functions identified in Question 2.2 demonstrate either odd or even symmetry.

Question 2.4. Determine all real roots of the functions identified in Question 2.2.

Question 2.5. Use plots of the functions identified in Question 2.2 to comment on the properties of each function. Your answer should include comment on the following properties.

i. Domain.
ii. Range.
iii. Asymptotes.
iv. Confirmation of the type of mapping identified in Question 2.1.

Question 2.6. Determine the domain and range of the function $f(x) = \sqrt[+]{x}$ on $\mathbb{R}$.

Question 2.7. Use an algebraic method to find intervals of z that solve the following inequalities. Confirm these with a computational plot.

a. $z^3 - 2z^2 - z > -2$
b. $z^2 + 2z + 5 \geq 0$
c. $-z^3 - 2z^2 - 2z > 0$
d. $z^2 - 4 \leq 0$

Question 2.8. Explain in words why the following statements are true.

a. A second-order polynomial must have either no real roots, two real roots, or one repeated real root.
b. An odd-ordered polynomial must always have at least one real root.

Question 2.9. Determine the parity of all possible products of two odd or even functions.

Question 2.10. Use a substitution method to find all possible real roots of the following functions.

a. $f(x) = e^{2x} + e^x - 30$
b. $g(z) = \cos^2(5z) - 4$
c. $h(y) = e^{6y} + e^{4y} - e^{2y} - 1$
d. $n(q) = \sin(2q^2)$
e. $m(p) = \ln\left(e^{2p} + 2e^p\right)$

Question 2.11. It is often convenient to define the following functions that are based on the standard circular functions.

$$\operatorname{cosec} x = \frac{1}{\sin x} \qquad \sec x = \frac{1}{\cos x} \qquad \cot c = \frac{1}{\tan x}$$

Explore each function and, in each case, comment on the following properties.

i. Domain
ii. Range
iii. Roots
iv. Symmetries
v. Asymptotes

Question 2.12. You are given that

$$f(x) = \sin x \quad g(x) = e^x \quad h(x) = \frac{2x+1}{2x^2+3}$$

State expressions for the following composite functions.

a. $(h \circ f)(x)$
b. $(g \circ h)(x)$
c. $(f \circ g \circ h)(x)$
d. $(g \circ f \circ g)(x)$

Question 2.13. Use algebraic techniques to determine the domain and range of the following functions over $\mathbb{R}$. Check your conclusions with a computational plot of each function.

a. $f(x) = \frac{2x-1}{2x^2+x-1}$
b. $g(z) = \ln\left(\frac{2z-1}{2z^2+z-1}\right)$
c. $h(y) = \sin\left(\frac{2y-1}{2y^2+y-1}\right)$
d. $j(p) = \exp\left(\frac{2p-1}{2p^2+p-1}\right)$

Question 2.14. Demonstrate that the following standard properties of logarithms are true.

a. $\log\left(a^b\right) \equiv b\log(a)$
b. $\log(ab) \equiv \log(a) + \log(b)$
c. $\log\left(\frac{a}{b}\right) \equiv \log(a) - \log(b)$

Question 2.15. Obtain $f^{-1}(x)$ for each of the following functions.

a. $f(x) = (x-2)^{\frac{1}{4}}$

b. $f(x) = \frac{x-2}{3x+9}$

c. $f(x) = e^{4x^2}$

d. $f(x) = \ln(x+6)$

Question 2.16. State the accumulation function for investments made at a compounding interest rate of 10% per annum. Use this function to determine the following quantities at this interest rate.

a. The accumulated value of \$100 after 17 years.

b. The accumulated value at $t = 10$ for a deposit of \$40 made at $t = 2$ and another of \$60 at $t = 7$.

c. The present value of a \$150,000 debt due 3 months in the future.

d. The net present value of cash flows \$40, \$50, and \$60 received at times $t = 10$, 11, and 12, respectively.

e. The accumulated value at $t = 15$ of the cash flows in part d.

Question 2.17. An investment opportunity requires an initial investment of £12m and will return £10m and £5m after 3 years 6 months and 7 years, respectively.

a. State the equation of value $f(i) = 0$ for this opportunity and determine all real roots of $f(i)$.

b. An investor is considering funding the investment with a loan and initially using the proceeds to repay the loan before generating a profit. If the prevailing interest rate on the loan is 7% per annum, should the investor use the loan? Explain your reasoning.

CHAPTER 3

Differential Calculus

Contents

Prerequisite knowledge	Intended learning outcomes

• Chapter 1
• Chapter 2
	• Recall and work with the fundamental definition of a derivative
	• Define and determine the following for simple functions
	  ◦ limits
	  ◦ continuity
	• Define and determine the following for standard functions
	  ◦ tangent lines
	  ◦ derivatives
	• Determine the derivatives of combinations of standard functions using
	  ◦ the product rule
	  ◦ the chain rule
	  ◦ the quotient rule
	• Determine the derivatives of simple inverse functions

In this chapter, we review the differential calculus that you may be familiar with from previous studies. The material can however be considered as a direct continuation of Chapter 2 and so you should not worry if you have little or no prior experience of calculus.

We begin by refining what we mean by *continuity*, a concept introduced in the previous chapter, and give a discussion of *limits*. Mathematical *derivatives* are then discussed within the context of these two ideas and from the perspective of calculating the gradients of functions. We then proceed to demonstrate how to obtain the derivatives of the common classes of functions discussed in the previous chapter, before moving on to a discussion of how to work with more complicated functions.

3.1 CONTINUITY

In Chapter 2, we referred regularly to whether a function was *continuous* or not when exploring its properties. At that stage we defined continuity as simply being that we could draw the function with a single line; that is, we would not have to take our pencil off the paper when sketching it. As we will see, continuity is crucial to the idea of mathematical derivatives[1] and it is worth taking some time to discuss continuity in a more formal sense. Key to the concept of continuity is that of a *limit* and a discussion of limits is our starting point.

3.1.1 Limits

The *limit* of a function describes the function's behavior as the independent variable approaches a particular value. It is crucial to understand that in the discussion of limits we consider the process of moving toward a point on the real number line, rather than evaluating a function directly at that point. If, as the independent variable x approaches the value c, the function $f(x)$ gets closer and closer to a single limiting value L, we say that "L is the limit of the function as x approaches c." Mathematically, we would write this as

$$\lim_{x \to c} f(x) = L$$

[1] It is assumed that readers have an interest in actuarial science and finance and so may be familiar with the word "derivative" in the financial sense. A financial derivative is something entirely different to a mathematical derivative and of course it is the mathematical derivative that we will be concerned with throughout most of this book on mathematical methods. You should assume that the term "derivative" refers to a mathematical derivative unless otherwise stated.

There is nothing special about the use of $f(x)$, c and L and we could equally consider

$$\lim_{y \to d} g(y) = C$$

for example, with obvious meaning.

Of course, when discussing a limit we can approach the particular (finite) point $x = c$ from values of x above c or below c. This distinction defines the *right-hand limit* and *left-hand limit*, respectively. In particular,

$$\lim_{x \to c^+} f(x) = L \quad \text{denotes the right-hand limit}$$

$$\lim_{x \to c^-} f(x) = L \quad \text{denotes the left-hand limit}$$

It should be clear that the superscript $^+$ indicates that we are approaching c from values of x greater than c, that is $x - c > 0$. Furthermore, the superscript $^-$ indicates that we are approaching c from values of x less than c, that is $c - x < 0$. We might also refer to these two cases as approaching x from *above* and *below*, respectively.

Right-hand and left-hand limits are illustrated in Figure 3.1. In particular, the function shown is such that we can have two situations.

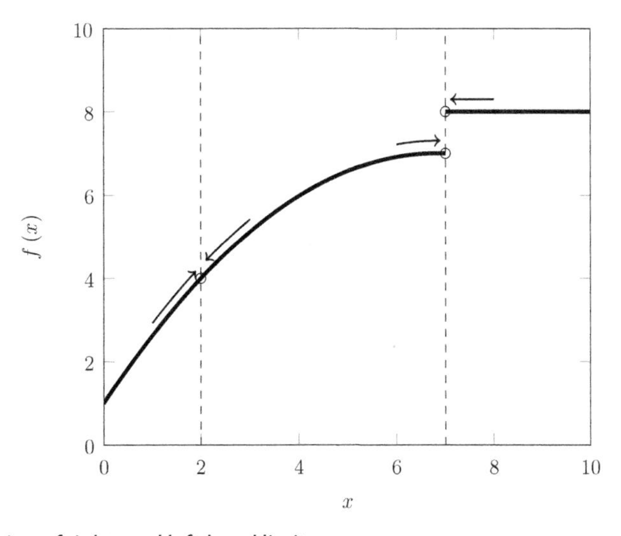

Figure 3.1 Illustration of right- and left-hand limits.

1.

$$\lim_{x \to 2^+} f(x) = \lim_{x \to 2^-} f(x) = 4$$

That is, the function takes the value 4 as the independent variable approaches $x = 2$ from either direction.

2.

$$\lim_{x \to 7^+} f(x) = 8$$

$$\lim_{x \to 7^-} f(x) = 7$$

That is, the function takes different values at $x = 7$ depending on how the independent variable approaches that point

The reason for this behavior is of course the "jump" in the function at $x = 7$. The jump is such that the limit of the function is ambiguous at $x = 7$ and we can go further and say that, because of this ambiguity, the limit *does not exist* at that point. In contrast, the function is better behaved everywhere else on the curve in the sense that the right- and left-hand limits are equal at every other point.

We therefore have a condition for a limit to exist:

$$\lim_{x \to c} f(x) = L \Leftrightarrow \lim_{x \to c^-} f(x) = L = \lim_{x \to c^+} f(x)$$

Expressed in words

the limit of $f(x)$ at $x = c$ is L if and only if both the right-hand and left-hand limits of $f(x)$ at $x = c$ equal L.

As we have seen before, the double-headed arrow indicates that the statement works in both directions. In particular,

i. that the limit exists, means that both the left-hand and right-hand limits are equal; and

ii. that both the left-hand and right-hand limits are equal means that the limit exists.

In practice, there are three general methods which can be used for calculating limits,

1. tabulating values of the function around $x = c$
2. plotting the function (either by hand or with a computer)
3. algebraic investigation

These three methods are explored in the following examples. Note that we have previously used the tabular method in Table 2.4 during our discussion of the "blow up" points (singularities) of rational functions. Note also that at this stage, it is reasonable to acknowledge that our knowledge of algebraic techniques may not be sufficient to reliably work with method 3 and you may wish to return to these examples after studying Chapter 5, for example.

EXAMPLE 3.1

Calculate the following limits.

a. $\lim_{x \to \pi} \frac{\cos x}{x}$

b. $\lim_{x \to 1} \frac{x-1}{\ln(x)}$

c. $\lim_{x \to 2} \frac{x^2-2x+2}{x-2}$

Solution

a. Table 3.1 gives values of the function as $x \to \pi^{\pm}$ and Figure 3.2(a) plots these values. In both approaches we see that the left-hand and right-hand limits are equal and take the value -0.3183, which is actually $-1/\pi$. Algebraically, we notice that neither the numerator nor denominator have any singular points or roots around $x = \pi$ and the limit can be evaluated directly. The three methods demonstrate that the limit exists and takes the value $\cos(\pi)/\pi = -1/\pi$.

b. Table 3.1 gives values of this function as $x \to 1^{\pm}$ and Figure 3.2(b) plots these values. In both approaches we see that the left-hand and right-hand limits are equal and take the value 1. Algebraically, we notice that the denominator is zero at $x = 1$ and so there is a possibility of a singular point there, however the numerator is also zero at this point. In this case, we may prefer to revert to rely on the tabulation method. A mathematical approach is discussed in Chapter 5.

c. Table 3.1 gives values of this function as $x \to 2^{\pm}$ and Figure 3.2(c) plots these values. In both approaches we see that the limit does not exist. Alternatively, from our knowledge of rational functions, we know that as $x = 2$ is a root of the denominator but not a root of the numerator, we therefore have a singular point.

Table 3.1 Tabular method for Example 3.1

x	4	3.5	3.2	3.16	3.145	3.142	3.142	3.1416
$f(x)$	−0.1634	−0.2676	−0.3120	−0.3164	−0.3180	−0.3183	−0.3183	−0.3183
x	2	2.5	2.75	3	3.1	3.14	3.141	3.1416
$f(x)$	−0.2081	−0.3205	−0.3361	−0.3300	−0.3223	−0.3185	−0.3184	−0.3183
x	1.5	1.2	1.1	1.01	1.001	1.0001	1.00001	1.000001
$g(x)$	1.2332	1.0970	1.0492	1.0050	1.0005	1.0000	1.0000	1.0000
x	0.5	0.8	0.9	0.99	0.999	0.9999	0.99999	0.999999
$g(x)$	0.7213	0.8963	0.9491	0.9950	0.9995	0.9999	1.0000	1.0000
x	3	2.5	2.1	2.01	2.001	2.0001	2.00001	2.000001
$h(x)$	5.0000	6.5000	22.1000	202.0100	2002	20,002	200,002	2,000,002
x	1.5	1.7	1.8	1.9	1.99	1.999	1.9999	1.99999
$h(x)$	−2.5000	−4.9667	−8.2000	−18.1000	−198.0100	−1998.0	−19,998	−199,998

$f(x) = \cos(x)/x$, $g(x) = (x-1)/\ln(x)$, and $h(x) = (x^2 - 2x + 2)/(x - 2)$.

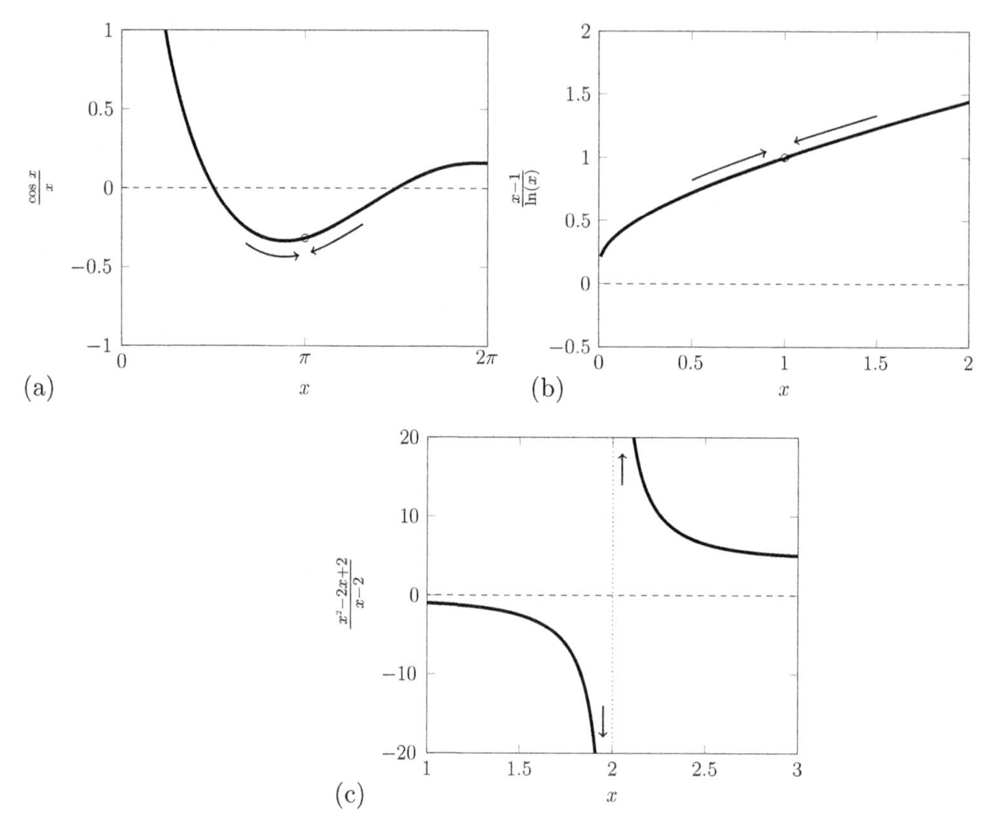

Figure 3.2 Illustrations of the limits in Example 3.1. (a) $x \to \pi^{\pm}$, (b) $x \to 1^{\pm}$, and (c) $x \to 2^{\pm}$.

Unless a computer is being used to plot the function, methods 1 and 2 are essentially the same. That is, one needs to evaluate the function at various points in order to generate the plot manually. Furthermore, the values used in Table 3.1 demonstrate that the resolution of the values chosen is important. If, for example, we were using the tabular method for the function in Example 3.1(c) and did not evaluate the function sufficiently close or at $x = 2$, we might have missed the existence of the discontinuity.

3.1.2 Continuity

If we think back to the intuitive definition of continuity, that is a curve is continuous if we can draw it with a single pencil line, we see that the functions in Figures 3.1 and 3.2(c) are not continuous at $x = 7$ and $x = 2$, respectively. These hints at the connection between the existence of limits and continuity and it should then be of no surprise that they are closely related. In the formal sense,

a function is continuous at point $x = c$ if the limit of $f(x)$ as x approaches c exists and has finite value $f(c)$.

Mathematically, the condition of continuity at point $x = c$ is stated as

$$\lim_{x\to c^-} f(x) = f(c) = \lim_{x\to c^+} f(x)$$

We explore this definition with some examples.

EXAMPLE 3.2

Determine whether the following function is continuous on the domain $\mathbb{R}$.

$$g(x) = \begin{cases} x^2 - 4 & \text{for } x \in (-\infty, 4] \\ 5 & \text{for } x \in (4, \infty) \end{cases}$$

Solution

The function is defined for all $\mathbb{R}$ and the quadratic polynomial and constant are clearly continuous where they are defined. The potential exception is then at $x = 4$ where we have

$$\lim_{x\to 4^-} h(x) = 12 \quad \text{and} \quad \lim_{x\to 4^+} h(x) = 5$$

The limit therefore does not exist at $x = 4$ and the function is not continuous at that point. We say that $g(x)$ has a *discontinuity* at $x = 4$. It is however continuous everywhere else.

EXAMPLE 3.3

Determine whether the following function is continuous on the domain $\mathbb{R}$.

$$h(x) = \begin{cases} x^3 - 2x + 1 & \text{for } x \in (-\infty, 1) \\ 2 & \text{for } x = 1 \\ x^3 - 2x + 2 & \text{for } x \in (1, \infty) \end{cases}$$

Solution

The cubic polynomial $x^3 - 2x + 2$ is continuous at all points it is defined. The question regarding the continuity of $g(x)$ is then around the point $x = 1$. We have $f(1) = 0$ and

$$\lim_{x\to 1^-} h(x) = 2 = \lim_{x\to 1^+} h(x)$$

Despite the left-hand and right-hand limits being equal at $x = 1$ (and so the limit exists), we note that their value does not equal $h(1)$. The function $h(x)$ therefore has a discontinuity at $x = 1$.

Example 3.3 illustrates why we distinguish between taking the limit at a point and evaluating the function at that point: the outcome of these operations may not be equal at a discontinuity. Of course, the above discussion also shows that there is no difference between the numerical value of a limit as $x \to c$ and a simple evaluation of the function at point $x = c$ if the function is continuous.

We can apply the idea of a discontinuity to the standard functions discussed in the previous chapter.

EXAMPLE 3.4

Explore the potential existence of discontinuities in the following functions.

a. $f(x) = \frac{x^3-11x^2+38x-40}{x-3}$

b. $g(x) = \tan(x)$

Solution

a. Recall that $f(x)$ is a rational function of order 3/1 and that rational functions have the potential for singular points (discontinuities) at the roots of the denominator unless these coincide with roots of the numerator. In this case, the denominator has a single root at $x = 3$ and we can easily demonstrate that this is *not* a root of the numerator. We therefore have a discontinuity at $x = 3$.

b. Recall that $g(x) = \frac{\sin(x)}{\cos(x)}$ and $\cos(x)$ has roots at $\frac{(2n+1)\pi}{2}$ for $n \in \mathbb{Z}$. These do not coincide with roots of $\sin(x)$ and so lead to discontinuities at those values of x.

EXAMPLE 3.5

Demonstrate the existence of the discontinuities found in Example 3.4 using left-hand and right-hand limits.

Solution

a. It is easily shown using a tabular method or a computational plot that

$$\lim_{x\to 3^-} f(x) = -\infty \quad \text{and} \quad \lim_{x\to 3^+} f(x) = \infty$$

The limit of $f(x)$ as $x \to 3$ does not therefore exist and the function has a discontinuity in the formal sense.

b. From Figure 2.20(c) we see that

$$\lim_{x\to \frac{(2n+1)\pi}{2}^-} g(x) = -\infty \quad \text{and} \quad \lim_{x\to \frac{(2n+1)\pi}{2}^+} g(x) = \infty \quad \forall n \in \mathbb{Z}$$

Again, we have demonstrated the existence of the discontinuities in the formal sense.

3.2 DERIVATIVES

Armed with knowledge of limits, we are able to proceed with a discussion of mathematical derivatives. As we shall see later, the concept of continuity is also important for derivatives but, for reasons of clarity, we initially choose to work with polynomial functions which of course are continuous over all $\mathbb{R}$. We link the concept of continuity and derivatives at the end of Section 3.2.1.

3.2.1 Gradients

The concept of a *gradient* may well be familiar to you. For example, in everyday life it is not unusual to see road signs warning us of "adverse gradients" which we interpret as a warning about the *slope* of the road ahead. The meaning of the term gradient is exactly the same in mathematics; that is, it is associated with slope. In particular, we will be concerned with the "slope of a function."

An intuitive definition of the gradient is that it gives a measure of how much vertical distance is obtained per unit of horizontal distance. For example, if a road climbs 10 m over a 40 m horizontal distance, it would be said to have a 1:4 gradient or, equivalently, a 25% gradient.

As a step toward a more mathematical definition of a gradient, it is convenient to think of the following expression

$$\text{gradient} = \frac{\text{vertical change}}{\text{horizontal change}} \tag{3.1}$$

The gradient of the road in question is then $10/40 = 0.25$. Of course, in practice the road would not be a perfect straight line and what we would actually calculate is the average gradient over the 40 m horizontal stretch. In mathematics, however, we can define perfectly straight lines and we begin our discussion of gradients with the perfectly straight lines defined by linear functions, that is first-order polynomials.

Using the notation Δ to be read as "the change in," f to represent the vertical distance, and x to represent the horizontal distance, we can express Eq. (3.1) in the more mathematical form of

$$\text{gradient} = \frac{\Delta f}{\Delta x} \tag{3.2}$$

You will recall from Eq. (2.9) that a linear function has the standard form

$$f(x) = a_1 x + a_0$$

This is often written as

$$f(x) = mx + c$$

where x is the independent variable, and m and c are parameters of the particular function. The interpretation of c is easily understood from the observation $f(0) = c$, that is c is simply the value the function takes at $x = 0$. The parameter c is sometimes referred to as the *y-intercept* which reflects that the straight line intercepts the vertical axis at value c.

You may recall that the parameter m is related to the gradient. To see this, we consider the change in the value of $f(x)$ over a fixed distance in x. For example, we note that $f(0) = c$ and $f(1) = m + c$, and use Eq. (3.2) to calculate the gradient as

$$\text{gradient} = \frac{f(1) - f(0)}{1 - 0} = \frac{m + c - c}{1} = m$$

EXAMPLE 3.6

State the gradient and y-intercept of the linear function $f(x) = 5x + 2$. Confirm your value for the gradient by calculating it between the following values.

a. $x = 10$ and $x = 15$.

b. $x = 1000$ and $x = 1250$.

c. $x = -100$ and $x = -10$.

Solution

Comparing the function $f(x) = 5x + 2$ to the standard form $f(x) = mx + c$ we immediately see that it has gradient 5 and it intercepts the vertical axis at $f(0) = 2$.

a. Since $f(10) = 52$ and $f(15) = 77$, the gradient is calculated as $(77 - 52)/(15 - 10) = 5$. This calculation is illustrated in Figure 3.3.

b. Since $f(1000) = 5002$ and $f(1250) = 6252$, the gradient is calculated as $(6252 - 5002)/(1250 - 1000) = 5$.

c. Since $f(-100) = -498$ and $f(-10) = -48$, the gradient is calculated as $(-498 + 48)/(-100 + 10) = 5$.

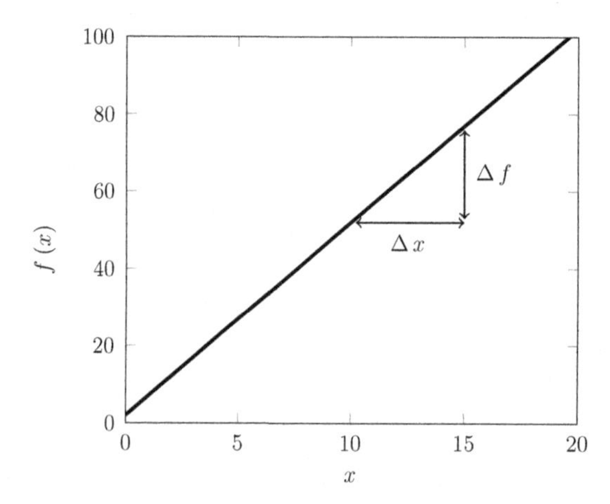

Figure 3.3 Calculation of the gradient of the straight line $f(x) = 5x + 2$ in Example 3.6(a).

EXAMPLE 3.7

State and interpret the gradient of the following straight-line functions

a. $f(x) = 2x + 100$
b. $g(x) = -5x + 12$
c. $h(x) = -2$

Solution

a. The gradient of $f(x)$ is 2. A plot of $f(x)$ would slope *upward*, that is, $f(x)$ increases as x increases.
b. The gradient of $g(x)$ is -5. A plot of $g(x)$ would slope *downward*, that is, $g(x)$ decreases as x increases.
c. The gradient of $h(x)$ is 0. A plot of $h(x)$ would be a *horizontal* line with $h(x) = -2$ for all x.

EXAMPLE 3.8

Determine the linear functions with the following properties.

a. Gradient 2 and $f(0) = 3$.
b. Gradient -1 and $f(2) = 6$.
c. Passes through $f(1) = 2$ and $f(6) = 12$.

Solution

In each case, we start with $f(x) = mx + c$ and are required to determine m and c from the information given.

a. We immediately see that $f(x) = 2x + 3$.
b. We have $f(x) = -x + c$ with c determined from $f(2) = 6 = -2 + c$. Therefore, $f(x) = -x + 8$.
c. We have $2 = m + c$ and $12 = 6m + c$ which are satisfied with $m = 2$ and $c = 0$. Therefore, $f(x) = 2x$.

As demonstrated in Example 3.6 and Figure 3.3, we can evaluate the gradient of a linear (or constant) function over any horizontal distance and obtain the same value. This is not however true of nonlinear functions. For example, the quadratic function $g(x) = x^2$ shown in Figure 3.4 clearly has a gradient that changes as we move away from $x = 0$ in either direction. The value of the gradient using Eq. (3.2) calculated between $x = 0$ and $x = 1$ will be different to that calculated between $x = 1$ and $x = 2$, for example. Using Eq. (3.2) for nonlinear functions merely returns the average gradient over the particular horizontal interval used. With this in mind, we need a method of defining the *instantaneous gradient* of a function at any particular value of x.

We define the instantaneous gradient of a function $f(x)$ at x to be

$$\lim_{\delta \to 0} \frac{f(x+\delta) - f(x)}{\delta} \tag{3.3}$$

It should be clear that this expression is essentially the same as computing Eq. (3.2) over a vanishingly small interval between x and $x + \delta$, where $\delta \in \mathbb{R}$. That is, rather than calculating an average gradient over an extended interval starting at x, we determine the gradient at a precise location x by calculating the limit of a ratio as the interval tends to zero size.

EXAMPLE 3.9

Use the expression for the instantaneous gradient (3.3) to determine the gradient of the following expressions at general position x.

a. $f(x) = 4$
b. $g(x) = 2x + 2$
c. $h(x) = x^2$

Solution

a. The instantaneous gradient is obtained from

$$\lim_{\delta \to 0} \frac{4 - 4}{\delta} = 0$$

This is as we would expect for a horizontal line.

b. The instantaneous gradient is obtained from

$$\lim_{\delta \to 0} \frac{2(x + \delta) + 2 - (2x + 2)}{\delta} = \lim_{\delta \to 0} \frac{2\delta}{\delta} = 2$$

This is as we would expect from a function of the form "$mx + c$."

c. The instantaneous gradient is obtained from

$$\lim_{\delta \to 0} \frac{(x + \delta)^2 - x^2}{\delta} = \lim_{\delta \to 0} \frac{x^2 + 2x\delta + \delta^2 - x^2}{\delta} = 2x$$

Example 3.9(c) demonstrates that the gradient of $h(x) = x^2$ is itself a function of x. We can test the result graphically using the concept of *tangents*. The tangent of a curve at position x is simply the straight line that "just touches" the curve at that point and shares the curve's instantaneous gradient. We can therefore form tangents at various points along a curve by fitting straight lines to the value of the function and its instantaneous gradient at the various positions.

EXAMPLE 3.10

Visually confirm the gradient function obtained in Example 3.9(c) by plotting $h(x) = x^2$ and its tangent lines at $x = -1$, $x = 0$, and $x = 2$.

Solution

- At $x = -1$ the gradient is $2x = -2$. The tangent must pass touch $h(-1) = 1$ and so $c = -1$. The tangent at this point has form $y = -2x - 1$.

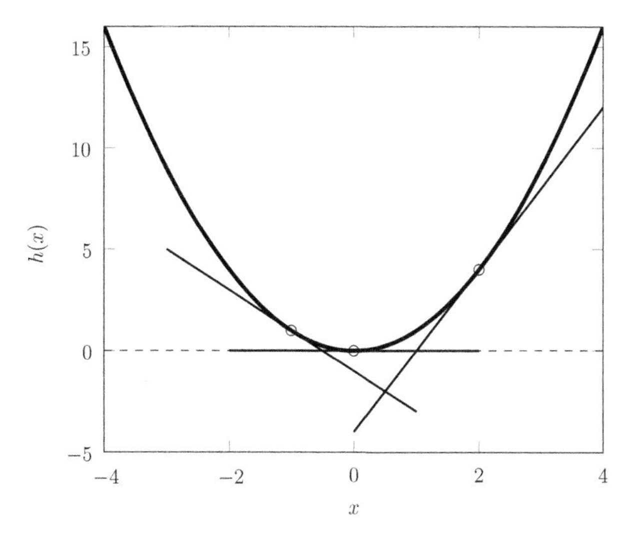

Figure 3.4 Visual confirmation of tangent lines to $h(x) = x^2$ calculated in Example 3.10.

- At $x = 0$, the gradient is $2x = 0$. The tangent must pass touch $h(0) = 0$ and so $c = 0$. The tangent at this point has form $y = 0$.
- At $x = 2$, the gradient is $2x = 4$ The tangent must pass touch $h(2) = 4$ and so $c = -4$. The tangent at this point has form $y = 4x - 4$.

These tangents are shown in Figure 3.4 where it can be seen that we have correctly calculated the instantaneous gradient at each point.

The above examples show that it is often possible to derive the gradient function of $f(x)$. For this reason it is common to say that the instantaneous gradient function (3.3) gives the *derivative* of $f(x)$ with respect to x. There are a number of standard ways to denote the derivative of $f(x)$, the three most common being $\frac{\mathrm{d}}{\mathrm{d}x}f(x)$, $\frac{\mathrm{d}f}{\mathrm{d}x}$, and $f'(x)$, and each is such that

$$\frac{\mathrm{d}}{\mathrm{d}x}f(x) \equiv \frac{\mathrm{d}f}{\mathrm{d}x} \equiv f'(x) \equiv \lim_{\delta \to 0} \frac{f(x+\delta) - f(x)}{\delta} \tag{3.4}$$

Of course, all this notation is flexible and we would denote the derivative of $g(y)$ with respect to y as $\frac{\mathrm{d}}{\mathrm{d}y}g(y)$, $\frac{\mathrm{d}g}{\mathrm{d}y}$, or $g'(y)$, for example. The three variations of notation will be used interchangeably and without comment in all that follows.

A very useful piece of notation is

$$\left.\frac{\mathrm{d}f}{\mathrm{d}x}\right|_a$$

which is interpreted as the derivative of $f(x)$ with respect to x evaluated at $x = a$. This has the obvious generalization to functions of the form $g(y)$, for example.

As the derivative is defined by a limit, it should be clear from our discussions in Section 3.1.1 that the limit is required to exist in order for the derivative to exist. This provides the connection between continuity and derivatives. In particular, if a function is *differentiable* at a point (that is its derivative exists), it must also be continuous at that point.

Now that we have defined what is meant by a derivative, it is useful to develop the general rules for determining the derivatives of the common classes of functions introduced in Chapter 2. It is possible to discuss the derivatives of polynomial, exponential, and sin and cos functions at this stage. However we require a little more work before we can discuss the derivatives of rational, logarithmic, and the tan function, and we return to these in Section 3.3.

3.2.2 Derivatives of polynomials

Example 3.9 suggests the following rules regarding the derivative of polynomials of order 0, 1, and 2.

$$\frac{\mathrm{d}}{\mathrm{d}x}x^0 = 0, \quad \frac{\mathrm{d}}{\mathrm{d}x}x = 1, \quad \frac{\mathrm{d}}{\mathrm{d}x}x^2 = 2x$$

In fact, the derivative of a general nth-order polynomial is

$$\frac{\mathrm{d}}{\mathrm{d}x}x^n = nx^{n-1} \tag{3.5}$$

It should be immediately clear that the above expressions for $n = 0$, 1, and 2 are consistent with this general rule, and we confirm that this is true for $n = 3$, 4, and 5 in the following example.

EXAMPLE 3.11

Use the definition of a derivative stated in Eq. (3.4) to confirm that the following statements are true.

a. $\frac{\mathrm{d}x^3}{\mathrm{d}x} = 3x^2$

b. $\frac{\mathrm{d}x^4}{\mathrm{d}x} = 4x^3$

c. $\frac{\mathrm{d}x^5}{\mathrm{d}x} = 5x^4$

Solution

We apply Eq. (3.4) in each case and use algebraic techniques to evaluate the resulting limit.

a.
$$\frac{\mathrm{d}x^3}{\mathrm{d}x} = \lim_{\delta\to 0}\frac{(x+\delta)^3 - x^3}{\delta} = \lim_{\delta\to 0}\left(3x^2 + 3\delta x + \delta^2\right) = 3x^2$$

b. $$\frac{dx^4}{dx} = \lim_{\delta\to 0} \frac{(x+\delta)^4 - x^4}{\delta} = \lim_{\delta\to 0} \left(4x^3 + 6\delta x^2 + 4\delta^2 x + \delta^3\right) = 4x^3$$

c. $$\frac{dx^5}{dx} = \lim_{\delta\to 0} \frac{(x+\delta)^5 - x^5}{\delta} = \lim_{\delta\to 0} \left(5x^4 + 10\delta x^3 + 10\delta^2 x^2 + 5\delta^3 x + \delta^4\right) = 5x^4$$

The rule stated in Eq. (3.5) is easily generalized for all $n \in \mathbb{R}$, although we do not prove this statement.

EXAMPLE 3.12

State the derivative of each of the following functions.

a. $f(x) = x^{10}$

b. $g(y) = y^{\pi}$

c. $h(z) = z^{a-1}$, where $a \in \mathbb{R}$ is a constant.

Solution

Using the general result (3.5), we can immediately write these derivatives as follows.

a. $f'(x) = 10x^9$

b. $g'(y) = \pi y^{\pi-1}$

c. $h'(z) = (a-1)z^{a-2}$

3.2.3 Derivatives of exponential functions

The exponential function $f(x) = e^x$ is unique in that its derivative is itself. That is

$$\frac{d}{dx}e^x = e^x$$

The interpretation of this is that the gradient of the exponential function at point x is simply the value of e^x, which is why the function demonstrates such rapid growth with moderate increases in x. See Figure 2.14, for example.

To demonstrate that this expression is true, we apply Eq. (3.4) to the function e^x as follows.

$$\frac{d}{dx}e^x = \lim_{\delta\to 0} \frac{e^{x+\delta} - e^x}{\delta} = e^x \times \lim_{\delta\to 0} \frac{e^{\delta} - 1}{\delta}$$

To proceed, it is then necessary to know the result

$$\lim_{\delta\to 0} \frac{e^{\delta} - 1}{\delta} = 1$$

It is possible to prove this using algebraic techniques, which we return to in Chapter 5, but for the moment it is sufficient to demonstrate the limit using the tabular method; this is done in Table 3.2. Using this limit, we see that

Table 3.2 Tabular method for determining that $\lim_{\delta\to 0}\frac{e^{\delta}-1}{\delta}=1$

δ	0.1	0.01	0.001	0.0001
$\frac{e^{\delta}-1}{\delta}$	1.0517	1.0050	1.0005	1.0001
δ	−0.1	−0.01	−0.001	−0.0001
$\frac{e^{\delta}-1}{\delta}$	0.9516	0.9950	0.9995	1.0000

$$\frac{d}{dx}e^x = \lim_{\delta\to 0}\frac{e^{x+\delta}-e^x}{\delta} = e^x\lim_{\delta\to 0}\frac{e^{\delta}-1}{\delta} = e^x$$

as required.

The limit demonstrated in Table 3.2 is a particular example of the general result

$$\lim_{\delta\to 0}\frac{a^{\delta}-1}{\delta} = \ln(a) \quad \text{for all } a>0$$

This general result enables one to determine the derivative of more general exponential functions. In particular,

$$\frac{d}{dx}a^x = \lim_{\delta\to 0}\frac{a^{x+\delta}-a^x}{\delta} = a^x\lim_{\delta\to 0}\frac{a^{\delta}-1}{\delta} = a^x\ln(a) \quad \text{for } a>0. \tag{3.6}$$

EXAMPLE 3.13

Use a tabular method to confirm that

$$\lim_{\delta\to 0}\frac{2^{\delta}-1}{\delta} = \ln(2)$$

Solution

δ	0.1	0.01	0.001	0.0001
$\frac{2^{\delta}-1}{\delta}$	0.7177	0.6956	0.6934	0.6932
δ	−0.1	−0.01	−0.001	−0.0001
$\frac{2^{\delta}-1}{\delta}$	0.6697	0.6908	0.6929	0.6931

We note that $\ln(2)\approx 0.6931$

EXAMPLE 3.14

Write down expressions for the following derivatives.

a. $\frac{d4^x}{dx}$

b. $\frac{d\pi^x}{dx}$

Solution

a. $\frac{d4^x}{dx} = 4^x\ln(4)$

b. $\frac{d\pi^x}{dx} = \pi^x\ln(\pi)$

3.2.4 Derivatives of circular functions

At this stage, it is possible to derive expressions for the derivative of the basic sin and cos functions. In particular,

$$\frac{\mathrm{d}\sin(x)}{\mathrm{d}x} = \cos(x) \tag{3.7}$$

$$\frac{\mathrm{d}\cos(x)}{\mathrm{d}x} = -\sin(x) \tag{3.8}$$

As might be expected, the derivation of Eq. (3.7) starts with

$$\frac{\mathrm{d}\sin(x)}{\mathrm{d}x} = \lim_{\delta\to 0} \frac{\sin(x+\delta) - \sin(x)}{\delta}$$

Progress is then made by expanding out the first sin term in the numerator using the "angle sum" identity stated in Appendix A. This leads to

$$\frac{\mathrm{d}\sin(x)}{\mathrm{d}x} = \lim_{\delta\to 0} \frac{\sin(x)\cos(\delta) + \cos(x)\sin(\delta) - \sin(x)}{\delta}$$

which, after some manipulation, gives

$$\frac{\mathrm{d}\sin(x)}{\mathrm{d}x} = \cos(x) \times \lim_{\delta\to 0} \frac{\sin(\delta)}{\delta} - \sin(x) \times \lim_{\delta\to 0} \frac{1-\cos(\delta)}{\delta}$$

The derivative is then seen to depend on the value of two further limits, which, using a tabular method or graphical method, can be shown to be

$$\lim_{\delta\to 0} \frac{\sin(\delta)}{\delta} = 1 \quad \lim_{\delta\to 0} \frac{1-\cos(\delta)}{\delta} = 0$$

You are asked to verify this in Question 3.4. Putting all the pieces together, we obtained Eq. (3.7).

The derivation of Eq. (3.8) can proceed along similar lines, however it is instructive to understand an alternative approach, based on the relationship between the sin and cos functions. In particular, we begin by recalling that

$$\cos(x) = \sin\left(x + \frac{\pi}{2}\right)$$

Taking derivatives of both sides, we have

$$\frac{\mathrm{d}}{\mathrm{d}x}\cos(x) = \frac{\mathrm{d}}{\mathrm{d}x}\sin\left(x + \frac{\pi}{2}\right)$$

It may be clear to you that the gradient of the function $\sin\left(x + \frac{\pi}{2}\right)$ is no different to the gradient of $\sin(x)$, that is, all we are doing is shifting the function in x, not changing its shape. With this idea in mind, we can use Eq. (3.7) to write

$$\frac{\mathrm{d}}{\mathrm{d}x}\cos(x) = \cos\left(x + \frac{\pi}{2}\right) \tag{3.9}$$

Note that $\cos\left(x + \frac{\pi}{2}\right) = -\sin(x)$, which leads to Eq. (3.8), as required.

The derivation of Eq. (3.8), as presented above, required some general reasoning regarding the shape of the shifted function that is not strictly rigorous. It is however possible to avoid such nonrigorous methods to demonstrate the derivative by using what we call the *chain rule*. This method is discussed in the next section.

3.3 DERIVATIVES OF MORE COMPLICATED FUNCTIONS

So far, we have been concerned only with the derivatives of basic functions with a single term. For example, we now know how to derive $f'(x)$ for $f(x) = x^4$, but not for $f(x) = 4x^4 - 2x^3 + x^2 - 1$ or $f(x) = \cos(x)\sin(x)$, for example. In this section, we discuss a broader collection of techniques that will significantly expand the types of functions that we can obtain the derivatives of.

3.3.1 Sums of functions

The operation of "taking a derivative" is linear. That is, when calculating $f'(x)$, Eq. (3.4) requires only linear powers of $f(x)$. For this reason, the derivative operator is distributive and it should be clear that, for a and b constants,

$$\frac{\mathrm{d}}{\mathrm{d}x}\left[af(x) + bg(x)\right] = a\frac{\mathrm{d}f}{\mathrm{d}x} + b\frac{\mathrm{d}g}{\mathrm{d}x} \tag{3.10}$$

This can be shown mathematically using Eq. (3.4)

$$\begin{aligned}\frac{\mathrm{d}}{\mathrm{d}x}\left[af(x) + bg(x)\right] &= \lim_{\delta\to 0}\frac{af(x+\delta) + bg(x+\delta) - (af(x) + bg(x))}{\delta}\\ &= a\times\lim_{\delta\to 0}\frac{f(x+\delta) - f(x)}{\delta} + b\times\lim_{\delta\to 0}\frac{g(x+\delta) - g(x)}{\delta}\\ &= a\frac{\mathrm{d}f}{\mathrm{d}x} + b\frac{\mathrm{d}g}{\mathrm{d}x}\end{aligned}$$

Of course, these arguments can be extended to the sum of any number of functions.

EXAMPLE 3.15

Write down the derivatives of the following functions.

a. $f(y) = 2y^3 - 4y^2 + y - 1$
b. $g(x) = 4\exp(x) - 5\sin(x)$
c. $h(z) = 10\cos(z) - 4\exp(z) + \sin(z) - z$

Solution

We can immediately write down the following derivatives using Eq. (3.10).

a. $f'(y) = 6y^2 - 8y + 1$

b. $g'(x) = 4\exp(x) - 5\cos(x)$

c. $h'(z) = -10\sin(z) - 4\exp(z) + \cos(z) - 1$

3.3.2 Product rule

From the discussions so far, it is not yet clear that there is a general rule for obtaining the derivatives of a product of functions. That is, it is unclear how to take the derivative of functions of the form $h(x) = f(x) \times g(x)$.

There is in fact a general rule called the *product rule*

$$\frac{\mathrm{d}}{\mathrm{d}x}\big(f(x)g(x)\big) = f(x) \times \frac{\mathrm{d}}{\mathrm{d}x}g(x) + g(x) \times \frac{\mathrm{d}}{\mathrm{d}x}f(x) \tag{3.11}$$

This expression can be extended to product of any number of functions in the obvious way. For example, it should be clear that

$$\big(f(x)g(x)h(x)\big)' = f'(x)g(x)h(x) + f(x)g'(x)h(x) + f(x)g(x)h'(x)$$

The proof of the product rule again follows from our fundamental definition of the derivative given in Eq. (3.4) and we demonstrate the method in the particular case of the product of two functions. We begin by stating that

$$\frac{\mathrm{d}}{\mathrm{d}x}(f(x)g(x)) = \lim_{\delta\to 0}\frac{f(x+\delta)g(x+\delta) - f(x)g(x)}{\delta}$$

and proceed by adding and subtracting the term $f(x+\delta)g(x)$ to the numerator. Using the property that the limit of a product is a product of a limit (see Question 3.3), some algebraic manipulation leads to

$$\begin{aligned}\frac{\mathrm{d}}{\mathrm{d}x}(f(x)g(x)) &= \lim_{\delta\to 0} f(x+\delta)\lim_{\delta\to 0}\frac{g(x+\delta) - g(x)}{\delta} + \lim_{\delta\to 0} g(x)\lim_{\delta\to 0}\frac{f(x+\delta) - f(x)}{\delta}\\ &= f(x)\frac{\mathrm{d}}{\mathrm{d}x}g(x) + g(x)\frac{\mathrm{d}}{\mathrm{d}x}f(x)\end{aligned}$$

Despite the relatively complicated derivation, the product rule is easy to use in practice.

EXAMPLE 3.16

Write down the derivatives of the following functions.

a. $f(x) = \exp(x)\cos(x)$

b. $g(y) = y^2\sin(y)$

c. $h(z) = z^4\exp(z)\cos(z)\sin(z)$

Solution

We use the product rule in each case and arrive at the following after some algebraic manipulation.

a. $f'(x) = \exp(x)(\cos x - \sin x)$

b. $g'(u) = y(2\sin y + y\cos y)$

c. $h'(z) = z^3 \exp(z)\left(4\cos(z)\sin(z) + z\cos(z)\sin(z) - z\sin^2(z) + z\cos^2(z)\right)$

Of course, given that $x^m \times x^n = x^{m+n}$, one should expect consistency between the rule we established for the differentiation of an $(m+n)$th-order polynomial and the product rule applied to the $x^m \times x^n$. This is demonstrated in the following example.

EXAMPLE 3.17

Using the polynomial functions $f(x) = x^n$, $g(x) = x^m$, and $h(x) = x^{n+m}$, demonstrate the consistency between the standard formula for calculating the derivative of a polynomial and the product rule.

Solution

- By Eq. (3.5), $h'(x) = (n+m)x^{n+m-1}$.
- By the product rule,
  $(f(x)g(x))' = nx^{n-1}x^m + mx^n x^{m-1} = (n+m)x^{n+m-1}$.

In principle, it is possible to derive an expression for the derivative of functions of the form $\cos^n(x)$, $\sin^n(x)$, and e^{nx} for $n \in \mathbb{N}$ using the product rule. However, as n increases, the method becomes cumbersome and, of course, it cannot cope with noninteger values of n. For these reasons it is useful to express such functions as composites with polynomial functions.

For example,

$$\cos^n(x) = (\cos x)^n = f(g(x)) = (f \circ g)(x) \tag{3.12}$$

where $g(x) = \cos(x)$ and $f(x) = x^n$ with $n \in \mathbb{R}$. The question then becomes one of deriving expressions for the derivatives of composite functions. This motivates discussion of the *chain rule.*

3.3.3 Chain rule

The chain rule (also known as the *function of a function rule*) enables one to determine the derivative of composite functions. In its simplest form, that is for a composite of two functions, the chain rule takes the form

$$\frac{\mathrm{d}}{\mathrm{d}x}(f \circ g)(x) = \frac{\mathrm{d}}{\mathrm{d}x}f(g(x)) = f'(g(x))g'(x) \tag{3.13}$$

Alternatively, we could write this using the "evaluated at" notation which perhaps gives a clearer statement of the rule.

$$(f \circ g)'(x) = \left.\frac{\mathrm{d}f(x)}{\mathrm{d}x}\right|_{g(x)} \left.\frac{\mathrm{d}g}{\mathrm{d}x}\right|_{x}$$

Essentially, the chain rule requires the use of a substitution to separate the composite into its constitute functions, and then proceeds by first taking the derivative of the "outer" function with respect to the substitute. For example, if we are required to differentiate the composite function in Eq. (3.12), we would begin by making the substitution $u = \cos x$ and continue as $f'(u) = \mathrm{d}u^n/\mathrm{d}u = nu^{n-1}$. However, an important point is that we wish to obtain the differential of the composite with respect to x, not u; we account for this by multiplying $f'(u)$ by $\mathrm{d}u/\mathrm{d}x$. This full process is illustrated in the following example.

EXAMPLE 3.18

Use the chain rule to derive an expression for

$$\frac{\mathrm{d}}{\mathrm{d}x}(\cos x)^n \text{ for } n \in \mathbb{N}$$

Solution

We use the substitution $u = \cos x$ and obtain

$$\left.\frac{\mathrm{d}}{\mathrm{d}u}u^n\right|_{u=\cos x} = \left.nu^{n-1}\right|_{u=\cos x} = n\cos^{n-1}x$$

The required derivative with respect to x is then, from Eq. (3.13),

$$\frac{\mathrm{d}}{\mathrm{d}x}(\cos x)^n = n\cos^{n-1}(x)\frac{\mathrm{d}u(x)}{\mathrm{d}x} = -n\sin(x)\cos^{n-1}(x)$$

Note that this result actually applies for any value $n \in \mathbb{R}$.

The chain rule can be used to obtain the derivatives of simple modifications to the *arguments* of the basic functions. For example, using Eq. (3.13), we can immediately write down expressions for the following:

$$\frac{\mathrm{d}}{\mathrm{d}x}\exp(2x) = 2\exp(2x)$$
$$\frac{\mathrm{d}}{\mathrm{d}x}\cos(10x - 4) = -10\sin(10x - 4)$$
$$\frac{\mathrm{d}}{\mathrm{d}y}\sin(6y - 1.2\pi) = 6\cos(6y - 1.2\pi)$$

We see that the action of a derivative on functions of the form $f(ax)$, where a is some constant, is to "bring out" the value of a. In contrast, derivatives of the functions of the form $f(x+a)$ have no effect. The general results are demonstrated as follows.

$$\frac{\mathrm{d}}{\mathrm{d}x}f(ax) = \left.\frac{\mathrm{d}f(x)}{\mathrm{d}x}\right|_{ax} \frac{\mathrm{d}}{\mathrm{d}x}(ax) = a\left.\frac{\mathrm{d}f(x)}{\mathrm{d}x}\right|_{ax} = af'(ax)$$

$$\frac{\mathrm{d}}{\mathrm{d}x}f(x+a) = \left.\frac{\mathrm{d}f(x)}{\mathrm{d}x}\right|_{x+a} \frac{\mathrm{d}}{\mathrm{d}x}(x+a) = \left.\frac{\mathrm{d}f(x)}{\mathrm{d}x}\right|_{x+a} = f'(x+a)$$

These observations illustrate the "stretching" and "translation" properties of modifying a function in these ways. In addition, Eq. (3.9) should now be immediately apparent without resorting to intuitive arguments about the "shape of functions" in the derivation of $\mathrm{d}\cos(x)/\mathrm{d}x$ presented in Section 3.2.4.

The chain rule can be extended to composite functions formed from a chain of any number of functions. In particular, for the composite of three functions we write

$$\frac{\mathrm{d}}{\mathrm{d}x}(f \circ g \circ h)(x) = \left.\frac{\mathrm{d}f(x)}{\mathrm{d}x}\right|_{g(h(x))} \left.\frac{\mathrm{d}g(x)}{\mathrm{d}x}\right|_{h(x)} \frac{\mathrm{d}h(x)}{\mathrm{d}x}$$

which can be extended to four or more functions in the obvious way.

EXAMPLE 3.19

If $f(x) = \exp(x)$, $g(x) = 4x$, and $h(x) = \sin x$, state the following composite functions and derive an expression for the derivative with respect to x in each case.

a. $(f \circ g)(x)$
b. $(g \circ h)(x)$
c. $(f \circ g \circ h)(x)$
d. $(g \circ h \circ f)(x)$

Solution

Expressed in full, the composite functions are

a. $(f \circ g)(x) = \exp(4x)$
b. $(g \circ h)(x) = 4\sin x$
c. $(f \circ g \circ h)(x) = \exp(4\sin x)$
d. $(g \circ h \circ f) = 4\sin(\mathrm{e}^x)$

Using the chain rule, the derivatives are

a. $(f \circ g)'(x) = \left.\frac{\mathrm{d}f(x)}{\mathrm{d}x}\right|_{g(x)} \left.\frac{\mathrm{d}g(x)}{\mathrm{d}x}\right|_{x} = 4\exp(4x)$

b. $(g \circ h)'(x) = \left.\frac{\mathrm{d}g(x)}{\mathrm{d}x}\right|_{h(x)} \left.\frac{\mathrm{d}h(x)}{\mathrm{d}x}\right|_{x} = 4\cos x$

c. $(f \circ g \circ h)'(x) = \left.\frac{\mathrm{d}f(x)}{\mathrm{d}x}\right|_{g(h(x))} \left.\frac{\mathrm{d}g(x)}{\mathrm{d}x}\right|_{h(x)} \left.\frac{\mathrm{d}h(x)}{\mathrm{d}x}\right|_{x} = 4\mathrm{e}^{4\sin x}\cos x$

d. $(g \circ h \circ f)'(x) = \left.\frac{\mathrm{d}g(x)}{\mathrm{d}x}\right|_{h(f(x))} \left.\frac{\mathrm{d}h(x)}{\mathrm{d}x}\right|_{f(x)} \left.\frac{\mathrm{d}f(x)}{\mathrm{d}x}\right|_{x} = 4\mathrm{e}^x\cos(4\mathrm{e}^x)$

EXAMPLE 3.20

Use the chain rule to find the gradient of the following function at $x = 1$.

$$f(x) = \mathrm{e}^{(\ln x)^2 - \ln x + 2}$$

Solution

We recognize that

$$f(x) = (l \circ m \circ n)(x)$$

where

$$l(x) = \mathrm{e}^x \quad m(x) = x^2 - x + 2 \quad \text{and} \quad n(x) = \ln x$$

The derivative is then

$$\begin{aligned}\frac{\mathrm{d}f}{\mathrm{d}x} &= \left.\frac{\mathrm{d}l(x)}{\mathrm{d}x}\right|_{m(n(x))} \left.\frac{\mathrm{d}m(x)}{\mathrm{d}x}\right|_{n(x)} \left.\frac{\mathrm{d}n(x)}{\mathrm{d}x}\right|_{x} \\ &= \exp\left((\ln x)^2 - \ln x + 2\right) \times (2\ln x - 1) \times \frac{1}{x} \\ &= \left(\frac{2\ln x - 1}{x^2}\right) \mathrm{e}^{(\ln x)^2 + 2}\end{aligned}$$

which reduces to $-\mathrm{e}^2 \approx -7.3891$ at $x = 1$.

3.3.4 Quotient rule

A very useful result for the taking derivatives of functions of the form $f(x)/g(x)$ can be found from combining the chain and product rules. This result is called the *quotient rule*, and is particularly useful when working with rational functions, for example. The quotient rule is stated as

$$\frac{\mathrm{d}}{\mathrm{d}x}\left(\frac{f(x)}{g(x)}\right) = \frac{f'(x)g(x) - f(x)g'(x)}{g^2(x)} \tag{3.14}$$

Writing the original function as $f(x)(g(x))^{-1}$ hints at why the quotient rule can be thought of as an application of both the product and chain rules. This is further demonstrated by the proof of the quotient rule, which is as follows.

$$\begin{aligned}\frac{\mathrm{d}}{\mathrm{d}x}\left(f(x)(g(x))^{-1}\right) &= \frac{\mathrm{d}f(x)}{\mathrm{d}x} \times (g(x))^{-1} + f(x) \times \frac{\mathrm{d}}{\mathrm{d}x}(g(x))^{-1} \\ &= \frac{\mathrm{d}f(x)}{\mathrm{d}x} \times (g(x))^{-1} - \frac{f(x)}{g^2(x)}\frac{\mathrm{d}g(x)}{\mathrm{d}x}\end{aligned}$$

$$= \frac{f'(x)g(x) - f(x)g'(x)}{g^2(x)}$$

It should be clear that the second line uses the result that

$$\frac{d}{dx}(g(x))^{-1} = \frac{1}{g^2(x)}g'(x)$$

which is an application of the chain rule. This proof demonstrates that one could choose to remember the form of Eq. (3.14) or, alternatively, rely on your knowledge of the product and chain rules.

The quotient rule is of course a very useful result for obtaining the derivatives of rational functions, which is why we have not been able to consider the derivatives of that class of standard functions until this point. The use of quotient rule is fairly straightforward in principle, although the algebra can get very complicated.

For example, in order to determine

$$\frac{d}{dx}\left(\frac{x^2 + x}{x - 3}\right)$$

we would identify $f(x) = x^2 + x$ and $g(x) = x - 3$, with $f'(x) = 2x + 1$ and $g'(x) = 1$. After some manipulation, Eq. (3.14) gives

$$\frac{d}{dx}\left(\frac{x^2 + x}{x - 3}\right) = \frac{(2x + 1) \times (x - 3) - (x^2 + x) \times 1}{(x - 3)^2}$$

$$= \frac{x^2 - 6x - 3}{(x - 3)^2}$$

Following our discussion in Section 3.2, it should be clear that this derivative does not exist at the singular point at $x = 3$.

EXAMPLE 3.21

Derive expressions for the derivatives of the following functions.

a. $f(z) = \frac{2z^2-2}{z+3}$

b. $g(x) = \frac{3x^3+2x^2-x+1}{2x^3+x^2+2}$

c. $h(y) = \frac{1}{2y^2-4}$

Solution

a. $f'(z) = \frac{2(z^2+6z+1)}{(z+3)^2}$

b. $g'(x) = \frac{-x^4+4x^3+13x^2+6x-2}{(2x^3+x^2+2)^2}$

c. $h'(y) = \frac{-y}{(y^2-2)^2}$

Note that some algebraic manipulation is required to obtain these results. The resulting derivatives do not exist where the functions $f(z)$, $g(x)$, and $h(y)$ are undefined.

Knowledge of the quotient rule also allows one to discuss the derivative of the basic tan function, which, as with the rational functions, we have been forced to delay until this point.

In particular,

$$\frac{\mathrm{d}}{\mathrm{d}x}\tan x = \frac{1}{\cos^2(x)} \tag{3.15}$$

for $x \neq n\pi$ with $n \in \mathbb{Z}$. This result is obtained as follows:

$$\begin{aligned}\frac{\mathrm{d}}{\mathrm{d}x}\tan x &= \frac{\mathrm{d}}{\mathrm{d}x}\left(\frac{\sin x}{\cos x}\right)\\ &= \frac{\cos x\cos x + \sin x\sin x}{\cos^2 x}\\ &= \frac{1}{\cos^2 x}\end{aligned}$$

where we have used the identity $\cos^2 x + \sin^2 x = 1$ from Appendix A. Together with Eqs. (3.7) and (3.8), Eq. (3.15) completes our statement of the derivatives of the basic circular functions.

With knowledge of the quotient, chain and product rules, we are now able to consider the derivatives of relatively complicated trigonometric functions, as in the following example.

EXAMPLE 3.22

Derive expressions for the derivatives of the following functions.

a. $f(x) = \tan(3x + 10)$
b. $g(x) = 1/\sin x$
c. $h(x) = 1/\cos x$
d. $l(x) = \tan x \sin x$

Solution

a. Using the chain rule and Eq. (3.15), $f'(x) = \frac{3}{\cos^2(3x+10)}$.
b. Using the chain rule and Eq. (3.7), $g'(x) = -\frac{\cos x}{\sin^2(x)} = \frac{-1}{\sin x \tan x}$.
c. Using the chain rule and Eq. (3.7), $h'(x) = -\frac{(-\sin x)}{\cos^2(x)} = \frac{\tan x}{\cos x}$.
d. Using the product rule and Eqs. (3.7) and (3.15),
$l'(x) = \frac{\sin x}{\cos^2 x} + \tan x \cos x = \frac{\tan x}{\cos x} + \sin x.$

Of course, the derivatives of the above functions do not exist where the $f(x)$, $g(x)$, $h(x)$, and $l(x)$ are undefined.

At this stage, it is useful to briefly review the trigonometric identities in Appendix A. In particular, note that standard notation exists to denote the reciprocal of the standard circular functions,

$$\text{cosec}(x) = \frac{1}{\sin x}, \quad \sec(x) = \frac{1}{\cos x}, \quad \cot x = \frac{1}{\cot x} \tag{3.16}$$

These functions are often extremely useful for representing complicated functions involving the basic circular functions in a compact way. The properties of the cosec, sec, and cot functions were explored in Question 2.11 and you are invited to review the results of that question now.

EXAMPLE 3.23

Use the expressions in Eq. (3.16) to simplify the answers in Example 3.22 where appropriate.

Solution

a. $f'(x) = 3\sec^2(3x + 10)$
b. $g'(x) = -\cos(x)\csc^2(x)$ or $-\cot(x)\csc(x)$
c. $h'(x) = \sin(x)\sec^2(x)$ or $\tan(x)\sec(x)$
d. $l'(x) = \sin(x)\left(1 + \sec^2(x)\right)$ or $\sin(x) + \tan(x)\sec(x)$

The solutions show that there are often many equivalent ways to represent the same function.

3.3.5 Derivatives of inverse functions

An obvious question might now be, given a function, is it possible to determine directly the derivative of its inverse function? The answer to this is that *sometimes* it is possible.

In particular, if an explicit expression of the inverse function exists and it is differentiable, the general rule is

$$\frac{\mathrm{d}}{\mathrm{d}x} f^{-1}(x) = \left(\left. \frac{\mathrm{d}f(x)}{\mathrm{d}x} \right|_{f^{-1}(x)} \right)^{-1} \tag{3.17}$$

which can be more compactly written as $1/f'(f^{-1}(x))$. The rigorous proof of this expression is beyond this scope of this book and we instead consider a more pragmatic approach to calculating the derivative of an inverse function. The results of this pragmatic method will then be used demonstrate that the above expression does indeed work in Example 3.26.

We introduce the method by obtaining the derivative of $\ln(x)$, which is the remaining basic function to discuss the derivative of, and is, of course, the inverse function of $\exp(x)$. We begin by taking the derivative of the function $y = \ln(x)$ with respect to y, noting that its inverse is $x(y) = \exp(y)$. This leads to

$$\frac{\mathrm{d}y}{\mathrm{d}y} = \frac{\mathrm{d}}{\mathrm{d}y} \ln(x)$$

$$\Rightarrow 1 = \frac{\mathrm{d}\ln(x)}{\mathrm{d}x}\frac{\mathrm{d}x}{\mathrm{d}y}$$

The right-hand side uses the chain rule and the left-hand side is simply that $\frac{\mathrm{d}y}{\mathrm{d}y} = 1$. In this case, $\frac{\mathrm{d}x}{\mathrm{d}y} = \exp(y) = x$ and so the above expression can be expressed in terms of x as

$$1 = \frac{\mathrm{d}\ln(x)}{\mathrm{d}x} \times x$$

which is rearranged to give

$$\frac{\mathrm{d}}{\mathrm{d}x}\ln(x) = \frac{1}{x} \tag{3.18}$$

It is clear that this expression is not defined at $x = 0$, but we note that $\ln x$ is also not defined at that point.

Equation (3.18) is a specific case of the following more general result.

$$\frac{\mathrm{d}}{\mathrm{d}x}\log_a(x) = \frac{1}{x\ln(a)} \quad \text{for } a > 0 \tag{3.19}$$

This expression can be derived in a similar way, as shown in the following example.

EXAMPLE 3.24

Derive the general rule expressed in Eq. (3.19).

Solution

We set $y = \log_a(x)$ which is known to have inverse $x(y) = a^y$. Taking derivatives of $y = \log_a(x)$ with respect to y and using the chain rule leads to

$$\frac{\mathrm{d}y}{\mathrm{d}y} = \frac{\mathrm{d}\log_a(x)}{\mathrm{d}y}$$

$$\Rightarrow 1 = \frac{\mathrm{d}\log_a(x)}{\mathrm{d}x}\frac{\mathrm{d}x}{\mathrm{d}y}$$

Recall from Eq. (3.6) that $\mathrm{d}x(y)/\mathrm{d}y = a^y\ln(x) = x\ln(x)$, and so we see that

$$\frac{\mathrm{d}}{\mathrm{d}x}\log_a(x) = \frac{1}{x\ln(a)}$$

as required.

EXAMPLE 3.25

Derive expressions for the following derivatives.

a. $\frac{\mathrm{d}}{\mathrm{d}x}\arcsin(x)$

b. $\frac{\mathrm{d}}{\mathrm{d}x}\arccos(x)$

c. $\frac{\mathrm{d}}{\mathrm{d}x}\arctan(x)$

Solution

a. If $y = \arcsin(x)$ and $x = \sin(y)$

$$\begin{aligned}\frac{\mathrm{d}y}{\mathrm{d}y} &= \frac{\mathrm{d}\arcsin(x)}{\mathrm{d}x}\frac{\mathrm{d}x}{\mathrm{d}y} \\ \Rightarrow 1 &= \frac{\mathrm{d}\arcsin(x)}{\mathrm{d}x}\cos(y) \\ &= \frac{\mathrm{d}\arcsin(x)}{\mathrm{d}x}\sqrt{1-x^2}\end{aligned}$$

where we have used the identity $\cos^2 y + \sin^2 y = 1 \Rightarrow \cos y = \sqrt{1-x^2}$. Hence

$$\frac{\mathrm{d}\arcsin(x)}{\mathrm{d}x} = \frac{1}{\sqrt{1-x^2}}$$

b. We set $y = \arccos(x)$ and $x = \cos(y)$ and obtain

$$\frac{\mathrm{d}\arccos(x)}{\mathrm{d}x} = \frac{-1}{\sqrt{1-x^2}}$$

c. We set $y = \arctan(x)$ and $x = \tan(y)$ and obtain

$$\frac{\mathrm{d}\arctan(x)}{\mathrm{d}x} = \frac{1}{\sqrt{1+x^2}}$$

EXAMPLE 3.26

Use Eq. (3.17) to write down the following results directly.

a. $\frac{\mathrm{d}}{\mathrm{d}x}\log_{10}(x) = \frac{1}{x\ln(10)}$

b. $\frac{\mathrm{d}}{\mathrm{d}x}\arcsin(x) = \frac{1}{\sqrt{1-x^2}}$

c. $\frac{\mathrm{d}}{\mathrm{d}x}\arctan(x) = \frac{1}{\sqrt{1+x^2}}$

Solution

a. If $f^{-1}(x) = \log_{10}(x)$ then $f(x) = 10^x$ and Eq. (3.17) leads to

$$\begin{aligned}\frac{\mathrm{d}}{\mathrm{d}x}\log_{10}(x) &= \left(\frac{\mathrm{d}}{\mathrm{d}x}10^x\bigg|_{\log_{10}(x)}\right)^{-1} \\ &= \left(10^{\log_{10}(x)}\ln 10\right)^{-1} \\ &= \frac{1}{x\ln 10}\end{aligned}$$

b. If $f^{-1}(x) = \arcsin x$ then $f(x) = \sin x$ and Eq. (3.17) leads to

$$\begin{aligned}\frac{\mathrm{d}}{\mathrm{d}x}\arcsin(x) &= \left(\frac{\mathrm{d}}{\mathrm{d}x}\sin x\bigg|_{\arcsin x}\right)^{-1}\\ &= (\cos(\arcsin x))^{-1}\\ &= \left(\sqrt{1-\sin^2(\arcsin x)}\right)^{-1}\\ &= \frac{1}{\sqrt{1-x^2}}\end{aligned}$$

c. If $f^{-1}(x) = \arctan x$ then $f(x) = \tan x$ and Eq. (3.17) leads to

$$\begin{aligned}\frac{\mathrm{d}}{\mathrm{d}x}\arctan(x) &= \left(\frac{\mathrm{d}}{\mathrm{d}x}\tan x\bigg|_{\arctan x}\right)^{-1}\\ &= \left(\sec^2(\arctan x)\right)^{-1}\\ &= \left(1+\tan^2(\arctan x)\right)^{-1}\\ &= \frac{1}{\sqrt{1+x^2}}\end{aligned}$$

3.4 ALGEBRAIC DERIVATIVES ON YOUR COMPUTER

Wolfram Alpha is a particularly powerful tool for performing algebraic derivatives. These can be performed using numerous commands, all of which are very intuitive. For example, an expression for

$$\frac{\mathrm{d}}{\mathrm{d}x}\cos\left(\mathrm{e}^{2x^2-1}\right)$$

is obtained using either of the commands

```
derivative cos(e^(2x^2-1))
```

or

```
dcos(e^(2x^2-1))/dx
```

Both commands result in

$$\frac{\mathrm{d}}{\mathrm{d}x}\cos\left(\mathrm{e}^{2x^2-1}\right) = -4\mathrm{e}^{2x^2-1}x\sin\left(\mathrm{e}^{2x^2-1}\right)$$

which you are invited to confirm by hand using the chain rule.

The derivatives of a function can be evaluated at particular points using intuitive extensions to these commands. For example, we can obtain

$$\left.\frac{\mathrm{d}}{\mathrm{d}x}\left(\mathrm{e}^{\frac{2x^2-1}{x^2+x+1}}\right)\right|_{x=1}$$

from the command

```
derivative e^((2x^2-1)/(x^2+x+1)) at x=1
```

This leads to the result $\sqrt[3]{\mathrm{e}}$.

It should now be clear that Wolfram Alpha understands many intuitive commands, often multiple commands for the same operation, and it is impossible to list every command the engine understands here. It is therefore worth spending some time exploring how Wolfram Alpha might be applied to the mathematics considered in this and previous chapters. You are now encouraged to do this, beginning with the following example. Note that the engine states its interpretation of the command given so you can confirm that it has actually understood you correctly.

EXAMPLE 3.27

Explore how Wolfram Alpha might be used solve the following problems.

a. Obtain the tangent line to $f(x) = x\cos\left(x^2+2\right)$ at $x = 0$.

b. Determine the parity of the inverse of $h(z) = \tan(z^3)$. Determine its range and an expression for its derivative.

c. If $g(y) = \tan\left(y^{2n}\right)$ for $n \in \mathbb{Z}$, determine an expression for $g'(y)$ for $n = 0, 1, 2, 3$.

d. If $m(x) = x^6 - 56x^4 + 784x^2 - 2304$, factorize this expression and its derivative.

Solution

a. The command

```
tangent line of x*cos(x^2+2) at x=0
```

results in the required line $y = x\cos(2)$. A plot is given so that we can visually confirm that this is correct.

b. The command

```
parity inverse tan(z^3)
```

returns that $\tan^{-1}\left(z^3\right)$ is odd. Furthermore, the commands

```
range inverse tan(z^3+1)
```

and

```
derivative inverse tan(z^3+1)
```

result in the range $\{y \in \mathbb{R} : -\frac{\pi}{2} < y < 0 \text{ or } 0 < y < \frac{\pi}{2}\}$ and derivative

$$\frac{\mathrm{d}}{\mathrm{d}z}\left(\tan^{-1}\left(z^2+1\right)\right) = \frac{3z^2}{(z^3+1)^2+1}$$

Plots are given so that the parity and range can be visually confirmed.

c. The command

```
derivative tan(y^(2n))
```

results in

$$\frac{d}{dy}\tan\left(y^{2n}\right) = 2ny^{2n-1}\sec^2(y^{2n})$$

and their value at each $n \in \mathbb{N}$.

d. The commands

```
factorise x^6-56x^4+784x^2-2304
```

and

```
factorise derivative x^6-56x^4+784x^2-2304
```

result in

$$(x-2)(x+2)(x-4)(x+4)(x-6)(x+6) \quad \text{and} \quad 2x(x^2-28)(3x^2-28)$$

respectively.

3.5 ACTUARIAL APPLICATION: THE FORCE OF INTEREST

We now return to the time value of money, as explained in the previous chapter, and demonstrate how we might begin to consider accumulations under interest rates that *vary* with time. As we shall see, this necessitates the use of differential calculus. In this section, we will derive a fundamental expression that connects the accumulation factor to an interest rate quantity through the use of a derivative with respect to time. We will return to the general solution of this expression in Chapter 7.

Before we begin to derive the expression, it is necessary to introduce two new actuarial concepts: *nominal interest rates* and the *principal of consistency*. We begin by assuming that interest rates are fixed and will generalize to varying interest rates once these new concepts are established.

We denote the (constant) *nominal rate of interest converted over a period of length h* by i_h per annum. This nominal rate is defined to be such that the *effective* compounding rate of interest over a period of length h is given by $h \times i_h$. For example,

- A nominal rate interest of 12% per annum *converted monthly* is denoted by $i_{1/12} = 12\%$ per annum and is such that the effective rate of interest is $\frac{12\%}{12} = 1\%$ *per month.*
- A nominal rate of 3% per annum *converted quarterly* is denoted by $i_{1/4} = 3\%$ per annum and is such that the effective rate is $\frac{3\%}{4} = 0.75\%$ per quarter.
- A nominal rate of 1% per annum *converted 2-yearly* is denoted by $i_2 = 1\%$ per annum and is such that the effective rate is $2 \times 1\% = 2\%$ per 2 years.

Nominal rates are often useful in practice, for example, when considering streams of payments paid less or more frequently than per annum. However, for all intents and purposes in this book, it is sufficient to understand what is meant by a nominal rate and how we might use it in an accumulation. It will prove necessary to incorporate nominal rates into accumulations in the discussion of varying interest rates later in this section.

EXAMPLE 3.28

Determine the required quantities given the stated constant nominal rates of interest.

a. The accumulated value of \$200 after 1 year under a nominal rate $i_{0.5} = 4\%$ per annum.

b. The accumulated value of \$100 after 2 years under a nominal rate $i_{0.25} = 5\%$ per annum.

c. The present value of \$10,000 due in 12 years under a nominal rate $i_3 = 1\%$ per annum.

Solution

In each case, we convert the nominal rate to an appropriate compounding "effective" rate and apply Eq. (2.22).

a. We are given the nominal rate converted every 6 months, $i_{0.5} = 4\%$ per annum, and so have an effective rate of $i = 2\%$ per 6 months. The accumulation after 1 year must be considered as two 6-month periods and we have an accumulated value of $200(1.02)^2 =$ \$208.08.

b. We are given the nominal rate converted every quarter (3 months), $i_{0.25} = 5\%$ per annum, and so have an effective rate of $i = 1.25\%$ per quarter. The accumulation after 2 years must be considered as eight quarters and we have an accumulated value of $100(1.0125)^8 \approx \$110.45$.

c. We are given the nominal rate converted every 3 years, $i_3 = 1\%$, and so have an effective rate of $i = 3\%$ per 3 years. The present value of the payment due in 12 years is therefore four periods of 3 years away and we have a present value of $10,000(1.03)^{-4} \approx \$8884.87$.

We now generalize our notation for the accumulation factor, as earlier defined in Eq. (2.22), to make clear both the investment and withdrawal times. This will simplify the following discussion of the principle of consistency. In particular, we now define the accumulated value at time t of a unit investment made at t_1 by $A(t_1, t)$. In the case that i is fixed, it is clear that,

$$A(t_1, t) = (1 + i)^{t-t_1} \tag{3.20}$$

and so is equal to $A(t_2 - t_1)$ in the notation of Eq. (2.22). Furthermore, if again i is constant, $A(0, t) \equiv A(t)$. The notation $A(t_1, t)$ may appear unusual to you to as it implies that A has two variables, t_1 and t. However, we understand t_1 to be a *fixed* time and so $A(t_1, t)$ is a function of t only. If $t > t_1$, it is clear that Eq. (3.20) is an accumulation factor; if $t < t_1$, Eq. (3.20) can be considered as a *discounting* factor. This is consistent with our previous discussion around $A(t)$ with positive and negative t.

Using this new notation, we see that $A(0, t_1) \times A(t_1, t_2)$ would give the accumulation of a unit investment made at time 0 to some time $t_1 > 0$, which is then immediately reinvested between t_1 and some later t_2. Since the payment of compound interest does not distinguish between prior interest paid and the initial investment made, it must be that

$$A(0, t_1)A(t_1, t_2) \equiv A(0, t_2)$$

That is, the two-stage accumulation to time t_2 results in exactly the same accumulation as the one-stage accumulation over the same time period. This is of course assuming that the same rates of interest are paid in each scenario. This expression is an example of the *principle of consistency*.

The principle of consistency, as stated here, can be generalized to show that the accumulation over any interval can be broken down into the accumulation of any number of subintervals. That is, if $t_0 < t_1 < \cdots < t_n$, then

$$A(t_0, t_n) \equiv A(t_0, t_1) \times A(t_1, t_2) \times \cdots \times A(t_{n-1}, t_n)$$

It should be clear that the principle also applies to a discounting process. Here the discounting factor from t_n to t_0 is given by

$$\frac{1}{A(t_0, t_n)} \equiv \frac{1}{A(t_0, t_1) \times A(t_1, t_2) \times \cdots \times A(t_{n-1}, t_n)}$$

We are now able to discuss what form the accumulation factor $A(0, t)$ must take when the interest rate is allowed to be a function of time, which is the aim of this section. To do this, it is necessary to incorporate the *force of interest*, $\delta(t)$, defined to be

$$\delta(t) = \lim_{h \to 0} i_h(t) \tag{3.21}$$

That is, the force of interest at time t is the nominal rate at time t converted on an ever-shorter time interval. Indeed $\delta(t)$ is often called the *continuously compounding* rate of interest, which refers to it being the nominal rate converted continuously. Note that both the force of interest and the nominal rate are now allowed to be functions of time. Of course, this does not make sense if the nominal rate is converted over a finite period of time within which it can vary. However, since we are looking at it converting near instantaneously in the limit $h \to 0$, we can assume that the associated effective rate $hi_h(t)$ is *constant* over this infinitesimally small interval.

Let us consider the accumulation of a unit investment between time t and $t + h$, where h is some (small) positive time period. It should be clear that,

$$A(t, t + h) = 1 + hi_h(t)$$

which follows directly from the definition of $i_h(t)$. Furthermore, the principle of consistency is such that

$$A(0, t + h) \equiv A(0, t)A(t, t + h) \Rightarrow A(t, t + h) = \frac{A(0, t + h)}{A(0, t)}$$

We can combine these two expressions and obtain

$$\frac{A(0, t + h)}{A(0, t)} = 1 + hi_h(t) \Rightarrow i_h(t) = \frac{A(0, t + h) - A(0, t)}{A(0, t)h}$$

and, as $h \to 0$, we have

$$\delta(t) = \lim_{h\to 0} i_h(t) = \lim_{h\to 0}\left(\frac{A(0,t+h)-A(0,t)}{A(0,t)h}\right)$$

Comparing this expression to Eq. (3.4), we find that

$$\delta(t) = \frac{1}{A(0,t)}\frac{\mathrm{d}}{\mathrm{d}t}A(0,t) = \frac{\mathrm{d}}{\mathrm{d}t}\ln A(0,t) \tag{3.22}$$

That is, when interest is allowed to be a function of time, the accumulation factor and the force of interest are connected by a derivative with respect to time.

Recall that our aim was to derive an expression for $A(0,t)$, or indeed $A(t_1,t)$, under varying interest rates. Equation (3.22) is progress toward this however we will have to wait until Chapters 6 and 7 to understand how to "reverse" this expression to make $A(0,t)$ the subject. Some insight into the process can be obtained if we revert back to the particular case that interest is constant and attempt to make sense of δ in terms of i. If $\delta(t) = \delta$, it should be clear that

$$A(0,t) = \mathrm{e}^{\delta t}$$

satisfies Eq. (3.22) as is confirmed by

$$\begin{aligned}\frac{\mathrm{d}}{\mathrm{d}t}\ln A(0,t) &= \frac{\mathrm{d}}{\mathrm{d}t}\ln \mathrm{e}^{\delta t}\\ &= \frac{\mathrm{d}}{\mathrm{d}t}(\delta t)\\ &= \delta\end{aligned}$$

Furthermore, since interest is again fixed, we know that $A(0,t) = (1+i)^t$. Therefore

$$\mathrm{e}^{\delta t} = (1+i)^t \Rightarrow \mathrm{e}^{\delta} = 1+i$$

In addition, it is clear that $A(t_1,t) = \mathrm{e}^{(t-t_1)\delta}$.

EXAMPLE 3.29

Compute the following under a fixed continuously compounding rate of interest of 8% per annum.

a. The accumulated value of \$10,000 after 4 years.

b. The present value of a \$100 due in 15 years.

c. The accumulated value at $t = 5$ of \$20,000 invested at $t = 4$. Time is measured in years.

d. The present value of \$10 due in 6 months and \$20 due in 25 months.

e. The equivalent effective annual rate of compounding interest. Interpret your answer.

Solution

We are given that $\delta = 0.08$ and so use the accumulation factor $A(t_1,t) = \mathrm{e}^{(t-t_1)\delta}$.

a. The accumulated value \$10,000 after 4 years is given by $10,000\mathrm{e}^{4\times 0.08} \approx \$13,771.28$.

b. The present value of a \$100 due in 15 years is given by $100\mathrm{e}^{-15\times 0.08} \approx \$30.12$.

c. The accumulated value at $t = 5$ of \$20,000 invested at $t = 4$ is given by $20,000e^{(5-4)\times 0.08} \approx \$21,665.74$.

d. The present value of \$10 due in 6 months and \$20 due in 25 months is given by $10e^{-0.5\times 0.08} + 20e^{-\frac{25}{12}\times 0.08} \approx \$26.54$.

e. The equivalent effective annual rate of compounding interest is i such that $1 + i = e^{0.08}$. Therefore, $i \approx 8.33\%$. We note that to achieve the same accumulation, a higher value of i is required. This is because $i \equiv i_1$ is "converted" at the end of each year, whereas δ is "converted" every instant. Therefore, under δ, interest is paid and begins to earn interest far more rapidly than under i.

We will return to the accumulation factors arising from a given $\delta(t)$ in Section 7.5. Until that point, it is only possible for us to use Eq. (3.22) to infer $\delta(t)$ from some given $A(0, t)$, which is of course much less useful. This is demonstrated in the following example.

EXAMPLE 3.30

The growth of a unit deposit in a bank account is given by the accumulation factor $e^{0.05t^2+0.02}$, where t is measured in years from the initial deposit. Determine the underlying force of interest that is being paid by this account.

Solution

We are given that $e^{0.05t^2+0.02t}$ and can use Eq. (3.22) to determine $\delta(t)$. In particular,

$$\begin{aligned}\delta(t) &= \frac{d}{dt}\ln\left(e^{0.05t^2+0.02t}\right)\\ &= \frac{d}{dt}\left(0.05t^2 + 0.02t\right)\\ &= 0.1t + 0.02\end{aligned}$$

3.6 QUESTIONS

The following questions are intended to test your knowledge of the material discussed in this chapter. Full solutions are available in Chapter 3 Solutions of Part III. You should use an algebraic approach unless otherwise stated.

Question 3.1. Determine the equation of the straight line through each of the following pairs of data points.

a. $(x, y) = (0, 0)$ and $(3, 3)$

b. $(x, y) = (-1, -5)$ and $(4, 0)$

c. $(x, y) = (\pi, 0)$ and $(\pi^2, -\pi)$

d. $(x, y) = (-10, -56)$ and $(-100, -129)$

Question 3.2. Determine whether the following functions are continuous at the locations indicated.

a. $f(x)$ at $x = 5$ where

$$f(x) = \begin{cases} x^3 + x^2 + 1 & \text{for } x \in (-\infty, -1] \\ x^3 + x^2 + 2 & \text{for } x \in (-1, \infty) \end{cases}$$

b. $g(y)$ at $y = 2$ where

$$g(y) = \begin{cases} y^5 - y^2 - 20 & \text{for } x \in (-\infty, 2] \\ y^3 & \text{for } x \in (2, \infty) \end{cases}$$

c. $h(z)$ at $z = 0$ where

$$h(z) = \begin{cases} e^{-2z} & \text{for } x \in (-\infty, 0) \\ 1.1 & \text{for } x = 0 \\ e^{-2z} & \text{for } x \in (0, \infty) \end{cases}$$

Question 3.3. You are given that, if

$$\lim_{x \to a} f(x) = L_1 \quad \text{and} \quad \lim_{x \to a} g(x) = L_2$$

then

$$\lim_{x \to a} (f(x)g(x)) = L_1 L_2$$

Use this result to demonstrate that

$$\lim_{x \to a} f^n(x) = \left[\lim_{x \to a} f(x) \right]^n$$

for $n = 1, 2, 3, \ldots$.

Question 3.4. Confirm the following limits using a tabular approach.

$$\lim_{\delta \to 0} \left(\frac{\sin(\delta)}{\delta} \right) = 1 \quad \text{and} \quad \lim_{\delta \to 0} \left(\frac{1 - \cos(\delta)}{\delta} \right) = 0$$

Question 3.5. Determine the tangent lines to the following functions at the location stated.

a. $m(p) = p^3 - 2p + \sin(2p)$ at $p = 2\pi$.

b. $n(q) = e^{-q} \left(\sin^2(q) + \cos^2(\pi q) \right)$ at $q = 0$.

Question 3.6. Use Eq. (3.4) to determine the derivatives of the following functions from first principles.

a. $f(x) = x^2 - 2x + 1$
b. $g(y) = e^{2y}$
c. $h(z) = \frac{1}{z^2}$
d. $m(p) = 4e^{p^2}$

Question 3.7. Determine the derivatives of the following expressions. State which approach you have used in each case.

a. $f(x) = \frac{x^2-2x+1}{x^4+25}$
b. $g(z) = \exp(\sin z)$
c. $h(y) = \cos^2 y - 4\cos y + 2$
d. $m(p) = \cos(p)\sin(e^p)$

Question 3.8. Explain how one might approach the following derivative. Determine it.

$$\frac{d}{dx}\left(\sin\left(e^{x^2-4}\right)\right)$$

Question 3.9. Determine the locations at which the instantaneous gradients of the following functions are zero. Interpret the meaning of these locations.

a. $f(z) = \frac{z^3}{3} - z + 5$
b. $g(y) = \cos(y) + 2$

Question 3.10. Derive expressions for the following derivatives.

a. $\frac{d}{dx}\arcsin(4x)$
b. $\frac{d}{dx}\arccos(x^2 - 1)$
c. $\frac{d}{dx}\arctan(e^x)$

Question 3.11. A bank account is such that it pays the following continuously compounding rate of interest

$$\delta = \begin{cases} 5\% \text{ for } t \in [0, 4] \\ 7\% \text{ for } t \in (4, 10] \end{cases}$$

where t is measured in years. Calculate the amount that should be invested now to cover a liability of \$5600 due in 7 years' time.

Question 3.12. Deposits in a bank account are known to accumulate with factor $A(0, t) = te^{0.05t^3}$, where t is measured in years. Determine the underlying force of interest paid by this account.

CHAPTER 4

Differential Calculus II

Contents

Prerequisite knowledge	Intended learning outcomes

• Chapter 1
• Chapter 2
• Chapter 3
	• Define and test for smoothness, using both the intuitive and strict definitions
	• Define and locate turning (stationary) points
	• Classify turning points (using general reasoning and the second-derivative test) as either
	  ◦ maxima
	  ◦ minima
	  ◦ points of inflection
	• Determine high-order derivatives
	• Determine the ranges of complicated functions

In this chapter, we discuss some applications of differential calculus that are relevant to the exploration and understanding of functions. In particular, we will consider the concept of *smoothness* and *turning points*. The discussion of both concepts necessarily requires an understanding of *higher-order derivatives* which we will discuss in the practical sense. At the end of this chapter, we return to an algebraic strategy for determining the range of a function; this is a direct continuation of the methods seen in Chapter 2.

4.1 AN INTRODUCTION TO SMOOTHNESS

Recall that we gave a very intuitive definition of *smoothness* in Chapter 2; in particular, we said that smoothness was simply the absence of corners or cusps on the plot of a function. In the fullest mathematical sense, a function f is said to be *smooth* over a domain if all *partial derivatives* of all possible *orders* are continuous on that domain. However, at this point, we are yet to consider *partial derivatives* (see Chapter 12) and *higher-order derivatives* (see Section 4.2), and so our immediate discussion of smoothness is necessarily limited.

We begin by defining smoothness as the property of a *continuous first derivative*. That is, a function $f(x)$ is smooth over domains where $f'(x)$ is continuous. This is a softer condition than the full requirement for smoothness but should be intuitively clear. Recall from Section 3.1 that a function $g(x)$ is continuous over domains where

$$\lim_{x \to c^-} g(x) = g(c) = \lim_{x \to c^+} g(x) \tag{4.1}$$

Therefore, function $f(x)$ is *smooth*, in our current softer sense, where

$$\lim_{x \to c^-} f'(x) = f'(c) = \lim_{x \to c^+} f'(x) \tag{4.2}$$

It should be immediately clear that since the derivative of a function does not exist at a discontinuous point, a discontinuous function cannot be smooth. For example, the function $\tan x$ is noncontinuous, nondifferentiable, and nonsmooth at $x = n\pi$ with $n \in \mathbb{Z}$; and the function $1/x$ is noncontinuous, nondifferentiable, and nonsmooth at $x = 0$. However, a continuous function can be nonsmooth and a discussion of such examples is our starting point.

Many of the functions we have considered to this point have in fact been smooth by the definition given above and it may be difficult to picture a nonsmooth function with corners or cusps. A useful class of functions for the discussion of smoothness is the class of *piecewise* functions. Piecewise functions are simply defined to be functions consisting of multiple subfunctions. For example,

$$l(x) = \begin{cases} x^2 - 4 & \text{for } x \in (-\infty, 4] \\ 4 & \text{for } x \in (4, \infty) \end{cases} \tag{4.3}$$

is a piecewise function with subfunctions $x^2 - 4$ and 4. A further example is the "absolute value" function

$$m(x) = |x| = \begin{cases} -x & \text{for } x \in (-\infty, 0) \\ x & \text{for } x \in [0, \infty) \end{cases} \tag{4.4}$$

From our discussion in Section 3.1, it should be clear that, in principle, there is no barrier to piecewise functions being continuous as the independent variable moves between subfunctions. For example, both $l(x)$ and $m(x)$ defined above are continuous by the property (4.1). That is their left-hand and right-hand limits are equal (i.e., the limit exists) at the crossover and the value of the function is equal to the limit in each case. This is not to say that all piecewise functions are continuous, of course, see Examples 3.2 and 3.3. The property of smoothness, even this initial softer definition, is however more restrictive.

It should be clear from Eq. (4.3) that

$$l'(x) = \begin{cases} 2x & \text{for } x \in (-\infty, 4] \\ 0 & \text{for } x \in (4, \infty) \end{cases}$$

and so

$$\lim_{x \to 4^-} l'(x) = 8 \quad \text{and} \quad \lim_{x \to 4^+} l'(x) = 0$$

Despite being continuous, $l(x)$ is *not* smooth at $x = 4$. Furthermore, Eq. (4.4) is such that

$$m'(x) = |x| = \begin{cases} -1 & \text{for } x \in (-\infty, 0) \\ 1 & \text{for } x \in [0, \infty) \end{cases}$$

and so

$$\lim_{x \to 0^-} l'(x) = -1 \quad \text{and} \quad \lim_{x \to 0^+} l'(x) = 1$$

Therefore, $m(x)$ is also continuous but *not* smooth at $x = 0$.

EXAMPLE 4.1

Determine whether the following functions are smooth (by the restricted definition 4.2) over $x \in \mathbb{R}$.

a. $f(x) = 2x^2 - 2x + 1$

b. $h(x) = \begin{cases} x & \text{for } x \in (-\infty, 1] \\ 2 & \text{for } x \in (1, \infty) \end{cases}$

c. $g(x) = \begin{cases} x^3 + 4 & \text{for } x \in (-\infty, 0] \\ 10x + 4 & \text{for } x \in (0, \infty) \end{cases}$

d. $k(x) = \begin{cases} x^3 & \text{for } x \in (-\infty, 0] \\ x^2 & \text{for } x \in (0, \infty) \end{cases}$

Solution

a. Both $f(x)$ and $f'(x) = 4x - 2$ are continuous from the properties of polynomials. The function $f(x)$ is therefore smooth over $\mathbb{R}$.

b. The subfunctions either side of $x = 1$ are smooth and continuous by the properties of polynomials. However, the left-hand and right-hand limits of $h(x)$ are not equal at $x = 1$ and so the function is not continuous there. The function cannot therefore be smooth at $x = 1$.

c. The subfunctions either side of $x = 0$ are smooth and continuous by properties of polynomials; $g(x)$ is also continuous at $x = 0$. However,

$$g'(x) = \begin{cases} 3x^2 & \text{for } x \in (-\infty, 0] \\ 10 & \text{for } x \in (0, \infty) \end{cases}$$

and so $g'(x)$ is not continuous at $x = 0$ (owing to different left-hand and right-hand limits). The function $g(x)$ is therefore not smooth at $x = 0$.

d. The subfunctions either side of $x = 0$ are smooth and continuous by properties of polynomials. We have

$$k'(x) = \begin{cases} 3x^2 & \text{for } x \in (-\infty, 0] \\ 2x & \text{for } x \in (0, \infty) \end{cases}$$

which is also continuous at $x = 0$. Function $k(x)$ is therefore smooth by the definition (4.2). We will return to this example in Example 4.4.

It is important to get an intuitive feel for how smoothness and nonsmoothness manifest in the visual sense. Previously, we defined smoothness as the absence of corners and cusps, and, with the mathematical definition of "continuous derivatives" in mind, this intuitive definition should now make some sense. These ideas are demonstrated in Figure 4.1 where the functions discussed in Example 4.1 are shown together. For example, clear corner can be seen at $x = 0$ in Figure 4.1(c).

4.2 HIGHER-ORDER DERIVATIVES

In the above discussion of smoothness, we were forced to consider the continuity of derivatives as functions in their own right. Indeed, the notion of a derivative was introduced in Chapter 3 as a function *derived* from another in a particular way. With this in mind, it is reasonable to expect to be able to study further properties of derivative functions. In order to do this, it may prove useful to have knowledge of the gradients of derivative functions. This, of course, prompts the concept of a "derivative of a derivative" of a function.

Working with function $f(x)$, the derivative of the derivative could be denoted in any of the following equivalent ways:

$$\frac{\mathrm{d}}{\mathrm{d}x}\left(\frac{\mathrm{d}f(x)}{\mathrm{d}x}\right) \equiv \frac{\mathrm{d}}{\mathrm{d}x}f'(x) \equiv \frac{\mathrm{d}^2}{\mathrm{d}x^2}f(x) \equiv f''(x)$$

The derivative of a derivative of $f(x)$ is often referred to as the *second-order derivative* or the *second derivative* of $f(x)$. The second derivative is a function itself and an explicit expression for it can be derived from the repeated application of the standard methods discussed in Chapter 3.

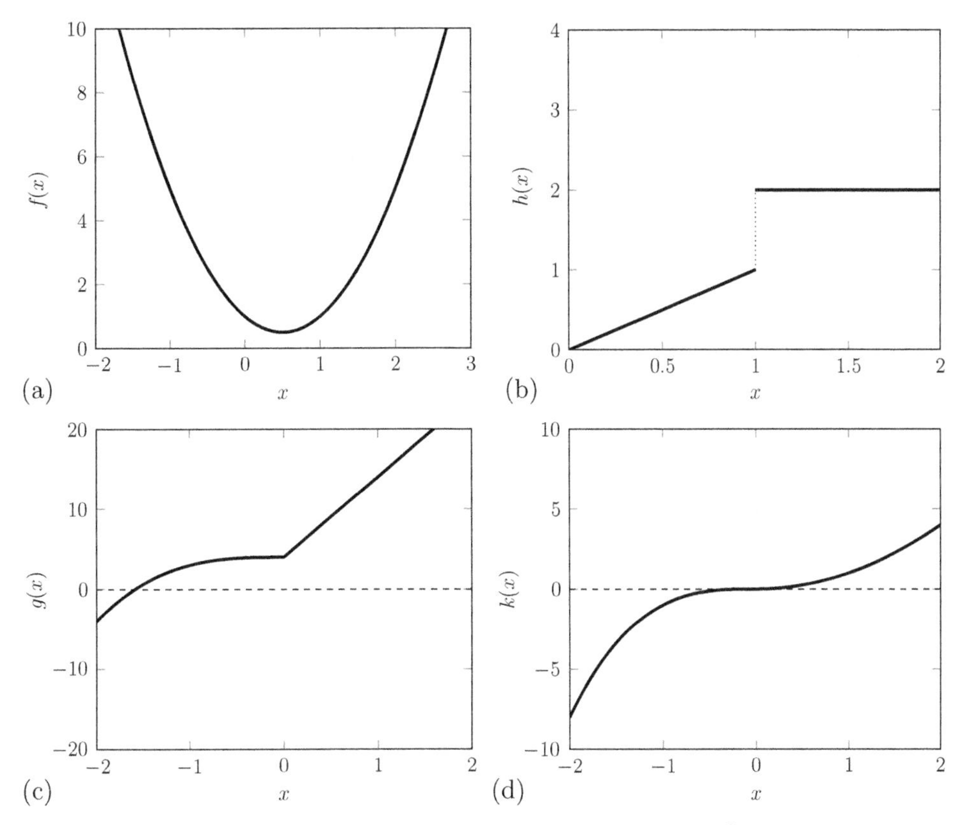

Figure 4.1 Illustrations of the functions in Example 4.1. (a) $f(x) = 2x^2 - 2x + 1$, (b) $h(x) = \begin{cases} x & \text{for } x \in (-\infty, 1] \\ 2 & \text{for } x \in (1, \infty) \end{cases}$, (c) $g(x) = \begin{cases} x^3 + 4 & \text{for } x \in (-\infty, 0] \\ 10x + 4 & \text{for } x \in (0, \infty) \end{cases}$, and (d) $k(x) = \begin{cases} x^3 & \text{for } x \in (-\infty, 0] \\ x^2 & \text{for } x \in (0, \infty) \end{cases}$.

EXAMPLE 4.2

Determine expressions for the second derivatives of the following functions.

a. $g(x) = 2x^2 + x - 2$

b. $h(x) = \sin(2x)$

c. $l(x) = x \exp(2x^2 - x)$

Solution

a. The first derivative is obtained as $g'(x) = 4x + 1$. The second derivative is then

$$\frac{\mathrm{d}}{\mathrm{d}x}(4x + 1) = 4$$

which we could write as $g''(x) = 4$.

b. The first derivative is $h'(x) = 2\cos(2x)$, and so $h''(x) = -4\sin(2x)$.

c. Using the chain and product rules gives

$$\begin{aligned} l'(x) &= \exp(2x^2 - x) + x(4x - 1)\exp(2x^2 - x) \\ &= (4x^2 - x + 1)\exp(2x^2 - x) \end{aligned}$$

Similarly,

$$\begin{aligned} l''(x) &= \frac{d}{dx}\left((4x^2 - x + 1)\exp(2x^2 - x)\right) \\ &= (16x^3 - 8x^2 + 13x - 2)\exp(2x^2 - x) \end{aligned}$$

Of course, in principle, one must not necessarily stop at second derivatives and we can define derivatives of any order while the function remains differentiable. The notation for higher-order derivatives continues as you might expect

$$\frac{d^3}{dx^3}f(x) \quad \frac{d^4}{dx^4}f(x) \quad \frac{d^5}{dx^5}f(x) \quad \cdots \quad \frac{d^n}{dx^n}f(x)$$

with $n \in \mathbb{N}^+$. The more compact notation does differ somewhat. For example, rather than representing the fifth derivative as $f'''''(x)$, we would write $f^{(5)}(x)$. More specifically, we would write increasing orders of the derivative of $f(x)$ as,

$$f'(x) \quad f''(x) \quad f'''(x) \quad f^{(4)}(x) \quad f^{(5)}(x) \quad \ldots \quad f^{(n)}(x)$$

Note that there is no definitive rule regarding at what n it is necessary to use the superscript (n) notation. For example, $f^{(1)}(x)$ would be understood as $f'(x)$, albeit unusual.

Deriving expressions for higher-order derivatives simply requires repeated use of standard derivative techniques.

EXAMPLE 4.3

Determine expressions for the fifth-order derivatives of the functions stated in Example 4.2.

Solution

We continue from the solution of Example 4.2.

a. The second derivative is $g''(x) = 4$ and so $g^{(n)} = 0$ for $n \geq 3$.

b.

$$\begin{aligned} h''(x) &= -4\sin(2x) \quad h^{(3)}(x) = -8\cos(2x) \\ h^{(4)}(x) &= 16\sin(2x) \quad h^{(5)}(x) = 32\cos(2x) \end{aligned}$$

c. Significant algebraic manipulation leads to

$$\begin{aligned} l''(x) &= (16x^3 - 8x^2 + 13x - 2)\exp(2x^2 - x) \\ l^{(3)}(x) &= (64x^4 - 48x^3 + 108x^2 - 37x + 15)\exp(2x^2 - x) \end{aligned}$$

$$l^{(4)}(x) = (256x^5 - 256x^4 + 736x^3 - 400x^2 + 313x - 52)\exp(2x^2 - x)$$

$$l^{(5)}(x) = (1024x^6 - 1280x^5 + 4480x^4 - 3360x^3 + 3860x^2 - 1321x + 365)\exp(2x^2 - x)$$

We will see an important mathematical application of second derivatives in the next section, namely, the classification of turning points. However, from the perspective of straightforward mathematical analysis, the use of derivatives beyond the second order is often limited to obtaining series expansions of functions, which we explore in Chapter 5, and the more formal definition of smoothness considered in the next section. In applied mathematics, higher-order derivatives may however have particular physical or economic interpretations and the ability to derive expressions for $f^{(n)}(x)$ is important.

Now that we are able to determine higher-order derivatives, we can broaden our understanding of smoothness yet further. In particular, in the strict sense, a single-variable function $f(x)$ is smooth if *all* derivatives are continuous. This is a clear extension of the practical condition defined by Eq. (4.2).

It should be clear from the patterns of higher-order derivatives obtained in Example 4.3(a)-(c) that the functions defined there are smooth in the fullest sense. In particular, $g^{(n)}(x)$ are zero for sufficient n, which is a continuous horizontal line; $h^{(n)}(x)$ are alternative between multiples of the basic circular functions, which we know to be continuous; and $l^{(n)}(x)$ are the product of polynomials of increasing order and an exponential, both being continuous. However, the function demonstrated to be smooth by condition (4.2) in Example 4.1(d) requires revisiting.

EXAMPLE 4.4

Investigate the smoothness of the following function first considered in Example 4.1(d).

$$k(x) = \begin{cases} x^3 & \text{for } x \in (-\infty, 0] \\ x^2 & \text{for } x \in (0, \infty) \end{cases}$$

Solution

We begin by writing down the first few derivatives

$$k'(x) = \begin{cases} 3x^2 & \text{for } x \in (-\infty, 0] \\ 2x & \text{for } x \in (0, \infty) \end{cases}$$

$$k''(x) = \begin{cases} 6x & \text{for } x \in (-\infty, 0] \\ 2 & \text{for } x \in (0, \infty) \end{cases}$$

$$k^{(3)}(x) = \begin{cases} 6 & \text{for } x \in (-\infty, 0] \\ 0 & \text{for } x \in (0, \infty) \end{cases}$$

$$k^{(4)}(x) = 0$$

In each case, the subfunctions are continuous either side of $x = 0$, but a question remains at $x = 0$. As discussed in Example 4.1(d), the left-hand and right-hand limits of $k'(x)$ are equal at $x = 0$ and are equal to $k'(0)$, the first derivative is therefore smooth. However, this is not the case for the second and third derivatives. The function $k(x)$ is therefore *not* smooth in the strictest sense at $x = 0$, despite being smooth by the softer first-derivative condition.

Example 4.4 demonstrates that the technical concept of smoothness is more than the absence of obvious corners and cusps in a plot of a function, which is of course a property of the first derivative only. For example, function $k(x)$ in Example 4.4 appears smooth, but is actually not smooth in the strict sense when we consider higher-order derivatives; this is demonstrated in Figure 4.2(d). However, it is often sufficient to use the softer first-derivative condition of smoothness and we will often do so throughout this book.

4.3 STATIONARY AND TURNING POINTS

A relatively simple application of derivatives, in addition to calculating gradients and testing for smoothness, is determining the existence of maximum and minimum values. It should be clear that, when they exist, a *global maximum* and *global minimum* determine the range of a function without resorting to computational plots as was done in Chapter 2. Furthermore, when the maximum and minimum values are only within finite intervals, i.e., are not global properties, they may have important consequences for the interpretation of the function in an applied context.

In this section, we discuss the concept of *stationary points*, which will naturally lead to the definition of *turning points* that include maximum and minimum values of a function. Collectively, these will add a further layer of algebraic tools for studying the behavior of functions. We then demonstrate the application of these methods to determining a function's range.

4.3.1 Stationary points

A stationary point of a function $f(x)$ is simply any point at which its instantaneous rate of change with respect to x is zero. If $x = s$ denotes a stationary point of $f(x)$, then the gradient of the function is zero at a stationary point and so $f'(s) = 0$. The tangent to the function at a stationary point therefore also has zero gradient and is parallel to the horizontal axis. A function can, in principle, have any number of stationary points (including none), depending on the function's particular properties.

EXAMPLE 4.5

Determine all stationary points of the following functions. Write down the equation of the tangent in each case.

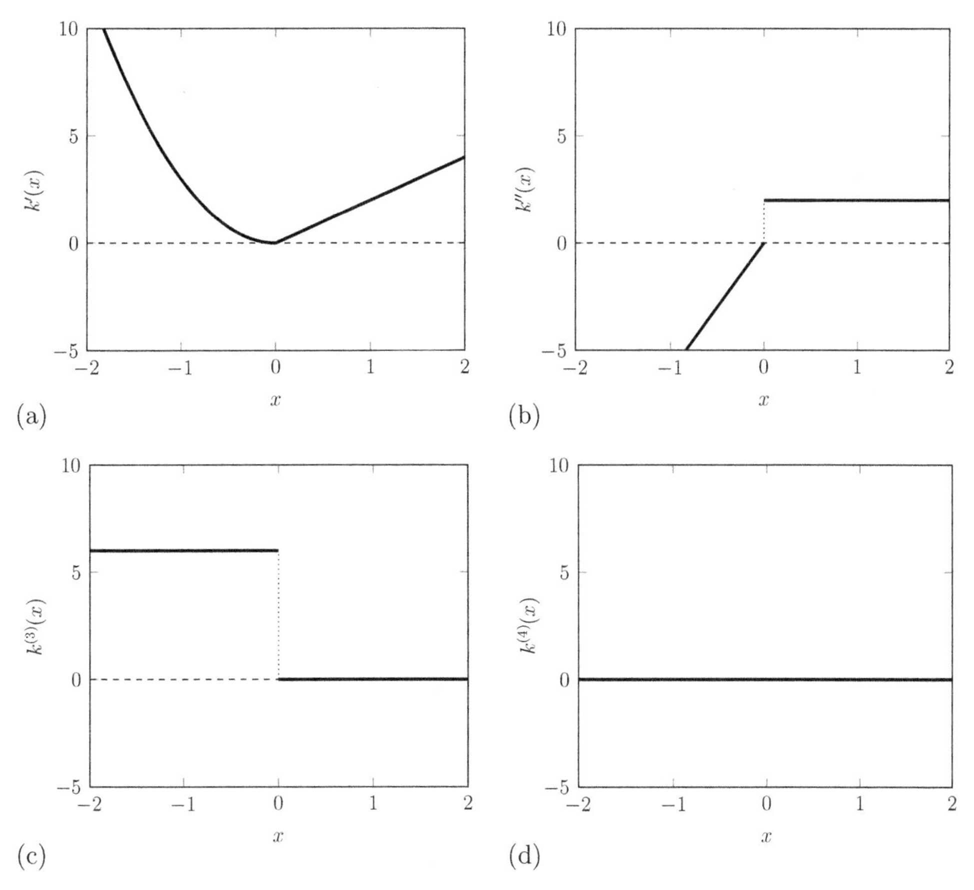

Figure 4.2 Higher-order derivatives of $k(x)$ from Example 4.4. (a) $k'(x) = \begin{cases} 3x^2 & \text{for } x \in (-\infty, 0] \\ 2x & \text{for } x \in (0, \infty) \end{cases}$, (b) $k''(x) = \begin{cases} 6x & \text{for } x \in (-\infty, 0] \\ 2 & \text{for } x \in (0, \infty) \end{cases}$, (c) $k^{(3)}(x) = \begin{cases} 6 & \text{for } x \in (-\infty, 0] \\ 0 & \text{for } x \in (0, \infty) \end{cases}$, and (d) $k^{(4)}(x) = \begin{cases} 0 & \text{for } x \in (-\infty, 0] \\ 0 & \text{for } x \in (0, \infty) \end{cases}$.

a. $h(x) = 3x + 2$
b. $k(x) = e^{x^2 - 2x + 2}$
c. $l(x) = \sin x$
d. $m(x) = x^3 + 2$

Solution

In each case, we begin by obtaining the first-derivative function and proceed to find its roots.

a. $h'(x) = 3$ implies there are no stationary points.

b. $k'(x) = 2(x-1)\exp(x^2 - 2x + 2)$ implies a single stationary point at $x = 1$. The equation of the tangent is $y = k(1) = \mathrm{e}$.
c. $l'(x) = \cos x$ implies an infinite number of roots at $x = (2n+1)\pi/2$ for $n \in \mathbb{Z}$. The tangents are then $y = 1$ for n even and $y = 1$ for n odd.
d. $m'(x) = 2x$ implies a single stationary point at $x = 0$. The equation of the tangent is simply $y = m(0) = 2$.

The stationary points and associated tangents of the functions in Example 4.5 are confirmed visually in Figure 4.3. We see three distinct types of stationary points: *maxima*, *minima*, and *points of inflection*.[1] In general, the classification of a stationary point at $x = s$ into one of the three categories is determined by the local behavior of the function around the point. In particular, maxima and minima can be defined as follows.

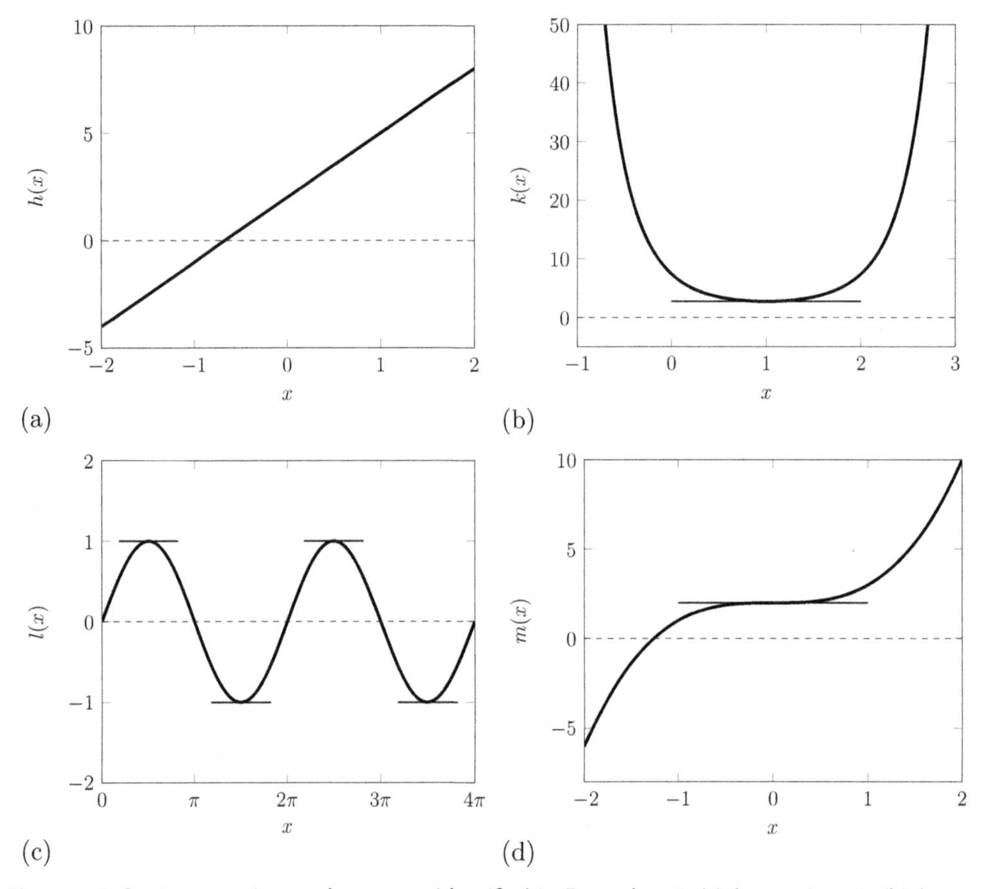

Figure 4.3 Stationary points and tangents identified in Example 4.5. (a) $h(x) = 3x + 2$, (b) $k(x) = \mathrm{e}^{x^2-2x+2}$, (c) $l(x) = \sin x$, and (d) $m(x) = x^3 + 2$.

[1] Note that some authors use the alternative spelling "inflexion."

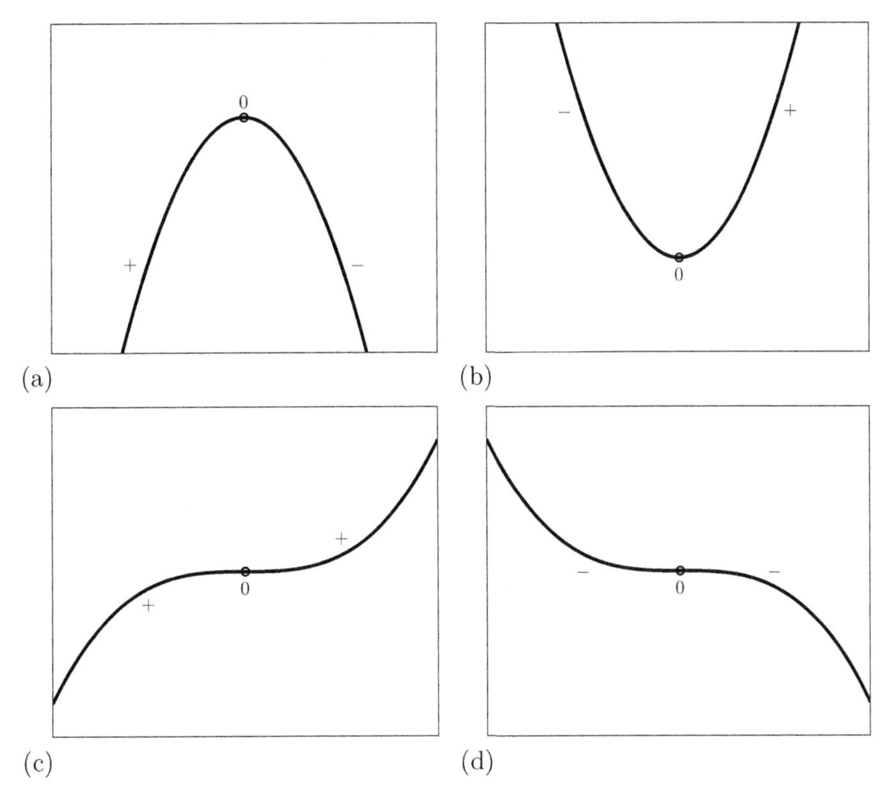

Figure 4.4 Classification of turning points and indication of local gradient. (a) Maximum, (b) minimum, (c) point of inflection I, and (d) point of inflection II.

- *Maximum*: as x approaches s from below and passes through, the gradient changes from positive, through zero to negative. See Figure 4.4(a).
- *Minimum*: as x approaches s from below and passes through, the gradient changes from negative, through zero to positive. See Figure 4.4(b).

A point of inflection can exist in one of two variations. These are described as follows.

- *Point of inflection I*: as x approaches s from below and passes through, the gradient changes from positive, through zero to positive. See Figure 4.4(c).
- *Point of inflection II*: as x approaches s from below and passes through, the gradient changes from negative, through zero to negative. See Figure 4.4(d).

The above classifications of stationary points can be thought of in terms of the function "changing direction," that is "turning," in different ways. For this reason, maxima, minima, and points of inflection are often referred to as *turning points*. In particular, a maximum indicates a *clockwise turning* with increasing x and a minimum indicates an *anticlockwise turning* with increasing x.

There is, however, a subtlety regarding the use of the term "point of inflection." In general terms, a point of inflection can refer to any point at which the sense of turning, that is clockwise or anticlockwise, changes. For example, there is a change in the sense of turning between a consecutive maximum and minimum on the sin function in Figure 4.3(c) and so there should be a point of inflection somewhere between the two. However, we know that there are no further stationary points between each consecutive maximum and minimum. It should therefore be clear that a point of inflection need not be a stationary point, which is in contrast to maxima and minima which are always stationary points. We could therefore say that a point of inflection need not be a stationary point, but a stationary point could be a point of inflection.

It is often the case that this subtly is ignored and the term "point of inflection" is used only to refer to particular types of stationary point; this will indeed be the case for the remainder of this book. The terms "stationary point" and "turning point" can then be used interchangeably. However, it is important to have some awareness of this subtly if you are reading around the subject.

The descriptions of the three types of stationary point given above provide a practical method for determining the classification. In particular, we can look at the local behavior of the derivative of the function around the stationary point to determine which classification it has. This is best illustrated with an example.

EXAMPLE 4.6

Classify the stationary point in each of the following functions.

a. $p(x) = x^2$

b. $q(x) = -x^2$

Solution

You should note that each function has a stationary point at $x = 0$. We then proceed to sample the properties of the first derivative either side of $x = 0$ in each case.

a. $p'(x) = 2x$. Since $p'(-0.1) < 0$ and $p'(0.1) > 0$ we have a minimum.

b. $q'(x) = -2x$. Since $q'(-0.1) > 0$ and $q'(0.1) < 0$ we have a maximum.

EXAMPLE 4.7

Classify all stationary points found in Example 4.5.

Solution

a. $g'(x) = 2x$ with a stationary point $x = 0$. Since $g'(-0.1) < 0$ and $g'(0.1) > 0$ we have a minimum.

b. $k'(x) = 2(x-1)\exp(x^2 - 2x + 2)$ with a stationary point at $x = 1$. Since $k'(0.9) < 0$ and $k'(1.1) > 0$ we have a minimum.

c. $l'(x) = \cos x$ with stationary points at $x = (4n-1)\pi/2$ which are minima, and at $x = (4n+1)\pi/2$ which are maxima; these classifications are clear from our knowledge of $\sin x$.

d. $m'(x) = 3x^2$ with a stationary point at $x = 0$. Since $m'(-0.1) > 0$ and $m'(0.1) > 0$ we have a point of inflection.

These classifications are confirmed in Figure 4.3.

In the above examples, we choose to look at behavior at positions $x \pm 0.1$. Of course, there is no definite rule for how close we should sample the function near a stationary point to determine its local behavior, but care should always be taken to ensure that the sampling remains local to the particular point in question. This is particularly important where there are multiple stationary points close together. For example, in Example 4.7(d), care should be taken to sample the derivative function sufficiently close to $x = \pi/2$ so that the properties of the other nearby stationary points are not picked up.

4.3.2 Classification via second derivatives

The sampling method to determine the classification of stationary points described above is cumbersome. With some thought, it is however possible to determine a classification method based on the second derivative of the function at the location of the stationary point.

In particular, as described previously, a maximum is such that the value of the local gradient *decreases* with increased x as we approach the stationary point from below. That is, the gradient of the gradient is negative near to the stationary point at $x = s$. Expressing this mathematically, we have

$$\left.\frac{\mathrm{d}}{\mathrm{d}x}f'(x)\right|_{x=s} < 0 \quad \text{that is} \quad \left.\frac{\mathrm{d}^2}{\mathrm{d}x^2}f(x)\right|_{x=s} < 0$$

at a maximum. Similarly, a minimum is such that the local gradient *increases* with increased x as we approach the stationary point from below and

$$\left.\frac{\mathrm{d}}{\mathrm{d}x}f'(x)\right|_{x=s} > 0 \Rightarrow \left.\frac{\mathrm{d}^2}{\mathrm{d}x^2}f(x)\right|_{x=s} > 0$$

The second derivative is therefore seen to be related to the direction of turning of the function. In particular, a negative second derivative implies an anticlockwise turning and a positive second derivative implies a clockwise turning. The situation is therefore more subtle for a point of inflection. We might at first expect that

$$\frac{\mathrm{d}}{\mathrm{d}x}f'(x) = 0 \Rightarrow f''(x) = 0$$

However, this is not a sufficient condition to identify a point of inflection and we are further required to demonstrate a change in sign of $f''(x)$ either side of the inflection point. The reason for this additional condition is to demonstrate the change from clockwise to anticlockwise (or vice versa) turning.

EXAMPLE 4.8

Use second derivatives to confirm the classifications in Example 4.7.

Solution

a. $g''(x) = 2 > 0$. The turning point at $x = 0$ is a minimum.

b. $k''(x) = 2(2x^2 - 4x + 2)\exp(x^2 - 2x + 2)$ and $k''(1) = 2e > 0$. The turning point at $x = 1$ is a minimum.

c. $l''(x) = -\sin x$ and $l''((4n-1)\pi/2) = 1 > 0$ which is a minimum, and $l''((4n+1)\pi/2) = -1 < 0$ which is a maximum.

d. $m''(x) = 6x$ and the second-derivative test at $x = 0$ is inconclusive. In this case, we are forced to revert to sampling the first derivatives either side of $x = 0$ and so conclude that we have a point of inflection.

EXAMPLE 4.9

Find and classify all stationary points of $m(x) = x^3 - 6x^2 + 12x - 9$.

Solution

We note that

$$m'(x) = 3(x-2)^2$$

which has a repeated root at $x = 2$. We therefore have a single stationary point for $m(x)$ at $x = 2$.

$$m''(x) = 6(x-2)$$

which is zero at $x = 2$. The second-derivative test is therefore inconclusive, however, we note that $m''(1.9) < 0$ and $m''(2.1) > 0$ and so we have a point of inflection at $x = 2$.

4.3.3 Application to ranges

It should be clear from the above discussion that the ability to determine stationary points is useful when exploring the properties of functions. However, the terms "maximum" and "minimum" refer to *local* properties of the function. For example, the value of the function at a maximum could in fact be smaller than the value of the function at a minimum. Stationary points do not then give information about the global properties of the function and we cannot simply state that the values of the local maximum and local minimum of a function determine the function's extreme values and range. Instead, it is typically necessary to use this information in conjunction with other properties of the function, for example, the asymptotic behavior as $x \to \pm\infty$ or the existence of singular points.

EXAMPLE 4.10

Find and identify all stationary points in the following functions. Use this information to help determine each function's range over $\mathbb{R}$.

a. $p(x) = x^3 - 4x^2 - x + 4$

b. $q(x) = x^4 - 4x^2$

c. $r(x) = \frac{4x^4-2x+1}{3x^2+1}$

d. $t(x) = \frac{x^2-2x+2}{x-3}$

Solution

a. *Stationary points*:
$p'(x) = 3x^2 - 8x - 1$ and so $p(x)$ has stationary points at $x_{\pm} = \frac{1}{3}(4 \pm \sqrt{19})$. We see that $p''(x_+) > 0$ and so is a minimum with $p(x_+) \approx -8.2088$, and $p''(x_-) < 0$ and so is maximum with $p(x_-) \approx 4.0607$.
Asymptotic properties:
The function is such that $\lim_{x\to\pm\infty} p(x) = \pm\infty$.
Singular points:
Polynomials do not have singular points.
Range:
Piecing this information together, it is possible for $p(x)$ to take any real value and we determine that $p(x)$ has range $\{p \in \mathbb{R}\}$.

b. *Stationary points*:
$q'(x) = 4x^3 - 8x$ and so $q(x)$ has stationary points at $x = 0$ and $x = \pm\sqrt{2}$. We see that $q''(0) < 0$ and so is a maximum with $q(0) = 0$, and $q''(\pm\sqrt{2}) > 0$ and so are minima with $q(\pm\sqrt{2}) = -4$.
Asymptotic properties:
The function is such that $\lim_{x\to\pm\infty} p(x) = \infty$.
Singular points:
Polynomials do not have singular points.
Range:
The upper limit of the range is unbounded, but the global lower limit is actually given by the local minima. The range is therefore $\{q \in \mathbb{R} : q \geq -4\}$.

c. *Stationary points*:
$r'(x) = \frac{2(12x^5+8x^3+3x^2-3x+1)}{(3x^2+1)^2}$ and we can use Goalseek or Wolfram Alpha, for example, to determine three turning points at $x \approx -0.3415$ (maximum with $r \approx 1.2871$), $x \approx -0.4542$, and $x \approx 0.5346$ (both minima with $r \approx 1.2840$ and 0.1386, respectively).
Asymptotic properties:
The function is such that $\lim_{x\to\pm\infty} p(x) = \infty$.
Singular points:
The denominator of the rational function has no real roots and so there are no singular points.
Range:
The upper limit of the range is unbounded, but the lower limit is given by the lowest minimum value. The range is therefore $\{r \in \mathbb{R} : 0.1386 \leq r < \infty\}$.

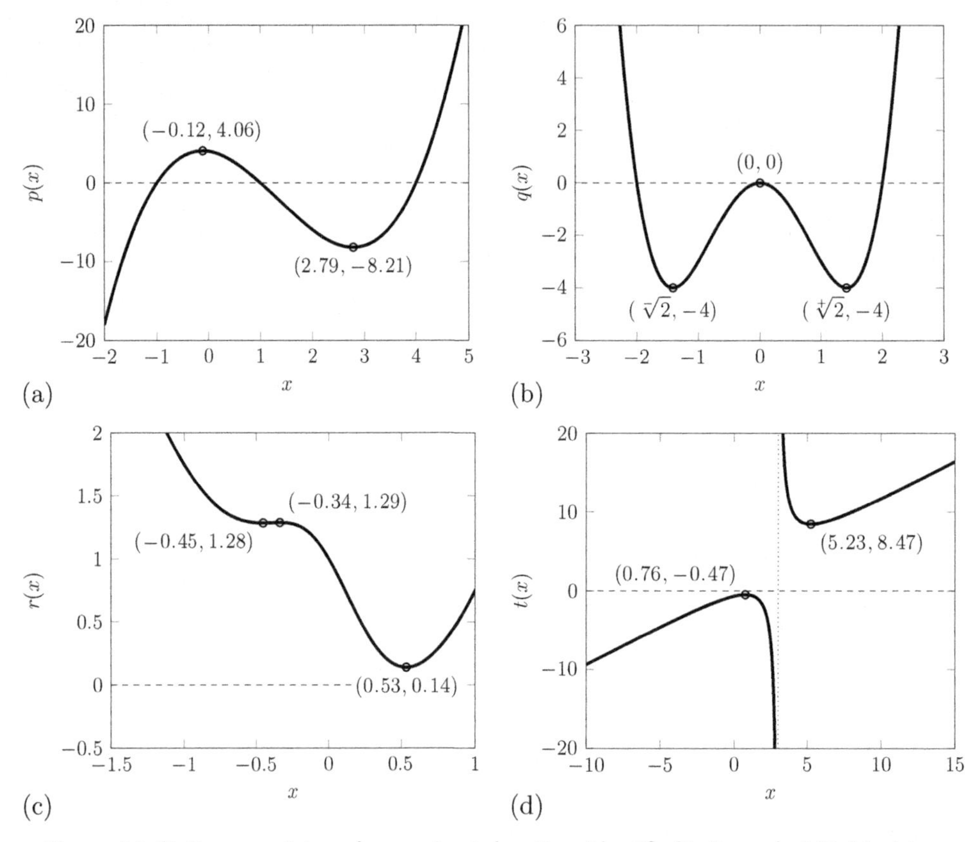

Figure 4.5 Stationary points and approximate locations identified in Example 4.10. (a) $p(x) = x^3 - 4x^2 - x + 4$, (b) $q(x) = x^4 - 4x^2$, (c) $r(x) = \frac{4x^4-2x+1}{3x^2+1}$, and (d) $t(x) = \frac{x^2-2x+2}{x-3}$.

d. *Stationary points*:
$t'(x) = \frac{x^2-6x+4}{(x-3)^2}$ and so $t(x)$ has stationary points at $x_{\pm} = 3 \pm \sqrt{5}$. We see that $t''(x_-) < 0$ and so is a maximum with $t(x_-) = 4 - 2\sqrt{5}$, and $t''(x_+) > 0$ and so is a minimum with $t(x_+) = 4 + 2\sqrt{5}$.
Asymptotic properties:
The function is such that $\lim_{x\to\pm\infty} p(x) = \pm\infty$.
Singular points:
The function has upward and downward asymptotes at $x = 3$ arising from a real root of the denominator.
Range:
We note that local minimum at x_+ has a greater value than the local maximum at x_-. In this case, this leaves a "hole" in the range. It is otherwise possible to obtain any other real value. The range is then $\{t \in \mathbb{R} : t \in (-\infty, 4 - 2\sqrt{5}] \cup [4 + 2\sqrt{5}, \infty)\}$.

These ranges can be confirmed in Figure 4.5 or using Wolfram Alpha's `range` command. The reader is invited to plot Figure 4.5(c) over $x \in [-0.5, -0.3]$ to confirm the classification of the stationary points at negative x.

In cases where the asymptotic behavior is such that the range is determined by the local properties of the function at finite values of x, the turning points can determine the range. The trick is of course to use all of the algebraic strategies we have developed so far in a flexible way. With this in mind, we revisit Example 2.14(a).

EXAMPLE 4.11

Use algebraic techniques to determine the range of

$$f(x) = \frac{3x^2 - 2x + 2}{2x^2 + x + 1}$$

Solution

The function has no discontinuities arising from real roots of the denominator and we note the asymptotic behavior

$$\lim_{x \to \pm\infty} f(x) = 1.5$$

The range of the function is therefore determined by the behavior of the function at finite values of x and it is instructive to investigate the stationary points.

$$f'(x) = \frac{7x^2 - 2x - 4}{(2x^2 + x + 1)^2} = 0 \quad \text{and so } x_{\pm} = \frac{1}{7}(1 \pm \sqrt{29})$$

These can be classified from the sign of $f''(x_{\pm})$ and some manipulation shows a maximum at x_- and a minimum at x_+. In particular,

$$f(x_-) = \frac{2}{7}(8 + \sqrt{29}) \quad \text{and} \quad f(x_+) = \frac{2}{7}(8 - \sqrt{29})$$

The asymptote is between these values, that is, $f(x_+) < 1.5 < f(x_-)$. The range of $f(x)$ is therefore determined by the local turning points which are actually global extreme values. The range is then given by $\{f \in \mathbb{R} : f \in [-\frac{2}{7}(-8 + \sqrt{29}), \frac{2}{7}(8 + \sqrt{29})]\}$. Note that this is consistent with the plot in Figure 2.12(a).

4.4 HIGHER-ORDER DERIVATIVES AND STATIONARY POINTS ON YOUR COMPUTER

It should be of no surprise that Wolfram Alpha can be used to determine higher-order derivatives of functions. For example, the easiest command to determine the second-order derivative of $l(x) = x \exp(2x^2 - x)$ from Example 4.2(c) is

```
d^2/dx^2 xe^(2x^2-x)
```

This leads to the expression determined in that example. The extension to higher-order derivatives follows as one might image and the reader is invited to confirm the higher-order derivatives of $l(x)$ as determined in Example 4.3(c). For example, the relevant command for the fifth derivative is

```
d^5/dx^5 xe^(2x^2-x)
```

Of course, the command, say,

```
solution (d/dx e^(x^2-2x+2) )
```

can be used to help determine the location ($x = 0$) of the stationary point of $k(x)$ appearing in Example 4.7(b). We could then proceed to evaluate the second derivative at this point with

```
d^2/dx^2 e^(x^2-2x+2) where x=0
```

to confirm that it is a minimum. However, Wolfram Alpha can also be used directly to locate and classify the turning points of a function. For example, the command

```
turning points e^(x^2-2x+2)
```

returns that there is a minimum at $x = 1$ with value e, a plot is also shown to confirm this.

EXAMPLE 4.12

You are given that

$$f(x) = \frac{x^8 - x^3 + x^2 - 1}{x^8 + 4}$$

Use Wolfram Alpha to determine the following.

a. $f^{(4)}(x)$.
b. The location and classification of any turning points of $f(x)$.
c. Any horizontal or vertical asymptotic behavior.
d. All real roots of $f^{(3)}(x)$.

Solution

a. The command

```
d^4/dx^4 (x^8-x^3+x^2-1)/(x^8+4)
```

returns a very complicated expression.

b. The command

```
turning points (x^8-x^3+x^2-1)/(x^8+4)
```

returns that the function has a maximum at $x \approx -1.68915$ (where $f(x) \approx 1.03803$) and a minimum at $x = 0$ (where $f(x) = -\frac{1}{4}$). A plot is shown to confirm these results.

c. The command

```
asymptotes (x^8-x^3+x^2-1)/(x^8+4)
```

returns that the function has a horizontal asymptote such that $\lim_{x\to\pm\infty} f(x) = 1$

d. The command

```
solve (d^3/dx^3 (x^8-x^3+x^2-1)/(x^8+4) )
```

returns six real roots location at

$$x \approx -2.17501 \quad x \approx 0.43063$$
$$x \approx -1.25408 \quad x \approx 0.97378$$
$$x \approx -0.86619 \quad x \approx 1.35800$$

4.5 ACTUARIAL APPLICATION: APPROXIMATING PRICE SENSITIVITIES

In this section, we consider a very pragmatic use of derivatives for approximating the value of a function in response to a small change in the independent variable. Although not explicitly stated in this section, the ideas are related to the *Taylor expansion* of a function that we will discuss in Chapter 5. You are invited to return again to this section and attempt to connect the two ideas once you have studied Chapter 5. Whilst the discussion of the ideas presented here are framed in terms of very actuarial applications, the same methods can be used with functions in any application.

Let us begin with an expression for the present value of a particular cash flow expressed as a function of the compounding interest rate, i,

$$f(i) = \frac{4}{(1+i)^3} + \frac{5}{(1+i)^5} + \frac{12}{(1+i)^7} + \frac{120}{(1+i)^{10}}$$

You might be able to read this expression as representing the present value of a cash flow of 4, 5, 12, and 120 at times $t = 3, 5, 7$, and 10, respectively, for some appropriate units of money and time. A possible interpretation of $f(i)$ is that it gives the maximum price an investor should be willing to pay for an investment that returns these cash flows if he demands at least a return of i% per annum. It should be clear that $f'(i)$ is then given by

$$f'(i) = -\frac{12}{(1+i)^4} - \frac{25}{(1+i)^6} - \frac{84}{(1+i)^8} - \frac{1200}{(1+i)^{11}}$$

These expressions are such that, at $i = 5\%$ per annum, we have

$$f(0.05) \approx 89.571 \quad \text{and} \quad f'(0.05) \approx -786.997$$

The question is now, can we use these values to approximate the value of $f(0.06)$, for example? That is, given we have the fair price at 5% per annum, can we estimate the price of the same investment if, instead, 6% per annum is required. The answer is, of course, yes we can.

In order to see how, we revisit the notion of a derivative as the local gradient. That is,

$$\frac{\Delta f}{\Delta i} \simeq \frac{df}{di}$$

where Δf is the change in the value of f in response to a change in i, denoted Δi. The approximation is more accurate as $\Delta i \to 0$. Therefore, if we begin at some interest rate i_0 and want to estimate the effect on f of moving to $i_1 = i_0 + h$, we have $\Delta i = i_1 - i_0 = h$, $\Delta f = f(i_1) - f(i_0)$ and so

$$f(i_1) \simeq f(i_0) + h \left.\frac{df}{di}\right|_{i_0} \tag{4.5}$$

In our particular example, $i_0 = 0.05$, $h = 0.01$, $f(i_0) \approx 89.571$, and $f'(0.05) \approx -786.997$. We can therefore estimate $f(0.06)$ from

$$f(0.06) \simeq 89.571 + 0.01 \times (-786.997) = 81.701$$

The actual value can be obtained by evaluating $f(i)$ at $i = 0.06$, and we find 82.083. The approximation is reasonably close to the actual value with an error that reflects that $h = 0.01 > 0$.

EXAMPLE 4.13

Use the information given above for $f(0.05)$ and $f'(0.05)$ to estimate the fair price for the same investment at the following rates. Comment on your answers.

a. $i = 5.1\%$ per annum
b. $i = 4\%$ per annum
c. $i = 8\%$ per annum

Solution

We have that $f(0.05) \approx 89.571, f'(0.05) \approx -786.997$ and, in each case, apply Eq. (4.5) for the appropriate h.

a. For $i = 5.1\%$ per annum, we have $h = 0.1\% = 0.001$ and so $f(0.051) \simeq 89.571 + 0.001 \times (-786.997) \approx 88.784$. The actual value is $f(0.051) \approx 88.788$.
b. For $i = 4\%$ per annum, we have $h = -1\% = -0.01$ and so $f(0.04) \simeq 89.571 - 0.01 \times (-786.997) \approx 97.441$. The actual value is $f(0.051) \approx 97.852$.
c. For $i = 8\%$ per annum, we have $h = 3\% = 0.03$ and so $f(0.08) \simeq 89.571 + 0.03 \times (-786.997) \approx 65.961$. The actual value if $f(0.08) \approx 69.163$.

We note that the approximation becomes increasing less accurate as $|h|$ increases. This is consistent with the observation that Eq. (4.5) is accurate in the limit $h \to 0$. Note also that the fair price reduces as a higher return is demanded. That is, since the receipts are fixed, a higher return is generated from a lower price.

At this stage you may think that this exercise is pointless as one could always just recalculate $f(i)$ at a different i. This is of course true in our motivating example. However, in practice, we may have information about the price of an investment and the derivative of that present value at some i, but have no actual knowledge of the function $f(i)$. In that case, it should be clear that the approach is particularly useful.

To understand how information about $f'(i)$ may be given in practice, it is useful to introduce the concept of the *effective duration* of an investment. In particular, the effective duration is a function of i, denoted $D(i)$ and such that

$$D(i) = -\frac{1}{f(i)}\frac{\mathrm{d}}{\mathrm{d}i}f(i)$$

Here $f(i)$ is the present value of the investment's proceeds, that is its price, at i. Note that $f(i) \neq 0$.

It should be immediately clear from the prior discussion that the effective duration is related to the sensitivity of the present value of changes in i. Indeed, we can write Eq. (4.5) in terms of $D(i)$ as

$$f(i_1) \simeq f(i_0)\,(1 - hD(i_0)) \tag{4.6}$$

EXAMPLE 4.14

For $i = 7\%$ per annum, an investment is known to have a price of \$123 and an effective duration of 7.4. Estimate the new price if the required return changes to

a. $i = 6.5\%$ per annum,
b. $i = 7.3\%$ per annum.

Solution

We apply Eq. (4.6) in both cases.

a. $i = 6.5\%$ per annum corresponds to $h = -0.005$ and so $f(0.065) \simeq 123 \times (1 + 0.005 \times 7.4) \approx \$127.55$.
b. $i = 7.3\%$ per annum corresponds to $h = 0.003$ and so $f(0.073) \simeq 123 \times (1 - 0.003 \times 7.4) \approx \$120.27$.

EXAMPLE 4.15

An investment project is expected to return \$3m in 2 years' time and \$4m in 4 years' time.

a. Determine the maximum price and effective duration of this project, if a return of $i = 3\%$ per annum is required.
b. If the required return changes to $i = 3.5\%$, use your answer in part (a) to estimate the new maximum price.

Solution

a. We begin by forming the present value of the project's proceeds. Working in money units of \$m and time units of years we have

$$f(i) = \frac{3}{(1+i)^2} + \frac{4}{(1+i)^4} \Rightarrow f'(i) = \frac{-6}{(1+i)^3} - \frac{16}{(1+i)^5}$$

and so $f(0.03) \approx 6.382$ and $f'(0.03) \approx -19.293$. We therefore compute the effective duration at $i = 0.03$ to be

$$D(i_0) = -\frac{1}{f(i_0)} f'(i_0) = \frac{19.293}{6.382} \approx 3.023$$

The maximum price is \$6.382m and the effective duration is 3.023 years.

b. We are now required to estimate $f(3.5\%)$ and can do so using the effective duration at 3%. That is, we apply Eq. (4.6)

$$f(0.035) \simeq f(0.03)\left(1 - 0.005 \times f'(0.03)\right) \approx 6.286$$

That is, the maximum price has reduced to \$6.286m. We can check this estimate by evaluating $f(3.5\%)$ directly. In particular, $f(0.035) \approx 6.286$.

There are a number of alternative definitions related to the effective duration of investment proceeds. A common variation is called simply the *duration*. The duration, $\tau(\delta)$, is very similar to the effective duration but is defined in terms of the force of interest δ rather that i. In particular,

$$\tau(\delta) = -\frac{1}{f(\delta)} \frac{\mathrm{d}}{\mathrm{d}\delta} f(\delta) \tag{4.7}$$

As we shall now see, the duration is particularly useful in practice.

EXAMPLE 4.16

Determine the duration of the investment proceeds in Example 4.15 at general δ. Interpret your expression.

Solution

We begin by determining the present value in terms of δ. Working in money units of \$m and time units of years, we find

$$f(\delta) = 3\mathrm{e}^{-2\delta} + 4\mathrm{e}^{-4\delta}$$

An expression for $f'(\delta)$ is then seen to be

$$f'(\delta) = -2 \times 3\mathrm{e}^{-2\delta} - 4 \times 4\mathrm{e}^{-4\delta}$$

The action of taking the derivative with respect to δ is to "bring down" the time at which each particular payment occurs. The duration is then given by

$$\tau(\delta) = \frac{2 \times 3\mathrm{e}^{-2\delta} + 4 \times 4\mathrm{e}^{-4\delta}}{3\mathrm{e}^{-2\delta} + 4\mathrm{e}^{-4\delta}}$$

The duration can therefore be interpreted as the weighted average timing of the proceeds, with the weighting given by each particular proceed's present value as a fraction of the total present value.

The interpretation given in Example 4.16 demonstrates that, in some sense, the noun "duration" makes sense for $\tau(\delta)$. That is, the duration gives a measure of the extent of an investment's proceeds in time. For example, for a fixed δ, an investment resulting in significant cash flows in the future will have a larger duration than an investment resulting in significant cash flows occurring very soon. This interpretation of τ arises purely from the properties of exponential functions. Note that τ is typically expressed in units of years.

EXAMPLE 4.17

Calculate the duration for each of the following investments at $\delta = 5\%$ per annum. Interpret your answers.

a. Purchase at $t = 0$ resulting in receipts of \$30 at $t = 1, 2, 3$, and 4.
b. Purchase at $t = 0$ resulting in receipts of \$30 at $t = 5, 6, 7$, and 8.
c. Purchase at $t = 0$ resulting in receipts of \$30 at $t = 1, 2, 5$, and 8.
Note that t is measured in years.

Solution

In each case, we begin by forming the present value expression of the proceeds as a function of δ and then apply Eq. (4.7).

a. We have

$$f(\delta) = 30\left(e^{-\delta} + e^{-2\delta} + e^{-3\delta} + e^{-4\delta}\right) \Rightarrow f(0.05) \approx 106.065$$
$$f'(\delta) = -30\left(e^{-\delta} + 2e^{-2\delta} + 3e^{-3\delta} + 4e^{-4\delta}\right) \Rightarrow f'(0.05) \approx -258.539$$

and so $\tau(0.05) = \frac{258.539}{106.065} \approx 2.438$ years.

b. We have

$$f(\delta) = 30\left(e^{-5\delta} + e^{-6\delta} + e^{-7\delta} + e^{-8\delta}\right) \Rightarrow f(0.05) \approx 86.839$$
$$f'(\delta) = -30\left(5e^{-5\delta} + 6e^{-6\delta} + 7e^{-7\delta} + 8e^{-8\delta}\right) \Rightarrow f'(0.05) \approx -559.029$$

and so $\tau(0.05) = \frac{559.029}{86.839} \approx 6.437$ years.

c. We have

$$f(\delta) = 30\left(e^{-\delta} + e^{-2\delta} + e^{-5\delta} + e^{-8\delta}\right) \Rightarrow f(0.05) \approx 99.156$$
$$f'(\delta) = -30\left(e^{-\delta} + 2e^{-2\delta} + 5e^{-5\delta} + 8e^{-8\delta}\right) \Rightarrow f'(0.05) \approx -360.524$$

and so $\tau(0.05) = \frac{360.524}{99.156} \approx 3.636$ years.

We note that the value of the duration appears to reflect the timings of the cash flows.

It should be clear that, for the same reasons discussed around the interpretation of $D(i)$, the duration $\tau(\delta)$ can also be considered as a measure of the sensitivity of the fair price of an investment to changes in the required δ. It may then seem that we have a problem: the duration has a useful interpretation as a measure of the time extent of the investment's proceeds, and the effective duration is useful because we might more naturally wish to understand the sensitivity of the price to changes in the required return expressed as an effective interest rate. Do we therefore need to know the value of both? The answer is no. In fact, it is perfectly possible to work only with the duration and still be able to infer the price sensitivity to changes in i. We consider this now.

Recall that, for fixed interest rates, $1 + i = e^{\delta}$. This means that we can consider i to be a function of δ, and also δ to be a function of i. In particular,

$$i(\delta) = e^{\delta} - 1 \quad \text{and} \quad \delta(i) = \ln(1 + i)$$

With this in mind, the duration can be rewritten using the *chain rule* as

$$\tau(\delta) = -\frac{1}{f(\delta)}\frac{d}{d\delta}f(\delta) = -\frac{1}{f(\delta(i))}\frac{di(\delta)}{d\delta}\frac{d}{di}f(\delta(i))$$

It is clear that $\frac{di}{d\delta} = e^{\delta} = (1 + i)$ and so

$$\tau(\delta) = -\frac{(1 + i)}{f(i)}\frac{df(i)}{di} = (1 + i)D(i)$$

We therefore see that, given the duration of an investment, we are able to use it to

1. determine a measure of the time extent of the investment's proceeds; and
2. obtain the effective duration and so investigate the sensitivity of the fair price to small changes in the required underlying interest rate i.

EXAMPLE 4.18

A fixed-interest investment has price £145 and duration 4.3 years at $i = 5.64\%$ per annum. Calculate an approximation to price in the following situations.

a. The required return, expressed as continuously compounding rate of interest, moves to 6% per annum.

b. The required return, expressed as compounding rate of interest, moves to 5% per annum.

Solution

a. We have a change in δ and so work in terms of duration. The initial value is $\delta_0 = \ln(1.0564) \approx 0.0549$ and it moves to $\delta_1 = 0.06$, therefore $\Delta\delta = h \approx 0.0051$. From the definition of duration, we see that

$$\tau(\delta) \simeq -\frac{1}{f(\delta)}\frac{\Delta f}{\Delta\delta} \Rightarrow f(0.06) \simeq f(0.0564)\,(1 - h\tau)$$

and so estimate that the new price is given by $145\,(1 - 0.0051 \times 4.3) \approx$ £143.82.

b. We have a change in i and so work in terms of effective duration. The initial value is $i_0 = 0.0564$ and it moves to $i_1 = 0.05$, therefore $\Delta i = h \approx -0.0064$. Furthermore, $D(i) = \frac{\tau}{(1+i)} = \frac{4.3}{1.0564}$ and so

$$f(0.05) \simeq f(0.0564)\,(1 - hD(0.0564)) = 145\left(1 + 0.0064 \times \frac{4.3}{1.0564}\right) \approx 148.78$$

We therefore estimate that the new price is £148.78.

4.6 QUESTIONS

The following questions are intended to test your knowledge of the material discussed in this chapter. Full solutions are available in Chapter 4 Solutions of Part III. You should use an algebraic approach unless otherwise stated.

Question 4.1. Demonstrate that the following functions are strictly smooth over $\mathbb{R}$.

a. $f(x) = \cos(2x)$
b. $h(y) = e^{y^2}$
c. $g(z) = z^2 - 2z + 2$
d. $k(p) = \frac{p^2}{p^2+2}$

Question 4.2. Identify the location of any turning points in the following functions.

a. $f(x) = \cos(2x)$
b. $h(y) = e^{y^2}$
c. $g(z) = z^2 - 2z + 2$
d. $k(p) = \frac{p^2}{p^2+2}$

Question 4.3. Classify each turning point found in the following functions. Use a computational plot to confirm your conclusions.

a. $f(x) = \cos(2x)$
b. $h(y) = e^{y^2}$
c. $g(z) = z^2 - 2z + 2$
d. $k(p) = \frac{p^2}{p^2+2}$

Question 4.4. Determine the third- and fourth-order derivatives of the following functions.

a. $f(x) = \cos(2x)$
b. $h(y) = e^{y^2}$
c. $g(z) = z^2 - 2z + 2$
d. $k(p) = \frac{p^2}{p^2+2}$

Question 4.5. Determine the range of each of the following functions.

a. $f(x) = \cos(2x)$
b. $h(y) = e^{y^2}$
c. $g(z) = z^2 - 2z + 2$
d. $k(p) = \frac{p^2}{p^2+2}$

Question 4.6. Determine the range of each of the following functions.

a. $f(x) = \frac{x^4-2x+4}{5x^4+2}$

b. $g(x) = \frac{x^4-2x+4}{5x^4-2}$

Comment on any differences between the ranges of these two very similar functions.

Question 4.7. An investor purchases an asset at time $t = 0$ for \$100. If the investor believes the market price of the asset will evolve according to the function

$$\$\left[150.4891 - (t-7)^2 - \frac{e^{5-t^2}}{100}\right]$$

determine the optimal time to sell the asset and the profit that he is expected to make. Note that t is measured in years.

Question 4.8. Due to staffing issues, the daily output of a fresh sandwich manufacturer based in New York varies over time and is modeled by the function

$$1000 \times \left[2 + \sin\left(\frac{2\pi}{365}t\right)\right] \text{ sandwiches}$$

Note that $t \in [0, 365]$ is measured in days from 1 January. The manufacturer knows that the lunch-time market in lower Manhattan has a near insatiable appetite for sandwiches and it is possible to sell any number at a unit price given by

$$\$\left[4 - \left(1 - \frac{2}{365}t\right)^2\right] \text{ per sandwich}$$

where again $t \in [0, 365]$ is measured in days from 1 January.

a. Determine the maximum output of sandwiches and the day on which this happens.

b. Determine the maximum price it is possible to charge and the day on which this is possible.

c. Determine the maximum daily revenue the sandwich manufacturer is able to generate and the day that this occurs. Comment on your answer.

d. If the manufacturer wishes to shut down for 1 day in the year for maintenance, which day should he choose?

You should assume that the manufacturer works 365 days a year.

Question 4.9. Compute the maximum price at $t = 0$, the effective duration, and the duration of the following projects if the required rate of return is 6% per annum (expressed as a compounding rate).

a. Annual proceeds received at the end of each of the next 5 years, starting at \$50 and inflating by 5% per annum.

b. Three-yearly proceeds each of £100 for a total of three payments (that is, after 3, 6, and 9 years), followed by £1000 received after 10 years.

Question 4.10. An investor is willing to pay \$1010 for an investment with duration 10 years. If these values are based on a required return of 3% per annum (expressed as a compounding rate), estimate the new price in the following situations.

a. The required return increases to 4% (expressed as a compounding rate).

b. The required return reduces to 2.5% (expressed as a continuously compounding rate).

CHAPTER 5

Sequences and Series

Contents

Prerequisite knowledge	**Intended learning outcomes**

• Chapter 1
• Chapter 2
• Chapter 3
• Chapter 4
	• Define and recognize series and sequences, including
	  ◦ finite and infinite series/sequence
	  ◦ arithmetic and geometric progressions
	• Distinguish between series and summations
	• Recognize patterns in simple series, including recursive examples
	• Manipulate and evaluate summations
	• Test a given summation for convergence
	• Determine Taylor and Maclaurin series of a given function and use for
	  ◦ function approximation
	  ◦ evaluation of limits

In this chapter, we study *sequences* and *series*. As we shall see, both concepts are very much connected. Series and sequences are interesting as mathematical concepts in their own right but are particularly useful in the discussion of cash-flow streams in actuarial and financial applications. This chapter is an opportunity to discuss Taylor and Maclaurin series, which are very useful concepts in the analysis of functions. As we shall see, Taylor series connects the material of this chapter with higher-order derivatives discussed in Chapter 4 and are particularly useful for evaluating limits.

5.1 SEQUENCES

As sequence is simply an ordered set of terms that are determined by a pattern or rule. The most obvious example of a sequence would be

$$1, 2, 3, 4, \ldots$$

This sequence starts at 1 and is defined by adding 1 to the previous term. Mathematically, we would say that the nth term of the sequence is n for $n \in \mathbb{N}^+$. Recall that in Chapter 1, we defined $\mathbb{N}$ to begin at 0 and $\mathbb{N}^+$ to begin at 1. It is then clear that the sequence continues as

$$5, 6, 7, 8, \ldots$$

A slightly more complicated example of a sequence would be

$$0, 3, 8, 15, \ldots$$

which has nth term given by $n^2 - 1$ for $n \in \mathbb{N}^+$. This sequence would continue as $24, 35, 48, 63, \ldots$.

We are careful to define a sequence as an "ordered set of *terms*" because a sequence need not necessarily be numerical. For example,

$$a, b, c, d, \ldots$$

is a perfectly well-defined sequence of terms, defined alphabetically. However, in this book, we are particularly concerned with numbers and so we will be restricted to numerical sequences in all that follows.

Although it is useful to define a (numerical) sequence in terms of a mathematical rule and a starting value, this is not always necessary and sequences can be presented simply as an ordered set of numbers. For example, the numerical sequences given above could be presented as

$$\{1, 2, 3, 4, \ldots\} \quad \text{and} \quad \{0, 3, 8, 15, \ldots\}$$

Note that we have used the notation $\{\}$, however a sequence defined like this is not actually a *set* in the strict sense. Recall that a set, as discussed in Chapter 1, is presented as a collection of *unique* members. In particular, a set need not be expressed in an order and members would not be repeated. For example, $\{1, -1, 1, -1, \ldots\}$ defines a sequence but does not form a set in the strict sense since the order is crucial and both terms 1 and -1 are repeated.

In practice, it is typically the case that, given a sequence, one would need to determine subsequent values of it; being able to identify the rule that defines a sequence is then crucial. No standard notation exists for the nth term of a sequence, but in what follows we will use x_n.

EXAMPLE 5.1

Determine the next four terms in the following numerical sequences. State a rule in each case for determining the nth term, where $n \in \mathbb{N}^+$.

a. $\{-1, -2, -3, -4, \ldots\}$
b. $\{1, 3, 5, 7, \ldots\}$
c. $\{1, 0, 1, 0, \ldots\}$
d. $\{10, 9, 8, 7, \ldots\}$
e. $\{2, 2, 2, 2, \ldots\}$

Solution

a. This is a sequence of negative integers, $\{-1, -2, -3, -4, -5, -6, -7, -8, \ldots\}$. We have $x_n = -n$ for $n \in \mathbb{N}^+$.
b. This is a sequence of odd natural numbers, $\{1, 3, 5, 7, 9, 11, 13, 15, \ldots\}$. We have $x_n = 2n - 1$ for $n \in \mathbb{N}^+$.
c. This is an alternating sequence of 1 and 0, $\{1, 0, 1, 0, 1, 0, 1, 0, \ldots\}$. We have $x_n = (1 - (-1)^n)/2$ for $n \in \mathbb{N}^+$.
d. This is a backwards sequence starting at 10, $\{10, 9, 8, 7, 6, 5, 4, 3, \ldots\}$. We have $x_n = 11 - n$ for $n \in \mathbb{N}^+$.
e. This is a flat sequence returning the value 2. We have $x_n = 2$ for $n \in \mathbb{N}^+$.

Example 5.1 illustrates that a sequence can be described as being *forwards*, *backwards*, or *alternating*. However, many sequences can also be considered as a mapping of the forward sequence $\{1, 2, 3, 4, \ldots\}$ via the particular rule x_n. This is true if the rule is of the form $x_n = f(n)$, that is some one-to-one or many-to-one mapping of n, as in each case of Example 5.1, but this is not necessarily always the case. For example, the well-known *Fibonacci Sequence*,

$$0, 1, 1, 2, 3, 5, 8, 13, 21, \ldots$$

has the rule $x_n = x_{n-1} + x_{n-2}$. The Fibonacci rule therefore operates on prior values of the sequence rather than some transformation of the underlying sequence $\{1, 2, 3, 4, \ldots\}$. Rules of this form are known as *recursive*.

EXAMPLE 5.2

Determine the recursive rule in each of the following sequences.

a. $\{0, 1, 1, 3, 5, 11, \ldots\}$
b. $\{0, 1, -1, 7, -13, 55, \ldots\}$

Solution

a. We have $x_n = 2x_{n-2} + x_{n-1}$.
b. We have $x_n = 6x_{n-2} - x_{n-1}$.

The sequences presented so far have been *infinite sequences*; that is, they continue indefinitely. In contrast, we could define a *finite sequence*; that is, a sequence of finite length.

EXAMPLE 5.3

For the following finite sequences, determine the rule x_n and value of m such that $n = 1, 2, 3, \ldots, m$.

a. $\{\frac{1}{2}, \frac{1}{4}, \frac{1}{6}, \ldots, \frac{1}{16}\}$
b. $\{-7, -5, -3, \ldots, 1\}$
c. $\{4, 7, 12, 19, \ldots, 10{,}003\}$
d. $\{1, \frac{4}{3}, \frac{3}{2}, \frac{8}{5}, \ldots, \frac{68}{35}\}$

Solution

a. We have $x_n = 1/2n$ for $n = 1, 2, 3, \ldots, 8$.
b. We have $x_n = n - 8$ for $n = 1, 2, 3, \ldots, 9$.
c. We have $x_n = n^2 + 3$ for $n = 1, 2, 3, \ldots, 100$.
d. We have $x_n = 2n/(n+1)$ for $n = 1, 2, 3, \ldots, 34$.

Sequences formed from the application of a recursive rule can also be either infinite or finite. In what follows, we will be predominantly be concerned with sequences defined by $x_n = f(n)$, that is by rules that are not recursive.

5.2 SERIES AND SUMMATIONS

5.2.1 Concepts and notation

A *series* arises from the sum of terms in a sequence. For example, a series could be formed from summing the terms in sequence $\{1, 2, 3, 4, \ldots\}$; that is, $1 + 2 + 3 + 4 + \cdots$. A *finite series* arises from summing a *finite* number of terms in either an infinite or a finite sequence. It should be clear that a finite series arising from a well-defined sequence will have a finite *value*.

EXAMPLE 5.4

Calculate the value of the finite series formed from the following sequences.

a. All terms of $\{10, 8, 6, 4, \ldots, 0\}$
b. The first five terms of $\left\{1, \frac{1}{2}, \frac{1}{3}, \frac{1}{4}, \ldots\right\}$

Solution

a. $10 + 8 + 6 + 4 + 2 + 0 = 30$

b. $1 + \frac{1}{2} + \frac{1}{3} + \frac{1}{4} + \frac{1}{5} = \frac{137}{60}$

In contrast, an *infinite series* arises from the sum of an *infinite* number of terms in an infinite sequence. The value of an infinite series could therefore be infinite or, in some cases, may converge to a finite value. We will return to the issue of *convergence* later in this chapter.

It is convenient to introduce *sigma notation* to avoid writing out a great number of terms when expressing a series. Sigma notation can be used to represent both finite and infinite series in terms of a sequence rule x_n for $n \in \mathbb{N}^+$. For example, the finite sequence formed from the first 10 terms of the infinite sequence defined by $x_n = 2n$ is $\{2, 4, 6, \ldots, 20\}$ and the associated series is denoted

$$\sum_{n=1}^{10} 2n = 2 \times 1 + 2 \times 2 + 2 \times 3 + \cdots + 2 \times 10$$

The notation $\sum_{n=1}^{10} 2n$ is read as "the sum of $2n$ for $n = 1$ to 10." Furthermore, the infinite series arising from the sequence defined by $x_n = n^2$ is denoted

$$1^2 + 2^2 + 3^2 + 4^2 \cdots = \sum_{n=1}^{\infty} n^2$$

This is read as "the sum of n^2 from $n = 1$ to infinity."

The *indexation* symbol, here n, is a dummy variable and any symbol can be used. For example, given a sequence defined by a general rule x_n, the associated series can be equivalently written as

$$\sum_{n=1}^{\infty} x_n \equiv \sum_{i=1}^{\infty} x_i \equiv \sum_{j=1}^{\infty} x_j \equiv \sum_{k=1}^{\infty} x_k$$

EXAMPLE 5.5

Express the following series in terms of sigma notation.

a. $2^1 + 2^2 + 2^3 + \cdots + 2^{100}$

b. $3 + 5 + 7 + 9 + 11 + \cdots$

c. $\frac{1}{11} + \frac{1}{12} + \frac{1}{13} + \cdots + \frac{1}{22}$

Solution

In each case, we first need to identify the sequence rule, x_n.

a. We have $x_n = 2^n$ for $n = 1, 2, \ldots, 100$ and so $\sum_{i=1}^{100} 2^i$

b. We have $x_n = 2n + 1$ for $n = 1, 2, \ldots$ and so $\sum_{j=1}^{\infty} (2j + 1)$

c. We have $x_n = \frac{1}{n+10}$ for $n = 1, 2, \ldots, 12$ and so $\sum_{k=1}^{12} \frac{1}{k+10}$

Up until now we have considered sequences and series in terms of simple numerical values; however, it is possible to extend the concepts to algebraic quantities. For example, the series $2^1 + 2^2 + 2^3 + \cdots$ and $3^1 + 3^2 + 3^3 + \cdots$ and both examples of the particular series

$$\sum_{n=1}^{\infty} x^n$$

with $x = 2$ and $x = 3$, respectively. Algebraic sequences and series are particularly important in actuarial science and financial applications.

As we shall see in later sections, the sigma notation is particularly powerful and it is useful to extend it to a broader class of objects called *summations* before proceeding. Summations are essentially the same as series, but without the restriction that they start at an indexing value of 1. For example, $\sum_{k=10}^{100}(x^{2k+1})$ is a summation but not a series because it starts at $k = 10$. It is essential to understand that if we remove the restriction that the indexing value starts at 1, we open up the possibility of allowing many, many equivalent representations of the same summations. This is considered in the next example.

EXAMPLE 5.6

Demonstrate that the following summations are equivalent.

$$\sum_{k=1}^{100} x^{k+10} \quad \text{and} \quad \sum_{i=11}^{110} x^i$$

Solution

Two approaches can be taken.

1. The first approach is to convince yourself that the terms in both summations are identical. In particular,

$$\sum_{k=1}^{100} x^{k+10} = x^{11} + x^{12} + x^{13} + \cdots + x^{110}$$

$$\sum_{i=11}^{110} x^i = x^{11} + x^{12} + x^{13} + \cdots + x^{110}$$

   and so the two are clearly identical.

2. Alternatively, we can attempt to use a transformation of the indexing variable to algebraically demonstrate that they are equivalent. In particular, if we set $k + 10 = j$, then $k = 1$ becomes $j = 1 + 10 = 11$ and $k = 100$ becomes $j = 100 + 10 = 110$. Therefore,

$$\sum_{k=1}^{100} x^{k+10} = \sum_{j=11}^{110} x^j$$

Since the indexing object is a dummy variable, we can simply replace j with i to demonstrate the equivalence.

The transformation process used here is an important skill and is revisited throughout this chapter.

It is important to realize that the index can take negative values. For example, the following summation is easily understood

$$\sum_{i=-3}^{3} 2i = -6 + (-4) + (-2) + 0 + 2 + 4 + 6$$

as is

$$\sum_{i=-6}^{3} i = -6 + (-5) + (-4) + (-3)$$

Essentially the index is understood to march upwards from the lower value toward the upper in unit steps.

5.2.2 Properties of summations

The sigma notion is important as it not only removes the cumbersome need to list individual terms in a summation but also enables us to think of summations as algebraic objects in their own right. With this in mind we now consider the algebraic properties of summations.

The fundamental properties of summations are stated mathematically as follows.

i. $\sum_{i=p}^{q} 1 = q - p + 1$

ii. $\sum_{i=p}^{q} cx_i = c\sum_{i=p}^{q} x_i$

iii. $\sum_{i=p}^{q} (x_i + y_i) = \sum_{i=p}^{q} x_i + \sum_{i=p}^{q} y_i$

where c is a constant, $p < q \in \mathbb{Z}$ and x_i and y_i represent rules (in terms of functions of i) that define sequences. A demonstration that these properties are indeed true is left as an exercise. Furthermore, it should be clear that Property iii. can be extended to the sum of any number of sequences.

EXAMPLE 5.7

Evaluate the following expressions using Properties i.-iii.

a. $\sum_{i=-10}^{10} 2$.

b. $\sum_{i=1}^{100} (2x_i - 4y_i)$ if $\sum_{i=1}^{100} x_i = 100$ and $\sum_{i=1}^{100} y_i = 20$.

c. $\sum_{i=-5}^{10} (10 - 5x_i)$ if $\sum_{i=-5}^{10} 5x_i = 6$.

Solution

a. Using Properties i. and ii., we have $\sum_{i=-10}^{10} 2 = 2\sum_{i=-10}^{10} 1 = 2 \times (10 + 10 + 1)$ $= 42$.

b. Using Properties ii. and iii., we have $\sum_{i=1}^{100}(2x_i - 4y_i) = \sum_{i=1}^{100}(2x_i) + \sum_{i=1}^{100}(-4y_i) = 2\sum_{i=1}^{100} x_i - 4\sum_{i=1}^{100} y_i = 2 \times 100 - 4 \times 20 = 120$.

c. Using Properties i., ii., and iii., we have $\sum_{i=-5}^{10}(10 - 5x_i) = \sum_{i=-5}^{10} 10 + \sum_{i=-5}^{10}(-5x_i) = 160 - 5\sum_{i=-5}^{10} x_i = 160 - 5 \times 6 = 130$.

Further properties can be derived from these basic properties. To begin with, let us consider the reverse of Property iii. For example, it should be clear that

$$\sum_{i=10}^{100} i^2 + \sum_{j=10}^{100} 2j^3 = \sum_{i=10}^{100} (2i^3 + i^2)$$

which can be generalized to the additional property

iv. $\sum_{i=p}^{q} x_i + \sum_{j=p}^{q} y_j = \sum_{i=p}^{q} (x_i + y_i)$.

This can be further generalized to the addition of any number of summations. However, it should be clear that, for example,

$$\sum_{i=10}^{100} i^2 + \sum_{j=1}^{100} 2j^3 \neq \sum_{i=10}^{100} (2i^3 + i^2)$$

The different indices of the two summations are of course the reason for this, and Property iv relies on the fact that p and q are the same for both summations.

One practical way around the problem of adding summations with different indices is to split the "largest summation." For example,

$$\begin{aligned}\sum_{i=10}^{100} i^2 + \sum_{j=1}^{100} 2j^3 &= \sum_{i=10}^{100} i^2 + \sum_{j=10}^{100} 2j^3 + \sum_{j=1}^{9} 2j^3 \\ &= \sum_{j=1}^{9} 2j^3 + \sum_{i=10}^{100} (2i^3 + i^2)\end{aligned}$$

In some cases, it may be possible to transform the index of one summation so that the addition can take place without splitting the largest summation. However, the ability to do this is particularly sensitive to the properties of the summations in question and it is often impossible for summations with a finite index. Both approaches are illustrated in the following example.

EXAMPLE 5.8

Use Properties i-iv to simplify the following expressions as far as possible.

a. $\sum_{i=2}^{\infty}(2i + 1) + \sum_{j=1}^{\infty} 2(j - 1)$

b. $\sum_{i=10}^{400} x^i + \sum_{j=10}^{410}(4i - 1)$

c. $\sum_{m=5}^{54} 2m + \sum_{j=1}^{50} 4j$

d. $\sum_{k=5}^{\infty} x^{k-10} + \sum_{j=1}^{\infty}(3j+1)x$

Solution

a. The simple form of both summations, particularly the infinite upper limits, lends itself to the transformation approach. In particular, if we substitute $i = k+1$ in the first summation we can transform the problem to

$$\sum_{k=2-1}^{\infty} (2(k+1)+1) + \sum_{j=1}^{\infty} 2(j-1) = \sum_{k=1}^{\infty}(2k+3) + \sum_{j=1}^{\infty} 2(j-1)$$
$$= \sum_{k=1}^{\infty}(4k+1)$$

b. There is no substitution to translate both the upper and the lower limits to common values and it necessary to consider splitting a summation. In particular,

$$\sum_{i=10}^{400} x^i + \sum_{j=10}^{410}(4j-1) = \sum_{i=10}^{400} x^i + \sum_{j=10}^{400}(4j-1) + \sum_{j=401}^{410}(4j-1)$$
$$= \sum_{i=10}^{400}(x^i + 4i - 1) + \sum_{j=401}^{410}(4j-1)$$
$$= -401 + \sum_{i=10}^{400}(x^i + 4i) + 4\sum_{j=401}^{410} j$$

We can go further and transform both summations to series as

$$= -401 + \sum_{n=1}^{391}(x^{n+9} + 4(n+9)) + 4\sum_{k=1}^{10}(k+400)$$
$$= 29{,}675 + 4\sum_{k=1}^{10} k + \sum_{n=1}^{391}(x^{n+9} + 4n)$$

c. Both the lower and the upper limits differ by 4 which implies that a transformation will be possible. In particular, if we set $m = i+4$ then

$$\sum_{m=5}^{54} 2m + \sum_{j=1}^{50} 4j = \sum_{i=1}^{50} 2(i+4) + \sum_{j=1}^{50} 4j$$
$$= 400 + 6\sum_{i=1}^{50} i$$

d. Both upper limits are infinite and a transformation method can be used. In particular, $k = i + 4$ leads to

$$\sum_{k=5}^{\infty} x^{k-10} + \sum_{j=1}^{\infty}(3j+1)x = \sum_{i=1}^{\infty} x^{i-6} + \sum_{j=1}^{\infty}(3j+1)x$$

$$= \sum_{i=1}^{\infty}(x^{i-6} + (3i+1)x)$$

5.3 EVALUATING SUMMATIONS

We have so far been concerned with manipulating summations rather than evaluating them. This is of course with the exception of the use of Property i which gives the evaluation of a very simple of summation. When considering the evaluation of a summation, it is necessary to distinguish between finite and infinite sums. In particular, it should be clear that the sum of a finite number of terms (each of finite value) has a finite evaluation, which is not necessarily the case for an infinite sum of finite values. We consider the issue of *convergence* of infinite summations in Section 5.3.3.

There are two approaches to evaluating a summation. The first approach is to use a computer to evaluate it directly. For example, one might use Excel's copy down and sum functions or particular commands in Wolfram Alpha. Alternatively, one could code a loop in VBA or C++.[1] Such approaches are certainly always possible for the sum of *finite* series; furthermore, experimentation using an increasingly large number of finite terms could lead to some insight into the convergence or otherwise of the sum of *infinite* terms. We do not, however, discuss these approaches in detail here. The second approach, that we will focus on, is to attempt to manipulate a summation into a standard form that has a known sum. You could consider this an algebraic approach, which is in contrast to the computational approach. We return to the use of Wolfram Alpha as a means of checking the results of an algebraic approach towards the end of this chapter.

5.3.1 Arithmetic and geometric progressions

In what follows, we will need to be able to classify a sequence (either finite or infinite) as either an *arithmetic progression* or a *geometric progression*. However, as we shall see, a sequence need not necessarily be either.

[1] This coding approach is particularly useful as the loop could be incorporated into a larger computational project; for example, a pension scheme valuation, which is not possible using Wolfram Alpha.

An arithmetic progression is a sequence in which consecutive terms differ by a *common difference*. For example, the sequence $\{4, 6, 8, 10, \ldots\}$ is an infinite arithmetic progression with *common difference* 2. Mathematically, we can express an arithmetical progression as

$$\{a, a + d, a + 2d, \ldots, a + (n - 1)d, \ldots\} \tag{5.1}$$

That is, the nth term is given by $x_n = a + (n - 1)d$ for $n \in \mathbb{N}^+$, where d is the common difference and a is the first term in the sequence.

In contrast, a geometric progression is a sequence in which consecutive terms are related by a *common factor*. For example, the sequence $\{1, 2, 4, 8, 16, \ldots\}$ is a geometric progression with *common factor* 2. Mathematically, we can express a geometric progression as

$$\{a, ar, ar^2, \ldots, ar^{n-1}, \ldots\} \tag{5.2}$$

That is, the nth term is given by $x_n = ar^{n-1}$ for $n \in \mathbb{N}^+$, where r is the common factor and a is the first term in the sequence.

Not all sequences can be classified as being either arithmetic or geometric progression. For example, the series defined by nth term $x_n = 1 + 4(n - 1) - 0.1^{n-1}$ can be considered as having two components, one arithmetic and one geometric. Furthermore, it may be difficult or indeed impossible to express a sequence defined with a recursive formula in terms of arithmetic and geometric components. The techniques that follow are therefore limited in their application, but the skills development from a solid practical understanding of them are very useful.

EXAMPLE 5.9

Classify the following as either arithmetic or geometric progressions (or some combination of both). State the starting value and common difference/factor in each case.

a. $\{10, 14, 18, 22, \ldots, 50\}$
b. $\{-1, 1, -1, 1, \ldots\}$
c. nth term given by $x_n = 0.2 \times 4^n$ for $n = 1, \ldots, 10$
d. nth terms given by $x_n = 2^n - 0.1n$ for $n = 1, 2, \ldots$.
e. $\{2, 5, 9, 15, \ldots, 16{,}413\}$

Solution

a. This is a finite arithmetic progression with common difference 4 and starting value 10. The nth term is given as $x_n = a + (n - 1)d = 10 + 4(n - 1)$ for $n = 1, \ldots, 11$.
b. This is an infinite geometric progression with common factor -1 and starting value -1. The nth term is given by $x_n = ar^{n-1} = (-1) \times (-1)^{n-1}$.
c. This is a finite geometric progression with common factor 4. However, it is not in the standard form of Eq. (5.2) and should be rewritten as $x_n = 0.8 \times 4^{n-1}$ for $n = 1, \ldots, 10$. The starting value is then clearly 0.8.
d. This can be considered as a combination of the two types of progression. The first component is $2^n = ar^{n-1} = 2 \times 2^{n-1}$ and so is an infinite geometric progression

with starting value 2 and common factor 2. The second component is $-0.1n = -0.1 + (-0.1)(n-1)$, an infinite arithmetical progression with starting value -0.1 and common difference -0.1.

e. Although it might not be immediately obvious, this is a finite series with nth term given by $x_n = 1 + 2(n-1) + 2^{n-1}$ for $n = 1, \ldots, 15$. The series is therefore formed from a finite arithmetic progression with starting value 1 and common difference 2, and a geometric progression with starting value 1 and common factor 2.

5.3.2 Evaluating finite sums

We begin by working toward a closed-form expression for the sum of an arithmetical progression of k terms. This expression will depend on the progression's starting value and common difference. We denote this sum as $S_{A,k}$, with the "A" indicating that it is the sum of an arithmetic progression and the k indicating the number of terms in the series. Using Eq. (5.1), we are seeking an expression for

$$S_{A,k} = \sum_{n=1}^{k} (a + (n-1)d)$$

Using the basic properties of summations, we can manipulate this as follows

$$\begin{aligned} S_{A,k} &= \sum_{n=1}^{k} (a + (n-1)d) \\ &= a\sum_{n=1}^{k} 1 + d\sum_{n=1}^{k} n - d\sum_{n=1}^{k} 1 \\ &= k(a-d) + d\sum_{n=1}^{k} n \end{aligned}$$

The problem therefore becomes one of finding an expression for $\sum_{n=1}^{k} n = 1 + 2 + 3 + \cdots + k$. An evaluation of this sum is easily obtained by adding the sequence to its reverse, term by term. That is,

$$2\sum_{n=1}^{k} n = \left\{ \begin{array}{lllll} 1 & +2 & +3 & +\cdots & +k \\ k & +k-1 & +k-2 & +\cdots & +1 \\ \hline (k+1) & +(k+1) & +(k+1) & +\cdots & +(k+1) \end{array} \right\} = k(k+1)$$

We therefore have

$$\sum_{n=1}^{k} n = \frac{1}{2}k(k+1) \tag{5.3}$$

This is a useful result in itself and can be used to complete the required the derivation of the closed-form expression,

$$S_{A,k} = \frac{k}{2}\left(2a + (k-1)d\right) \tag{5.4}$$

This expression can be used to evaluate any summation that can be manipulated into the standard form of a finite arithmetic progression, as given in Eq. (5.1).

EXAMPLE 5.10

Use an algebraic method to evaluate the following summations.

a. $\sum_{n=1}^{10}(50 + 4(n-1))$

b. $4 + 3 + \cdots - 15$

c. $\sum_{n=1}^{200} 10n$

d. $\sum_{n=16}^{25}(2 + n)$

Solution

a. This is already in the standard form of the sum of an arithmetic progression with $a = 50$, $d = 4$, and $k = 10$. Equation (5.4) leads to $S_{A,10} = 5 \times (100 + 9 \times 4) = 680$.

b. The nth term in the sequence begin summed is given by $x_n = 4 - (n-1)$ for $n = 1, \ldots, 20$. It is therefore an arithmetic progression with $a = 4$, $d = -1$, and $k = 20$, and so $S_{A,20} = 10 \times (8 - 19 \times 1) = -110$.

c. We can write this as $\sum_{n=1}^{200}(10 + 10(n-1))$ which is a sequence formed from a standard arithmetic progression with $a = 10$, $d = 10$, and $k = 200$. The sum is then obtained as $S_{A,200} = 100 \times (20 + 199 \times 10) = 201{,}000$.

d. This is the sum of an arithmetic progression with common difference 1, but some manipulation is required to express it in standard form. Using the substitution $i = n - 15$, we proceed as

$$\sum_{n=16}^{25}(2+n) = \sum_{i=1}^{10}(i+17)$$
$$= \sum_{i=1}^{10}(18 + 1(i-1))$$

and so have a standard form of the sum of an arithmetic progression with $a = 18$, $d = 1$, and $k = 10$. The sum is therefore $S_{A,10} = 5 \times (36 + 9) = 225$.

EXAMPLE 5.11

Determine an expression for the following series for general $k \in \{2, 3, 4, \ldots\}$.

$$\sum_{n=1}^{k} \ln(4^n)$$

Evaluate

$$\sum_{n=1}^{10} \ln(4^n) \quad \text{and} \quad \sum_{n=10}^{100} \ln(4^n)$$

Solution

Using the properties of logarithms and some further manipulation, we can rewrite the series as

$$\begin{aligned}\sum_{n=1}^{k} \ln(4^n) &= \sum_{n=1}^{k} n \ln(4) \\ &= \ln(4) \sum_{n=1}^{k} n \\ &= \frac{\ln(4)}{2} k(k+1)\end{aligned}$$

Note that we have also used Eq. (5.3). We can then determine that

$$\sum_{n=1}^{10} \ln(4^n) = \frac{\ln(4)}{2} \times 10 \times 11 \approx 76.246$$

Note that the second summation requires further manipulation. In particular, we need to express the summation as a series. This can be done by setting $n = i + 9$ and so

$$\begin{aligned}\sum_{n=10}^{100} \ln(4^n) &= \sum_{i=1}^{91} \ln(4^{i+9}) \\ &= \ln(4) \left(\sum_{i=1}^{91} i + 9 \sum_{i=1}^{91} 1 \right) \\ &= \ln(4) \left(\frac{91 \times 92}{2} + 9 \times 91 \right) \\ &\approx 6938.4033\end{aligned}$$

Note that we have again used Eq. (5.3).

We now consider the sum of a finite geometric progression. This is denoted $S_{P,k}$, such that

$$S_{G,k} = \sum_{n=1}^{k} ar^{n-1}$$

where a is the starting value, r the common factor, and $k \in \mathbb{N}^+$. A simple way to obtain the closed-form expression is to subtract the sequence formed by $rS_{G,k}$ from $S_{G,k}$. In particular, we have

$$\left\{\begin{array}{r} S_{G,k} \\ -rS_{G,k} \\ \hline S_{G,k}-rS_{G,k} \end{array}\right\} = \left\{\begin{array}{llllll} a & +ar & +ar^2 & +\ldots & +ar^{k-1} \\ -ar & -ar^2 & -ar^3 & -\ldots & -ar^k \\ \hline a & & & & -ar^k \end{array}\right\}$$

which leads to

$$S_{G,k} = \frac{a(1-r^k)}{1-r} \tag{5.5}$$

This expression enables one to calculate the sum of a finite number of terms in a geometric expression when $r \neq 1$. This restriction on r reflects that Eq. (5.5) is a rational function with $r = 1$ a root of the denominator.

EXAMPLE 5.12

Use an algebraic method to evaluate the following summations.

a. $\sum_{n=1}^{5} 10 \times 2^{n-1}$

b. $1 + 10 + 100 + \cdots + 100{,}000$

c. $\sum_{n=1}^{500} \left(\frac{1}{2}\right)^n$

d. $\sum_{n=-10}^{25} 0.4^{n+1}$

Solution

a. This is already in the standard form of the sum of a geometric progression with $a = 10$, $r = 2$, and $k = 5$. Equation (5.5) then leads to $S_{G,5} = 10 \times \frac{(1-2^5)}{1-2} = 310$.

b. The nth term in the sequence can be represented by $x_n = 1 \times 10^{n-1}$ for $n = 1, \ldots, 6$. It is then a geometric progression with $a = 1$, $r = 10$, and $k = 6$, and $S_{G,6} = \frac{1-10^6}{1-10} = 111{,}111$.

c. We can write this as $\sum_{n=1}^{200} 0.5 \times 0.5^{n-1}$ which is the sum of a standard geometric progression with $a = 0.5$, $r = 0.5$, and $k = 500$. The sum is then obtained as $S_{P,500} = \frac{0.5\times(1-0.5^{500})}{1-0.5} = 1$.

d. This is the sum of a geometric progression with common ratio 0.4, but some manipulation is required to express it in standard form. Using the substitution $i = n + 11$ we proceed as

$$\sum_{n=-10}^{25} 0.4^{n+1} = \sum_{i=1}^{36} 0.4^{i-10}$$
$$= \sum_{i=1}^{36} 0.4^{-9} \times 0.4^{i-1}$$

and so have a standard form with $a = 0.4^{-9}$, $r = 0.4$, and $k = 36$. The value is $S_{G,36} = \frac{0.4^9\times(1-0.4^{36})}{1-0.4} \approx 6357.83$.

EXAMPLE 5.13

Evaluate the following summations.

a. $\sum_{n=1}^{15}(2n+2^n)$

b. $\sum_{n=10}^{57}\left(3+0.1n+(0.9)^{-\frac{n}{2}}\right)$

Solution

a. We note that

$$\begin{aligned}\sum_{n=1}^{15}(2n+2^n) &= \sum_{n=1}^{15}2n+\sum_{n=1}^{15}2^n \\ &= \sum_{n=1}^{15}(2+2(n-1))+\sum_{n=1}^{15}2\times 2^{n-1}\end{aligned}$$

which is seen to be the sum of a standard arithmetic progression (with $a=2$, $d=2$, and $k=15$) and a standard geometric progression (with $a=2$, $r=2$, and $k=15$). These two components can be evaluated separately as

$$\begin{aligned}S_{A,15}+S_{G,15} &= \frac{15}{2}\times(4+14\times 2)+\frac{2\times(1-2^{15})}{1-2} \\ &= 240+65{,}534 \\ &= 65{,}774\end{aligned}$$

b. We can manipulate this into the sum of an arithmetic and a geometric progression,

$$\begin{aligned}\sum_{n=10}^{57}\left(3+0.1n+(0.9)^{-\frac{n}{2}}\right) &= \sum_{i=1}^{48}\left(3+0.1(i+9)+(0.9)^{-\frac{i+9}{2}}\right) \\ &= \sum_{i=1}^{48}(4+0.1(i-1))+\sum_{i=1}^{48}\left((0.9)^{-5}\times\left(\frac{1}{\sqrt{0.9}}\right)^{i-1}\right) \\ &= \frac{48}{2}(2\times 4+47\times 0.1)+\frac{(0.9)^{-5}\times\left(1-(0.9)^{-24}\right)}{1-1/\sqrt{0.9}} \\ &\approx 665.98\end{aligned}$$

using the standard expressions for the sums.

5.3.3 Evaluating infinite sums

We now consider the sum of a sequence of infinitely many terms. Intuitively you might expect that the sum of infinitely many terms in either an arithmetic progression or a geometric progression has an infinite value; however, this is not always the case. An infinite sequence that sums to a finite value is called a *convergent* series, as opposed to a *divergent* series that has an infinite sum.

To motivate the idea of a convergent series, we begin by considering how we might construct an infinite series starting from a finite value of its sum. That is, let us begin by considering the problem the "wrong way around":

Consider the finite portion of the real number line between 0 and 1. It is clear that we can divide this line into half to give two intervals both of width 1/2. We divide the second interval in half again to give three intervals of length 1/2, 1/4, and 1/4. We then divide the last interval in half yet again to give four intervals of length 1/2, 1/4, 1/8, and 1/8. We can continue this process indefinitely. That is, we can easily imagine the process of dividing a finite interval into infinitely many sub-intervals. The process is demonstrated in Figure 5.1.

From the mathematical perspective, we can express the total length of the interval between 0 and 1 at each stage of this process as

$$\begin{aligned}
1 &= \frac{1}{2} + \frac{1}{2} \\
1 &= \frac{1}{2} + \frac{1}{4} + \frac{1}{4} \\
1 &= \frac{1}{2} + \frac{1}{4} + \frac{1}{8} + \frac{1}{8} \\
&\vdots \\
1 &= \frac{1}{2} + \frac{1}{2^2} + \frac{1}{2^3} + \cdots + \frac{1}{2^n}
\end{aligned}$$

The expression as $n \to \infty$ can be written in terms of summation notation as

$$1 = \sum_{i=1}^{\infty} \frac{1}{2^i} = \frac{1}{2} \times \sum_{i=1}^{\infty} \left(\frac{1}{2}\right)^{i-1} \tag{5.6}$$

This is therefore an example of a *convergent* infinite geometric progression with starting value $a = 1/2$ and common factor $r = 1/2$.

In general, we use the notation S_∞ for an infinite sum. For example, $S_{A,\infty}$ denotes the sum of an infinite arithmetic progression and $S_{G,\infty}$ denotes the sum of an infinite geometric progression.

It should be immediately clear that the sum of an infinite arithmetic progression is *always* divergent because it involves the sum of infinitely many values of a *finite common difference*. This can be demonstrated algebraically by taking the limit as $k \to \infty$

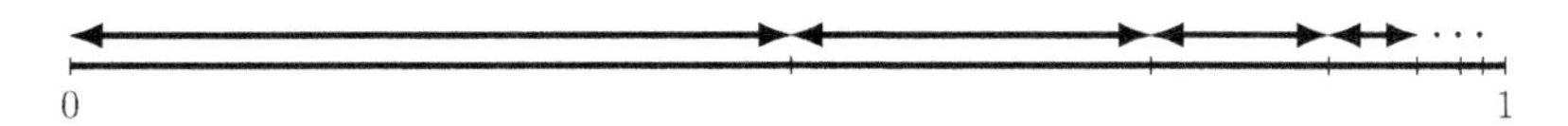

Figure 5.1 Repeated bisection of the interval [0, 1].

in Eq. (5.4). In particular,

$$\begin{aligned} S_{A,\infty} &= \lim_{k\to\infty} S_{A,k} \\ &= \lim_{k\to\infty} \left\{ \frac{k}{2}\left(2a + (k-1)d\right) \right\} \\ &= \infty \end{aligned}$$

As demonstrated above, a geometric progression can, however, be convergent and we now turn to investigating what conditions are required for this convergence. As expressed by Eq. (5.5), the sum of a finite geometric progression depends on the starting point, a, and the common ratio, r. It also should be clear from the form of that expression that the convergence or divergence will be determined by r. In particular, some manipulation leads to

$$\lim_{k\to\infty} S_{G,k} = \lim_{k\to\infty} \left\{ \frac{a(1-r^k)}{1-r} \right\} = \begin{cases} \frac{a}{1-r} & \text{for } |r| < 1 \\ \infty & \text{for } |r| \geq 1 \end{cases}$$

We therefore see that a geometric progression is convergent if it has a common ratio such that $|r| < 1$. This is summarized as

$$S_{G,\infty} = \frac{a}{1-r} \iff |r| < 1 \qquad (5.7)$$

Revisiting the motivating example of Eq. (5.6), we have $a = 1/2$ and $r = 1/2$ and so, by Eq. (5.7), the series is convergent and has value

$$\sum_{k=1}^{\infty} \frac{1}{2^k} = \frac{1/2}{1-1/2} = 1$$

This value is of course that expected from the original construction of the problem.

EXAMPLE 5.14

Determine whether each of the following are convergent or divergent. Calculate the converged value of the sum where appropriate.

a. $1 + 3 + 5 + \cdots$

b. $5 + 5/3 + 5/3^2 + \cdots$

c. $\sum_{j=1}^{\infty} 0.01 \times 2^j$

d. $\sum_{j=1}^{\infty} \left\{ \left(-\frac{1}{5}\right)^j + 2(j-1) \right\}$

e. $\sum_{k=1}^{\infty} 0.01 \times (-0.2)^k$

Solution

a. This is an infinite arithmetic progression with common difference of 2 and so is divergent.

b. This is an infinite geometric progression with $a = 5$ and $r = 1/3$. The series is convergent since $|r| < 1$ and has value given by Eq. (5.3.3).

$$S_{G,\infty} = \frac{5}{1 - 1/3} = 7.5$$

c. This is an infinite geometric progression with $|r| > 1$. It is therefore divergent.

d. This can be considered as a combination of an independent infinite geometric and an infinite arithmetic progressions. The infinite arithmetic component is divergent irrespective of the geometric component and so the series is divergent.

e. This is a geometric progression with $a = 0.01$ and $r = -0.2$. The series is convergent since $|r| < 1$ and has value given by Eq. (5.7).

$$S_{G,\infty} = \frac{0.01}{1 + 0.2} = \frac{1}{120}$$

Of course the infinite series that arise in practice may not be in the standard form of Eq. (5.7). In such cases, it may be necessary to manipulate the summation to get it into the standard form. This is explored in Example 5.15.

EXAMPLE 5.15

Demonstrate that the following series are convergent and determine their value.

a. $\sum_{k=1}^{\infty}(-0.99)^{k+10}$

b. $\sum_{j=1}^{\infty} 2\left(\frac{1}{10}\right)^{5j}$

c. $\sum_{n=10}^{\infty}(-1)^n(0.2)^{n-7}$

d. $\sum_{i=-100}^{\infty} 10 \times (0.95)^{2i}$

Solution

In each case, the form of the summation is not consistent with a direct application of Eq. (5.7). Some further manipulation is needed before the standard expression can be applied.

a. We note that $\sum_{k=1}^{\infty}(-0.99)^{k+10} = (-0.99)^{11}\sum_{k=1}^{\infty}(-0.99)^{k-1}$ and see that this is the standard form of an infinite geometric progression with $a = (-0.99)^{11}$ and $r = -0.99$. The sequence is convergent since $|r| < 1$. The full series is evaluated as

$$\frac{(-0.99)^{11}}{1 + 0.99} \approx -0.4499$$

b. We note that $\sum_{j=1}^{\infty} 2\left(\frac{1}{10}\right)^{5j} = \sum_{j=1}^{\infty} 2 \times (1/10)^5 \times \left(\left[\frac{1}{10}\right]^5\right)^{j-1}$ and see that this is in a standard form of an infinite progression with $a = 2 \times (1/10)^5$ and $r = (1/10)^5$. The sequence is convergent since $|r| < 1$ and the sum is evaluated as

$$\frac{2 \times (1/10)^5}{1 - (1/10)^5} = \frac{2}{99{,}999}$$

c. We begin by using the substitution $i = n - 9$ which addresses the indexing of the sum such that $\sum_{n=10}^{\infty}(-1)^n(0.2)^{n-7} = \sum_{i=1}^{\infty}(-1)^{i+9}(0.2)^{i+2} = (0.2)^3\sum_{i=1}^{\infty}(-0.2)^{i-1}$.

This is easily seen as a standard infinite geometric progression with $a = (0.2)^3$ and $r = -0.2$ and so is convergent. The series is then evaluated as

$$\frac{0.2^3}{1+0.2} = \frac{1}{150}$$

d. We begin by using the substitution $k = i + 101$ which addresses the indexing of the sum. Some further manipulation leads to

$$\sum_{i=-100}^{\infty} 10 \times (0.95)^{2i} = \sum_{k=1}^{\infty} \frac{10}{0.95^{200}} (0.95^2)^{k-1}$$

which is in the standard form of an infinite geometric progression with $a = 10 \times (0.95)^{-200}$ and $r = 0.95^2$. The sequence is convergent and the series is evaluated as

$$\frac{10 \times (0.95)^{-200}}{1 - 0.95^2} \approx 2{,}926{,}000$$

5.4 TAYLOR AND MACLAURIN SERIES

5.4.1 Concepts

A very useful application of ideas developed in this chapter is the *Taylor series* of a function, often alternatively known as the *Taylor expansion* of a function. As we shall see, the Taylor series of a function is a polynomial representation of the function, with coefficients obtained from the function's properties at a particular value of the independent variable. In theory, the polynomial expansion would have an infinite number of terms, but in practice it is typically possible to *truncate* the infinite expansion to a finite number of terms. Obtaining a finite-order polynomial that approximates a function with sufficient accuracy is often the practical motivation of using Taylor expansions; we will also see some further uses later in this section.

Let us consider a general, smooth function $f(x)$ that is not necessarily a polynomial. The Taylor expansion of $f(x)$ about the point $x = a$ is defined as

$$f(x) = \sum_{n=0}^{\infty} \frac{f^{(n)}(a)}{n!} (x-a)^n \tag{5.8}$$

Note that the symbol "!" denotes the *factorial* operation which operates on integers, such that $n! = n \times (n-1) \times \cdots \times 2 \times 1$. For example, $4! = 4 \times 3 \times 2 \times 1 = 24$ and $6! = 720$. Note that by convention, $0! = 1$. Recall that the property of smoothness at $x = a$ means that all derivatives, $f^{(n)}(a)$, exist and have finite value. The Taylor expansion of $f(x)$ therefore has the form

$$f(x) = c_0 + c_1(x-a) + c_2(x-a)^2 + \cdots$$

which is clearly a polynomial in increasing powers of $(x - a)$. The coefficients are given by $c_n = \frac{f^{(n)}(a)}{n!}$. Equation (5.8) is stated without proof and the interested reader is invited to explore *Taylor polynomials* for an insight into the development of the Taylor series. Our focus in this book is the application of Taylor series.

The use of the sigma notation in Eq. (5.8) to represent a polynomial may be new to you. However, it should be clear from the discussion in this chapter and Chapter 2 that any polynomial function can be represented as

$$\sum_{n=0}^{\infty} c_n x^n$$

where c_n is the coefficient of the nth-order term. A cubic polynomial, for example, would then have $c_3 \neq 0$ and $c_n = 0$ for $n \geq 4$.

EXAMPLE 5.16

Determine the Taylor expansions about the point $x = 1$ of the following functions

a. $f(x) = \exp(x)$

b. $g(x) = \ln(1 + x)$ for $x > -1$

c. $h(x) = x^2$

Solution

We apply Eq. (5.8) at $a = 1$ in each case.

a. We note that $f^{(n)}(x) = f(x)$ and so $f^{(n)}(1) = \mathrm{e}^1$. The Taylor expansion about $x = 1$ is then,

$$\begin{aligned}\exp(x) &= \mathrm{e}^1\left(1 + (x-1) + \frac{1}{2!}(x-1)^2 + \frac{1}{3!}(x-1)^3 + \cdots\right)\\ &= \mathrm{e}^1\left(x + \frac{1}{2!}(x-1)^2 + \frac{1}{3!}(x-1)^3 + \cdots\right)\end{aligned}$$

b. We have $g'(x) = 1/(x+1)$, $g''(x) = -1/(x+1)^2$, $g'''(x) = 2/(x+1)^3$ from which the values of $g^{(n)}(1)$ can be determined as $g(1) = \ln(2)$, $g'(1) = 1/2$, $g''(1) = -1/4$, $g'''(x) = 1/4$. The Taylor expansion about $x = 1$ is then,

$$\ln(1 + x) = \ln(2) + \frac{1}{2}(x-1) - \frac{1}{8}(x-1)^2 + \cdots$$

c. We have $h'(x) = 2x$, $h''(x) = 2$, $h^{(n)}(x) = 0$ for $n \geq 3$ from which the values of $h^{(n)}(1)$ can be determined as $h(x) = 1$, $h'(1) = 2$, $h''(x) = 2$, $h^n(x) = 0$ for $n \geq 3$. The Taylor expansion about $x = 1$ is then,

$$x^2 = 1 + 2(x-1) + (x-1)^2$$

This expression is easily shown to simplify to x^2.

It is common to say that the expansion is *centered* on the value of a used to calculate the coefficients. Of course the value of a must be chosen in advance and is often determined by one's intentions for the expansion, as we shall see in later examples. The expansion resulting from the particular case that $a = 0$ is often known as the *Maclaurin series* or *Maclaurin Expansion.*

EXAMPLE 5.17

Determine the Maclaurin series of the functions in Example 5.16.

Solution

In each case, we apply Eq. (5.8) at $a = 0$ using the derivative functions obtained in Example 5.16.

a. The Maclaurin series is,

$$\exp(x) = 1 + x + \frac{1}{2!}x^2 + \frac{1}{3!}x^3 + \cdots$$

b. The Maclaurin series is,

$$\ln(1 + x) = x - \frac{x^2}{2} + \frac{x^3}{3} + \cdots$$

c. The Maclaurin series is,

$$x^2 = x^2$$

This result is what we might expect for a simple polynomial.

Examples 5.16 and 5.17 demonstrate the practical result that Taylor expansions of polynomial functions have a finite number of terms and are identical to the original polynomial, although some manipulation may be needed to show this. In addition, they show that the Taylor expansions of nonpolynomial functions are in general infinite series.

5.4.2 Convergence and function approximation

Equation (5.8) shows that a general Taylor series has no common factor and so cannot be considered as a geometric progression. The question of what conditions are needed to be met for a Taylor expansion to converge for any particular x is then difficult to answer in general. Indeed, it is necessary to consider convergence on a case-by-case basis using the *ratio test*.

The ratio test is a test for the convergence of an infinite series of the form

$$\sum_{n=0}^{\infty} b_n$$

where each term b_n is a real[2] number. The test is also sometimes known as the *Cauchy ratio test* or the *d'Alembert's ratio test*. The ratio test involves finding L such that

$$L = \lim_{n \to \infty} \left| \frac{b_{n+1}}{b_n} \right|$$

That is, L is the ratio of successive terms far along the series. The condition for convergence of the series is then that $L < 1$. If $L > 1$ the series does not converge and if $L = 1$ the test is inconclusive.

Note that the ratio test is consistent with the test used to demonstrate the convergence of a geometric progression, that is the common ratio of a geometric progression has absolute value less than unity. The ratio test is a powerful tool as it can be used in cases where there is no common ratio and this is our motivation here.

When the ratio test is used for Taylor series, b_n is a function of x such that

$$b_n = \frac{f^{(n)}(a)}{n!}(x-a)^n$$

The ratio test therefore often allows one to determine an interval of x such that the Taylor series of function $f(x)$ is convergent. This is best illustrated with an example.

EXAMPLE 5.18

Determine the interval of x such that the Maclaurin expansions of the following functions are convergent.

a. $\exp x$
b. $\ln(1+x)$

Solution

The Maclaurin expansions were calculated in Example 5.17(a) and (b) and we work from those.

a.
$$\exp(x) = 1 + x + \frac{1}{2!}x^2 + \frac{1}{3!}x^3 + \cdots = \sum_{n=0}^{\infty} \frac{x^n}{n!}$$

We therefore have $b_n = x^n/n!$ and

$$L = \lim_{n \to \infty} \left| \frac{b_{n+1}}{b_n} \right| = \lim_{n \to \infty} \left| \frac{x}{n+1} \right|$$

which is zero for all finite x. The ratio test therefore shows that the Maclaurin expansion for $\exp(x)$ is convergent for all real x.

[2] The ratio test can also be used for complex series, although this is beyond the scope of this chapter. Complex numbers will be considered in Chapter 8.

b.

$$\ln(1+x) = x - \frac{x^2}{2} + \frac{x^3}{3} + \cdots = \sum_{n=0}^{\infty} \frac{(-1)^n}{n+1} x^{n+1}$$

We therefore have $b_n = \frac{(-1)^n}{n+1} x^{n+1}$ and

$$L = \lim_{n\to\infty} \left| \frac{b_{n+1}}{b_n} \right| = \lim_{n\to\infty} \left| -x \left\{ \frac{1+1/n}{1+2/n} \right\} \right| = |x|$$

The ratio test therefore shows that the Maclaurin expansion of $\ln(1+x)$ is convergent only for $x \in (-1, 1)$.

Whether a particular expansion converges is closely related to the practical use of truncated Taylor expansions: namely that they are used a means of approximating functions close to $x = a$. Such approximations are of significant practical importance and are often the motivation for working with Taylor expansions.

By way of illustration, consider again the function $f(x) = \ln(1+x)$. Logarithm functions are difficult to work with and it may be that we would, for whatever reason, prefer to approximate it with a polynomial function in the region around $x = 0$. An obvious way to do this would be to use the Maclaurin expansion

$$\ln(1+x) = \sum_{n=0}^{\infty} \frac{(-1)^n}{n+1} x^{n+1}$$

Simply replacing the function with the convergent but still infinite series is clearly impractical. It would be much better to investigate how many terms of the expansion are required to give a level of accuracy appropriate with our particular needs. We know this Maclaurin series to be convergent for $x \in (-1, 1)$ which contains the region of interest around $x = 0$, see Example 5.18(b); each addition term of the series could therefore be interpreted as an improvement on the accuracy of a polynomial approximation of $\ln(1+x)$. That is, each additional term, with a higher power of x, will contribute toward the convergence onto the accurate value of $\ln(1+x)$ for any $x \in (-1, 1)$. To illustrate this, consider the following polynomial approximations of the function formed from an increasing number of terms in the Maclaurin series,

$$f_1(x) = x$$

$$f_2(x) = x - \frac{x^2}{2}$$

$$f_3(x) = x - \frac{x^2}{2} + \frac{x^3}{3}$$

$$\vdots$$

Table 5.1 Approximations for $\ln(1 + 0.5) = 0.4055$ formed by increasing numbers of terms in the Maclaurin expansion $f_n(x)$ evaluated at $x = 0.5$

n	$f_n(0.5)$	n	$f_n(0.5)$
1	0.5000	7	0.4058
2	0.3750	8	0.4053
3	0.4167	9	0.4055
4	0.4010	10	0.4054
5	0.4073	11	0.4055
6	0.4047	12	0.4055

The actual value the function at $x = 0.5$ is $\ln(1 + 0.5) = 0.4055$ to four decimal places, and this can be compared against the value of each $f_n(0.5)$ as shown in Table 5.1. It should be clear that additional terms are seen to move the approximation toward $f(x) = 0.4055$ which is consistent with the series being convergent for this x. Similar behavior can be seen for all $x \in (-1, 1)$ in Figure 5.2 where $\ln(1 + x)$ is plotted along with the first four approximations.

Of course if a series is not convergent at a particular x, we cannot expect additional terms in the series to better approximate the actual value of the function at that point. For example, we know that the Maclaurin expansion of $\ln(1 + x)$ is not convergent at $x = 1.5$, say, and it is evident from Figure 5.2 that increasing orders of approximation

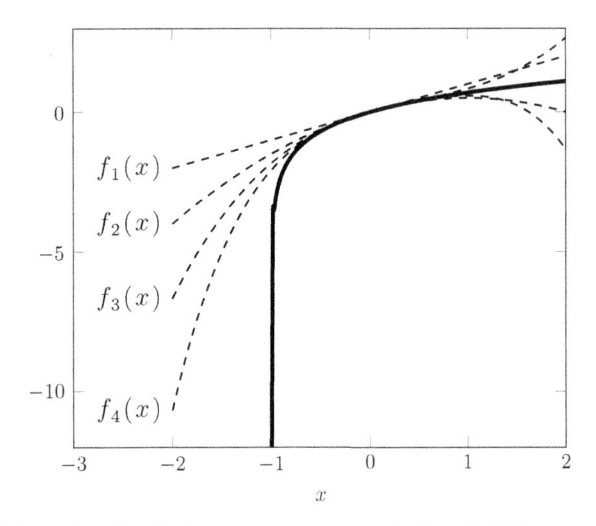

Figure 5.2 A plot of the nth-order Maclaurin expansions of $\ln(1 + x)$, $f_n(x)$, against $\ln(1 + x)$.

are not moving toward the actual value of $\ln(1+1.5) = 0.9163$. The reader is invited to demonstrate this by forming a table equivalent to Table 5.1 at $x = 1.5$.

The above discussion of the nonconvergence of the Maclaurin series of $\ln(1+x)$ for $x \notin (-1, 1)$ does not mean that it is impossible to approximate the function outside of this interval using polynomial approximations. Instead, it is necessary to work with an alternative expansion centered elsewhere. For example, we note that the function exists and is smooth at $x = 1.5$ and so the Taylor expansion centered there will also exist and can be used to approximate the function close to $x = 1.5$. This is considered in the following example.

EXAMPLE 5.19

Develop a third-order polynomial that approximates $\ln(1+x)$ close to $x = 1.5$. Use this expression to approximate the function at $x = 1.3$ and 1.6.

Solution

The Taylor series of $\ln(1+x)$ centered at $a = 1.5$ can be shown to be

$$\ln(1+x) = \ln(2.5) + \sum_{n=1}^{\infty} \frac{(-1)^{n+1}}{n \times 2.5^n}(x-1.5)^n$$

The third-order polynomial approximation is then

$$\ln(1+x) \simeq \ln(2.5) + \frac{2}{5}(x-1.5) - \frac{2}{25}(x-1.5)^2 + \frac{8}{375}(x-1.5)^3 \tag{5.9}$$

The ratio test applied to the full Taylor expansion leads to the condition for convergence that $|0.4(x-1.5)| < 1$ (see Question 5.5) and so it is possible to use this expansion to approximate the function at $x = 1.3$ and $x = 1.6$. In fact, Eq. (5.9) returns $\ln(1+1.3) \simeq 0.8531$ and $\ln(1+1.6) \simeq 0.9483$ and these values can be compared to $\ln(1+1.3) = 0.8329$ and $\ln(1+1.6) = 0.9555$, respectively.

Let us now consider the function $\exp(x)$ which was shown in Example 5.18 to have a universally convergent Maclaurin expansion. It is important to realize that although the Maclaurin expansion of $\exp(x)$ is convergent for all x, it will not necessarily be the best expansion to use to approximate $\exp(x)$ for x away from $x = 0$. Instead, if an approximation is required close to $x = 10$ for example, it would be better to use a polynomial approximation based on the Taylor expansion centered on $x = 10$. This concept is related to the *speed of convergence* of an approximation, the analysis of which is beyond the scope of this text on fundamental methods. For our purposes, the speed of convergence can be understood as being measured by the number of terms required in a polynomial expansion to achieve the required degree of accuracy. The concept is neatly illustrated in the following example.

EXAMPLE 5.20

Investigate the speed of convergence of polynomial approximations of $\exp(x)$ at locations in the interval $x \in [8, 12]$ formed from

a. the Maclaurin expansion and

b. the Taylor expansion centered on $x = 10$.

Solution

We begin by deriving the relevant expansions. After some work we find,

a. $\exp(x) = \sum_{n=0}^{\infty} \frac{x^n}{n!}$

b. $\exp(x) = e^{10} \sum_{n=0}^{\infty} \frac{(x-10)^n}{n!}$

A straight-forward method of investigating the speed of convergence is to consider the number of terms needed in each expansion to obtain the correct value of the function to two decimal places at various $x \in [8, 12]$. The results (generated by Excel) are summarized in Table 5.2. Both methods are seen to converge to the correct value, but the Maclaurin series takes significantly more terms in the expansion to achieve this. It should also be clear that the distance of x from the value of a used as the center is crucial to the speed of convergence.

Table 5.2 The number of terms required in (a) the Maclaurin expansion and (b) the Taylor expansion centered at $x = 10$ for obtaining $\exp(x)$ correct to two decimal places in Example 5.20

x	n, Case a.	n, Case b.
8	25	14
9	26	9
9.5	29	8
9.9	29	4
10.1	30	4
10.5	31	7
11	32	9
12	35	13

EXAMPLE 5.21

Derive a third-order polynomial that approximates the following functions over the interval $x \in [2, 4]$. Visually test the accuracy of your approximations.

a. $f(x) = \sin(x)$

b. $g(x) = \exp(x^2/10)$

Solution

We choose to find the Taylor expansions of the functions centered at $x = 3$, the midpoint of the interval of concern. Some work shows that the truncated Taylor expansions are

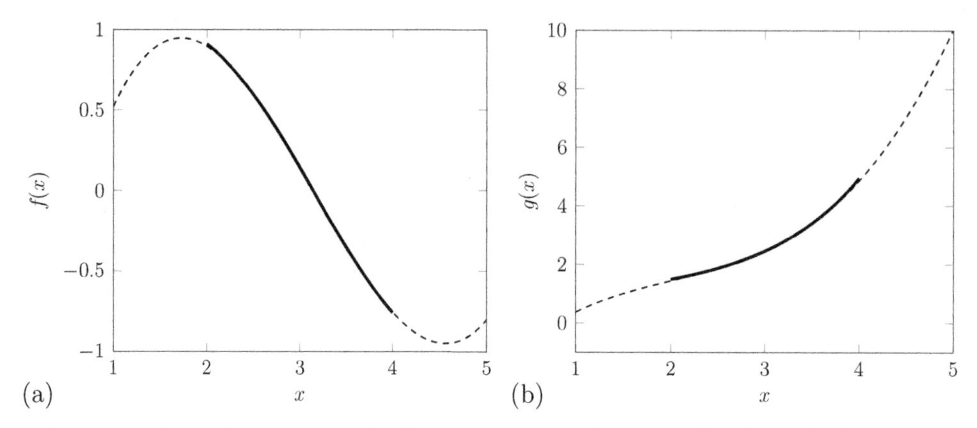

Figure 5.3 Comparisons over $x \in [2,4]$ between the actual functions (bold) and the third-order approximations centered on $x = 3$ (dashed) obtained in Example 5.21. (a) $f(x) = \sin(x)$. (b) $g(x) = \exp(x^2/10)$.

a. $f(x) \simeq \sin(3) + \cos(3)(x-3) - \frac{1}{2}\sin(3)(x-3)^2 - \frac{1}{6}\cos(3)(x-3)^3$

b. $g(x) \simeq e^{9/10}\left(1 + \frac{3}{5}(x-3) + \frac{7}{25}(x-3)^2 + \frac{12}{125}(x-3)^3\right)$

These approximations are compared to the actual functions in Figure 5.3. In both cases, remarkable agreement and no convergence issues are seen over the interval $[2,4]$.

5.4.3 Further uses

In the previous section, we saw that truncated Taylor expansions can be used to approximate functions. Although this is a significant practical use of the concept, the approximation of functions is not the only practical use of Taylor series. We discuss other useful applications here, for example, we will see how Taylor series can be used to investigate the properties of functions, including the evaluation of limits and the investigation of symmetries, and also to evaluate higher-order derivatives.

We begin with a discussion of limits and return to the limit

$$\lim_{x\to 0}\left(\frac{\sin x}{x}\right) = 1 \tag{5.10}$$

that we first encountered in Chapter 2. At that time we were unable to compute this limit algebraically and we were forced to use the long-winded approach of evaluating the function at various x approaching 0. As we will now see, the use of Taylor series provides an algebraic means of evaluating such limits. The power of a Taylor-series approach is that it allows one to express any smooth function as a polynomial and our previous discussions should have convinced you that polynomials are easy to work with.

For example, we know that the numerator of the function in Eq. (5.10) can be expanded as

$$\sin x = x - \frac{x^3}{6} + \frac{x^5}{120} + \cdots$$

and so we can immediately determine a polynomial expression for $\sin(x)/x$ as

$$\frac{\sin x}{x} = 1 - \frac{x^2}{6} + \frac{x^4}{120} + \cdots$$

It is then clear from this expression that

$$\lim_{x \to 0} \frac{\sin x}{x} = \lim_{x \to 0} \left(1 - \frac{x^2}{6} + \frac{x^4}{120} + \cdots\right)$$
$$= 1$$

These ideas are further explored in the following examples.

EXAMPLE 5.22

Use Taylor series to evaluate the following limits.

a. $\lim_{x \to 0} \frac{\sin x^2 - x^2}{x^6}$

b. $\lim_{x \to 0} \frac{e^{4x} - 1}{\sin x}$

c. $\lim_{x \to 1} \frac{\ln x}{\sqrt{x} - 1}$

d. $\lim_{x \to 3\pi} \frac{\tan x}{x - 3\pi}$

Solution

a. We need the limit at $x = 0$ and so use a Maclaurin series. In particular,

$$\sin x^2 - x^2 = x^2 - \frac{x^6}{6} + \frac{x^{10}}{120} + \cdots - x^2$$
$$= -\frac{x^6}{6} + \frac{x^{10}}{120} + \cdots$$

and so

$$\frac{\sin x^2 - x^2}{x^6} = \frac{1}{6} + \frac{x^4}{120} + \cdots$$

From this we can immediately see that

$$\lim_{x \to 0} \frac{\sin x^2 - x^2}{x^6} = -\frac{1}{6}$$

b. Again we use Maclaurin expansions of the constituent functions. In particular,

$$e^{4x} - 1 = 1 + 4x + 8x^2 + \cdots - 1$$
$$= 4x + 8x^2 + \cdots$$

and so

$$\frac{e^{4x}-1}{\sin x} = \frac{4x + 8x^2 + \cdots}{x - \frac{x^3}{6} + \cdots}$$
$$= \frac{4 + 8x + \cdots}{1 - \frac{x^2}{6} + \cdots}$$

From this we immediately see that

$$\lim_{x \to 0} \frac{e^{4x}-1}{\sin x} = 4$$

c. It will prove useful to consider Taylor expansions centered on $a = 1$. In particular,

$$\ln x = (x-1) - \frac{1}{2}(x-1)^2 + \frac{1}{3}(x-1)^3 + \cdots$$

and

$$\sqrt{x} - 1 = 1 + \frac{x-1}{2} - \frac{(x-1)^2}{8} + \cdots - 1$$
$$= \frac{x-1}{2} - \frac{(x-1)^2}{8} + \cdots$$

Therefore,

$$\frac{\ln x}{\sqrt{x}-1} = \frac{(x-1) - \frac{1}{2}(x-1)^2 + \frac{1}{3}(x-1)^3 + \cdots}{\frac{x-1}{2} - \frac{(x-1)^2}{8} + \cdots}$$
$$= \frac{1 - \frac{1}{2}(x-1) + \frac{1}{3}(x-1)^2 + \cdots}{\frac{1}{2} - \frac{x-1}{8} + \cdots}$$

and we see that

$$\lim_{x \to 1} \frac{\ln x}{\sqrt{x}-1} = 2$$

d. We use Taylor expansions centered on $a = 3\pi$. In particular,

$$\tan x = (x - 3\pi) + \frac{1}{3}(x-3\pi)^3 + \frac{2}{15}(x-3\pi)^5 + \cdots$$

and so

$$\frac{\tan x}{x - 3\pi} = \frac{(x-3\pi) + \frac{1}{3}(x-3\pi)^3 + \frac{2}{15}(x-3\pi)^5 + \cdots}{x - 3\pi}$$
$$= 1 + \frac{1}{3}(x-3\pi)^2 + \frac{2}{15}(x-3\pi)^4 + \cdots$$

We therefore see that

$$\lim_{x\to 3\pi} \frac{\tan x}{x - 3\pi} = 1$$

In addition to determining limits, Taylor expansions can be used to determine if a function possesses any symmetry. Recall from Chapter 2 that an even function is one such that $f(-x) = f(x)$ and an odd function is such that $f(-x) = -f(x)$; that is, odd and even functions exhibit some symmetry about the vertical axis at $x = 0$. It should be clear that if $f(x)$ possesses this symmetry about $x = 0$, then the Maclaurin expansion of $f(x)$ will also possess this symmetry. Going further, the Maclaurin series of an even function must consist of *even* powers of x, and the Maclaurin series of an odd function must consist of *odd* powers of x. Maclaurin series therefore provide an algebraic means of determining the symmetry of a function.

EXAMPLE 5.23

Determine whether the following functions possess any symmetry about $x = 0$.

a. $g(x) = \cos x - x^2$

b. $h(x) = \sin x^3 - x^6$

c. $k(x) = \sinh(x^2) = \frac{1}{2}\left(e^{x^2} - e^{-x^2}\right)$

Solution

In each case, we are looking for symmetry about $x = 0$ and so use Maclaurin expansions.

a. We have $\cos x = 1 - \frac{x^2}{2} + \frac{x^4}{24} + \cdots$, which consists of even powers of x and demonstrates that $\cos x$ is an even function. Furthermore,

$$\cos x - x^2 = 1 - \frac{3x^2}{2} + \frac{x^4}{24} + \cdots$$

and we see that $g(x)$ is an even function.

b. We have $\sin x^3 = x^3 - \frac{x^9}{6} + \frac{x^{15}}{120} + \cdots$, which consists of odd powers of x and demonstrates that $\sin x^3$ is an odd function. Furthermore,

$$\sin x^3 - x^6 = x^3 - x^6 - \frac{x^9}{6} + \frac{x^{15}}{120} + \cdots$$

and we see that $h(x)$ consists of both odd and even powers of x and so does not possess any symmetry about $x = 0$.

c. We have

$$e^{x^2} = 1 + x^2 + \frac{x^4}{2} + \cdots$$

$$e^{-x^2} = 1 - x^2 + \frac{x^4}{2} + \cdots$$

which demonstrates that both are even functions. Furthermore,

$$\frac{1}{2}\left(e^{x^2} - e^{-x^2}\right) = x^2 + \frac{x^6}{6} + \cdots$$

and we see that $k(x)$ is an even function.

We can go further and investigate the symmetry of a function about other vertical lines. For example, the function $f(x) = (x-1)^2$ is not even about $x = 0$, but possesses an even-type symmetry about $x = 1$. The symmetry about a particular $x = a$ is immediately obvious from polynomials formed of powers of $(x - a)$, for example, $(x - 4)$ has an odd-type symmetry about $x = 4$, as does $(x-4) + (x-4)^3$. However, such symmetries can be less obvious for nonpolynomial functions. Taylor series provide a convenient means of algebraically investigating symmetries of nonpolynomial functions; this is demonstrated in the following example.

EXAMPLE 5.24

Determine whether the following functions are symmetrical about the stated value of x.

a. $l(x) = \cos\left(x^2 - 4x + 4\right)$ about $x = 2$.

b. $m(x) = \sin\left(x^3 - 3x^2 + 3x - 1\right)$ about $x = 1$.

c. $n(x) = \frac{\ln(x^2+4)}{x-10}$ about $x = 10$.

Solution

a. We require the Taylor expansion centered on $a = 2$.

$$\cos(x^2 - 4x + 4) = 1 - \frac{1}{2}(x-2)^4 + \frac{1}{24}(x-2)^8 + \cdots$$

which consists of even powers of $(x - 2)$ and so $l(x)$ has an even-type symmetry about $x = 2$.

b. We require the Taylor expansion centered on $a = 1$.

$$\sin(x^3 - 3x^2 + 3x - 1) = (x-1)^3 - \frac{1}{6}(x-1)^9 + \frac{1}{120}(x-1)^{15} + \cdots$$

which consists of odd powers of $(x - 1)$ and so $m(x)$ has an odd-type symmetry about $x = 1$.

c. We require the Taylor expansion centered on $a = 10$.

$$\frac{\ln(x^2+4)}{x-10} = \frac{\ln(104)}{x-10} + \frac{5}{26} - \frac{3(x-10)}{338} + \cdots$$

which consists of a mix of powers of $(x - 10)$ and so $n(x)$ has no symmetry about that $x = 10$. In addition, the first term of the Taylor expansion shows that there is a singular point at $x = 10$.

The final application of Taylor series that we will consider here is in relation higher-order derivatives. In principle, we know from Chapter 4 that it is possible to determine $f^{(n)}(0)$ for any smooth function $f(x)$ for any $n \in \mathbb{N}^+$ by repeatedly taking the derivative of $f(x)$ then $f'(x)$, etc., the required number of times. For example, if $f(x) = \mathrm{e}^{4x^3}$ we have

$$
\begin{aligned}
f'(x) &= 12x^2\mathrm{e}^{4x^3} \\
f''(x) &= 24\mathrm{e}^{4x^3}(6x^4 + x) \\
f'''(x) &= 24\mathrm{e}^{4x^3}(72x^6 + 36x^3 + 1) \\
&\vdots
\end{aligned}
$$

However, how can one efficiently determine an expression for $f^{(90)}(x)$, for example? Although some patterns in the derivatives of particular $f(x)$ can be exploited, we cannot quickly arrive at expressions for higher-order derivatives of general $f(x)$. It is, however, sometimes possible to efficiently determine the *value* of $f^{(n)}(a)$ using a Taylor-series approach.

Let us consider $f(x) = \mathrm{e}^{4x^2}$. Using the standard expansion $\mathrm{e}^x = \sum_{k=0}^{\infty} \frac{x^k}{k!}$ it is easy to determine that the Maclaurin expansion of $f(x)$ is

$$f(x) = \mathrm{e}^{4x^3} = 3\sum_{k=0}^{\infty} \frac{4^k(x^3)^k}{k!}$$

Now we can equate this to Eq. (5.8) with $a = 0$ and see that

$$3\sum_{k=0}^{\infty} \frac{4^k(x^3)^k}{k!} = \sum_{n=0}^{\infty} \frac{f^{(n)}(0)}{n!}x^n$$

The value of $f^{(90)}(0)$ would therefore appear on the right-hand side when $n = 90$, which, by equating the coefficients of the relevant power of x, corresponds to $k = 90/3 = 30$ on the left-hand expression. We therefore have

$$3 \times \frac{4^{30}}{30!} = \frac{f^{(90)}(0)}{90!}$$

and so $f^{(90)}(0) = 3 \times 4^{30} \times \frac{90!}{30!}$. This result can be confirmed using Wolfram Alpha.

This Taylor-series approach for evaluating the nth derivative of $f(x)$ relies on the existence of an x^n term in the Taylor expansion; this is not always the case and the approach is therefore of limited practical use. For example, we have seen that odd functions will not have any even-powered terms in their Maclaurin expansion and so $f^{(n)}(0)$ of an odd function cannot be evaluated in this way for even n.[3] The issue is not, however, restricted to odd/even functions and we have seen many examples where particular powers of x do not appear in the Maclaurin expansion of a function.

EXAMPLE 5.25

Use a Taylor-expansion approach to evaluate the following

a. the 112th derivative of $g(x) = \cos(x^2)$ at $x = 0$

b. the 545th derivative of $h(x) = xe^{x^2}$ at $x = 0$

c. the 46th derivative of $m(x) = \ln(4x^2 + 1)$ at $x = 0$

d. the 346th derivative of $p(x) = \sin\left(x^3\right)$ at $x = 0$.

Solution

a. We determine the Maclaurin expansion as

$$\sum_{k=0}^{\infty} \frac{(-1)^k (x^2)^{2k}}{(2k)!} = \sum_{n=0}^{\infty} \frac{g^{(n)}(0)}{n!} x^n$$

After equating coefficients of relevant powers of x we see that $n = 128$ corresponds to $k = 28$. Therefore,

$$\frac{(-1)^{28}}{(2 \times 28)!} = \frac{g^{(112)}(0)}{112!} \quad \Rightarrow \quad g^{(112)}(0) = \frac{112!}{56!}$$

b. We determine the Maclaurin expansion as

$$\sum_{k=0}^{\infty} \frac{x^{2k+1}}{k!} = \sum_{n=0}^{\infty} \frac{h^{(n)}(0)}{n!} x^n$$

After equating coefficients of relevant powers of x we see that $n = 545$ corresponds to $k = 272$. Therefore,

$$\frac{1}{272!} = \frac{h^{(545)}(0)}{545!} \quad \Rightarrow \quad h^{(545)}(0) = \frac{545!}{272!}$$

c. We determine the Maclaurin expansion as

$$-\sum_{k=1}^{\infty} \frac{(-4)^k (x^2)^k}{k} = \sum_{n=0}^{\infty} \frac{m^{(n)}(0)}{n!} x^n$$

After equating coefficients of relevant powers of x we see that $n = 46$ corresponds to $k = 23$. Therefore,

$$\frac{4^{23}}{23} = \frac{m^{(46)}(0)}{46!} \quad \Rightarrow \quad m^{(46)}(0) = \frac{4^{23} \times 46!}{23}$$

[3] Of course this does not mean that the even derivatives of an odd function do not exist.

d. We determine the Maclaurin expansion as

$$\sum_{k=0}^{\infty} \frac{(-1)^k}{(2k+1)!} x^{6k+3} = \sum_{n=0}^{\infty} \frac{p^{(n)}(0)}{n!} x^n$$

The left-hand expression possesses only odd powers of x and so we are unable to use the Taylor-series approach to calculate $p^{(346)}(0)$, or indeed $p^{(n)}(0)$ for any even n.

Similar ideas can often be applied with Taylor series centered on $x = a$ to determine $f^{(n)}(a)$, the only difference being that the coefficients of $(x - a)^n$ are equated.

We will return to an additional use of Taylor expansions in the next chapter.

5.5 SERIES AND SUMMATIONS ON YOUR COMPUTER

As in previous chapters, much of the material covered here can be further explored with Wolfram Alpha. The engine is also a useful means of checking answers that you may have obtained from pen-and-paper approaches. You are invited to spend a little time exploring commands that are relevant to the material in this chapter.

A key aim of this chapter has been the evaluation of series and summations. Wolfram Alpha's command `sum` is of relevance to this. For example, the finite series

$$\sum_{i=1}^{100} (i^2 - 2i)$$

can be evaluated using the command

```
sum i^2-2i between i=1 and 100
```

Wolfram Alpha reports back its interpretation of our request, which we see to be correct, and also returns the numerical value as 328,250.

The same command can be used to verify or derive closed-form expressions for summations for more general series. For example, the command

```
sum i^2-2i between i=1 and k
```

can be used to determine that

$$\sum_{i=1}^{k} (i^2 - 2i) = \frac{1}{6} k(k+1)(2k-5)$$

The commands can be easily modified to deal with summations, that is a series that starts at some indexing value not equal to 1. For example, the command

```
sum 0.2k between k=-10 and 5
```

returns the result that

$$\sum_{k=-10}^{5} 0.2k = -8$$

The sums of infinite sequences can also be evaluated using the commands `infty` or `infinity` for infinity. Recall that the issue of convergence was of particular importance for infinite summations and Wolfram Alpha will state if a requested infinite sum is not convergent. For example, the command

```
sum 0.2j between j=1 and infty
```

returns that the sum $\sum_{j=1}^{\infty} 0.2j$ does not converge. However, the command

```
sum 0.2^j between j=1 and infty
```

returns that

$$\sum_{j=1}^{\infty} 0.2^j = 0.25$$

The engine also explicitly states that this is a convergent geometric series and gives a closed-form expression of the *partial sum*, that is the equivalent sum over a finite number of terms,

$$\sum_{j=1}^{n} 0.2^j = -0.25(0.2^n - 1)$$

Wolfram Alpha can also be used to determine Taylor and Maclaurin expansions of functions. For example, the Taylor series of $f(x) = \sin\left(x^2 - 2\right)$ about $x = 1$ is obtained from the command

```
Taylor expansion sin(x^2-2) about x=1
```

Maclaurin expansions can also be obtained. This can be done either from the command

```
Taylor expansion sin(x^2-2) about x=0
```

or

```
Maclaurin expansion sin(x^2-2)
```

EXAMPLE 5.26

Use Wolfram Alpha to perform the following.

a. Evaluate $\sum_{i=50}^{150} (i^2 + 2\sin(\pi i))$.

b. Evaluate $\sum_{i=-10}^{10} \sin\left(\frac{\pi i}{2}\right)$.

c. Evaluate the sum of the first 20 terms of a geometric progression with starting value 10 and common ratio 1.2.

d. Determine a closed-form expression for $\sum_{i=n}^{k}(2i-\pi)$ for $k > n \in \mathbb{N}^+$.

e. Determine the Taylor expansion of $f(y) = e^{y^2-3}$ about $y = 1$.

Solution

a. We use the command

```
sum i^2+2sin(pi*i) between i=50 and 150
```

and obtain that

$$\sum_{i=50}^{150}(i^2 + 2\sin(\pi i)) = 1{,}095{,}850$$

b. We use the command

```
sum sin(i*pi/2) between i=-10 and 10
```

and obtain that

$$\sum_{i=-10}^{10} \sin\left(\frac{\pi i}{2}\right) = 0$$

c. We use the command

```
sum 10*1.2^(i-1) between i=1 and 20
```

and obtain that

$$\sum_{i=1}^{20} 10 \times 1.2^{i-1} \approx 1866.88000$$

d. We use the command

```
sum (2i-pi) between i=n and k
```

and obtain that

$$\sum_{i=n}^{k}(2i-\pi) = (k-n+1)(k+n-\pi)$$

e. We use the command

```
Taylor expansion of e^(y^2-3) about y=1
```

and obtain that

$$g(y) = \frac{1}{e^2}\left(1 + 2(y-1) + 3(y-1)^2 + 10(y-1)^3 + \cdots\right)$$

5.6 ACTUARIAL APPLICATION: ANNUITIES

In our previous discussions of the time value of money, we have been limited to single or small numbers of individual cash flows. However, in practice, real investment opportunities and business ventures typically consist of many, many cash flows. With this in mind, Actuaries have developed standard notation for the properties of many basic "building blocks" that can be used to describe mathematically much more complicated cash flow structures. The most fundamental building block is called a *level annuity* and the process for determining its present value is a nice application of the ideas and concepts described in this chapter. While there are many such building blocks, we will be concerned only with level annuities and related issues here.

A *unit level n-year annuity payable in arrears* is a precise description of the most fundamental "building block." In particular, it represents a stream of payments each of 1 (i.e., level and of unit amount), made at the end of each year (i.e., in arrears) for n years. The particular case of a 5-year annuity is shown in Figure 5.4; note that a total of 5 is paid in five equal installments at the end of each year.

The present value (at $t = 0$) of an n-year unit annuity paid in arrears is denoted $a_{\overline{n}|}$ in standard actuarial notation. It should be clear that, from first principles, the value of $a_{\overline{5}|}$ is obtained from the sum of the present value of each individual payment. That is, for some fixed compounding interest rate i,

$$a_{\overline{5}|} = \frac{1}{(1+i)} + \frac{1}{(1+i)^2} + \frac{1}{(1+i)^3} + \frac{1}{(1+i)^4} + \frac{1}{(1+i)^5}$$

which is easily identified as a finite geometric progression. In the notation of Section 5.3, the progression has an initial value $a = \frac{1}{(1+i)}$ and a common ratio $r = \frac{1}{(1+i)}$. The sum of this is therefore given by

$$S_{G,5} = \frac{a(1-r^5)}{1-r} = \frac{1}{(1+i)} \frac{1-(1+i)^{-5}}{1-\frac{1}{(1+i)}}$$

and, after a little manipulation, we find that

$$a_{\overline{5}|} = \frac{1-(1+i)^{-5}}{i}$$

This process can be generalized to give the present value of an n-year unit annuity paid in arrears as

$$a_{\overline{n}|} = \frac{1-(1+i)^{-n}}{i} \tag{5.11}$$

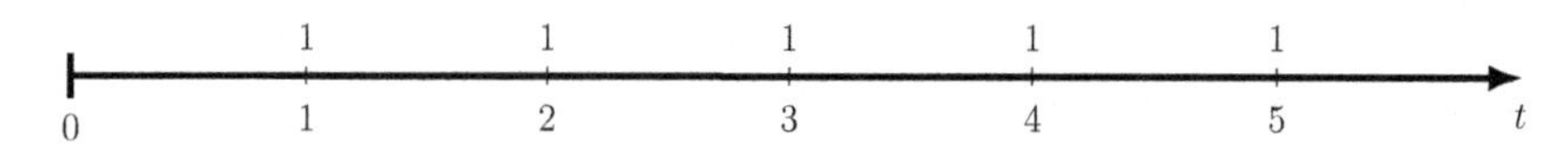

Figure 5.4 The cash flows in a 5-year unit level annuity payable in arrears.

EXAMPLE 5.27

Calculate the present value of the following cash flows if $i = 6\%$ per annum.

a. \$1 payable at the end of each of the next 3 years.

b. \$1000 payable at the end of each of the next 50 years.

Solution

Using appropriate units, both cash flows become examples of n-year unit annuities payable in arrears and Eq. (5.11) can be used.

a. Working in units of \$1, we have,

$$a_{\overline{3}|} = \frac{1-(1+i)^{-3}}{i} = \frac{1-(1.06)^{-3}}{0.06} = 2.67301$$

That is, the cash flow has a present value of \$2.67.

b. Working in units of \$1000, we have,

$$a_{\overline{50}|} = \frac{1-(1+i)^{-50}}{i} = \frac{1-(1.06)^{-50}}{0.06} = 15.76186$$

That is, the cash flow has a present value of \$15,761.86.

Clearly real cash flows are rarely of unit amounts. It is therefore useful to realize that Eq. (5.11) can be scaled to give the present value of any level payment stream payable in arrears. For example, if an amount X is paid at the end of each year, the present value of the associated n-year annuity is given by

$$Xa_{\overline{n}|} = X\left(\frac{1-(1+i)^{-n}}{i}\right)$$

EXAMPLE 5.28

Calculate the present value of the following cash flows if $i = 2\%$ per annum.

a. £45 payable at the end of each of the next 10 years.

b. £14,563 payable at the end of each of the next 25 years.

Solution

Both cash flows are examples of n-year annuities payable in arrears.

a. We have,

$$45a_{\overline{10}|} = 45\left(\frac{1-(1+i)^{-10}}{i}\right) = 45\left(\frac{1-(1.02)^{-10}}{0.02}\right) \approx 404.22$$

That is, the cash flow has a present value of £404.22

b. We have,

$$14{,}563a_{\overline{25}|} = 14{,}563\left(\frac{1-(1+i)^{-25}}{i}\right) = 14{,}563\left(\frac{1-(1.02)^{-25}}{0.02}\right) \approx 284{,}320.10$$

That is, the cash flow has a present value of £284,320.10.

EXAMPLE 5.29

An investment project is expected to return fixed dividends of \$10,000 per annum at the end of each of the next 15 years. Calculate the maximum price to pay for this investment if an investor is seeking an annual yield of at least 5% per annum.

Solution

The price P equals the present value, at $i = 0.05$, of the dividend payments. That is,

$$P = 10{,}000 a_{\overline{15}|} = 10{,}000 \left(\frac{1 - (1.05)^{-15}}{0.05} \right) \approx 103{,}796.58$$

Therefore, if the investor pays \$103,796.58 for this investment, he would obtain an annualized yield of 5% per annum. If he pays less than this price, the yield will be higher. P is therefore the maximum he should pay to secure a yield of at least 5% per annum.

The approach taken to derive $a_{\overline{n}|}$ can be extended to determine the present values of payments in arrears that, rather than being fixed, increase at some fixed percentage. For example, consider the cash flow illustrated in Figure 5.5. In this case, the amount paid starts at 1 but increases by 4% each year. It should be clear that $a_{\overline{n}|}$ does not apply in this case because the cash flows are not level. Instead we are forced to begin at first principles and write the present value of the cash flow stream as

$$\frac{1}{(1+i)} + \frac{1.04}{(1+i)^2} + \frac{1.04^2}{(1+i)^3} + \frac{1.04^3}{(1+i)^4}$$
$$= \frac{1}{1.04} \left[\frac{1.04}{1+i} + \left(\frac{1.04}{1+i} \right)^2 + \left(\frac{1.04}{1+i} \right)^3 + \left(\frac{1.04}{1+i} \right)^4 \right]$$

We now note that the terms in the square brackets on the right-hand side give another example of a finite geometric progression. This particular progression has initial value $a = \frac{1.04}{1+i}$ and common factor $r = \frac{1.04}{1+i}$ and its sum can be evaluated for some value of i. This is shown in the following example.

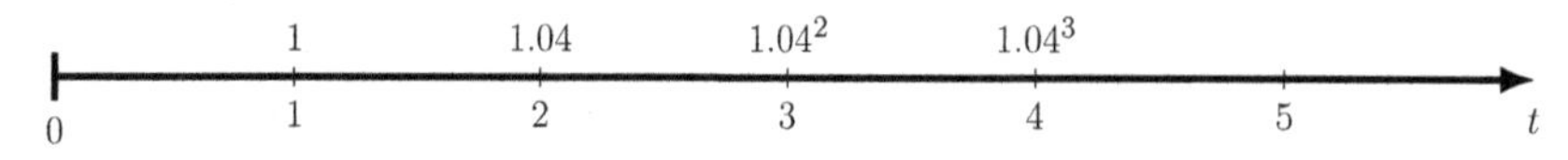

Figure 5.5 Cash flows for a 4-year payment stream with payments increasing at a fixed 4% per annum.

EXAMPLE 5.30

Determine the present value of the cash flow given in Figure 5.5 if $i = 6\%$ per annum.

Solution

We have $i = 0.06$ and so, in the previous expression, the starting value and common factor in the bracketed progression are both $a = r = 1.04/1.06 \approx 0.9811$. Using the standard expression for $S_{PG,4}$, we find the present value of the cash flow to be,

$$\frac{1}{1.04}\left(\frac{0.9811 \times (1 - 0.9811^4)}{1 - 0.9811}\right) \approx 3.67$$

In practice, not all cash flows will begin in the first year. For example, it is possible to conceive of a *deferred* level cash flow with first payment at, say, $t = 4$ as shown in Figure 5.6. It should be clear that is a deferred 7-year unit level annuity. The present value will be related to $a_{\overline{7}|}$; however, this alone would give the present value of a payment stream as at $t = 3$ rather than $t = 0$. In order to get the present value at $t = 0$, it is necessary to discount $a_{\overline{7}|}$ a further 3 years and we obtain the present value

$$\frac{1}{(1+i)^3}a_{\overline{7}|}$$

Note that we wish to work in terms of an annuity paid in arrears; therefore the first payment at $t = 4$ is considered to be paid at the end of the year beginning at $t = 3$.

EXAMPLE 5.31

Calculate the present value at $t = 0$ of a 10-year payment stream consisting of \$500 per year. The first payment is at $t = 5$, the second is at $t = 6$, etc. The interest rate is fixed at $i = 7\%$ per annum and time is measured in years.

Solution

We can consider the payment stream to be a 10-year annuity of \$500 per annum paid in arrears and deferred for 4 years. The present value is therefore given by

$$\frac{500}{1.07^4}a_{\overline{10}|} = \frac{500}{1.07^4}\left(\frac{1 - (1.07)^{-10}}{0.07}\right) \approx 2679.13$$

That is, the payment stream has a present value of \$2679.13.

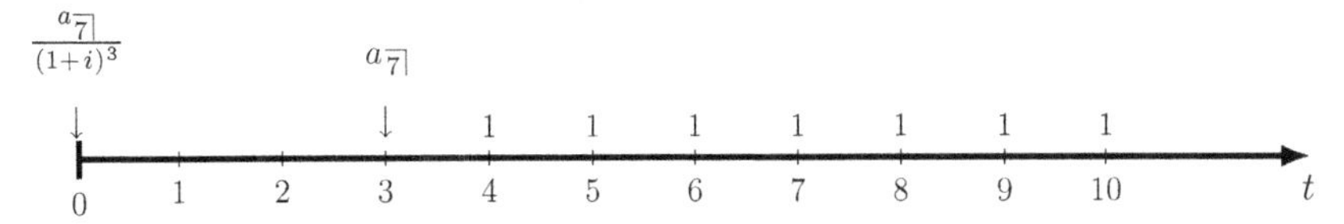

Figure 5.6 A deferred level cash flow.

A full discussion of the other "building blocks" for the analysis of more complicated cash flows is beyond the scope of this book and we do not discuss any other annuities beyond the fundamental $a_{\overline{n}|}$. However, we demonstrate some further practical uses of $a_{\overline{n}|}$ in the following examples.

EXAMPLE 5.32

Under the *discounted cash flow model*, the price of a single share in a company is given by the present value of all future expected dividends. Assuming that a particular share is expected to pay a fixed dividend of \$2 per annum in perpetuity (i.e., for ever) with the first payment due in exactly 1 year's time, calculate the maximum price an investor should pay if he requires a yield of at least 4% per annum.

Solution

It should be clear that the cash flow is a level annuity paid in arrears and so the maximum price will be given by $2a_{\overline{n}|}$ evaluated at $i = 0.04$. The problem, however, it that the value of n is unclear. The question states that the share is expected to pay in perpetuity, that is we can consider $n \to \infty$. While this might appear to be an unrealistic assumption, it is often made and is very useful in practice. The required price is therefore given by

$$2 \lim_{n\to\infty} a_{\overline{n}|} = 2a_{\overline{\infty}|}$$

Recall that $a_{\overline{n}|}$ is the sum of a geometric progression with common factor $r = \frac{1}{1+i}$. Furthermore, we know that $i > 0$ and so $|r| < 1$; the progression is therefore convergent. That is, $a_{\overline{\infty}|}$ is expected to have a finite value for all $i > 0$. In fact, the finite value is given by

$$a_{\overline{\infty}|} = \lim_{n\to\infty} \left(\frac{1-(1+i)^{-n}}{i} \right) = \frac{1}{i}$$

and so, in this case, the value of the share price is given by

$$P = \frac{2}{i} = \frac{2}{0.04} = 50$$

We therefore conclude that, if an investor wishes to obtain a yield of at least 4% per annum, the maximum price he should pay for the share is \$50.

A level unit annuity paid for ever is typically called a *perpetuity*. We would therefore say that share in this example has been assumed to be a perpetuity of \$2 per annum paid in arrears.

EXAMPLE 5.33

An investment is expected to result in payments of \$5000 at times $t = 1, 2, 3,$ and 4 and \$6000 at times $t = 5, 6,$ and 7, where t is measured in years. Calculate the maximum price an investor should pay for this investment if he requires a yield of at least 5.67% per annum.

Solution

The cash flows for this investment are illustrated in Figure 5.7. We see that the problem naturally falls into two-level annuities: a 4-year annuity in arrears of 5 per annum and a 3-year annuity in arrears of 6 per annum but deferred for 4 years. The combined present value therefore gives the price as

$$P = 5a_{\overline{5}|} + \frac{6}{(1+i)^4} a_{\overline{3}|}$$

This can be evaluated at $i = 0.0567$ to give $P \approx 33.765$. That is, for the yield required, the maximum price for the investment would be approximately \$33,765.

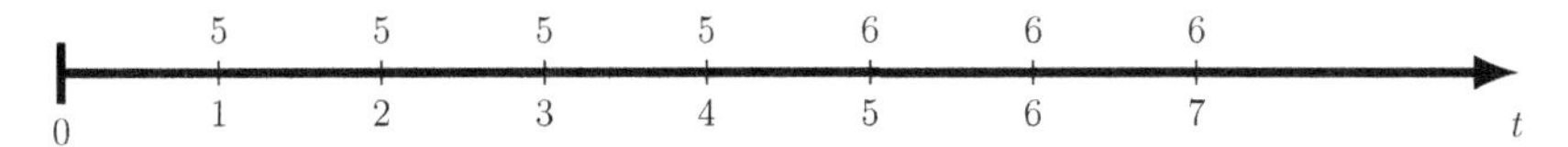

Figure 5.7 The cash flows resulting from the investment in Example 5.33. Monetary units of \$1000 have been used.

EXAMPLE 5.34

A project requires an initial investment of \$100,000 and returns \$12,000 at the end of each of the next 10 years. Calculate the annualized yield obtained on this project.

Solution

This case is different to those in the above examples in that we are required to obtain i, rather determine the net present value for some given i. Working in units of \$1000, we can equate the present values of the inflows and outflows to obtain,

$$100 = 12a_{\overline{10}|} \quad \Rightarrow \quad \frac{100}{12} = \frac{1-(1+i)^{-10}}{i}$$

The problem is now reduced to finding the value of i that satisfies this equation. We begin by noting that a total of $10 \times \$12{,}000 = \$120{,}000$ is obtained from an investment of \$10,000 and so we should expect a positive yield. Strategies for obtaining this by hand are explored in Chapter 13 and in the interim we note that either Excel's Goalseek function or Wolfram Alpha can be used. In particular, using Wolfram Alpha and the command

```
solve 100/12 = (1-(1+i)^(-10))/i
```

leads to $i \approx 3.46\%$ per annum. Note that Wolfram Alpha asks to confirm that i should be interpreted as variable rather than the imaginary number discussed in Chapter 8.

5.7 QUESTIONS

The following questions are intended to test your knowledge of the material discussed in this chapter. Full solutions are available in Chapter 5 Solutions of Part III. You should use an algebraic approach unless otherwise stated.

Question 5.1. State the rule x_n for $n \in \mathbb{N}^+$ for the following sequences. Use this rule to extend the sequences by a further three terms.

a. $\left\{1, \frac{1}{4}, \frac{1}{9}, \frac{1}{16}, \ldots\right\}$
b. $\{3, 6, 11, 18, \ldots\}$
c. $\{-2, 4, -8, 16, \ldots\}$
d. $\{2, 7, 8, 13, \ldots\}$

Question 5.2. You are given that

$$\sum_{i=1}^{100} (0.1i)^2 = \frac{6767}{2} \quad \text{and} \quad \sum_{i=1}^{10} (-0.2)^i = -\frac{1{,}627{,}604}{9{,}765{,}625}$$

Determine values for the following summations.

a. $\sum_{i=1}^{100} (1 + (0.2i)^2)$
b. $\sum_{k=1}^{100} (-0.2)^{i+10}$
c. $\sum_{i=1}^{20} (-0.2)^i$
d. $\sum_{j=-100}^{-1} j^2 + \sum_{k=6}^{15} (-0.2)^{k-6}$

Question 5.3. State whether the following series have a finite or infinite value. Calculate the finite value where appropriate.

a. $\sum_{i=1}^{10} 10 \times 1.2^{i-1}$
b. $\sum_{i=1}^{15} (5 + 2i)$
c. $\sum_{i=1}^{\infty} \frac{\pi}{2} \times \left(\frac{1}{2}\right)^i$
d. $\sum_{i=1}^{\infty} 2n$

Question 5.4. The following rules give the nth term in an infinite sequence, where $n \in \mathbb{N}^+$. Determine which associated infinite series are convergent.

a. $x_n = n + 1$
b. $x_n = 2^{n+1}$
c. $x_n = (-1)^n 0.99^{n-2}$
d. $x_n = 2n + 0.2^n$

Question 5.5. Determine the range of x over which the Taylor expansion of $\ln(1 + x)$ about $x = 1.5$ will converge. Note that this result was stated but not proved in Example 5.19.

Question 5.6. Determine the following limit algebraically.

$$\lim_{\delta \to 0} \left(\frac{\cos \delta - 1}{\delta} \right)$$

Note that this limit was considered using a tabular approach in Question 3.4.

Question 5.7. Determine the Maclaurin expansions of the following functions and determine the range of x for which each expansion is convergent.

a. $f(x) = e^{2x^3}$

b. $g(x) = \frac{\sin x}{x}$

Question 5.8. The notation $\ddot{a}_{\overline{n}|}$ is used to denote the present value of an n-year level unit annuity paid in *advance* (i.e., at the *start* of each year). Derive an expression for $\ddot{a}_{\overline{n}|}$ in terms of some fixed compounding interest rate i.

Question 5.9. An investment guarantees dividends of £10,000 at times $t = 1, 2, \ldots, 5$, followed by £20,000 increasing at a fixed rate of 5% per annum at times $t = 6, 7, \ldots, 10$. Determine the maximum price an investor should pay to secure an annualized yield of 6% per annum. Note that t is measured in years.

Question 5.10. In return for a capital investment of $500,000, a manufacturing company is able to increase production and generate an additional $20,000 at the end of each of all future years, increasing with inflation. Assuming that inflation will remain at a constant 4% per annum for ever more and that production can continue indefinitely, estimate the annualized yield obtained from the capital investment.

CHAPTER 6

Integral Calculus I

Contents

Prerequisite knowledge	Intended learning outcomes

• Chapter 1
• Chapter 2
• Chapter 3
• Chapter 4
• Chapter 5
	• Define and recognize an indefinite integral
	• Recognize indefinite integration as the reverse operation of differentiation
	• Determine the indefinite integrals of the standard functions
	• Determine the indefinite integrals of more complication functions, using
	  ◦ the change of variables approach
	  ◦ integration by parts

In this chapter, we introduce the concept of *integration*. The approach taken here is to consider integration from two different perspectives which, although will lead to fundamentally the same mathematical results, will give alternative insights into this branch of calculus. The two approaches will necessarily lead to us distinguishing between *indefinite* and *definite* integrals. Indefinite integrals are the subject of this chapter and definite integrals are considered in Chapter 7.

The chapter begins by presenting integration as the reverse operator of differentiation; we then determine the indefinite integrals of the standard classes of functions introduced in Chapter 2. Strategies for determining the indefinite integrals of more complicated functions that are likely to appear in practice are then developed.

6.1 INDEFINITE INTEGRALS OF BASIC FUNCTIONS

Perhaps the easiest way to approach *integration* is to consider it as the reverse operation of differentiation. We have already seen reverse operations: division is the reverse operation of multiplication, and subtraction is the reverse operation of addition, for example. Introducing integration as the reverse operation of differentiation is entirely analogous.

To demonstrate this we consider $g(x) = x^4$ and note that

$$\frac{\mathrm{d}g}{\mathrm{d}x} = 4x^3$$

If integration is the reverse operation of differentiation we would expect the *integral* of $g'(x) = 4x^3$ to return the original $g(x)$. However, things are slightly more subtle as we shall now see.

The first thing to discuss is the standard notation used to denote integration. In the same way that we need to state the variable about which a derivative is taken, we need to indicate the variable about which an integral is performed. With this in mind, the full notation for an integral of function $g(x)$, say, with respect to x is

$$\int g(x)\,\mathrm{d}x$$

The quantity being integrated, in this case $g(x)$, is called the *integrand* and the integration symbol $\int$ can be understood as an elongated "S" standing for "sum." The relationship between integrals and sums will become clear in Chapter 7. We therefore write the integral of $g'(x) = 4x^3$ with respect to x as

$$\int 4x^3\,\mathrm{d}x \tag{6.1}$$

While at this stage it may seem pointless to explicitly state that we are integrating a single-variable function with respect to its single variable, the notation is important when integrating constants (where there is no variable, see Example 6.2) and in *multi-variate* calculus which we consider in Chapter 11.

To evaluate the integral in Eq. (6.1) we define integration as the reverse operation of differentiation and restate the integral as the question "what function is differentiated with respect to x to give $4x^3$?" The properties of polynomials immediately lead to the answer x^4. However, we also note that the differential of $x^4 + 2$ is $4x^3$, as is the differential of $x^4 - 100$. In fact, the differential of any function of the form $x^4 + c$ is $4x^3$ for all values of constant c. The resulting integral is therefore not precisely known and we write

$$\int 4x^3 \, dx = x^4 + c$$

This requirement for an arbitrary constant, c, means that the integral is *indefinite*. The need for the constant can be intuitively understood if we recall that the derivative of a function gives its gradient. Clearly the gradient of a function is a property of its shape and is immune to shifting the function upwards or downwards, which is of course the action of adding a constant.

The above discussion regarding the connection between the derivatives and the indefinite integrals can be stated mathematically as

$$\frac{df(x)}{dx} = f'(x) \quad \Longleftrightarrow \quad \int f'(x) \, dx = f(x) + c \tag{6.2}$$

EXAMPLE 6.1

Determine the following indefinite integrals.

a. $\int x^2 \, dx$
b. $\int y^{9.2} \, dy$
c. $\int z^{-5} \, dz$

Solution

We use Eq. (6.2) in each case.

a. Note that x^3 is differentiated with respect to x to give $3x^2$ and so we must have

$$\int x^2 \, dx = \frac{x^3}{3} + c$$

b. Note that $y^{10.2}$ is differentiated with respect to y to give $10.2y^{9.2}$ and so we must have

$$\int y^{9.2} \, dy = \frac{y^{10.2}}{10.2} + c$$

c. Note that z^{-4} is differentiated with respect to z to give $-4z^{-5}$ and so we must have

$$\int z^{-5} \, dz = -\frac{z^{-4}}{4} + c$$

The results of Example 6.1 suggest a general rule for integrating polynomial terms of order n,

$$\int x^n \, dx = \frac{x^{n+1}}{n+1} + c \tag{6.3}$$

This can be summarized in words as

the integral of a polynomial term is obtained by adding 1 to the power and dividing by the new power.

The rule extends to negative values of n and so it can also be considered a rule for very basic rational function terms with constant numerator. Of course, we cannot divide by 0 and so we must restrict the rule to exclude the particular case that $n = -1$; we will return to the integral of $1/x$ in Section 6.2. Equation (6.3) is, however, perfectly applicable in the case of $n = 0$ and gives the rule for integrating a constant.

EXAMPLE 6.2

Determine the following indefinite integrals.

a. $\int 4\,dx$

b. $\int -2\,dy$

c. $\int 7\,dz$

Solution

We can write the integrand, 4, as either $4x^0$, $4y^0$, or $4z^0$ and so, by Eq. 6.3 with $n = 0$, the integrals are as follows.

a. $\int 4\,dx = 4x + c$

b. $\int -2\,dy = -2y + c$

c. $\int 7\,dz = 7z + c$

It should be clear from our knowledge of differentiation that constant multipliers can be brought outside of the integration and that integration is *distributive* across addition. That is

$$\int af(x)\,dx = a\int f(x)\,dx \quad \text{and}$$

$$\int \left(f(x) + g(x) + \cdots\right) dx = \int f(x)\,dx + \int g(x)\,dx + \cdots$$

Here, a is any constant and "$\cdots$" denotes that the expression can be extended to the sum of any number of functions in the integrand. These properties can be combined to give the more general result that

$$\int \left(af(x) + bg(x) + \cdots\right)\,dx = a\int f(x)\,dx + b\int g(x)\,dx + \cdots \tag{6.4}$$

where b is another constant. If the individual functions in Eq. (6.4) are considered as simple polynomial terms, the rule provides the means to integrate more general polynomial functions.

EXAMPLE 6.3

Determine the following indefinite integrals.

a. $\int \left(3x^2 + 2x + 1\right)\, \mathrm{d}x$

b. $\int \left(2\sqrt{x} + \frac{2}{x^2}\right)\, \mathrm{d}x$

c. $\int \left(\frac{2}{\sqrt[4]{x}} + x^{-4} + x^{\pi}\right)\, \mathrm{d}x$

d. $\int \left(7x^{-1/2} + 5x^4 + 3x^3 + 2x + 17\right)\, \mathrm{d}x$

Solution

Using Eqs. (6.4) and (6.3) we obtain the following.

a. We proceed term by term and obtain

$$\int (3x^2 + 2x + 1)\, \mathrm{d}x = x^3 + x^2 + x + c$$

b. We proceed term by term and obtain

$$\int \left(2\sqrt{x} + \frac{2}{x^2}\right) \mathrm{d}x = \int (2x^{1/2} + 2x^{-2})\, \mathrm{d}x = \frac{4}{3}x^{3/2} - 2x^{-1} + c$$

c. We proceed term by term and obtain

$$\int \left(\frac{2}{\sqrt[4]{x}} + x^{-4} + x^{\pi}\right) \mathrm{d}x = \int (2x^{-1/4} + x^{-4} + x^{\pi})\, \mathrm{d}x = \frac{8}{3}x^{3/4} - 3x^{-3} + \frac{x^{1+\pi}}{1+\pi} + c$$

d. We proceed term by term and obtain

$$\int (7x^{-1/2} + 5x^4 + 3x^3 + 2x + 17)\, \mathrm{d}x = 14x^{1/2} + x^5 + \frac{3}{4}x^4 + x^2 + 17x + c$$

In each case, an arbitrary constant technically arises from the integration of each term, but these are combined into a single constant.

Note that the product rule for differentiation (see Section 3.3.2) means that the distributive rule for addition does not extend to products of functions. That is, in general,

$$\int f(x) \times g(x)\, \mathrm{d}x \neq \int f(x)\, \mathrm{d}x \times \int g(x)\, \mathrm{d}x \tag{6.5}$$

We will return to integrands formed from products of functions in Section 6.3.

EXAMPLE 6.4

Demonstrate Eq. (6.5) with $f(x) = x$ and $g(x) = x^2$.

Solution

Ignoring the arbitrary constants, we have

$$\int f(x)\, \mathrm{d}x = \int x\, \mathrm{d}x = \frac{x^2}{2} \quad \text{and} \quad \int g(x)\, \mathrm{d}x = \int x^2\, \mathrm{d}x = \frac{x^3}{3}$$

and

$$\int f(x)g(x)\,dx = \int x^3\,dx = \frac{x^4}{4} \neq \frac{x^2}{2} \times \frac{x^3}{3}$$

Of course the discussion to this point is not sufficient to integrate general rational functions of the type considered in Section 2.3.2; that is functions formed from general polynomial numerator and denominator. For the purposes of integration, it is often convenient to consider rational functions as the product of two functions and we return to this more complicated situation in Section 6.3.

The results of Chapter 3 can be used with Eq. (6.2) to directly infer the indefinite integrals of many of the other standard functions. In particular, it should be immediately clear that

$$\frac{d}{dx}(e^x) = e^x \quad \Longleftrightarrow \int e^x\,dx = e^x + c \tag{6.6}$$

$$\frac{d}{dx}(\sin x) = \cos x \quad \Longleftrightarrow \int \cos x\,dx = \sin x + c \tag{6.7}$$

$$\frac{d}{dx}(\cos x) = -\sin x \quad \Longleftrightarrow \int \sin x\,dx = -\cos x + c \tag{6.8}$$

$$\frac{d}{dx}(\tan x) = \sec^2 x \quad \Longleftrightarrow \int \sec^2 x\,dx = \tan x + c \tag{6.9}$$

The integration of an exponential function with general base $a > 0$ can also be performed with reference to previous results and Eq. (3.6). That is,

$$\frac{d}{dx}a^x = a^x \ln(a) \quad \Longleftrightarrow \quad \int a^x\,dx = \frac{1}{\ln a}a^x + c \tag{6.10}$$

An alternative approach to generating this integral is discussed at the end of Section 6.2.

6.2 CHANGE OF VARIABLES APPROACH

It is often necessary to integrate standard functions with stretched and shifted arguments. For example, we might wish to integrate a function of the form $f(4x + 1)$ where $f(x)$ is one of the standard functions we have already seen. This motivates the idea of *changing variables* within an integration. This approach will also prove useful in the more complicated integrals of Section 6.3.

Consider the indefinite integral

$$\int (4x+1)^2 \, dx \tag{6.11}$$

The integrand can be considered as $f(4x+1)$ where $f(x) = x^2$. From our previous discussions we know that $\int f(x)\, dx = x^3/3$, but how can we link Eq. (6.11) to this standard integral? The answer is to *change the variable*; in particular, if we define $u = 4x + 1$ then the integrand will be u^2 which we can immediately integrate with respect to u. This is not the complete picture, however, as we need to understand how to modify the integral from being with respect to x to being with respect to u. In order to do this we note that

$$\frac{du}{dx} = \frac{d}{dx}(4x+1) = 4$$

and so, in some sense, we see that $du \equiv \frac{dx}{du}\, du = 4\, dx$. Note that a more general justification of this process is considered in Section 6.3. The integral can now be rewritten and computed as

$$\int (4x+1)^2 \, dx = \int u^2 \underbrace{dx}_{\frac{1}{4} du} = \int u^2 \frac{1}{4}\, du = \frac{u^3}{4 \times 3} + c = \frac{1}{12}(4x+1)^3 + c$$

EXAMPLE 6.5

Determine the following indefinite integrals using a suitable change of variables.

a. $\int \exp(16x - 2)\, dx$

b. $\int (4y + 11)^{-2}\, dy$

c. $\int \cos(5z + 0.1)\, dz$

d. $\int \sec^2(10u + 1)\, du$

Solution

a. If $u = 16x - 2$, $du \equiv 16\, dx$ and we proceed as

$$\int \exp(16x-2)\, dx = \int \exp u \underbrace{dx}_{\frac{1}{16} du} = \int \frac{e^u}{16}\, du = \frac{e^u}{16} + c = \frac{e^{16x-2}}{16} + c$$

b. If $u = 4y + 11$, $du \equiv 4\, dy$ and we proceed as

$$\int (4y+11)^{-2}\, dy = \int u^{-2} \underbrace{dy}_{\frac{1}{4} dy} = \int \frac{u^{-2}}{4}\, du = \frac{u^{-1}}{-4} + c = -\frac{(4y+11)^{-1}}{4} + c$$

c. If $u = 5z + 0.1$, $du \equiv 5\,dz$ and we proceed as

$$\int \cos(5z + 0.1)\,dz = \int \cos u \underbrace{dz}_{\frac{1}{5}du} = \int \frac{\cos u}{5}\,du = \frac{\sin u}{5} + c = \frac{\sin(5z + 0.1)}{5} + c$$

d. If $s = 10u + 1$, $ds \equiv 10\,du$ and we proceed as

$$\int \sec^2(10u + 1)\,du = \int \sec^2(s) \underbrace{du}_{\frac{1}{10}ds} = \int \frac{\sec^2(s)}{10}\,ds = \frac{\tan s}{10} + c = \frac{\tan(10u + 1)}{10} + c$$

Of course the use of "u" for the new variable is purely arbitrary and any letter can be used.

The results of Example 6.5 can be summarized with a general rule

$$\int f(ax + b)\,dx = \frac{1}{a}\int f(u)\,du \quad \text{where } u(x) = ax + b \tag{6.12}$$

Note that the change of variables approach will not work on an integral of a standard function with a nonlinear argument; for example, the integral of $\sin(2x^2 + 2)$. Essentially it is the constant gradient of the argument that is being exploited in this approach. The reader can quickly understand this by trying to perform this integral with the substitution $u = 2x^2 + 2$. In particular, the integral element dx becomes $2x\,du = 2\sqrt{0.5(u - 2)}\,du$ which is of no help.

In addition to performing integrals of the type in Eq. (6.12), the change of variables approach is useful to determine the integrals of other standard functions. In particular, the derivative of $\ln x$ given in Eq. (3.18) enables us to consider the integral of $\int x^{-1}\,dx$, which is the special case of Eq. (6.3) for $n = -1$. That is

$$\frac{d}{dx}\ln x = \frac{1}{x} \quad \Longleftrightarrow \quad \int \frac{1}{x}\,dx = \ln x + c$$

This statement is, however, only true if $x > 0$ as the logarithm is only defined for *positive* arguments. If we wish to obtain the indefinite integral of $1/x$ for $x < 0$ it is necessary to rewrite the integrand such that the resulting ln function will have a positive argument. This is done by introducing a negative sign to the numerator and denominator and making the substitution $u = -x$. Noting that $du \equiv -\,dx$, we proceed as

$$\int \frac{-1}{-x}\,dx = \int \frac{1}{u}\,du = \ln u = \ln(-x)$$

for $x < 0$. The result for positive and negative x can be summarized as

$$\int \frac{1}{x}\,dx = \ln|x| + c \tag{6.13}$$

Furthermore, the change of variables approach provides an alternative method to determine the integral of a^x for $a > 0$. It should be clear from Chapter 2 that the following expression is true

$$a^x = e^{x \ln a}$$

If we define $u = x \ln a$, $du \equiv \ln a \, dx$ and we can proceed as

$$\int a^x \, dx = \int e^{x \ln a} \, dx = \int e^u \underbrace{dx}_{\frac{1}{\ln a} du} = \int \frac{e^u}{\ln a} \, du = \frac{e^u}{\ln a} + c = \frac{a^x}{\ln a} + c$$

which is identical to result in Eq. (6.10).

Of course we are still missing the integrals of standard functions $\ln x$ and $\tan x$. These functions are not amenable to the techniques discussed to this point and we return to them at the end of the next section.

6.3 INDEFINITE INTEGRALS OF PRODUCTS OF FUNCTIONS

So far we have determined the indefinite integrals of a number of standard functions using a process relying on us recognizing the derivatives of other functions. We have also made some preliminary use of the change of variables approach to compute the derivatives of standard functions but with arguments of the form $ax + b$. However, these techniques are very limited in practice and it is necessary to consider a much broader collection of strategies for computing more complicated integrals. In this section, we consider a number of approaches for integrating functions that can be considered as the product of other functions, that is indefinite integrals of the form

$$\int f(x)g(x) \, dx$$

The approach to be taken for any particular integrand of this form will depend on the particular $f(x)$ and $g(x)$; an important skill is recognizing the correct approach needed.

6.3.1 Change of variables

We begin by returning to the change of variables method which is useful when $g(x)$ and $f(x)$ can be related through a derivative of an appropriately chosen new variable. For example, consider the integral

$$\int 2x(x^2 - 6)^{10} \, dx$$

It is of course possible to expand $(x^2 - 6)^{10}$ as a polynomial of order 20 and multiply through by $2x$ leading to a polynomial of order 21 that can be easily integrated term by term. However, such an approach is time consuming and prone to error. Instead, it is useful to recognize that the integrand is in fact the product of two functions, $g(x) = 2x$ and $f(x) = (x^2 - 6)^{10}$. Furthermore, we note that the new variable $u = x^2 - 6$ is such that $u'(x) = 2x = g(x)$. This enables the integral to be rewritten with the effective substitution $\mathrm{d}u \equiv 2x\,\mathrm{d}x$ and proceeds as

$$\int 2x(x^2-6)^{10}\,\mathrm{d}x = \int u^{10} \underbrace{2x\,\mathrm{d}x}_{\mathrm{d}u} = \int u^{10}\,\mathrm{d}u = \frac{u^{11}}{11} + c = \frac{1}{11}(x^2-6)^{11} + c$$

The approach is extremely useful when there exists some connection between the derivative of the substituting variable, u, and $g(x)$. As a further simple example, consider

$$\int 7x^5(5x^6+\pi)^{10}\,\mathrm{d}x$$

We immediately note that the derivative of the argument $5x^6 + \pi$ is $30x^5$ which can be exploited in a change of variables approach. In particular, let $u = 5x^6 + \pi$ and so $\mathrm{d}u \equiv 30x^5\,\mathrm{d}x$, then

$$\int 7x^5(5x^6+\pi)^{10}\,\mathrm{d}x = \int u^{10} \underbrace{7x^5\,\mathrm{d}x}_{\frac{7\,\mathrm{d}u}{30}} = \int \frac{7u^{10}}{30}\,\mathrm{d}u = \frac{7u^{11}}{330} + c = \frac{7}{330}\left(5x^6+\pi\right)^{11} + c$$

As a demonstration of the variety of integrals than can be approached in this way, consider now the indefinite integral

$$\int \frac{\ln x}{x}\,\mathrm{d}x$$

We note that

$$\frac{\mathrm{d}}{\mathrm{d}x}\ln x = \frac{1}{x}$$

and so we can use the new variable $u = \ln x$ which is such that $\mathrm{d}u \equiv \frac{1}{x}\,\mathrm{d}x$. The integration then proceeds as

$$\int \ln x \underbrace{\frac{1}{x}\,\mathrm{d}x}_{\mathrm{d}u} = \int u\,\mathrm{d}u = \frac{u^2}{2} + c = \frac{1}{2}(\ln x)^2 + c$$

Furthermore, consider

$$\int \sin x \cos^3 x(2+\cos x)\,\mathrm{d}x$$

The crucial point here is that the substitution of $u = \cos x$ still leaves a complicated integrand, but it is one that can be simplified with the new variable to give something manageable. In particular, we have $\mathrm{d}u \equiv -\sin x\,\mathrm{d}x$ and so

$$\int \sin x \cos^3 x(2+\cos x)\,\mathrm{d}x = \int -\cos^3 x(2+\cos x)\underbrace{(-\sin x)\,\mathrm{d}x}_{\mathrm{d}u} = -\int (2u^3+u^4)\,\mathrm{d}u$$

$$= -\frac{u^4}{2} - \frac{u^5}{5} + c = -\frac{\cos^4 x}{2} - \frac{\cos^5 x}{5} + c$$

EXAMPLE 6.6

Use an appropriate change of variable to determine the following indefinite integrals.

a. $\int \sqrt[4]{x}\sqrt{2+x^{5/4}}\,\mathrm{d}x$

b. $\int \sin y \cos^5 y\,\mathrm{d}y$

c. $\int \sec^4 z \tan^6 z\,\mathrm{d}z$

d. $\int u^2 \exp(u^3-5)\,\mathrm{d}u$

Solution

a. We use $u = 2 + x^{5/4}$, $\mathrm{d}u \equiv \frac{5}{4}x^{1/4}$ and so

$$\int \sqrt[4]{x}\sqrt{2+x^{5/4}}\,\mathrm{d}x = \int \sqrt{u}\,\underbrace{x^{1/4}\,\mathrm{d}x}_{\frac{4}{5}\mathrm{d}u}$$

$$= \int \frac{4u^{1/2}}{5}\,\mathrm{d}u + c = \frac{8}{15}u^{3/2} = \frac{8}{15}(2+x^{5/4})^{3/2} + c$$

b. We use $u = \cos y$, $\mathrm{d}u \equiv -\sin y\,\mathrm{d}y$ and so

$$\int \sin y \cos^5 y\,\mathrm{d}y = \int -u^5\underbrace{(-\sin y)\,\mathrm{d}y}_{\mathrm{d}u} = -\int u^5\,\mathrm{d}u = -\frac{u^6}{6} + c = -\frac{\cos^6 y}{6} + c$$

c. We use $u = \tan z$, $\mathrm{d}u \equiv \sec^2 z\,\mathrm{d}z$ and so

$$\int \sec^4 z \tan^6 z\,\mathrm{d}z = \int u^6 \underbrace{\sec^2 z}_{1+\tan^2 z = 1+u^2} \underbrace{\sec^2 z\,\mathrm{d}z}_{\mathrm{d}u}$$

$$= \int (u^6 + u^8)\,\mathrm{d}u = \frac{u^7}{7} + \frac{u^9}{9} + c = \frac{\tan^7 z}{7} + \frac{\tan^9 z}{9} + c$$

d. We use $s = u^3 - 5$, $\mathrm{d}s \equiv 3u^2\,\mathrm{d}u$ and so

$$\int u^2 \exp u^3 - 5\,\mathrm{d}u = \int \mathrm{e}^s \underbrace{u^2\,\mathrm{d}u}_{\frac{1}{3}\mathrm{d}s} = \int \frac{\mathrm{e}^s}{3}\,\mathrm{d}s = \frac{\mathrm{e}^s}{3} + c = \frac{1}{3}\exp(u^3-5) + c$$

EXAMPLE 6.7

Use the change of variable approach to determine the indefinite integral of $\tan x$.

Solution

It is convenient to express the integral as

$$\int \tan x \, dx = \int \frac{\sin x}{\cos x} \, dx$$

which shows that using $u = \cos x$ will work. In particular, we note that $du \equiv -\sin x \, dx$ and so

$$\int \tan x \, dx = \int \frac{\sin x}{\cos x} \, dx = \int \frac{-1}{u} \underbrace{(-\sin x) \, dx}_{du} = -\ln |u| + c$$
$$= -\ln |\cos x| + c \tag{6.14}$$

Note that the absolute value of u is required from Eq. (6.13).

6.3.2 Integration by parts

Although the change of variable method can be very useful, not all integrands can be expressed as $u'(x)f'(u)$ with some appropriately chosen $u(x)$ and the method has limited use. This motivates the introduction of an additional approach for integrals of products, called *integration by parts*.

Integration by parts is related to the product rule for the derivatives of products, as discussed in Section 3.3.2. Recall that the product rule is such that

$$\frac{d}{dx}(u(x)v(x)) = u(x)\frac{dv}{dx} + v(x)\frac{du}{dx}$$

This can be rewritten as

$$v(x)\frac{du}{dx} = \frac{d}{dx}(u(x)v(x)) - u(x)\frac{dv}{dx}$$

Therefore, in situations where an integrand can be expressed as $v(x)u'(x)$ for some appropriate $v(x)$ and $u(x)$, we are able to express this as

$$\int v(x)\frac{du}{dx} \, dx = \int \frac{d}{dx}(u(x)v(x)) \, dx - \int u(x)\frac{dv}{dx} \, dx$$
$$= u(x)v(x) - \int u(x)\frac{dv}{dx} \, dx \tag{6.15}$$

As we shall see in the following examples, this approach can be particularly useful.

The key to integration by parts is to choose carefully from the integrand which factor to use as $v(x)$ and which as $u'(x)$. Of course, $u'(x)$ should be easily integrated to obtain $u(x)$ required in the right-hand side of Eq. (6.15), and $u(x)v'(x)$ should also be easily integrated. As with the change of variable method, confidence with integration by parts requires practice.

We demonstrate its use with

$$\int x \sin x \, dx$$

This integral is clearly not a candidate for the change of variable approach; however, it is amenable to integration by parts. We begin by comparing the integral with Eq. (6.15). Note that if we choose $v = x$ and $u' = \sin x$ then $v'(x) = 1$ and $u(x) = -\cos x$ which enables us to proceed as

$$\int \underbrace{x}_{v(x)} \underbrace{\sin x}_{u'(x)} \, dx = \underbrace{-x \cos x}_{u(x)v(x)} - \int \underbrace{(-\cos x)}_{u(x)v'(x)} \, dx = -x \cos x$$
$$+ \int \cos x \, dx = -x \cos x + \sin x + c$$

Note that the full integral is not completed until the intermediate integral on the right-hand side is performed. The introduction of the arbitrary constant is therefore left until this very final step.

EXAMPLE 6.8

Use integration by parts to perform the indefinite integral

$$\int 2x e^{-4x} \, dx$$

Solution

If we write $v(x) = 2x$ and $u'(x) = \exp(-4x)$ then $v'(x) = 2$ and $u(x) = -0.25 \exp(-4x)$. Using these in Eq. (6.15) leads to

$$\int 2x e^{-4x} \, dx = -\frac{x e^{-4x}}{2} - \int \frac{1}{2} e^{-4x} \, dx = -\frac{1}{8} e^{-4x} (4x + 1) + c$$

The intermediate integral in the integration by parts approach does not necessarily have to be performed directly and it may be that a change of variable or indeed further uses of the integration by parts is required. To demonstrate this more complicated situation consider

$$\int 2x^2 \cos x \, dx$$

We set $v(x) = 2x^2$ and $u'(x) = \cos x$ which are such that $v'(x) = 4x$ and $u(x) = \sin x$. The integration then proceeds as

$$\int 2x^2 \cos x \, dx = 2x^2 \sin x - \int 4x \sin x \, dx$$

The intermediate integral requires further use of integration by parts and leads to

$$-\int 4x \sin x \, \mathrm{d}x = 4x \cos x - 4 \sin x$$

This then enables us to complete the desired integral as

$$\int 2x^2 \cos x \, \mathrm{d}x = 2(x^2 - 2) \sin x + 4x \cos x + c$$

It may seem that fundamentally the integration by parts approach relies on $v^{(n)}(x)$ returning a constant for some finite (preferably small) n. However, this is not entirely true as the method can also sometimes be used when repeated derivatives of $v(x)$ have a cyclical property, for example, with $v(x) = \exp x$ or $\sin x$, say. When $u'(x)$ also demonstrates cyclical properties on repeated integration, it is possible to exploit this and successfully use integration by parts. For example, consider

$$\int \mathrm{e}^x \cos x \, \mathrm{d}x$$

If $v(x) = \mathrm{e}^x$ and $u'(x) = \cos x$ then $v'(x) = \mathrm{e}^x$ and $u(x) = \sin x$ and

$$\int \mathrm{e}^x \cos x \, \mathrm{d}x = \mathrm{e}^x \sin x - \int \mathrm{e}^x \sin x$$

The intermediate integrate can be approached using integration by parts to give

$$\int \mathrm{e}^x \sin x \, \mathrm{d}x = -\mathrm{e}^x \cos x + \int \mathrm{e}^x \cos x \, \mathrm{d}x$$

The original integral can now be expressed as

$$\int \mathrm{e}^x \cos x \, \mathrm{d}x = \mathrm{e}^x \sin x + \mathrm{e}^x \cos x - \int \mathrm{e}^x \cos x \, \mathrm{d}x$$

This can be rearranged to give that

$$\int \mathrm{e}^x \cos x \, \mathrm{d}x = \frac{1}{2} (\mathrm{e}^x \sin x + \mathrm{e}^x \cos x) + c$$

EXAMPLE 6.9

Use integration by parts to compute the following indefinite integrals.

a. $\int \mathrm{e}^{2x}(4x - 1) \, \mathrm{d}x$

b. $\int y^3 \mathrm{e}^{-y} \, \mathrm{d}y$

c. $\int \mathrm{e}^z(z^2 - 1) \, \mathrm{d}z$

d. $\int \mathrm{e}^s \sin s \cos s \, \mathrm{d}s$

Solution

a. We set $v(x) = 4x - 1$ and $u'(x) = e^{2x}$ which are such that $v'(x) = 4$ and $u(x) = 0.5e^{2x}$. Integration by parts then gives

$$\int e^{2x}(4x-1)\,dx = \frac{1}{2}e^{2x}(4x-1) - 2\int e^{2x}\,dx$$
$$= e^{2x}\left(2x - \frac{3}{2}\right) + c$$

b. We set $v(y) = y^3$ and $u'(y) = e^{-y}$ which are such that $v'(y) = 3y^2$ and $u(y) = -e^{-y}$. Integration by parts then gives

$$\int y^3 e^{-y}\,dy = -y^3 e^{-y} + 3\int y^2 e^{-y}\,dy$$

The intermediate integral is obtained by integration by parts as

$$3\int y^2 e^{-y}\,dy = -3y^2 e^{-y} + 6\int y e^{-y}\,dy$$

This second intermediate integral is again obtained by integration by parts to give

$$6\int y e^{-y}\,dy = -6(y-1)e^{-y}$$

Back substituting the intermediate integrals leads to the desired answer

$$\int y^3 e^{-y}\,dy = -e^{-y}(y^3 + 3y^2 + 6y + 6) + c$$

c. We set $v(z) = z^2 - 1$ and $u'(z) = e^z$ which are such that $v'(z) = 2z$ and $u(z) = e^z$. Integration by parts then gives

$$\int e^z(z^2-1)\,dz = e^z(z^2-1) - 2\int z e^z\,dz$$

The intermediate integral is computed using integration by parts to give

$$\int e^z(z^2-1)\,dz = e^z(z-1)^2 + c$$

d. This example is included as a warning. You might naturally proceed by setting $v(s) = e^s$, therefore $v'(s) = e^s$, and $u'(s) = \sin s \cos s$. However, $u(s)$ is difficult to determine and requires a change of variable approach with new variable $g(s) = \cos s$ before we can form the integration by parts formula.

This approach has clearly become very complicated and it is worth taking a step back to see if a simpler approach is possible. We note from Appendix A that a simpler form of the original integral is possible using double-angle identities. In particular, using $\sin s \cos s = \sin(2s)/2$, the integral is rewritten as

$$\int e^s \sin s \cos s \, ds = \frac{1}{2}\int e^s \sin(2s) \, ds$$

which is an example of a cyclical integral. If $u'(s) = e^s$ and $v(s) = \sin(2s)$, then $u(s) = e^s$ and $v'(s) = 2\cos(2s)$ and integration by parts leads to

$$\frac{1}{2}\int e^s \sin(2s) \, ds = \frac{1}{2}e^s \sin(2s) - \int e^2 \cos(2s) \, ds$$

The intermediate integral proceeds in similar way to give

$$\frac{1}{2}\int e^s \sin(2s) \, ds = \frac{1}{2}e^s \sin(2s) - \left[e^2 \cos(2s) + 2\int e^s \sin(2s)\right]$$

which can be rearranged to give the final result that

$$\int e^s \sin s \cos s \, ds = \frac{e^s}{10}(\sin(2s) - 2\cos(2s))$$

At this stage, it is possible to return to the integral of $f(x) = \ln x$ and the other remaining standard functions. The integral of $\ln x$, for example, is of course not amenable to a change in variable and, at first glance, it may not appear amendable to integration by parts. However, recall from Chapter 3 that we can differentiate this function with ease. This gives some hint as to why integration by parts could be used in this and other cases. This is demonstrated in the following example.

EXAMPLE 6.10

Use integration by parts to determine the following indefinite integrals of standard functions.

a. $\int \ln x \, dx$
b. $\int \arcsin \, dx$
c. $\int \arccos \, dx$
d. $\int \arctan \, dx$

Solution

a. We begin by rewriting the integral as

$$\int 1 \times \ln x \, dx$$

and set $v(x) = \ln x$ and $u'(x) = 1$. These choices are such that $v'(x) = 1/x$ and $u(x) = x$ and integration by parts leads to

$$\int \ln x \, dx = x \ln x - \int \frac{x}{x} \, dx = x(\ln x - 1) + c$$

b. The derivative of the integrand is known from Example 3.25 and we use the same trick as in part a. In particular, we consider

$$\int 1 \times \arcsin x \, \mathrm{d}x$$

and set $v(x) = \arcsin x$ and $u'(x) = 1$. These choices are such that $v'(x) = (1-x^2)^{-1/2}$ and $u(x) = x$ and integration by parts leads to

$$\int \arcsin x \, \mathrm{d}x = x \arcsin x - \int x \left(1 - x^2\right)^{-1/2}$$

The intermediate integral can be computed with the substitution $s = 1 - x^2$, leading to

$$\int \arcsin x \, \mathrm{d}x = x \arcsin x + \sqrt{1-x^2} + c$$

c. This proceeds as in part b. to give

$$\int \arccos x \, \mathrm{d}x = x \arcsin x - \sqrt{1-x^2} + c$$

d. This proceeds as in parts b. and c. to give

$$\begin{aligned}\int 1 \times \arctan \, \mathrm{d}x &= x \arctan x - \int \frac{x}{1+x^2} \, \mathrm{d}x \\ &= x \arctan x - \frac{1}{2} \ln(1+x^2) + c\end{aligned}$$

6.4 INDEFINITE INTEGRALS OF RATIONAL FUNCTIONS

It has been necessary to leave the integration of rational functions until this relatively late stage. Despite being easily defined as the ratio of two polynomial functions, great variation in the form of rational functions is possible. Unlike polynomial functions, which are always easily integrated, the approach to integrating a rational function depends on the particular form of the rational function in hand. For this reason there is no standard result for the indefinite integral of a general rational function; instead, it is necessary to consider each integral on its own merits.

For example, consider the integral

$$\int \frac{2x+1}{x^2+x-10} \, \mathrm{d}x$$

Here, it should be immediately obvious that a simple change of variable approach is needed with $u = x^2 + x - 10$ and $\mathrm{d}u \equiv (2x+1)\,\mathrm{d}x$. This is because the numerator is related (in fact identical) to the derivative of the denominator. The integral then proceeds as

$$\int \frac{2x+1}{x^2+x-10}\,dx = \int \frac{1}{u}\underbrace{(2x+1)\,dx}_{du} = \int \frac{1}{u}\,du = \ln\left|x^2+x-10\right| + c$$

Similarly, the integral

$$\int \frac{10x^4+6x^2}{(x^5+x^3+1)^{12}}\,dx$$

is such that the numerator is proportional to the derivative of the denominator and a change of variable to $u = x^5 + x^3 + 1$ will work. We have that $du \equiv (5x^4 + 3x^2)\,dx$ and so

$$\int \frac{10x^4+6x^2}{(x^5+x^3+1)^{12}}\,dx = \int \frac{2}{u^{12}}\underbrace{(5x^4+3x^2)\,dx}_{du}$$

$$= \int \frac{2}{u^{12}}\,du = -\frac{2}{11(x^5+x^3+1)^{11}} + c$$

However, the following integral is not amenable to a change of variable approach,

$$\int \frac{x+3}{(x^2+2x-8)}\,dx$$

It is possible to express the integrand as a *partial fraction* and this gives an alternative approach. In particular,

$$\int \frac{x+3}{(x^2+2x-8)}\,dx = \int \frac{5}{6(x-2)}\,dx + \int \frac{1}{6(x+4)}\,dx$$

$$= \frac{5}{6}\ln|x-2| + \frac{1}{6}\ln|x+4| + c$$

EXAMPLE 6.11

Determine the following indefinite functions.

a. $\int \frac{x+4}{x+2}\,dx$

b. $\int \frac{35x^6+25x^4}{(x^7+x^5-\pi)^5}\,dx$

c. $\int \frac{2x+4}{3x^2+6x-24}\,dx$

Solution

a. Here, we note that the numerator can be written as $x + 4 = x + 2 + 2$ and so the integral can be restated as

$$\int \frac{x+4}{x+2}\,dx = \int \left(1 + \frac{2}{x+2}\right)\,dx = x + 2\ln|x+2| + c$$

b. Here, we note that a new variable $u = x^7 + x^5 - \pi$ is such that $du = (7x^5 + 5x^4)\,dx$. Therefore

$$\int \frac{35x^6 + 25x^4}{(x^7 + x^5 - \pi)^5}\,dx = \int \frac{1}{u^5}\underbrace{(35x^6 + 25x^4)\,dx}_{5\,du} = \int \frac{5}{u^5}\,du = -\frac{5}{4(x^7 + x^5 - \pi)^4} + c$$

c. Here, we use partial fractions to rewrite the integral as

$$\int \frac{35x^6 + 25x^4}{(x^7 + x^5 - \pi)^5}\,dx = \int \frac{2}{9}\left(\frac{1}{x+4} + \frac{2}{x-2}\right)dx = \frac{2}{9}(\ln|x+4| + 2\ln|x-2|) + c$$

Although we cannot derive a standard expression for the integral of a general rational function, we can, however, exploit the standard expression for the derivative of the arctan function, for example. From the result in Example 3.25 we know that

$$\int \frac{1}{x^2+1}\,dx = \arctan x + c \tag{6.16}$$

and from this result we can determine a general strategy for obtaining integrals of rational functions of the particular form

$$\int \frac{1}{ax^2 + bx + c}\,dx \tag{6.17}$$

where a, b, and c are constants. The approach involves manipulating the denominator of the integrand to look more like that in Eq. (6.16).

The algebraic approach required to manipulate the denominator is called *completing the square*, which you may be familiar with from the manipulation of quadratic expressions. The aim when completing the square is to express a quadratic in the form $(x+p)^2 + q^2$. For example,

$$x^2 + 2x + 5 \equiv (x+1)^2 + 4$$

This identity is easily seen to be true by multiplying out $(x+1)^2$ and adding 4. In order to obtain the completed-square form of a general quadratic, we note that whatever appears added to the x in the square term will appear multiplied by $2x$ after multiplying out. Therefore, the first term of the completed square form of $x^2 + ax + b$ is $(x + a/2)^2$. The additional term is then found by introducing the appropriate constant to ensure the left-hand and right-hand sides are equal.

EXAMPLE 6.12

Obtain the completed-square form of the following quadratics.

a. $x^2 - 8x + 18$

b. $x^2 + 2x + 4$

c. $x^2 - 4x + 10$

Solution

a. The coefficient of x is -8, therefore we need $(x-4)^2$. This is multiplied out to give $x^2 - 8x + 16$ and so a further $+2$ is required. Therefore

$$x^2 - 8x + 18 \equiv (x-4)^2 + 2$$

b. The coefficient of x is 2, therefore we need $(x+1)^2$. This is multiplied out to give $x^2 + 2x + 1$ and so a further $+3$ is required. Therefore

$$x^2 + 2x + 4 \equiv (x+1)^2 + 3$$

c. The coefficient of x is -4, therefore we need $(x-2)^2$. This is multiplied out to give $x^2 - 4x + 4$ and so a further $+6$ is required.. Therefore

$$x^2 - 4x + 10 \equiv (x-2)^2 + 6$$

Now that we are able to complete the square we turn to evaluating integrals of the form in Eq. (6.17). The reason for completing the square is of course to express the integrand in a manner consistent with that in the standard result of Eq. (6.16). For example,

$$\int \frac{1}{x^2 + 2x + 5}\,\mathrm{d}x = \int \frac{1}{(x+1)^2 + 4}\,\mathrm{d}x = \frac{1}{4}\int \frac{1}{\left(\frac{x+1}{2}\right)^2 + 1}\,\mathrm{d}x$$

If we then use the change of variable $u = (x+1)/2$ so $\mathrm{d}u \equiv \mathrm{d}x/2$, we have

$$\frac{1}{4}\int \frac{1}{\left(\frac{x+1}{2}\right)^2 + 1}\,\mathrm{d}x = \frac{1}{4}\int \frac{1}{u^2+1}\underbrace{\mathrm{d}x}_{2\,\mathrm{d}u} = \frac{1}{2}\int \frac{1}{u^2+1}\,\mathrm{d}u$$

The integral is then easily obtained from Eq. (6.16) as

$$\int \frac{1}{x^2 + 2x + 5}\,\mathrm{d}x = \frac{1}{2}\arctan\left(\frac{x+1}{2}\right) + c$$

EXAMPLE 6.13

Determine the following indefinite integrals using the completing the square method.

a. $\int \frac{1}{x^2 - 8x + 18}\,\mathrm{d}x$

b. $\int \frac{1}{y^2 + 2y + 4}\,\mathrm{d}y$

c. $\int \frac{1}{2z^2 - 8z + 20}\,\mathrm{d}z$

Solution

The results of Example 6.12 are used in each case.

a.

$$\int \frac{1}{x^2-8x+18}\,dx = \int \frac{1}{(x-4)^2+2}\,dx = \frac{1}{2}\int \frac{1}{\left(\frac{x-4}{\sqrt{2}}\right)^2+1}\,dx$$

$$= \frac{1}{\sqrt{2}}\arctan\left(\frac{x-4}{\sqrt{2}}\right)+c$$

b.

$$\int \frac{1}{y^2+2y+4}\,dy = \int \frac{1}{(y+1)^2+3}\,dy = \frac{1}{3}\int \frac{1}{\left(\frac{y+1}{\sqrt{3}}\right)^2+1}\,dy$$

$$= \frac{1}{\sqrt{3}}\arctan\left(\frac{y+1}{\sqrt{3}}\right)+c$$

c. We first need to factor out the 2 multiplying the z^2 term. Then we can proceed by completing the square.

$$\int \frac{1}{2z^2-8z+20}\,dy = \frac{1}{2}\int \frac{1}{z^2-4z+10} = \frac{1}{2}\int \frac{1}{(z-2)^2+6}\,dz$$

$$= \frac{1}{12}\int \frac{1}{\left(\frac{z-2}{\sqrt{6}}\right)^2+1}\,dz = \frac{1}{2\sqrt{6}}\arctan\left(\frac{z-2}{\sqrt{6}}\right)+c$$

The results of Example 6.13 can be summarized in the general result

$$\int \frac{1}{ax^2+bx+c}\,dx = \frac{2}{\sqrt{4ac-b^2}}\arctan\left(\frac{2ax+b}{\sqrt{4ac-b^2}}\right)+c \tag{6.18}$$

The reader is invited to derive this in Question 6.7. The expression of course is only applicable when $b^2 < 4ac$, otherwise the square root of a negative number is required which is not permitted in the domain of real numbers.

An equivalent approach using the standard derivatives of arcsin and arccos functions can be used in a number of other special cases. In particular, we see from Example 3.25 that

$$\int \frac{1}{\sqrt{1-x^2}}\,dx = \arcsin(x)+c \quad \text{and} \quad \int \frac{-1}{\sqrt{1-x^2}}\,dx = \arccos(x)+c$$

These expressions are clearly useful in conjunction with a completing the square approach for a further class of integrands. It is important to note that

$$\int \frac{-1}{\sqrt{1+x^2}}\,dx = -\int \frac{1}{\sqrt{1+x^2}}\,dx$$

and so either form could be used. This is illustrated in the following example.

EXAMPLE 6.14

Determine the following indefinite integral

$$\int \frac{-1}{\sqrt{7-2x-x^2}}\,dx$$

Solution

We proceed by using the completing the square approach on the quadratic inside the square root

$$\int \frac{-1}{\sqrt{7-2x-x^2}}\,dx = \frac{1}{2\sqrt{2}}\int \frac{-1}{\sqrt{1-\left(\frac{x+1}{8}\right)^2}}\,dx = \arccos\left(\frac{x+1}{2\sqrt{2}}\right) + c_1$$

Alternatively,

$$\int \frac{-1}{\sqrt{7-2x-x^2}}\,dx = \frac{-1}{2\sqrt{2}}\int \frac{1}{\sqrt{1-\left(\frac{x+1}{8}\right)^2}}\,dx = -\arcsin\left(\frac{x+1}{2\sqrt{2}}\right) + c_2$$

6.5 INDEFINITE INTEGRALS ON YOUR COMPUTER

Essentially this chapter has been concerned entirely with a single concept: indefinite integration. Our discussion extended a number of pages after the concept was introduced in Section 6.1; however, much of that discussion was concerned with strategies for performing integrals using *algebraic* approaches. Such a lengthy discussion is not, however, required when using Wolfram Alpha to determine integrals directly. That is, we do not have to state that the integral should be performed using a particular change of variable or by integration by parts, rather we just need to state a function and that it is to be integrated.

The command for performing integrals directly in Wolfram Alpha is `integrate` which can also be shortened to `int`. For example, the integral of $\frac{1}{x^2+4}$ is obtained from the command

```
int 1/(x^2+4) dx
```

which returns

$$\frac{1}{2}\tan^{-1}\left(\frac{x}{2}\right) + \text{constant}$$

While practicing integration by parts by hand, you might also wish to explore the use of the command

```
integrate by parts x sin(x) dx
```

This command reflects our desire to use Wolfram Alpha as a learning tool rather than a replacement for proper understanding.

EXAMPLE 6.15

Use Wolfram Alpha to perform the following indefinite integrals.

a. $\int x^2 e^{x^3}\,dx$

b. $\int y^2 \cos(2y)\,dy$

Solution

a. We use the command

```
int x^2 e^(x^3) dx
```

and obtain that

$$\int x^2 e^{x^3}\,dx = \frac{e^{x^3}}{3} + \text{constant}$$

Note that the strategy for performing this integral algebraically is to use the change of variable $u = x^3$.

b. We use the command

```
int y^2 cos(2y) dy
```

and determine that

$$\int y^2 \cos(2y)\,dy = \frac{1}{4}\left((2y^2 - 1)\sin(2y) + 2y\cos(2y)\right) + \text{constant}$$

Note that the strategy for performing this integral algebraically is to use integration by parts twice, beginning with $u = x^2$ and $v' = \cos(2y)$.

6.6 QUESTIONS

The following questions are intended to test your knowledge of the material discussed in this chapter. Full solutions are available in Chapter 6 Solutions of Part III. You should use an algebraic approach unless otherwise stated.

Question 6.1. Perform the following indefinite integrals.

a. $\int \cos(5x)\,dx$

b. $\int 10\sin\left(0.7 + \frac{y}{2}\right)\,dy$

c. $\int \left(e^{5z} + \ln(3z)\right)\,dz$

d. $\int \left(\text{cosec}^2(3p) - \sec^2(8p)\right)\,dp$

e. $\int \sin(\pi q)\cot(\pi q)\,dq$

Question 6.2. Perform the following indefinite integrals.

a. $\int x^3(x^4 - 2)^3\,dx$

b. $\int \cos(y) e^{\sin y}\,dy$

c. $\int \frac{2z}{z^2+5}\,dz$

d. $\int \tan^4(p)\sec^2(p)\,\mathrm{d}p$

e. $\int q^n\sqrt{1+q^{n+1}}\,\mathrm{d}q$ for $n \in \mathbb{N}^+$

Question 6.3. Perform the following indefinite integrals.

a. $\int x\sin x\,\mathrm{d}x$

b. $\int y^2 e^{6y}\,\mathrm{d}y$

c. $\int e^{2z}\sin z\,\mathrm{d}z$

d. $\int p\sqrt{1+p}\,\mathrm{d}p$

e. $\int \sin(q)\cos(q)\,\mathrm{d}q$

Question 6.4. Perform the following indefinite integrals.

a. $\int \frac{4}{1+x^2}\,\mathrm{d}x$

b. $\int \frac{-2}{y^2+4y+5}\,\mathrm{d}y$

c. $\int \frac{1}{\sqrt{4-8z-4z^2}}\,\mathrm{d}z$

d. $\int \frac{6}{25p^2-5p+1}\,\mathrm{d}p$

e. $\int \frac{-2}{q^2+4q-5}\,\mathrm{d}q$

Question 6.5. Using the methods stated, demonstrate that the following indefinite integral is correct.

$$\int \sec^3 x\,\mathrm{d}x = \frac{1}{2}\sec x\tan x + \frac{1}{2}\ln|\sec x + \tan x| + c$$

a. Integration by parts. You are given that $\int \sec x\,\mathrm{d}x = \ln|\sec x + \tan x|$.

b. The substitution $u = \sin x$.

Question 6.6. Derive the standard expression for the roots of a quadratic polynomial $a_1x^2 + a_2x + a_3$ for $a_i \in \mathbb{R}$ and $a_1 \neq 0$.

Hint. Begin by obtaining the completed-square form of the polynomial.

Question 6.7. Derive the general result,

$$\int \frac{1}{a_1x^2 + a_2x + a_3}\,\mathrm{d}x = \frac{2}{\sqrt{4a_1a_3 - a_2^2}}\arctan\left(\frac{2a_1x + a_2}{\sqrt{4a_1a_3 - a_2^2}}\right) + c$$

Note that this was stated in Eq. (6.18).

Question 6.8. Perform the following indefinite integral and comment on your result,

$$\int \sum_{k=0}^{\infty} \frac{(-1)^k x^{2k}}{(2k)!}\,\mathrm{d}x$$

Hint. Maclaurin series.

CHAPTER 7

Integral Calculus II

Contents

Prerequisite knowledge	Intended learning outcomes

• Chapter 1
• Chapter 2
• Chapter 3
• Chapter 4
• Chapter 5
• Chapter 6
	• Define and recognize a definite integral
	• Recognize definite integration as a process for determining areas under curves
	• Determine the definite integrals of standard and more complication functions, using, as appropriate,
	  ◦ the change of variables approach
	  ◦ integration by parts
	• Calculate the area between curves and also bounded by curves

In this chapter, we discuss *definite integrals* that arise from considering integration as the limit of a *summation process*. This approach to integration, which is distinct from that presented in Chapter 6, will allow us to determine areas bounded by functions and will give an alternative insight into the integration process. As we shall see, the mathematical procedures required for performing a definite integral are very closely related to those for performing an indefinite integral; it is assumed that the reader has a good understanding of Chapter 6.

7.1 DEFINITE INTEGRALS

Consider a function $f(x)$ defined on the domain of real numbers; that is, $x \in \mathbb{R}$. The function can of course be thought of as defining a curve $y = f(x)$ and this is shown in Figure 7.1. The aim of the following discussion is to construct a mathematical method

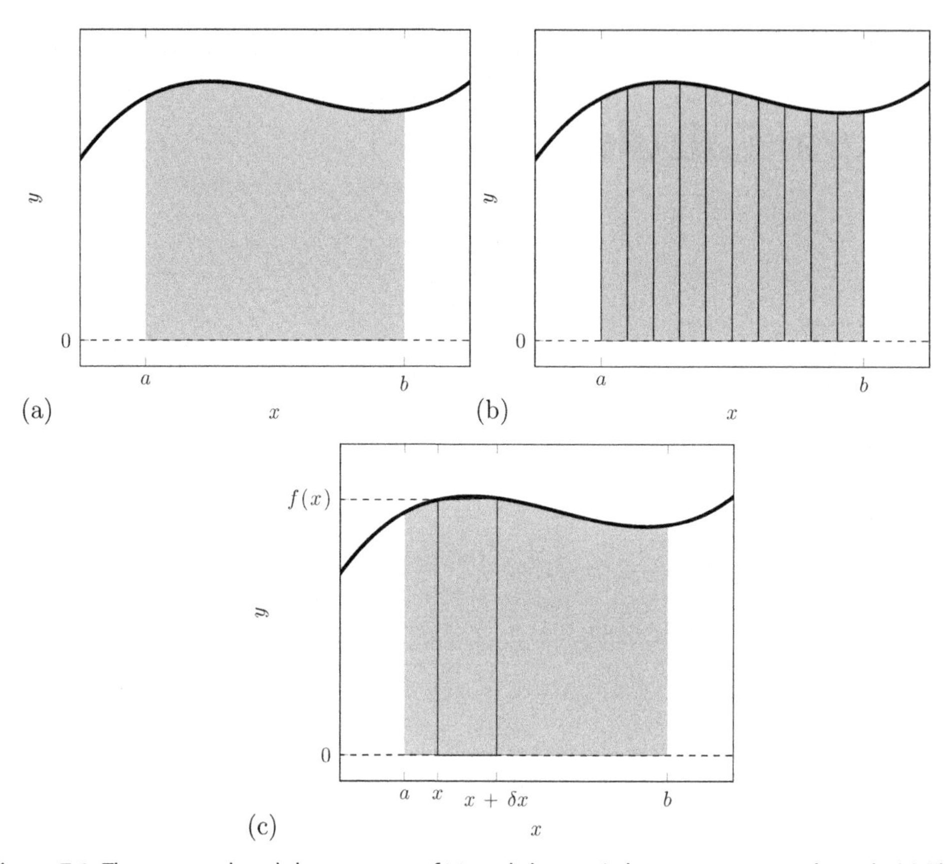

Figure 7.1 The area enclosed the curve $y = f(x)$ and the x-axis between $x = a$ and $x = b$. (a) The required area. (b) The area as the sum of rectangular elements. (c) A single element placed at some position x.

for computing the area that is enclosed between the curve and the x-axis, horizontally bounded by $x = a$ and $x = b$. Here, $a < b \in \mathbb{R}$ are constants and the required area is shown as the shaded region in Figure 7.1(a). While we are not imposing any particular form on $f(x)$ at this stage, we are restricted to functions that are continuous over the interval $x \in [a, b]$ and will begin by assuming that it remains *above* the x-axis for all $x \in [a.b]$.

An *approximation* to the area indicated can be obtained by dividing the region into a number of rectangular elements, each of equal width δx, and summing the contribution from each; this is illustrated in Figure 7.1(b). A particular rectangular element is shown in more detail in Figure 7.1(c). This element has its lower left corner placed at some $x_i \in [a, b - \delta x]$ and has height given by the function at that x, that is $y_i = f(x_i)$. If we denote the contribution to the total area A given by this element as δA_i, it should be immediately clear that

$$\delta A_i = y_i \delta x \tag{7.1}$$

Similar elements can be used over the interval, with each x_i placed at $x_i = \{a, a + \delta x, \ldots, b - \delta x\}$. The total area is then approximated by

$$A \approx \sum_{i=1}^{n} \delta A_i = \sum_{i=1}^{n} y_i \delta x = \sum_{i=1}^{n} f(x_i) \delta x \tag{7.2}$$

When δx has finite size, Eq. (7.2) represents only an approximation to A because each rectangular element fails to capture the addition area above the rectangle when $f(x) < f(x + \delta x)$ (i.e., the curve has positive gradient), and over-estimates the area when $f(x) > f(x + \delta x)$ (i.e., the curve has negative gradient); this can be seen in Figure 7.1(c). The approximation does, however, become increasingly accurate as δx reduces to 0. The limiting case is

$$A = \lim_{\delta x \to 0} \left(\sum_{i=1}^{n} f(x_i) \delta x \right)$$

and this expression defines the *definite integral* of $f(x)$ between $x = a$ and b. We denote this as

$$\int_a^b f(x)\, \mathrm{d}x = \lim_{\delta x \to 0} \left(\sum_{i=1}^{n} f(x_i) \delta x \right) \tag{7.3}$$

It is natural to ask, given our knowledge of indefinite integrals from Chapter 6, why Eq. (7.3) can be considered an integral? The answer is related to Eq. (7.1) which can be expressed as

$$\frac{\delta A}{\delta x} = y$$

In the limit that $\delta x \to 0$, which is of course crucial to the construction of Eq. (7.3), it should be clear that this becomes the *derivative* of A with respect to x. That is

$$\frac{\mathrm{d}A}{\mathrm{d}x} = y$$

This can be interpreted as

$$A = \int y\, \mathrm{d}x = \int f(x)\, \mathrm{d}x$$

and compared directly to Eq. (7.3).

Although it is not a formal argument, this discussion is intended to demonstrate that, in addition to being the reverse operation of differentiation, integration can be considered as a *summation process* to determine the area "under" a function. Indeed this is the origin of the symbol $\int$ which is an elongated "S" denoting a sum.

Recall that indefinite integrals arise from defining integration as the reverse process of a derivative. That is, given an *unknown* function's gradient, we can algebraically integrate the gradient function to determine the unknown function. However, the gradient is a property of the shape of the function with changes in x, not its particular position on the y-axis. It is therefore necessary to include an arbitrary constant in the statement of the result of the integral; that is, the result is indefinite. In contrast, a definite integral is not considered as an algebraic operation to determine an unknown function, rather it gives the definite area under a *known* function between two horizontal bounds. However, it is often possible to use the algebraic operations discussed in Chapter 6 to evaluate a definite integral, as we shall now see.

Consider the definite integral of the function $f(x) = 4x^3$ between $a = 1$ and $b = 1.5$. Note that this was the example used to introduce indefinite integrals in Section 6.1. The integral is written as

$$\int_1^{1.5} 4x^3 \, dx$$

and is interpreted as the area under the curve defined by $y = 4x^3$ bounded by $x = 1$ and $x = 1.5$, as shown in Figure 7.2. The indefinite integral of $4x^3$ is $x^4 + c$ and this can be used in a natural way to compute the definite integral as

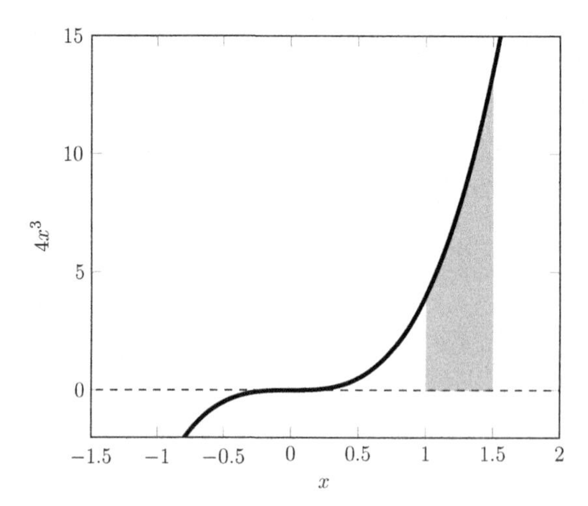

Figure 7.2 Graphical interpretation of the definite integral $\int_1^{1.5} 4x^3 \, dx$.

$$\int_{1}^{1.5} 4x^3\,dx = (1.5^4 + c) - (1^4 + c) = 1.5^4 - 1^4 \approx 4.0625$$

We note that the arbitrary constant cancels and is redundant to our evaluation of the definite integral. A more efficient way to denote the procedure of performing the definite integral is

$$\int_{1}^{1.5} 4x^3\,dx = [x^4]_1^{1.5} = 1.5^4 - 1^4 \approx 4.0625$$

The algebraic process for computing a general definite integral between $x = a$ and $x = b$ is summarized as

$$\int_a^b f(x)\,dx = \int_a^b g'(x)\,dx = [g(x)]_a^b = g(b) - g(a) \tag{7.4}$$

Here $f(x) = g'(x)$, and $g(x)$ is obtained from the algebraic integration of $f(x)$, that is the indefinite integral neglecting the arbitrary constant. Equation (7.4) is often referred to as the *fundamental theorem of calculus*.

As was briefly discussed in Chapter 6, it is not possible to perform all indefinite integrals algebraically. For this same reason, we are unable to determine the definite integrals of all continuous functions using the algebraic approach presented so far. Instead, it is often necessary to resort to numerical methods and these are considered in Chapter 13. In this chapter, however, we will remain concerned with definite integrals that can be evaluated algebraically.

EXAMPLE 7.1

Evaluate and sketch an interpretation of each of the following definite integrals.

a. $\int_1^2 (3x^2 + 2x + 2)\,dx$

b. $\int_3^4 \ln x\,dx$

c. $\int_{-3}^{-2} \frac{1}{x^2}\,dx$

d. $\int_1^2 e^x\,dx$

Solution

Each algebraic integral can be performed easily using the techniques of Chapter 6 before Eq. (7.4) is applied.

a. $\int_1^2 (3x^2 + 2x + 2) = [x^3 + x^2 + 2x]_1^2 = 12$

b. $\int_3^4 \ln x\,dx = [x(\ln x - 1)]_3^4 \approx 1.2493$

c. $\int_{-3}^{-2} x^{-2}\,dx = \left[-\frac{1}{x}\right]_{-3}^{-2} = \frac{1}{6}$

d. $\int_1^2 e^x\,dx = [e^x]_1^2 \approx 4.6701$

Graphical interpretations of these integrals are shown in Figure 7.3.

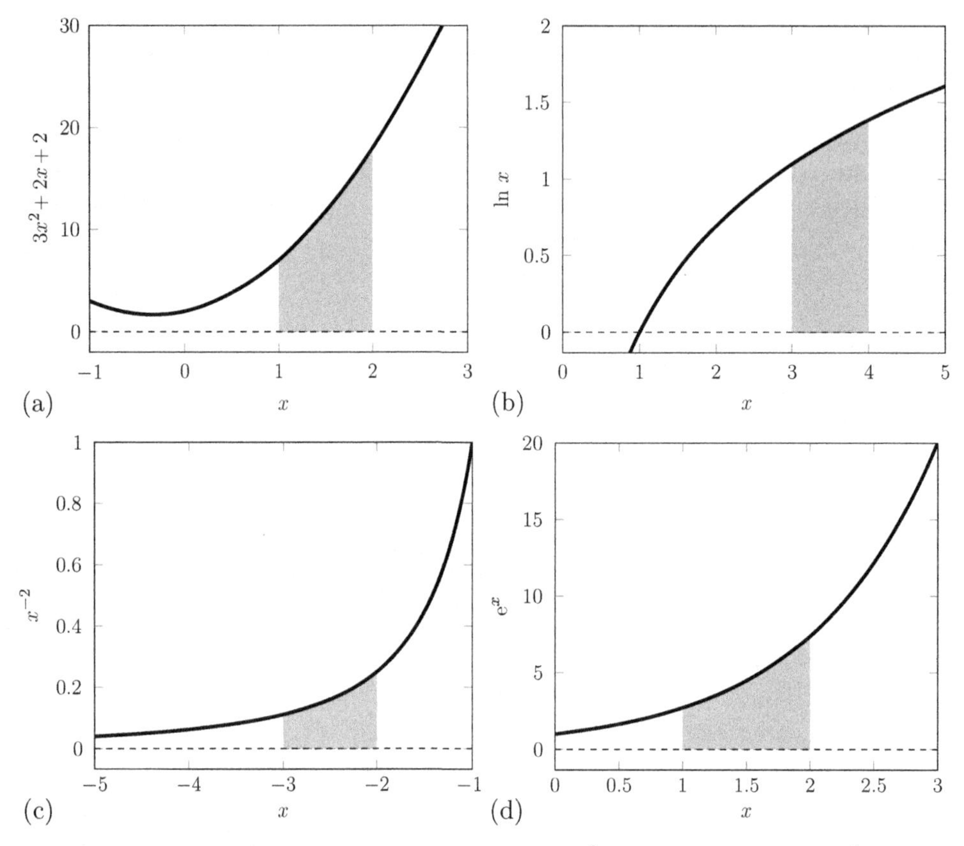

Figure 7.3 Definite integrals of Example 7.1. (a) $\int_1^2 (3x^2 + 2x + 2)\,\mathrm{d}x$. (b) $\int_3^4 \ln x\,\mathrm{d}x$. (c) $\int_{-3}^{-2} \frac{1}{x^2}\,\mathrm{d}x$. (d) $\int_1^2 \mathrm{e}^x\,\mathrm{d}x$.

The graphical interpretation of definite integrals intuitively shows the following property of definite integrals to be true

$$\int_a^b f(x)\,\mathrm{d}x = \int_a^{a_1} f(x)\,\mathrm{d}x + \int_{a_1}^b f(x)\,\mathrm{d}x$$

for $a \leq a_1 \leq b$. Informally this states that an area can be evaluated as the sum of two adjoining areas. Indeed, this can of course be extended to any number of separate areas as

$$\int_a^b f(x)\,\mathrm{d}x = \int_a^{a_1} f(x)\,\mathrm{d}x + \int_{a_1}^{a_2} f(x)\,\mathrm{d}x + \int_{a_2}^{a_3} f(x)\,\mathrm{d}x + \cdots + \int_{a_i}^b f(x)\,\mathrm{d}x \quad (7.5)$$

for $a \leq a_1 \leq a_2 \leq \cdots \leq a_i \leq b$.

EXAMPLE 7.2

Explore Eq. (7.5) using the definite integral

$$\int_0^{10} (x^2 + 2)\,\mathrm{d}x$$

Solution

The definite integral is easily computed as

$$\int_0^{10} (x^2 + 2)\,\mathrm{d}x = \left[\frac{x^3}{3} + 2x\right]_0^{10} = \frac{1060}{3}$$

We can choose to split the integral at $x = 5$, for example. In this case, we have

$$\begin{aligned}\int_0^{5} (x^2 + 2)\,\mathrm{d}x + \int_5^{10} (x^2 + 2)\,\mathrm{d}x &= \frac{155}{3} + \frac{905}{3} \\ &= \frac{1060}{3}\end{aligned}$$

Alternatively, splitting at $x = 3$ and $x = 7$ gives

$$\begin{aligned}\int_0^{3} (x^2 + 2)\,\mathrm{d}x + \int_3^{7} (x^2 + 2)\,\mathrm{d}x + \int_7^{10} (x^2 + 2)\,\mathrm{d}x &= 15 + \frac{340}{3} + 225 \\ &= \frac{1060}{3}\end{aligned}$$

This is demonstrated visually in Figure 7.4.

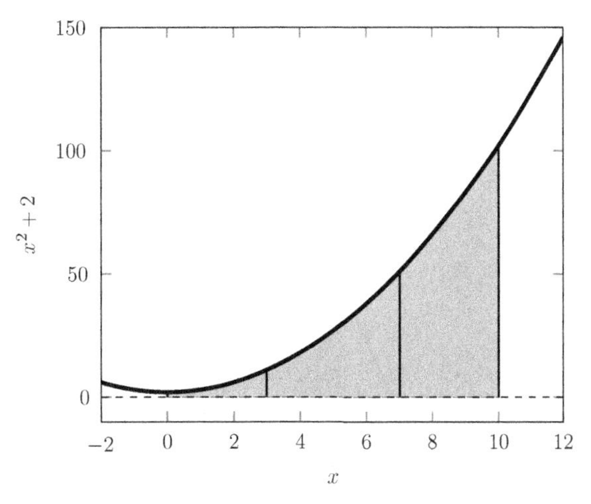

Figure 7.4 $\int_0^{10}(x^2+2)\,\mathrm{d}x = \int_0^{3}(x^2+2)\,\mathrm{d}x + \int_3^{7}(x^2+2)\,\mathrm{d}x + \int_7^{10}(x^2+2)\,\mathrm{d}x$.

We have so far been restricted to integrands that are continuous over the domain of integration. In particular, for the definite integral to exist, the integrand must not have any vertical asymptotes between $x = a$ and $x = b$. The reason for this should be intuitively clear from the graphical interpretation we have been using to discuss definite integrals. For example, the integrand in Example 7.1(c), $f(x) = x^{-2}$, is clearly not continuous at $x = 0$ and the area under the curve that includes this point is infinite. We can demonstrate this using a domain that spans $x = 0$, that is $a < 0 < b$ and splitting the integration either side of $x = 0$. In particular, if we have $a < a_1 < 0 < b_1 < b$ can write the sum of two definite integrals

$$\int_a^{a_1} x^{-2}\,\mathrm{d}x + \int_{b_1}^{b} x^{-2}\,\mathrm{d}x$$

which is such that

$$\int_a^{b} x^{-2}\,\mathrm{d}x = \lim_{a_1 \to 0} \int_a^{a_1} x^{-2}\,\mathrm{d}x + \lim_{b_1 \to 0} \int_{b_1}^{b} x^{-2}\,\mathrm{d}x$$

The right-hand side of this expression can be evaluated to give

$$\frac{1}{a} - \frac{1}{b} - \lim_{a_1 \to 0} \frac{1}{a_1} + \lim_{b_1 \to 0} \frac{1}{b_1}$$

which is clearly undefined.

To this point all definite integrals have resulted in a positive value, which is consistent with the idea that they represent an area. However, this is not necessarily always the case. For example,

$$\int_{-2}^{2} (x^2 - 4)\,\mathrm{d}x = \left[\frac{x^3}{3} - 4x\right]_{-2}^{2} = -\frac{10}{3}$$

But how do we interpret this "negative area"? The visual interpretation of this particular definite integral, shown in Figure 7.5(a), clearly shows the distinction between this area

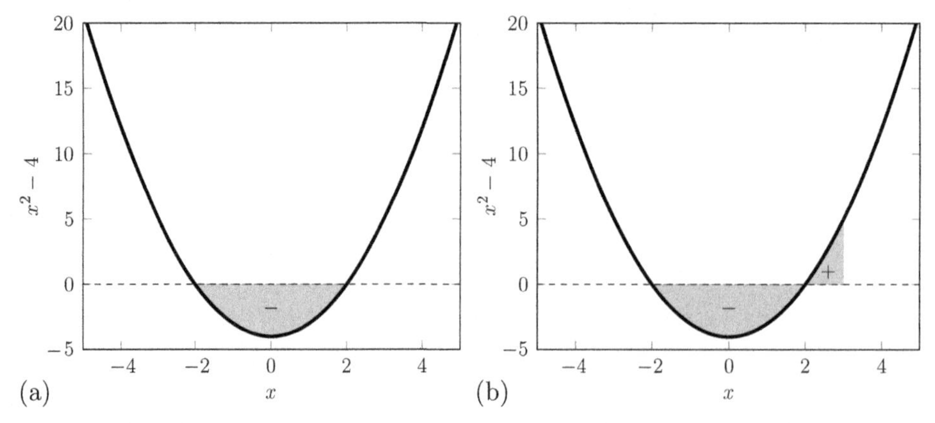

Figure 7.5 An interpretation of positive and negative areas. (a) $\int_{-2}^{2}(x^2 - 4)\,\mathrm{d}x$. (b) $\int_{-2}^{3}(x^2 - 4)\,\mathrm{d}x$.

and the "positive area" shown in Figure 7.4, say. That is, the area between the curve and the x-axis is negative when the curve is *below* the x-axis.

Of course not all definite integrals are represented by areas that are contained entirely on one side of the x-axis. For example, $\int_{-2}^{3}(x^2 - 4)\,dx$ has portions below and above the x-axis, as demonstrated in Figure 7.5(b). This definite integral can be evaluated by partitioning the integral into its negative and positive components

$$\begin{aligned}\int_{-2}^{3}(x^2 - 4)\,dx &= \int_{-2}^{2}(x^2 - 4)\,dx + \int_{2}^{3}(x^2 - 4)\,dx \\ &= -\frac{32}{3} + \frac{7}{3} = -\frac{25}{3}\end{aligned}$$

The overall result is negative because there is a larger contribution from below the x-axis than there is from above. The partitioning of this integral relied on prior knowledge of the roots and shape of the function, but this is not necessary. In practice, one would typically evaluate the integral directly as

$$\int_{-2}^{3}(x^2 - 4)\,dx = \left[\frac{x^3}{3} - 4x\right]_{-2}^{3} = -\frac{25}{3}$$

These results have implications for the definite integrals of *odd* and *even* functions. Recall that if $f(x)$ is odd then $f(x) = -f(x)$, and any definite integral between bounds symmetric about $x = 0$ is zero. In contrast, if $g(x)$ is even then $g(x) = g(-x)$, and any definite integral between bounds symmetric about $x = 0$ is nonzero. More specifically,

$$\begin{array}{ll}\int_{-a}^{a} f(x)\,dx = 0 & f(x) \text{ is an odd function} \\ \int_{-a}^{a} g(x)\,dx = 2\int_{0}^{a} g(x)\,dx & g(x) \text{ is an even function}\end{array} \tag{7.6}$$

This is demonstrated in Figure 7.6. These results can be extended to other functions that have odd and even symmetry about any x, as shown in the following example. We return briefly to the definite integrals of symmetric functions in Example 7.5.

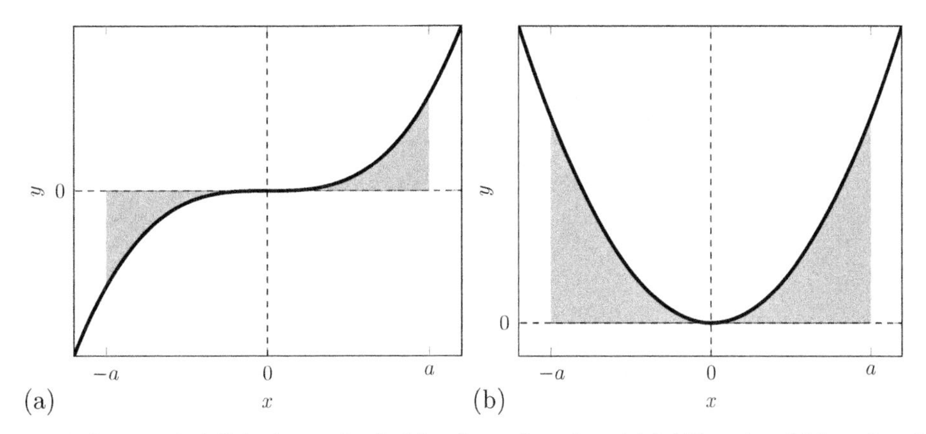

Figure 7.6 Symmetric definite integrals of odd and even functions. (a) Odd function. (b) Even function.

EXAMPLE 7.3

Plot and evaluate the following definite integrals.

a. $\int_{-1}^{7}(x^2 - 6x + 9)\,\mathrm{d}x$

b. $\int_{-2}^{4}(z^3 - 3z^2 + 3z - 1)\,\mathrm{d}z$

Solution

The plots are shown in Figure 7.7 and in both cases we see symmetries about some nonzero x than can be exploited.

a. The definite integral has even symmetry about $x = 3$ and we would expect a (positive) nonzero result. Exploiting this symmetry we can write

$$\int_{-1}^{7}(x^2 - 6x + 9)\,\mathrm{d}x = 2\int_{3}^{7}(x^2 - 6x + 9)\,\mathrm{d}x = \frac{128}{3}$$

The symmetry is also evident from the factorization of the integrand $(x - 3)^2$ and that the bounds can be written as 3 ± 4.

b. The definite integral has odd symmetry about $z = 1$ and is therefore 0. The symmetry is also evident from the factorization of the integrand $(z - 1)^3$ and that the bounds can be written as 1 ± 3.

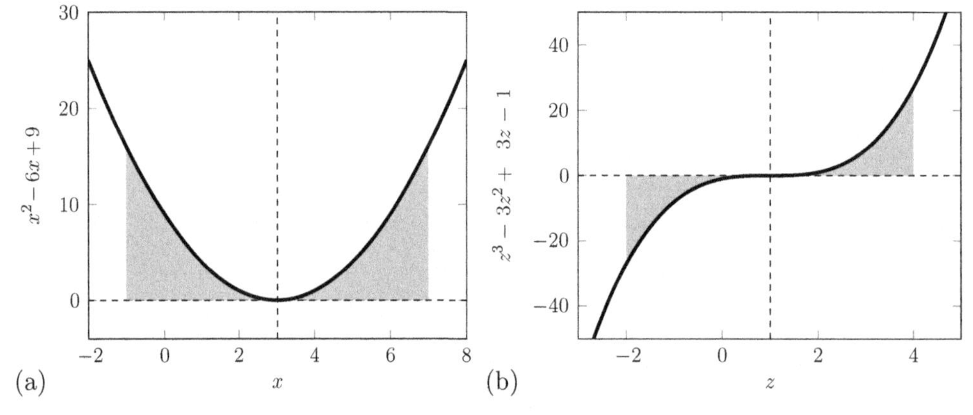

Figure 7.7 Plots of the definite integrals in Example 7.3. (a) $\int_{-1}^{7}(x^2 - 6x + 9)\,\mathrm{d}x$. (b) $\int_{-2}^{4}(z^3 - 3z^2 + 3z - 1)\,\mathrm{d}z$.

We end this section on the general properties of definite integrals with the result

$$\int_{a}^{b} f(x)\,\mathrm{d}x = -\int_{b}^{a} f(x)\,\mathrm{d}x \tag{7.7}$$

Although this is not evident from the visual interpretation of a definite integral, the result should be clear from the algebraic process of performing the integral. In particular, working with the expression (7.4), we have that

$$\int_a^b g'(x)\,dx = g(b) - g(a) = -(g(a) - g(b)) = -\int_b^a g'(x)\,dx$$

For example, we have already seen that $\int_{-2}^{3}(x^2 - 4)\,dx = -25/3$ and it is easily shown that

$$\int_3^{-2}(x^2 - 4)\,dx = \frac{25}{3}$$

7.2 INTEGRATION STRATEGIES

Recall that there are a number of approaches for performing the algebraic integration of integrands that can be considered as products of functions. These were discussed in Chapter 6 and include the *change of variables* and *integration by parts* approaches. In this section, we demonstrate how these approaches can be used to determine definite integrals.

7.2.1 Change of variables

It should be clear that the change of variable approach is needed to perform the indefinite integral

$$\int x(x^2 + 5)^2\,dx$$

We use the substitution $u = x^2 + 5$, which is such that $du \equiv 2x\,dx$, and proceed as

$$\int x(x^2 + 5)^2\,dx = \int u^2 \underbrace{x\,dx}_{\frac{du}{2}} = \frac{1}{2}\int u^2\,du = \frac{1}{6}(x^2 + 5)^3 + c$$

A definite integral of the same integrand proceeds along very similar lines; however, there is a subtly with regard the bounds that we clarify now.

Consider now the definite integral

$$\int_{x=-1}^{x=2} x(x^2 + 5)^2\,dx$$

which has been written to emphasize that the bounds are over x. As above, we require a change in variable, from x to u, but this must also be reflected in the bounds. In particular, if we use $u = x^2 + 5$, the definite integral proceeds as

$$\int_{x=-1}^{x=2} x(x^2 + 5)^2\,dx = \frac{1}{2}\int_{u=(-1)^2+5}^{u=2^2+5} u^2\,du = \left[\frac{u^3}{6}\right]_6^9 = \frac{171}{2}$$

Note carefully that the upper and lower bounds were also transformed to be in terms of u using $u = x^2 + 5$.

EXAMPLE 7.4

Use the change of variables approach to determine the following definite integrals.

a. $\int_1^2 \frac{x+1}{x^2+2x+1}\,\mathrm{d}x$

b. $\int_0^{\pi/2} \cos y \sin^2 y\,\mathrm{d}y$

c. $\int_0^{0.5} z\,\mathrm{e}^{z^2}\,\mathrm{d}z$

Solution

a. We use $u = x^2 + 2x + 1$ which is such that $\mathrm{d}u \equiv 2(x+1)\,\mathrm{d}x$ and proceed as

$$\int_{x=1}^{x=2} \frac{1}{u} \underbrace{(x+1)\,\mathrm{d}x}_{\frac{\mathrm{d}u}{2}} = \frac{1}{2}\int_{u=4}^{u=9} \frac{\mathrm{d}u}{u} = \frac{1}{2}\left[\ln u\right]_4^9 = \ln\left(\frac{3}{2}\right)$$

b. We use $u = \sin y$ which is such that $\mathrm{d}u \equiv \cos y\,\mathrm{d}y$ and proceed as

$$\int_{y=0}^{y=\pi/2} u^2 \underbrace{\cos y\,\mathrm{d}y}_{\mathrm{d}u} = \int_{u=0}^{u=1} u^2\,\mathrm{d}u = \left[\frac{u^3}{3}\right]_0^1 = \frac{1}{3}$$

c. We use $u = z^2$ which is such that $\mathrm{d}u \equiv 2z\,\mathrm{d}z$ and proceed as

$$\int_{z=0}^{z=0.5} \mathrm{e}^u \underbrace{z\,\mathrm{d}z}_{\frac{\mathrm{d}u}{2}} = \frac{1}{2}\int_{u=0}^{u=0.25} \mathrm{e}^u\,\mathrm{d}u = \frac{1}{2}[\mathrm{e}^u]_0^{0.25} \approx 0.14201$$

EXAMPLE 7.5

Use a change of variables approach to prove Eq. (7.6) for odd and even functions.

Solution

Consider a function $f(x)$ that is either odd or even. The symmetric definite integral $f(x)$ between $x = -a$ and $x = a$ can be partitioned at $x = 0$ to give two components, I_1 and I_2. That is

$$\int_{-a}^{a} f(x)\,\mathrm{d}x = \underbrace{\int_{-a}^{0} f(x)\,\mathrm{d}x}_{I_1} + \underbrace{\int_{0}^{a} f(x)\,\mathrm{d}x}_{I_2}$$

Using the change of variables $u = -x$ in I_1, we can proceed as

$$\begin{aligned}-\int_{a}^{0} f(-u)\,\mathrm{d}x + \int_0^a f(x)\,\mathrm{d}x &= \int_0^a f(-u)\,\mathrm{d}u + \int_0^a f(x)\,\mathrm{d}x \\ &= \int_0^a \{f(-x) + f(x)\}\,\mathrm{d}x\end{aligned}$$

where in the last line we have simply relabeled the dummy variable u back to x and regrouped the integral. In the particular case that $f(x)$ is an odd function, $f(-x) = -f(x)$

and we have

$$\int_{-a}^{a} f(x)\,\mathrm{d}x = 0$$

In contrast, if $f(x)$ is an even function, $f(-x) = f(x)$ and we have

$$\int_{-a}^{a} f(x)\,\mathrm{d}x = 2\int_{0}^{a} f(x)\,\mathrm{d}x$$

These expressions are consistent with Eq. (7.6), as required.

7.2.2 Integration by parts

The use of *integration by parts* for definite integrals is exactly as one would expect. In particular, there are no changes of variables and the integration limits are unchanged throughout the process. The method for definite integrals is summarized by the following expression which is a simple modification to Eq. (6.15)

$$\int_{a}^{b} v(x)\frac{\mathrm{d}u}{\mathrm{d}x}\,\mathrm{d}x = [u(x)v(x)]_a^b - \int_{b}^{a} u(x)\frac{\mathrm{d}v}{\mathrm{d}x}\,\mathrm{d}x \tag{7.8}$$

This is best illustrated with examples.

EXAMPLE 7.6

Evaluate

$$\int_{1}^{3} x\,\mathrm{e}^{x}\,\mathrm{d}x$$

Solution

It should be immediately clear that this is amenable to integration by parts and direct application of Eq. (7.8) with $v(x) = x$, $v'(x) = 1$, and $u'(x) = \mathrm{e}^x = u(x)$ leads to

$$\begin{aligned}\int_{1}^{3} x\,\mathrm{e}^{x}\,\mathrm{d}x &= [x\,\mathrm{e}^x]_1^3 - \int_{1}^{3} \mathrm{e}^x\,\mathrm{d}x \\ &= [x\,\mathrm{e}^x]_1^3 - [\mathrm{e}^x]_1^3 \\ &= 2\,\mathrm{e}^3 \approx 40.1711\end{aligned}$$

EXAMPLE 7.7

Evaluate

$$\int_{\pi}^{2\pi} \mathrm{e}^x \cos x\,\mathrm{d}x$$

Solution

You may recognize this as a cyclical example of integration by parts. We set $v(x) = \mathrm{e}^x = v'(x)$, $u'(x) = \cos x$, and $u(x) = \sin x$ and obtain, after some manipulation,

$$\int_{\pi}^{2\pi} e^x \cos x \, dx = [e^x \sin x + e^x \cos x]_{\pi}^{2\pi} - \int_{\pi}^{2\pi} e^x \cos x \, dx$$

The algebraic process was considered in detail in Section 6.3.2 and repeating this process gives

$$\int_{\pi}^{2\pi} e^x \cos x \, dx = \frac{1}{2} [e^x \sin x + e^x \cos x]_{\pi}^{2\pi}$$
$$= \frac{e^{\pi}}{2} \left(1 + e^{\pi}\right)$$

EXAMPLE 7.8

Use integration by parts to compute the following definite integrals. Note that the equivalent indefinite integrals were considered in Example 6.9.

a. $\int_{-1}^{1} e^{2x}(4x - 1)\, dx$
b. $\int_{1}^{2} y^3 e^{-y}\, dy$
c. $\int_{-3}^{-2} e^{z}(z^2 - 1)\, dz$
d. $\int_{0}^{\pi/2} e^{s} \sin s \cos s\, ds$

Solution

The indefinite integrals are taken directly from Example 6.9.

a. $\int_{-1}^{1} e^{2x}(4x - 1)\, dx = \left[e^{2x}\left(2x - \frac{3}{2}\right)\right]_{-1}^{1} \approx 4.1682$
b. $\int_{1}^{2} y^3 e^{-y}\, dy = \left[-e^{-y}\left(y^3 + 3y^2 + 6y + 6\right)\right]_{1}^{2} \approx 0.7433$
c. $\int_{-3}^{-2} e^{z}(z^2 - 1)\, dz = \left[e^{z}(z^2 - 1)^2\right]_{-3}^{-2} \approx 0.4214$
d. $\int_{0}^{\pi/2} e^{s} \sin s \cos s\, ds = \left[\frac{e^{s}}{10}\left(\sin(2s) - 2\cos(2s)\right)\right]_{0}^{\pi/2} \approx 1.1621$

7.3 AREA BETWEEN CURVES

As we have seen, definite integrals can be used to find the area between a curve and the horizontal axis. It is also possible to extend this idea to determine the area between two curves. For example, consider the area between the curves defined by functions $f(x) = x$ and $g(x) = x^2$ bounded between $x = 0$ and $x = 1$. We denote area under the curve of $f(x)$ by A_f and the area under the curve of $g(x)$ as A_g, as shown in Figure 7.8. It should be clear that $A_f > A_g$ and the area between the curves is $A = A_f - A_g$. We have

$$A_f = \int_0^1 f(x)\, dx = \frac{1}{2} \quad A_g = \int_0^1 g(x)\, dx = \frac{1}{3}$$

and so $A = 1/6$.

Consider now the area between the same functions but bounded between $x = 1$ and $x = 2$; this area is shown in Figure 7.9. In this case, we see that $A_g > A_f$ and the required area is given by $A = A_g - A_f$. That is

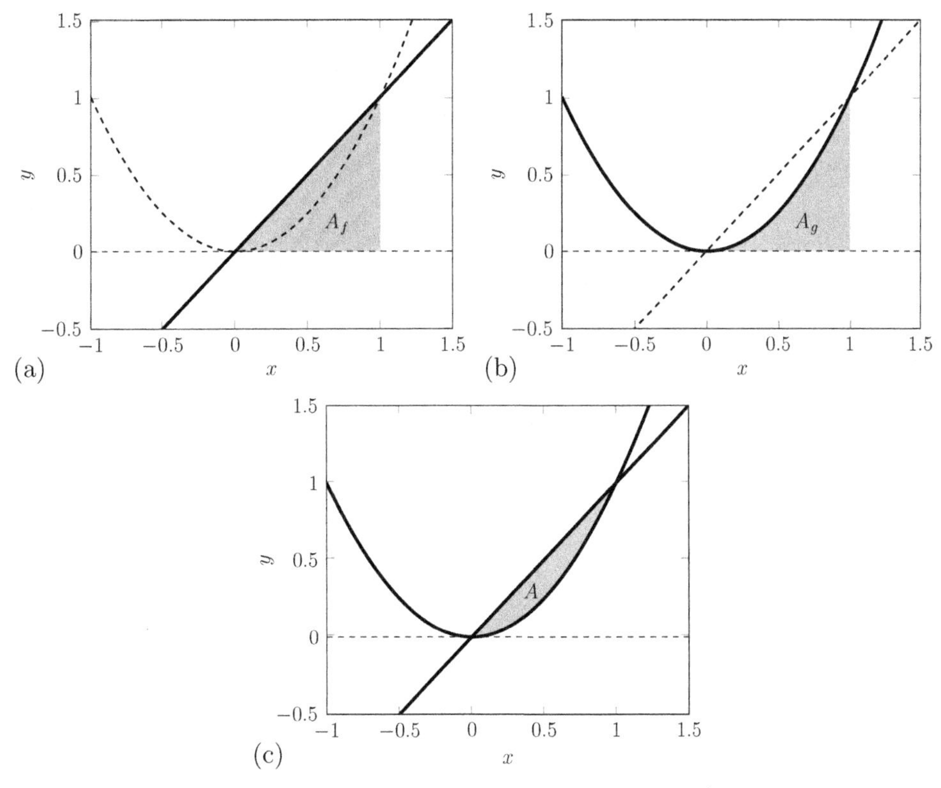

Figure 7.8 Obtaining the area between the curves $f(x) = x$ and $g(x) = x^2$ for $x \in [0, 1]$. (a) The area A_f under the curve of $f(x) = x$. (b) The area A_g under the curve of $g(x) = x^2$. (c) The area $A = A_f - A_g$ between the curves of $f(x)$ and $g(x)$.

$$A = \int_1^2 x^2 \, dx - \int_1^2 x \, dx$$

$$= \frac{7}{3} - \frac{3}{2} = \frac{5}{6}$$

The important observation here is that using the expression $A_f - A_g$ as we did for $x \in [0, 1]$ would have yielded a *negative* area, despite it being contained in the upper half plane. Although one might at first think that this could be easily rectified at the interpretation stage, it does demonstrate the problem that could arise when working out the area in the extended interval $x \in [0, 2]$, for example. That is, without careful partitioning of the integral, the contribution from $x \in [0, 1]$ would offset the contribution from $x \in (1, 2]$. For example, consider now the area between the two curves over the interval $x \in [0, 2]$. One might be tempted to naively compute that

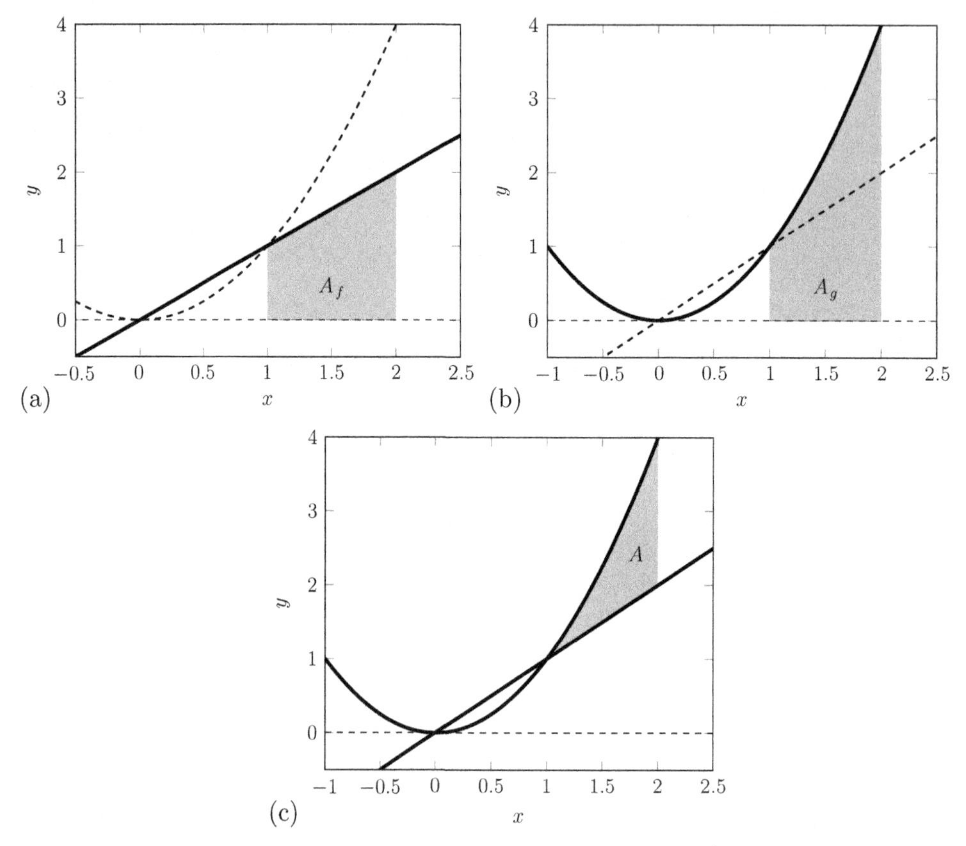

Figure 7.9 Obtaining the area between the curves $f(x) = x$ and $g(x) = x^2$ for $x \in [1, 2]$. (a) The area A_f under the curve of $f(x) = x$. (b) The area A_g under the curve of $g(x) = x^2$. (c) The area $A = A_g - A_f$ between the curves of $f(x)$ and $g(x)$.

$$\int_0^2 x\,dx - \int_0^2 x^2\,dx = \int_0^2 (x - x^2)\,dx = -\frac{2}{3}$$

which is clearly incorrect for the total area contained between the two curves. However, by partitioning the integral into distinct intervals where $x > x^2$ and $x < x^2$ we can obtain the correct area. Note that the two curves cross at $x = 1$, with $x > x^2$ for $x \in [0, 1]$ and $x < x^2$ for $x \in [1, 2]$. This partitioning then gives

$$\int_0^1 (x - x^2)\,dx + \int_1^2 (x^2 - x)\,dx = \left[\frac{x^2}{2} - \frac{x^3}{3}\right]_0^1 + \left[\frac{x^3}{3} - \frac{x^2}{2}\right]_1^2 = 1$$

which is the correct area between the curves over this interval.

This leads to the general conclusion that, in order to calculate the area between two curves, it is first necessary to determine any points of intersection and partition the total integral between these points (if they exist). The total area is obtained from the contribution from each partition calculated by

$$\begin{cases} \int_{a_i}^{a_{i+1}} (g(x) - f(x))\, \mathrm{d}x & \text{over the intervals where } g(x) > f(x) \\ \int_{a_i}^{a_{i+1}} (f(x) - g(x))\, \mathrm{d}x & \text{over the intervals where } g(x) < f(x) \end{cases}$$

where $\{a_i\}$ denotes consecutive points of intersection. In using this approach, we are necessarily forcing each contribution to the total area to be positive. The total area between two curves is therefore always *positive*, even if the area of interest lies beneath the x-axis. This is demonstrated in the following examples.

EXAMPLE 7.9

In each case, determine the area between the stated curves for x bounded between 0 and $\pi/2$.

a. $f(x) = \sin x$ and $g(x) = \cos x$

b. $f(x) = -\sin x$ and $g(x) = -\cos x$

Solution

a. It should be clear that $\sin x$ and $\cos x$ intersect when $\tan x = 1$, that is $x = \pi/4$ on this interval. Furthermore, $\cos x > \sin x$ for $x \in [0, \pi/4)$ and $\cos x < \sin x$ for $x \in (\pi/4, \pi/2]$. The area is therefore determined by

$$A = \int_0^{\pi/4} (\cos x - \sin x)\, \mathrm{d}x + \int_{\pi/4}^{\pi/2} (\sin x - \cos x)\, \mathrm{d}x \approx 0.8284$$

This is illustrated in Figure 7.10(a).

b. This case is identical to part a. but reflected in the x-axis. The intersection still occurs at $x = \pi/4$, but $-\sin x > -\cos x$ for $x \in [0, \pi/4)$ and $-\sin x < -\cos x$ for $x \in (\pi/4, \pi/2]$. The area is therefore again determined by

$$A = \int_0^{\pi/4} (-\sin x + \cos x)\, \mathrm{d}x + \int_{\pi/4}^{\pi/2} (\sin x - \cos x)\, \mathrm{d}x \approx 0.8284$$

We note that this is identical to the area determined in part a. despite the areas existing either side of the x-axis. This is illustrated in Figure 7.10(b).

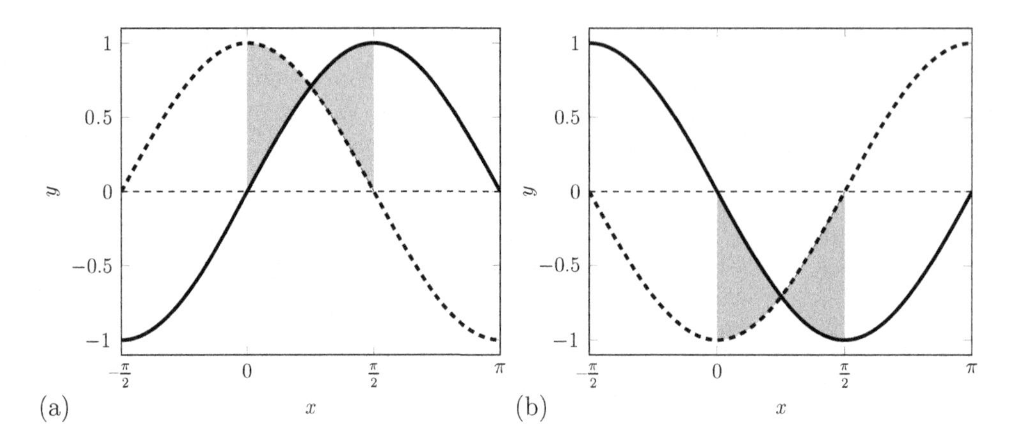

Figure 7.10 Graphical representation of Example 7.9. (a) Area between $\sin x$ and $\cos x$ (dashed) bounded between $x = 0$ and $\frac{\pi}{2}$. (b) Area between $-\sin x$ and $-\cos x$ (dashed) bounded between $x = 0$ and $\frac{\pi}{2}$.

EXAMPLE 7.10

Determine the area between the curves $f(x) = x^2 + 2$ and $g(x) = x + 4$ bounded between $x = -2$ and 3.

Solution

We begin by determining any points of intersection. These are such that

$$x^2 + 2 = x + 4 \Rightarrow x^2 - x - 2 = 0 \quad \text{and so } x = -1, 2$$

These points are within the integration limits and so it is necessary to partition the total area between

$$\begin{cases} x \in [-2, -1] & \text{where } x^2 + 2 > x + 4 \\ x \in [-1, 2] & \text{where } x^2 + 2 < x + 4 \\ x \in [2, 3] & \text{where } x^2 + 2 > x + 4 \end{cases}$$

The area is therefore given by

$$A = \int_{-2}^{-1} (x^2 + 2 - x - 4)\,\mathrm{d}x + \int_{-1}^{2} (x + 4 - x^2 - 2)\,\mathrm{d}x \int_{2}^{3} (x^2 + 2 - x - 4)\,\mathrm{d}x = \frac{49}{6}$$

EXAMPLE 7.11

Determine the area *bounded* by the functions

a. $f(x) = x$ and $g(x) = 4x^2$

b. $f(x) = 2$ and $g(x) = 1 - x^2$

Solution

Note the use of the word *bounded*. Rather than determining the area between the curves over a particular interval, we are required to determine the finite area enclosed between the intersection of the two curves (if any).

a. It should be clear that the two curves intersect at $x = 0$ and $x = 0.25$ and bound a finite area over $x \in [0, 0.25]$. The required area is that between the curves over this interval, as seen in Figure 7.11(a). Note that $x > 4x^2$ for $x \in [0, 0.25]$ and so the bounded area is obtained as

$$\int_0^{0.25} (x - 4x^2)\, \mathrm{d}x = \frac{1}{96}$$

b. In this case, the two functions do not intersect and no region is bounded between the two curves. Although the area *between* the curves over any stated interval would be nonzero, the *bounded* area is undefined. This is seen in Figure 7.11(b).

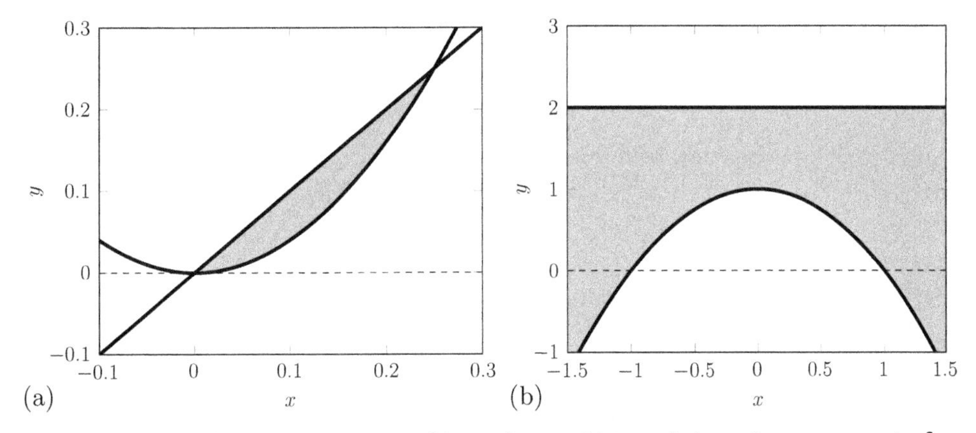

Figure 7.11 Graphical representation of Example 7.11. (a) Bounded area between x and $4x^2$. (b) Unbounded area between 2 and $1 - x^2$.

7.4 DEFINITE INTEGRALS ON YOUR COMPUTER

The relevant command for a definite integral in Wolfram Alpha is a simple modification to that for an indefinite integral. For example, the value of

$$\int_2^5 (x^2 + 2x)\, \mathrm{d}x$$

can be obtained from the command

```
int x^2+2x dx between x=2 and 5
```

and Wolfram Alpha returns that this integral has a value of 60.

As we know from the discussion in this chapter, the ordering of the integration limits is important and you should be careful to define them correctly. In particular, you should state the lower limit first. For example, the command

```
int x^2+2x dx between x=5 and 2
```

returns a value of −60. The reason for this is of course that

$$\int_b^a f(x)\,dx = -\int_a^b f(x)\,dx$$

and the engine has actually been instructed to compute

$$\int_5^2 (x^2 + 2x)\,dx$$

Wolfram Alpha can also be used to determine the area between curves and the area bounded by curves. The commands for these are entirely intuitive. For example, the area between curves x^2 and $-x$ bounded between $x = -2$ and 1 is obtained using the command

```
area between x^2 and -x between x=-2 and 1
```

The output demonstrates that Wolfram Alpha proceeds to partition the relevant integral between the points of intersection, as we would by hand. The numerical answer returned is $\frac{11}{6}$.

In contrast, the command

```
area between x^2 and -x
```

returns a numerical answer of $\frac{1}{6}$. Again, in response to this command, Wolfram Alpha proceeds as we would on paper: it determines the points of intersection and identifies the associated *bounded* areas.

EXAMPLE 7.12

Use Wolfram Alpha to compute the following.

a. $\int_{-8}^{10} \frac{x^5}{x^2+5}\,dx$

b. The area between $f(x) = x^2$ and $g(x) = 2 - 4x^3$ bounded between $x = -2$ and 5.

c. The area bounded between the two functions $h(x) = 2x^2$ and $k(x) = 4 - x$

Solution

a. We use the command

```
int x^5/(x^2+5) between x=-8 and 10
```

and obtain that

$$\int_{-8}^{10} \frac{x^5}{x^2+5}\,\mathrm{d}x \approx 1391.2$$

b. We use the command

```
area between x^2 and 2-4x^3 between x=-2 and 5
```

and obtain a numerical result of approximately 676.094.

c. We use the command

```
area between 2x^2 and 4-x
```

and obtain a numerical result of approximately 7.8988.

7.5 ACTUARIAL APPLICATION: THE FORCE OF INTEREST AS A FUNCTION OF TIME

Recall from our discussion in Chapter 3 that the general force of interest $\delta(t)$ (also known as the continuously compounding rate of interest) and the associated accumulation factor $A(0, t)$ are connected via Eq. (3.22). That is,

$$\delta(t) = \frac{1}{A(0,t)}\frac{\mathrm{d}}{\mathrm{d}t}A(0,t) = \frac{\mathrm{d}}{\mathrm{d}t}\ln A(0,t)$$

We had previously been unable to "reverse" this expression and so could only infer the underlying $\delta(t)$ that would have led to a given $A(0, t)$. In practice, however, it is much more useful to be able to determine $A(0, t)$ from a given $\delta(t)$. This is the subject of this short section on an example use of definite integrals in actuarial science.

It should be clear that Eq. (3.22), restated above, can be integrated with respect to t and this will takes us a step toward obtaining $A(0, t)$ in terms of $\delta(t)$. In particular,

$$\int_0^t \delta(s)\,\mathrm{d}s = \int_0^t \frac{\mathrm{d}}{\mathrm{d}s}\ln A(0,s)\,\mathrm{d}s$$

$$\Rightarrow \ln A(0,t) - \ln A(0,0) = \int_0^t \delta(s)\,\mathrm{d}s$$

Note that we have relabeled the time variable in any integrands to avoid confusion with the t in the upper integration limit.

It is important to note that $A(0, 0) \equiv 1$, which reflects that there has been no time within which a unit investment could have grown. We therefore have $\ln A(0, 0) \equiv 0$ and can write that

$$A(0,t) = \exp\left(\int_0^t \delta(t)\,\mathrm{d}t\right)$$

This expression is the desired result. That is, we are able to determine the accumulation factor $A(0, t)$ under the action of a general force of interest $\delta(t)$. This result can be generalized with very little effort to show that

$$A(t_1, t_2) = \exp\left(\int_{t_1}^{t_2} \delta(s)\, ds\right) \tag{7.9}$$

EXAMPLE 7.13

Compute the following accumulations under the given $\delta(t)$ defined on $t \in [0, \infty]$ and measured in years.

a. The accumulated value at time $t = 1$ of a unit investment made at $t = 0$ under $\delta(t) = \delta$ per annum.

b. The accumulated value at $t = 10$ of £100 invested at $t = 0$ under $\delta(t) = 0.005t$ per annum.

c. The accumulated value at $t = 5$ of £56,000 invested at $t = 2$ under $\delta(t) = 0.01(1 + t^2)$ per annum.

Solution

In each case, we are required to use Eq. (7.9).

a. The accumulated value at time $t = 1$ of a unit investment made at $t = 0$ under $\delta(t) = \delta$ is given by

$$A(0, 1) = \exp\left(\int_0^1 \delta\, dt\right) = e^{\delta}$$

We note that it is consistent with the result determined in Chapter 3 for constant δ.

b. The accumulated value at $t = 10$ of £100 invested at $t = 0$ under $\delta(t) = 0.005t$ is given by

$$\begin{aligned} 100A(0, 10) &= 100 \exp\left(\int_0^{10} 0.005t\, dt\right) \\ &= 100 \exp\left(\left[\frac{0.005}{2} t^2\right]_0^{10}\right) \\ &= 100\, e^{0.25} \\ &\approx £128.40 \end{aligned}$$

c. The accumulated value at $t = 5$ of £56,000 invested at $t = 2$ under $\delta(t) = 0.01(1 + t^2)$ is given by

$$\begin{aligned} 56{,}000A(2, 5) &= 56{,}000 \exp\left(\int_2^5 0.01(1 + t^2)\, dt\right) \\ &= 56{,}000 \exp\left(\left[0.01\left(t + \frac{t^3}{3}\right)\right]_2^5\right) \\ &= 56{,}000\, e^{21/50} \\ &\approx 85{,}229.85 \end{aligned}$$

As we know from previous discussions of the time value of money, the discounting factor that is used to discount a unit amount due at some t_2 back to $t_1 < t_2$ is given by $1/A(t_1, t_2)$. Equation (7.9) can therefore be used to determine the discounting factor applicable under a varying force of interest. In particular,

$$\frac{1}{A(t_1, t_2)} = \exp\left(-\int_{t_1}^{t_2} \delta(s)\, ds\right) \tag{7.10}$$

EXAMPLE 7.14

Compute the following under the given $\delta(t)$ defined on $t \in [0, \infty]$ and measured in years.

a. The present value at $t = 0$ of \$550 due at time $t = 2$ under $\delta(t) = 0.02t^3$ per annum.

b. The present value at $t = 5$ of \$10,000 due at time $t = 10$ under $\delta(t) = 0.05 + 0.001t^2$ per annum.

Solution

a. The present value at $t = 0$ of \$550 due at time $t = 2$ under $\delta(t) = 0.02t^3$ is given by

$$\begin{aligned}
\frac{550}{A(0,2)} &= 550 \exp\left(-\int_0^2 0.02t^3\, dt\right) \\
&= 550 \exp\left(\left[-\frac{t^4}{200}\right]_0^2\right) \\
&= 550\, e^{-0.08} \\
&\approx \$507.71
\end{aligned}$$

b. The present value at $t = 5$ of \$10,000 due at time $t = 10$ under $\delta(t) = 0.05 + 0.001t^2$ is given by

$$\begin{aligned}
\frac{10{,}000}{A(5,10)} &= 10{,}000 \exp\left(-\int_5^{10} \left(0.05 + 0.001t^2\right) dt\right) \\
&= 10{,}000 \exp\left(\left[-0.05t - \frac{0.001t^3}{3}\right]_5^{10}\right) \\
&\approx 10{,}000\, e^{-0.5416} \\
&\approx \$5,817.78
\end{aligned}$$

7.6 QUESTIONS

The following questions are intended to test your knowledge of the material discussed in this chapter. Full solutions are available in Chapter 7 Solutions of Part III. You should use an algebraic approach unless otherwise stated.

Question 7.1. Compute the following definite integrals.

a. $\int_3^5 (x^3 - 2x + 1)\,dx$
b. $\int_0^1 y^2(y^3 + 2)^5\,dy$
c. $\int_4^5 z\,e^{-2z}\,dz$
d. $\int_{-\frac{\pi}{2}}^0 \sin(p)\sqrt{\cos p}\,dp$
e. $\int_9^{10} \frac{q^2}{4q^3-1}\,dq$

Question 7.2. Compute the following definite integrals and interpret your answer.

a. $\int_0^\infty e^{-4y}\,dy$
b. $\int_{-\infty}^2 e^{-\pi x}\,dx$

Question 7.3. Compute

$$\int_0^1 \frac{1}{5 + (x - 2)^2}\,dx$$

Question 7.4. You are given that

$$f(x) = |x| \quad \text{and} \quad g(x) = x^2 - 2$$

a. Calculate the area A_a between the two curves bounded between $x = -5$ and 5.
b. Calculate the area A_b bounded between the two curves.

Question 7.5. Compute the area between the following curves. Illustrate your answers.

a. $f(x) = x^3$ and $g(x) = x^2$ bounded between $x = 0$ and 0.5.
b. $f(x) = x^3$ and $g(x) = x^2$ bounded between $x = -1$ and 5.

Question 7.6. Calculate the areas bounded between the following curves. Illustrate your answers.

a. $f(x) = 2x^2$ and $g(x) = 4 - x$
b. $h(y) = y^2$ and $k(y) = \pm\sqrt{y}$
c. $l(z) = 1$ and $m(z) = 2 - z^2$

Question 7.7. The equation $x = y^2 - 3$ implicitly defines a curve on the x-y plane. Calculate the area enclosed between this curve and $y = x - 1$.

Question 7.8. Determine the accumulated amount at $t = 12$ of an investment of \$500 made at $t = 0$ under the action of the following force of interest.

$$\delta(t) = \begin{cases} 4\% & \text{per annum for } t \in [0, 5) \\ 5\% & \text{per annum for } t \in [5, 9) \\ 0.6t\% & \text{per annum for } t \in [9, 20] \end{cases}$$

where t is measured in years.

Question 7.9. Determine the present value of \$100 due at time $t \in [0, 20]$ under the force of interest stated in Question 7.8.

Question 7.10. An n-year continuously paid unit annuity is a continuous payment stream paid at a constant rate such that a total of \$1 is received each year. Determine an expression for the present value (at $t = 0$) of an n-year continuously paid unit annuity under a constant force of interest δ.

PART II

Further Mathematics

CHAPTER 8

Complex Numbers

Contents

Prerequisite knowledge	**Intended learning outcomes**
• Chapter 1 • Chapter 2 • Chapter 5	• Define and recognize real, imaginary, and complex numbers • Add, subtract, multiply, and divide complex numbers • Produce Argand diagrams • Use Euler's formula to move between Cartesian and polar forms • Give a geometrical interpretation of simple operations on complex numbers • Recall the connection between the real circular and complex exponential functions; use this to derive standard trigonometric identities

Part I of this book was concerned entirely with real numbers. This is, all quantities were part of the number system $\mathbb{R}$ in which any number can be represented as a point on a number line. In this chapter, we broaden our number system to the set of complex numbers, $\mathbb{C}$. As we shall see, complex numbers are a much broader number system, indeed $\mathbb{R} \subset \mathbb{C}$. While at first sight complex numbers may appear to have limited use, they are in fact a very powerful mathematical tool. This chapter is intended to give a solid understanding of the properties of complex numbers and a glimpse of their power in applied mathematics.

8.1 IMAGINARY AND COMPLEX NUMBERS

An important property of real numbers is that the square of any real number is *positive*. For example, $2^2 = 4$ and also $(-2)^2 = 4$, and these lead to the reverse operation $\sqrt{4} = \pm 2$. The consequence of this is that the square root of a *negative* number does not exist. More accurately, we should say that the square root of a negative number does not exist *in the set of real numbers*, $\mathbb{R}$. This is our motivation for defining *imaginary numbers*.

We define the basic imaginary number as $\mathrm{i} = \sqrt[+]{-1}$, and note that the square root of any negative, real number can now be written as some real multiple of i. For example

$$\sqrt{-4} = \sqrt{-1 \times 4} = \sqrt{-1} \times \sqrt{4} = \pm 2\mathrm{i}$$

Similarly $\sqrt{-9} = \pm 3\mathrm{i}$ and $\sqrt{-2} = \sqrt{2}\mathrm{i}$, for example.[1] The quantity i can be thought of as a mathematical object in its own right and is such that, for example,

$$\begin{aligned}
\mathrm{i}^2 &= \sqrt{-1} \times \sqrt{-1} = -1 \\
\mathrm{i}^3 &= \mathrm{i} \times \mathrm{i}^2 = -\mathrm{i} \\
\mathrm{i}^4 &= \mathrm{i}^2 \times \mathrm{i}^2 = -1 \times -1 = 1 \\
\mathrm{i}^5 &= \mathrm{i} \times \mathrm{i}^4 = \mathrm{i} \\
&\vdots
\end{aligned}$$

Furthermore,

$$\frac{1}{\mathrm{i}} = \frac{\mathrm{i}}{\mathrm{i}^2} = \frac{\mathrm{i}}{-1} = -\mathrm{i}$$

EXAMPLE 8.1

Express the following in terms of i.

a. $\sqrt{-m^2}$ for $m \in \mathbb{R}$

b. i^{10}

c. $(a\mathrm{i})^7$ for $a \in \mathbb{R}$

d. $(-\pi)^{101/2}$

[1] Note that $\sqrt{-1}$ is sometimes denoted by j, although we will use i throughout this book.

Solution

a. $\sqrt{-m^2} = \sqrt{-1} \times \sqrt{m^2} = \pm m\mathrm{i}$

b. $\mathrm{i}^{10} = (\mathrm{i}^2)^5 = (-1)^5 = -1$

c. $(a\mathrm{i})^7 = a^7\mathrm{i}^7 = a^7 \times \left(\mathrm{i}^2\right)^3 \times \mathrm{i} = -a^7\mathrm{i}$

d. $(-\pi)^{101/2} = \pi^{101/2} \times \mathrm{i}^{101} = \pm\sqrt{\pi^{101}} \times (-1)^{50} \times \mathrm{i} = \pm\sqrt{\pi^{101}}\mathrm{i}$

In an analogous way to real numbers, imaginary numbers can be thought of as existing on a continuous number line. For example, 3i, 3.14i, and πi are all examples of imaginary numbers. Crucially, the imaginary number line is a *different* number line to the real number line, that is, real and imaginary numbers are distinct. However, it is possible to move between the real and imaginary number lines using, for example, the operations in Example 8.1. We will return to an interesting geometrical interpretation of real and imaginary numbers in Section 8.4 which will shed light on such transformations.

In contrast to both purely real and purely imaginary numbers, a *complex number* has both real and imaginary components. In general, a complex number is written as

$$z = a + b\mathrm{i} \tag{8.1}$$

where $a, b \in \mathbb{R}$. The quantities a and b are said to be the real and imaginary components of z, respectively. We might denote this as

$$\mathrm{Re}(a + b\mathrm{i}) = a \quad \text{and} \quad \mathrm{Im}(a + b\mathrm{i}) = b$$

EXAMPLE 8.2

Classify the following numbers as either real, imaginary, or complex.

a. 4

b. 3i

c. $10 + \mathrm{i}$

d. $-5 - \pi\mathrm{i}$

Solution

a. 4 is a real number.

b. 3i is an imaginary number.

c. $10 + \mathrm{i}$ is a complex number.

d. $-5 - \pi\mathrm{i}$ is a complex number.

We also note that real and imaginary numbers are contained in the set complex numbers. All are therefore complex numbers too.

Unlike purely real and purely imaginary numbers that exist on one-dimensional number lines, complex numbers are two dimensional, that is they have two distinct components and can be thought of as existing on the *complex plane*. The complex plane is a useful way of picturing the complex number system $\mathbb{C}$, which is in clear contrast the

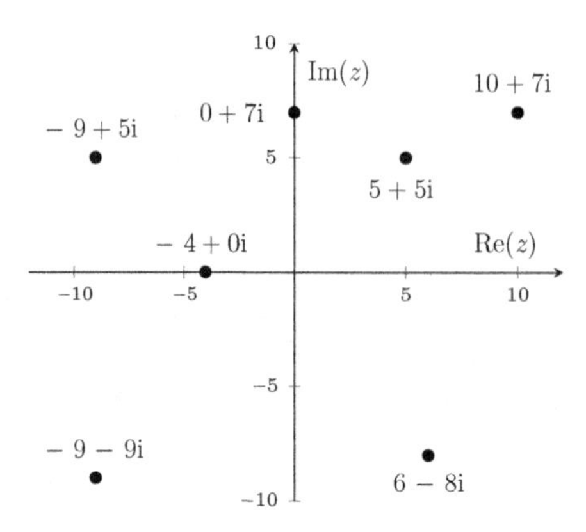

Figure 8.1 The complex plane (Argand diagram).

one-dimensional real number system $\mathbb{R}$. Note that it is convention to indicate the real line as the horizontal axis of the complex plane and the imaginary line as the vertical axis. If $a = 0$ in Eq. (8.1) the complex number is purely imaginary and exists on the vertical axis, and if $b = 0$ the complex number is purely real and exists on the horizontal real line. This explains why we consider the set of real numbers to be a subset of the set of complex numbers, that is $\mathbb{R} \subset \mathbb{C}$.

The complex plane and a number of complex numbers are illustrated in Figure 8.1. Note that such a graphical representation of a complex number is referred to as an *Argand diagram*. The analogy to a Cartesian coordinate system should be clear with the x- and y-axes of the Cartesian system being equivalent for the real and imaginary axes of an Argand diagram.

8.2 SIMPLE OPERATIONS ON COMPLEX NUMBERS

We now consider how one can work with complex numbers, in particular how we add, subtract, multiply, and divide them. We return to raising complex numbers to powers in Section 8.5.

Adding and subtracting complex numbers is a relatively simple process as the real and imaginary parts remain independent and can be considered entirely separately. For example, we can add $z_1 = 5 + 7\text{i}$ to $z_2 = 3 + \text{i}$ by treating the real and imaginary components as independent. This is

$$\begin{aligned} z_1 + z_2 &= (5 + 7\text{i}) + (3 + \text{i}) \\ &= \underbrace{(5 + 3)}_{\text{Re}(z_1)+\text{Re}(z_2)} + \underbrace{(7 + 1)}_{\text{Im}(z_1)+\text{Im}(z_2)}\text{i} \\ &= 8 + 8\text{i} \end{aligned}$$

Subtraction follows in the obvious way

$$\begin{aligned} z_1 - z_2 &= (5+7\mathrm{i}) - (3+\mathrm{i}) \\ &= \underbrace{(5-3)}_{\mathrm{Re}(z_1)-\mathrm{Re}(z_2)} + \underbrace{(7-1)}_{\mathrm{Im}(z_1)-\mathrm{Im}(z_2)}\mathrm{i} \\ &= 2+6\mathrm{i} \end{aligned}$$

Multiplication is, however, more complicated as the independence of the real and imaginary components is not conserved; for example, we know that $\mathrm{i} \times \mathrm{i} = -1$, that is, the product of two imaginary numbers is a real number. The product $z_1 \times z_2$ should therefore be approached with care and is actually akin to "multiplying out" in algebra. In particular,

$$\begin{aligned} z_1 \times z_2 &= (5+7\mathrm{i})(3+\mathrm{i}) \\ &= 5 \times (3+\mathrm{i}) + 7\mathrm{i} \times (3+\mathrm{i}) \\ &= 15 + 5\mathrm{i} + 21\mathrm{i} + 7\mathrm{i}^2 \\ &= 15 - 7 + (5+21)\mathrm{i} \\ &= 8 + 26\mathrm{i} \end{aligned}$$

The operations of addition, subtraction, and multiplication can be extended to three or more complex numbers in the obvious way.

EXAMPLE 8.3

Determine the following for $z_3 = -2 + 6\mathrm{i}$, $z_4 = 20 + 0.3\mathrm{i}$, and $z_5 = -\pi + 15\mathrm{i}$.

a. $z_3 + z_4 + z_5$
b. $z_3 + z_4 - z_5$
c. $4z_3z_4$
d. $z_3z_4z_5$

Solution

a. $z_3 + z_4 + z_5 = (-2 + 20 - \pi) + (6 + 0.3 + 15)\mathrm{i} = 18 - \pi + 21.3\mathrm{i}$
b. $z_3 + z_4 - 2z_5 = (-2 + 20 + 2\pi) + (6 + 0.3 - 2 \times 15)\mathrm{i} = 18 + 2\pi - 23.7\mathrm{i}$
c. $4z_3z_4 = 4 \times (-2 + 6\mathrm{i}) \times (20 + 0.3\mathrm{i}) = 4 \times (-40 - 0.6\mathrm{i} + 120\mathrm{i} - 1.8) = -167.2 + 477.6\mathrm{i}$
d. $z_3z_4z_5 = (-2 + 6\mathrm{i})(20 + 0.3\mathrm{i})(-\pi + 15\mathrm{i}) = (-41.8 + 119.4\mathrm{i})(-\pi + 15\mathrm{i}) = 41.8\pi - 1791 - (119.4\pi + 627)\mathrm{i}$

Dividing complex numbers is slightly more complicated than multiplying them. For example, consider

$$\frac{z_1}{z_2} = \frac{5+7\mathrm{i}}{3+\mathrm{i}}$$

It is not possible to perform this division directly as the denominator has two independent components (the real and the imaginary components). Progress could be made, however, if the denominator could be manipulated to be entirely real. In order to do this we need to first discuss the concept of *conjugation*.

We begin by defining the *complex conjugate* of general complex number $z = a + bi$ to be $z^* = a - bi$. As we shall see in Section 8.3, complex numbers often arise in complex conjugate pairs, that is $a \pm bi$, and we will return to this concept later. For the moment, however, it is sufficient to note the effect of multiplying z by its complex conjugate z^*

$$zz^* = (a + bi)(a - bi) = a^2 + b^2 + (ab - ba)\mathrm{i} = a^2 + b^2$$

For example, it is easy to see that

$$(1 + 4\mathrm{i})(1 - 4\mathrm{i}) = 17$$

$$(4 - 2\mathrm{i})(4 + 2\mathrm{i}) = 20$$

This fact can be exploited in the division z_1/z_2 as follows

$$\begin{aligned}\frac{z_1}{z_2} = \frac{5 + 7\mathrm{i}}{3 + \mathrm{i}} &= \frac{5 + 7\mathrm{i}}{3 + \mathrm{i}} \times \frac{3 - \mathrm{i}}{3 - \mathrm{i}} \\ &= \frac{1}{10}(5 + 7\mathrm{i})(3 - \mathrm{i}) \\ &= \frac{11}{5} - \frac{8}{5}\mathrm{i}\end{aligned}$$

That is, to divide by a complex number, one must first multiply both the numerator and denominator by the complex conjugate of the denominator. This expresses the denominator as a purely real number and the division can proceed as a multiplication. The general expression for computing the quotient of two complex numbers is therefore written as

$$\frac{z_1}{z_2} = \frac{z_1 \times z_2^*}{z_2 \times z_2^*} \tag{8.2}$$

Of course, setting $z_1 = 1$ shows that this expression can be used to take the reciprocal of a complex number.

EXAMPLE 8.4

Simplify the following expressions.

a. $\frac{2-6\mathrm{i}}{1+5\mathrm{i}}$

b. $\frac{1-5\mathrm{i}}{\mathrm{i}}$

c. $\frac{1}{1+2\mathrm{i}}$

d. $\frac{(3+2\mathrm{i})(4+4\mathrm{i})}{(2-3\mathrm{i})^2}$

Solution

Using Eq. (8.2), we obtain

a. $\frac{2-6i}{1+5i} = \frac{2-6i}{1+5i} \times \frac{1-5i}{1-5i} = \frac{-28-16i}{26} = -\frac{14}{13} - \frac{8}{13}i$

b. $\frac{1-5i}{i} = \frac{1-5i}{i} \times \frac{-i}{-i} = -5 - i$

c. $\frac{1}{1+2i} = \frac{1}{1+2i} \times \frac{1-2i}{1-2i} = \frac{1}{5} - \frac{2}{5}i$

d. $\frac{(3+2i)(4+4i)}{(2-3i)^2} = \frac{4+20i}{-5-12i} = \frac{4+20i}{-5-12i} \times \frac{-5+12i}{-5+12i} = \frac{-260-52i}{169} = -\frac{20}{13} - \frac{4}{13}i$

Note that part b. could also have be obtained from $\frac{1-5i}{i} = \frac{1}{i} - 5\frac{i}{i} = -5 - i$

8.3 COMPLEX ROOTS OF REAL POLYNOMIAL FUNCTIONS

The properties of polynomial functions were first discussed in Chapter 2. You may recall from that discussion that polynomials of order n can have at most n real roots, although there is no requirement that they have exactly n real roots. We return to this property now in the context of complex numbers.

In all that follows we assume that we have a real polynomial function; that is, the coefficients $a_0, a_1, \ldots, a_n$ in Eq. (2.9) are real, but we allow the independent variable to be complex.[2]

8.3.1 Quadratic polynomials

Consider a general quadratic function (polynomial function of order 2) of the form

$$f(z) = a_2 z^2 + a_1 z + a_0 \tag{8.3}$$

You will recall from your prior studies that the two roots of this quadratic function are given by

$$z = \frac{-a_1 \pm \sqrt{a_1^2 - 4a_0 a_2}}{2a_2} \tag{8.4}$$

At this stage it is useful to define the *discriminant* of a quadratic of form Eq. (8.3) to be

$$\text{discriminant} = a_1^2 - 4a_2 a_0$$

that is, the discriminant is the value to be square rooted in Eq. (8.4).

When the coefficients of a polynomial are such the discriminant is positive, the quadratic has two real roots. However, if the discriminant is negative, the roots of the quadratic are determined by the square root of a negative number and so do not exist as real numbers. We visualize these situations in Figure 8.2 with the functions

[2] The results can be readily extended to $a_0, a_1, \ldots, a_n \in \mathbb{C}$ but this would take us beyond the scope of our discussion.

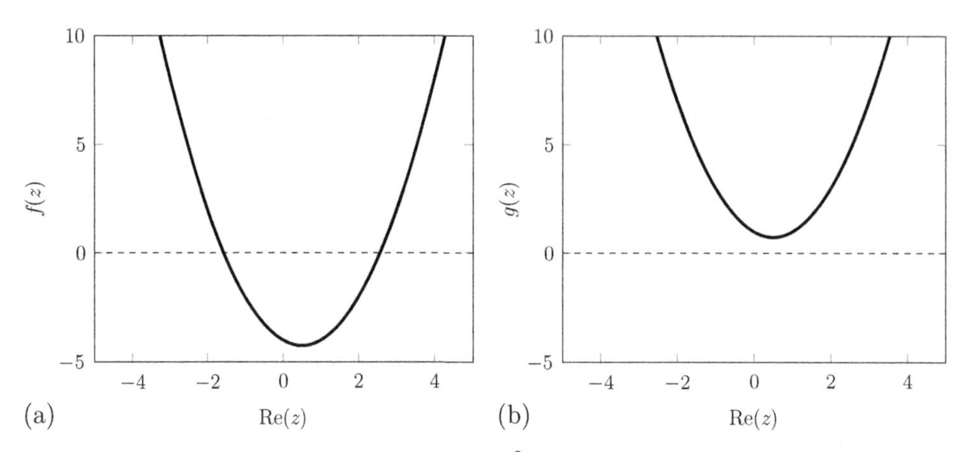

Figure 8.2 Roots of quadratic expressions. (a) $f(z) = z^2 - z - 4$ for $z \in \mathbb{R}$ has two real roots. (b) $g(z) = z^2 - z + 1$ for $z \in \mathbb{R}$ has no real roots.

$f(z) = z^2 - z + 4$ and $g(z) = z^2 - z + 1$. The discriminate of $f(x)$ is $17 > 0$ and it has two real roots. In contrast the discriminant of $g(x)$ is $-3 < 0$ and it has no real roots. Note that if the discriminant is zero, then there is a single repeated root and the curve touches the horizontal axis. This is the clearly the boundary between having two real roots and no real roots.

Now that we are aware of complex numbers, we can make more progress in cases where the discriminant is negative. Equation (8.4) shows that, although there are no real roots, there in fact two complex roots that exist as a *complex-conjugate pair*. For example, $g(x)$ has complex roots given by

$$z = \frac{+1 \pm \sqrt{1-4}}{2} = \frac{1}{2} \pm \frac{\sqrt{3}}{2}\mathrm{i}$$

The roots can be verified by evaluating the function at these complex values

$$\left(\frac{1}{2} + \frac{\sqrt{3}}{2}\mathrm{i}\right)\left(\frac{1}{2} + \frac{\sqrt{3}}{2}\mathrm{i}\right) - \left(\frac{1}{2} + \frac{\sqrt{3}}{2}\mathrm{i}\right) + 1 = \frac{1}{4} + \frac{\sqrt{3}}{2}\mathrm{i} - \frac{3}{4} - \frac{1}{2} - \frac{\sqrt{3}}{2}\mathrm{i} + 1 = 0$$

$$\left(\frac{1}{2} - \frac{\sqrt{3}}{2}\mathrm{i}\right)\left(\frac{1}{2} - \frac{\sqrt{3}}{2}\mathrm{i}\right) - \left(\frac{1}{2} - \frac{\sqrt{3}}{2}\mathrm{i}\right) + 1 = \frac{1}{4} - \frac{\sqrt{3}}{2}\mathrm{i} - \frac{3}{4} - \frac{1}{2} + \frac{\sqrt{3}}{2}\mathrm{i} + 1 = 0$$

We therefore see that a polynomial function of order 2 has two roots in the complex plane. These are either two real, a single repeated real root, or two complex roots.

EXAMPLE 8.5

Obtain and interpret the discriminant of the following quadratic functions. Determine the two roots in each case.

a. $h(z) = z^2 + z + 1$
b. $j(z) = z^2 + 4z + 1$

c. $k(z) = z^2 - 5z + 1$
d. $l(z) = 2z^2 + z + 2$
e. $m(z) = 4z^2 - 8z + 4$

Solution

a. The discriminant is -3 and we expect two complex roots. These are given by $z = -\frac{1}{2} \pm \frac{\sqrt{3}}{2}\mathrm{i}$.

b. The discriminant is 12 and we expect two real roots. These are given by $z = -2 \pm \sqrt{3}$.

c. The discriminant is 21 and we expect two real roots. These are given by $z = \frac{5}{2} \pm \frac{\sqrt{21}}{2}$.

d. The discriminant is -15 and we expect two complex roots. These are given by $z = -\frac{1}{4} \pm \frac{\sqrt{15}}{4}\mathrm{i}$.

e. The discriminant is 0 and we expected a repeated real root. This is given by $z = 1$.

8.3.2 Cubic polynomials

Recall our discussion in Chapter 2 regarding real cubic polynomials. In particular, recall that the asymptotic behavior of a cubic (i.e., its behavior for large positive and negative values of the real independent variable) is such that it must have at least one real root. With this in mind it should be clear that if the real root is z_1, a general cubic can be factorized as

$$a_2z^2 + a_1z + a_0 = (z - z_1)(pz^2 + qz + r)$$

for some $p, q, r \in \mathbb{R}$. The interpretation of this expression then follows from our discussion of the roots of quadratic expressions. In particular, in addition to the real root at z_1, a cubic could have either two real roots, a single repeated real root, or two complex roots existing as a complex conjugate pair. The discriminant of the quadratic factor of course determines which case it actually is for any given polynomial.

For example, consider the following three cubic functions.

$$f(z) = z^3 - 2z^2 - z + 2, \quad g(z) = z^3 + z^2 - z - 1, \quad h(z) = z^3 + z - 2$$

It should be clear that, in each case, $z = 1$ is a root and so $(z - 1)$ is a factor. After some manipulation it is possible to show that the functions can be rewritten as

$$f(z) = (z - 1)(z^2 - z - 2), \quad g(z) = (z - 1)(z^2 + 2z + 1),$$
$$h(z) = (z - 1)(z^2 + z + 2)$$

The discriminant of the quadratic factors are 9, 0, and -7, respectively, which indicates that, in addition to the real root at $z = 1$, $f(z)$ has two real roots, $g(z)$ has a repeated root, and $h(z)$ has two complex roots. This can be observed in Figure 8.3.

This discussion is not intended to develop a method for determining the roots of a general cubic polynomial. Rather it is intended to give an insight into the roots theoretical possible for a cubic function.

The crucial result to take from this discussion is that a real polynomial of order 3 has three roots in the complex plane. This is consistent with our previous result that a real

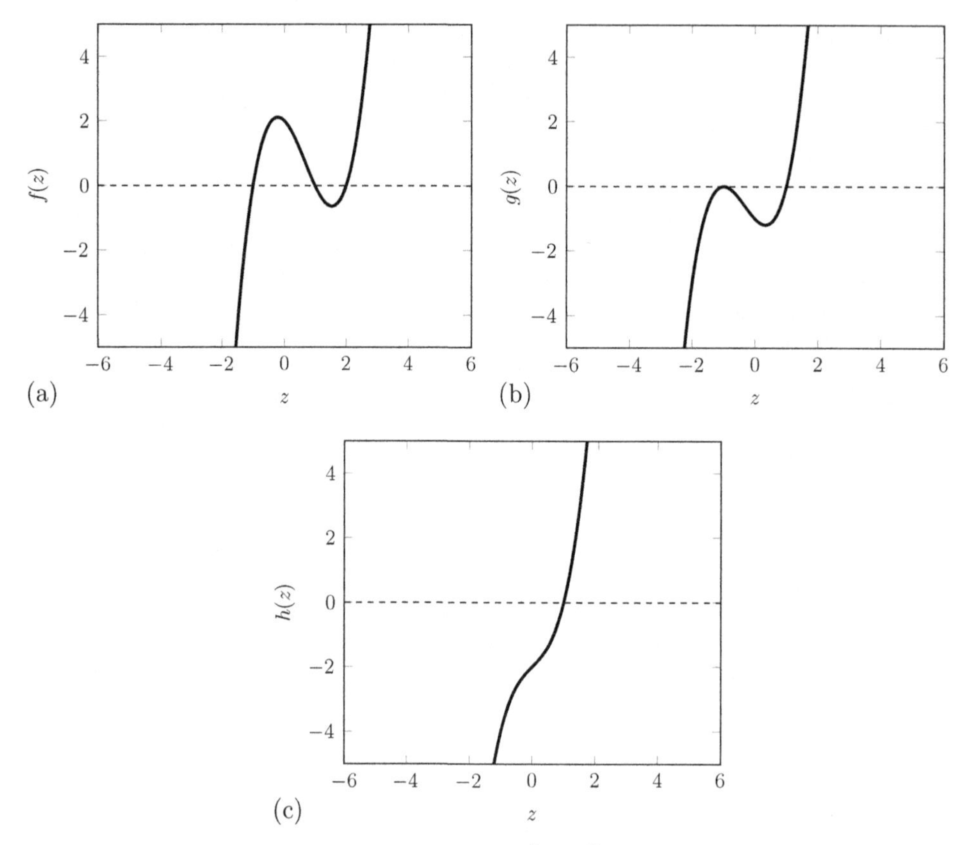

Figure 8.3 Roots of quadratic expressions. (a) $f(z) = z^3 - 2z^2 - z + 2$ for $z \in \mathbb{R}$ has three real roots. (b) $g(z) = z^3 + z^2 - z + 1$ for $z \in \mathbb{R}$ has three real roots, one repeated. (c) $h(z) = z^3 + z - 2$ for $z \in \mathbb{R}$ has one real root.

polynomial of order 2 has two roots (indeed the result for cubics relies on the result for quadratics). It is possible to generalize these two observations to the result that

a real polynomial of order n has n roots in the complex plane.

This is stated without formal proof, but it should be of no surprise that this is true.

EXAMPLE 8.6

Determine all roots of the function $f(z) = z^3 - 1$ for $z \in \mathbb{C}$.

Solution

One might be tempted to simply write down that $f(z)$ has a single root at $z = 1$, but this is not the full story as the function is a polynomial of order 3 and so has three roots in the complex plane. We know that $(z - 1)$ is a factor and some manipulation shows that

$$f(z) = (z - 1)\left(z^2 + z + 1\right)$$

The quadratic is easily solved to give the two further roots $z = -\frac{1}{2} \pm \frac{\sqrt{3}}{2}\mathrm{i}$. We return to the cube root of unity in Example 8.14.

EXAMPLE 8.7

Use your knowledge of quadratic functions to determine all roots in the complex plane of the quartic function $g(z) = z^4 - 1$.

Solution

We make the substitution $u = z^2$ which leads to $g(u) = u^2 - 1$. This has roots $u = \pm 1$. In terms of z, we have $z^2 = \pm 1$, which leads to a total of four roots

$$z = \pm\sqrt{1} \text{ and } \pm \mathrm{i}$$

This particular quartic function is therefore seen to have four roots in the complex plane. The reader is invited to confirm that these are indeed roots. Note that this approach is in contrast to the factorization approach used previously which would require one to determine the factorization $g(z) = (z-1)(z+1)(z^2+1)$ and is much more difficult.

8.4 ARGAND DIAGRAMS AND THE POLAR FORM

8.4.1 Polar form

We have previously discussed the necessity of thinking of complex numbers as existing on a two-dimensional complex plane, as illustrated in Figure 8.1. Until now, we have chosen to define a complex number as a point on the complex plane given by its real and imaginary components and this led to the idea of an Argand diagram that uses co-ordinates analogous to a two-dimensional Cartesian plane. Although this analogy is useful for working with complex numbers under the action of addition and subtraction, where the independence of the real and imaginary components is preserved, it is not very useful for operations such as multiplication and division where the independence is not preserved. To this end, we now consider an alternative geometrical interpretation of complex numbers, the *polar form*.

The polar form still considers a complex number as existing on a two-dimensional complex plane and so still requires two numbers to define it. However, rather than defining a horizontal and vertical position, the polar form specifies an angle measured from the positive horizontal axis, called the *argument*, and a radial distance from the origin, called the *modulus*. This is illustrated in Figure 8.4.

Clearly the modulus, r, and argument, θ, are related to the real and imaginary components of a complex number through projections onto the horizontal and vertical axes. That is,

$$a = r\cos\theta \quad \text{and} \quad b = r\sin\theta$$

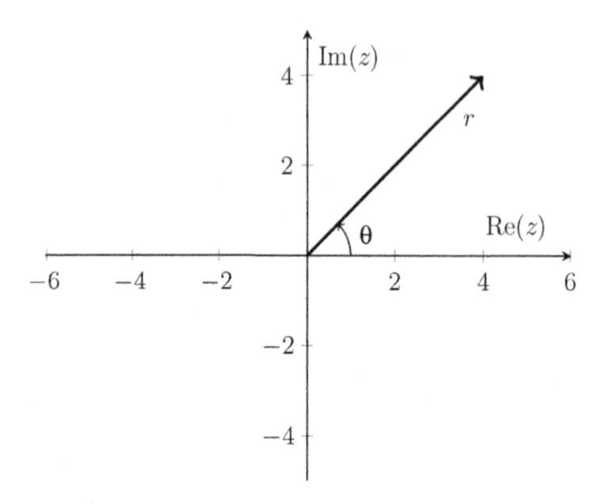

Figure 8.4 Argand diagram for polar form of complex number $z = 4 + 4\mathrm{i}$.

and so any complex number can be represented as

$$a + b\mathrm{i} = r(\cos\theta + \mathrm{i}\sin\theta) \tag{8.5}$$

where

$$r = |a + b\mathrm{i}| = \sqrt{a^2 + b^2} = \sqrt{zz^*} \tag{8.6}$$

$$\theta = \mathrm{Arg}(a + b\mathrm{i}) = \arctan\left(\frac{b}{a}\right) \tag{8.7}$$

Given the cyclical nature of the argument, it is often convenient to work with the *principal argument*. This is defined as the unique value of θ under Eq. (8.7) such that $-\pi < \theta \leq \pi$ (in radians). Any complex number in the upper half plane of the Argand diagram, that is, has positive imaginary component, has a positive principal argument; any complex number with negative imaginary component has a negative argument. The general argument and the principal argument are often distinguished by the use of $\arg(z)$ and $\mathrm{Arg}(z)$, respectively. It should be clear that $\arg(z) = \{\mathrm{Arg}(z) + 2k\pi : k \in \mathbb{Z}\}$.

EXAMPLE 8.8

Express the following complex numbers in polar form. Illustrate each complex number on an Argand diagram.

a. $z = 4$
b. $z = 2\mathrm{i}$
c. $z = 2 + 4\mathrm{i}$
d. $z = -1 + 2\mathrm{i}$
e. $z = -1 - 3\mathrm{i}$
f. $z = 2 - 5\mathrm{i}$

Solution

a. We have $r = 4$, $\mathrm{Arg}(4) = 0$ and so $z = 4(\cos 0 + \mathrm{i}\sin 0)$.
b. We have $r = 2$, $\mathrm{Arg}(4) = \frac{\pi}{2}$ and so $z = 2(\cos\frac{\pi}{2} + \mathrm{i}\sin\frac{\pi}{2})$.

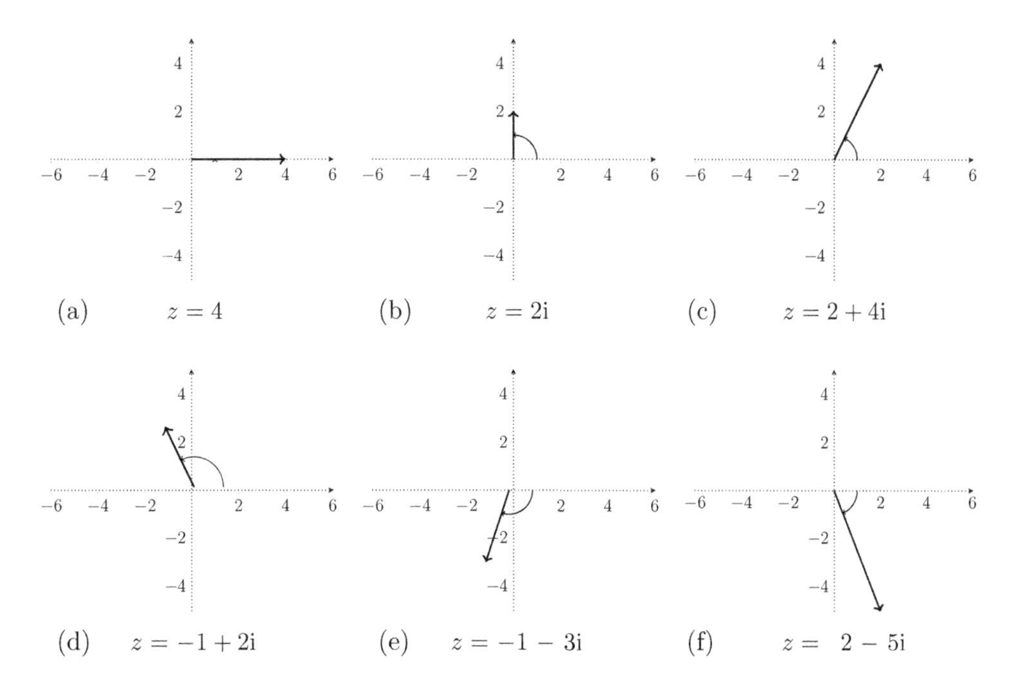

Figure 8.5 Argand diagrams for complex numbers in Example 8.8. (a) $z = 4$. (b) $z = 2\mathrm{i}$. (c) $z = 2 + 4\mathrm{i}$. (d) $z = -1 + 2\mathrm{i}$. (e) $z = -1 - 3\mathrm{i}$. (f) $z = 2 - 5\mathrm{i}$.

c. We have $r = \sqrt{2^2 + 4^2} = 2\sqrt{5}$, $\mathrm{Arg}(2 + 4\mathrm{i}) = \arctan 2 \approx 1.107$ and so $z = 2\sqrt{5}(\cos 1.107 + \mathrm{i} \sin 1.107)$.

d. We have $r = \sqrt{1^2 + 2} = \sqrt{5}$, $\mathrm{Arg}(-1 + 2\mathrm{i}) = \arctan(-2) = \pi - \arctan 2 \approx 2.034$ and so $z = \sqrt{5}(\cos 2.034 + \mathrm{i} \sin 2.034)$.

e. We have $r = \sqrt{1^2 + 3^2} = \sqrt{10}$, $\mathrm{Arg}(-1 - 3\mathrm{i}) = \arctan(-3) = \arctan(3) - \pi \approx -1.893$ and so $z = \sqrt{10}(\cos(-1.893) + \mathrm{i} \sin(-1.893))$.

f. We have $r = \sqrt{2^2 + 5^2} = \sqrt{29}$, $\mathrm{Arg}(2 - 5\mathrm{i}) = \arctan\left(-\frac{5}{2}\right) \approx -1.190$ and so $z = \sqrt{29}(\cos(-1.190) + \mathrm{i} \sin(-1.190))$.

The Argand diagrams are shown in Figure 8.5.

The advantage of using the polar method to work with complex numbers is that the operations of multiplication and division can be thought of as acting on the argument and modulus of a complex number; the operations are then easily pictured in terms of *rotating* and *stretching* in the complex plane. For example, the effect of multiplying a positive real number by i is to change the argument of that real number such that it rotates $\frac{\pi}{2}$ to the imaginary line. This avoids the difficultly under the Cartesian analogy of moving between independent axes.

The disadvantage of the polar form, as it appears here, should be apparent from Eq. (8.5): it involves familiarity with the trigonometric identities discussed in Chapter 2 and Appendix A. We return to a much more useful version of the polar form in Section 8.5 which will remove this problem. However, it is important that you are first able to operate with complex numbers in this cumbersome but intuitive form. In particular, our aim in this section is to derive standard expressions for multiplication and division in the complex plane that can be interpreted geometrically.

8.4.2 Multiplication and division of complex numbers

We begin by looking at two general complex numbers in polar form

$$z_1 = r_1(\cos\theta_1 + \mathrm{i}\sin\theta_1)$$
$$z_2 = r_2(\cos\theta_2 + \mathrm{i}\sin\theta_2)$$

The product $z_1 z_2$ is then

$$\begin{aligned} z_1 z_2 &= r_1(\cos\theta_1 + \mathrm{i}\sin\theta_1)\, r_2(\cos\theta_2 + \mathrm{i}\sin\theta_2) \\ &= r_1 r_2(\cos\theta_1\cos\theta_2 - \sin\theta_1\sin\theta_2 + \mathrm{i}(\cos\theta_1\sin\theta_2 + \sin\theta_1\cos\theta_2)) \end{aligned}$$

In order to proceed, we note the following identities from Appendix A

$$\cos\theta_1\cos\theta_2 - \sin\theta_1\sin\theta_2 \equiv \cos(\theta_1 + \theta_2)$$
$$\cos\theta_1\sin\theta_2 + \sin\theta_1\cos\theta_2 \equiv \sin(\theta_1 + \theta_2)$$

and immediately see that

$$z_1 z_2 = r_1 r_2(\cos(\theta_1 + \theta_2) + \mathrm{i}\sin(\theta_1 + \theta_2)) \tag{8.8}$$

This expression can be interpreted in terms of the stretching and rotating effects of multiplication in the complex plane. That is, the effect of multiplying z_1 by z_2 is to rotate z_1 anticlockwise about the origin such that it has an argument equal to the sum of θ_1 and θ_2, and also stretch its modulus by a factor r_2.

For example, consider $z_1 = 1 + \mathrm{i}$ and $z_2 = 1 + 2\mathrm{i}$. It is easy to show that the polar forms of these complex numbers are

$$z_1 = \sqrt{2}\left(\cos\frac{\pi}{4} + \mathrm{i}\sin\frac{\pi}{4}\right) \quad \text{and} \quad z_2 = \sqrt{5}(\cos 1.107 + \mathrm{i}\sin 1.107)$$

Using Eq. (8.8) we can immediately write down their product as,

$$z_1 z_2 = \sqrt{10}\left(\cos\left(\frac{\pi}{4} + 1.107\right) + \mathrm{i}\sin\left(\frac{\pi}{4} + 1.107\right)\right) \approx \sqrt{10}(\cos 1.892 + \mathrm{i}\sin 1.892)$$

This can be seen in Figure 8.6.

Note that since we have previously defined θ in the polar form to be the principal argument, that is $-\pi < \theta < \pi$, care should be taken to ensure that $\theta_1 + \theta_2$ is expressed in its principal form. For example, if two complex numbers, z_3 and z_4, have $\theta_3 = \frac{3\pi}{4}$ and $\theta_3 = \frac{\pi}{2}$, the product $z_3 z_4$ has argument $\theta_3 + \theta_4 = \frac{5\pi}{4} > \pi$. Although this is a valid argument for $z_3 z_4$, it is not the principal argument. The principal argument is $-\frac{3\pi}{4}$.

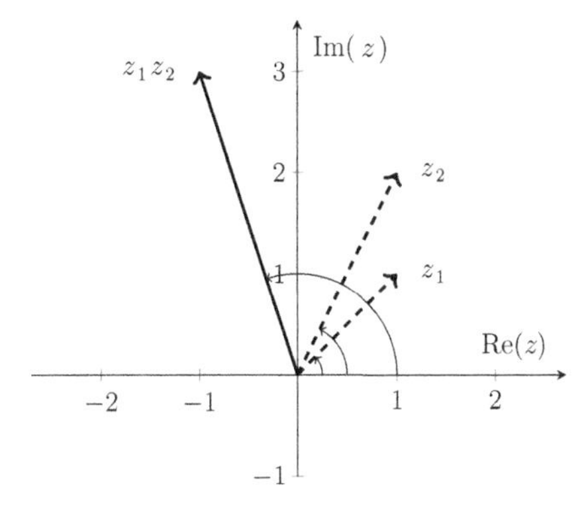

Figure 8.6 Effect of multiplication in the complex plane, $z_1 = 1 + \mathrm{i}$ and $z_2 = 1 + 2\mathrm{i}$.

EXAMPLE 8.9

Write down the principal argument and modulus of the product z_1z_2 for

a. $z_1 = 2\left(\cos\frac{\pi}{3} + \mathrm{i}\sin\frac{\pi}{3}\right)$ and $z_2 = 4\left(\cos\frac{\pi}{6} + \mathrm{i}\sin\frac{\pi}{6}\right)$

b. $z_1 = \sqrt{3}\left(\cos\frac{\pi}{2} + \mathrm{i}\sin\frac{\pi}{2}\right)$ and $z_2 = \sqrt{5}\left(\cos\frac{2\pi}{3} + \mathrm{i}\sin\frac{2\pi}{3}\right)$

c. $z_1 = 2 + \mathrm{i}$ and $z_2 = z_1$

d. $z_1 = 2 + 10\mathrm{i}$ and $z_2 = -1 + \mathrm{i}$

Solution

a. Using Eq. (8.8), we can immediately write down that z_1z_2 has modulus $r = 8$ and argument $\theta = \frac{\pi}{2}$ which is the principal argument.

b. We immediately write down that z_1z_2 has modulus $r = \sqrt{15}$ and argument $\frac{7\pi}{6}$ which translates to a principal argument of $\theta = \frac{-5\pi}{6}$.

c. We determine that $r_1 = r_2 = \sqrt{3}$ and $\theta_1 = \theta_2 = 0.464$ and so the product has modulus 3 and argument $\theta \approx 0.928$ which is the principal argument.

d. We determine that $r_1 = \sqrt{104}$, $\theta_1 \approx 1.373$, $r_2 = \sqrt{2}$, and $\theta_2 = \frac{3\pi}{4}$ and so the product has modulus $r = 4\sqrt{13}$ and the argument is 3.729 which translates to a principal argument of $\theta \approx -2.554$.

Consider now the quotient of the two general complex numbers

$$\frac{z_1}{z_2} = \frac{r_1\left(\cos\theta_1 + \mathrm{i}\sin\theta_1\right)}{r_2\left(\cos\theta_2 + \mathrm{i}\sin\theta_2\right)}$$

As previously discussed, progress can made by multiplying the numerator and denominator by the complex conjugate of the denominator, z_2^*. This leads to,

$$\begin{aligned}\frac{z_1}{z_2} &= \frac{r_1\left(\cos\theta_1 + \mathrm{i}\sin\theta_1\right)}{r_2\left(\cos\theta_2 + \mathrm{i}\sin\theta_2\right)} \times \frac{r_2\left(\cos\theta_2 - \mathrm{i}\sin\theta_2\right)}{r_2\left(\cos\theta_2 - \mathrm{i}\sin\theta_2\right)} \\ &= \frac{r_1}{r_2} \times \frac{\cos\theta_1\cos\theta_2 + \sin\theta_1\sin\theta_2 + \mathrm{i}\left(\sin\theta_1\cos\theta_2 - \cos\theta_1\sin\theta_2\right)}{\cos^2\theta_2 + \sin^2\theta_2}\end{aligned}$$

Noting the following identities from Appendix A,

$$\cos\theta_1 \cos\theta_2 + \sin\theta_1 \sin\theta_2 \equiv \cos(\theta_1 - \theta_2)$$
$$\cos\theta_1 \sin\theta_2 - \sin\theta_1 \cos\theta_2 \equiv \sin(\theta_1 - \theta_2)$$
$$\cos^2\theta_2 + \sin^2\theta_2 \equiv 1$$

we see that

$$\frac{z_1}{z_2} = \frac{r_1}{r_2}\left(\cos(\theta_1 - \theta_2) + \mathrm{i}\sin(\theta_1 - \theta_2)\right) \tag{8.9}$$

This can again be interpreted in terms of the stretching and rotating effects of division in the complex plane. That is, the effect of dividing z_1 by z_2 is to rotate z_1 clockwise so that it has an argument equal to $\theta_1 - \theta_2$, and to change its modulus by a dividing factor of r_2.

For example, consider $z_1 = 5 + 5\mathrm{i}$ and $z_2 = 4\mathrm{i}$ which have polar forms

$$z_1 = 5\sqrt{2}\left(\cos\frac{\pi}{4} + \mathrm{i}\sin\frac{\pi}{4}\right) \quad \text{and} \quad z_2 = 4\left(\cos\frac{\pi}{2} + \mathrm{i}\sin\frac{\pi}{2}\right)$$

Using Eq. (8.9) we can immediately write down that

$$\frac{z_1}{z_2} = \frac{5\sqrt{2}}{4}\left(\cos\left(-\frac{\pi}{4}\right) + \mathrm{i}\sin\left(-\frac{\pi}{4}\right)\right)$$

This can be seen in Figure 8.7.

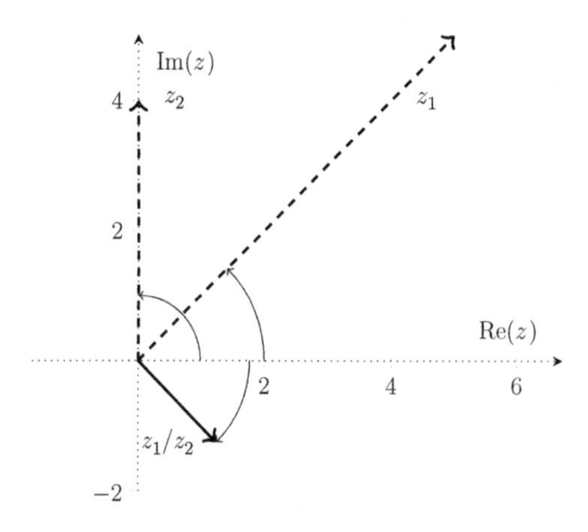

Figure 8.7 The effect of division in the complex plane, $z_1 = 5 + 5\mathrm{i}$ and $z_2 = 4\mathrm{i}$.

EXAMPLE 8.10

Write down the principal argument and modulus of the quotient z_1/z_2 for

a. $z_1 = 2\left(\cos\frac{\pi}{3} + \mathrm{i}\sin\frac{\pi}{3}\right)$ and $z_2 = 4\left(\cos\frac{\pi}{6} + \mathrm{i}\sin\frac{\pi}{6}\right)$

b. $z_1 = \sqrt{3}\left(\cos\frac{\pi}{2} + \mathrm{i}\sin\frac{\pi}{2}\right)$ and $z_2 = \sqrt{5}\left(\cos\frac{2\pi}{3} + \mathrm{i}\sin\frac{2\pi}{3}\right)$

c. $z_1 = 2 + \mathrm{i}$ and $z_2 = z_1$

d. $z_1 = 2 + 10\mathrm{i}$ and $z_2 = -1 + \mathrm{i}$

Solution

We note that these complex numbers are the same as in Example 8.9.

a. Using Eq. (8.9), we immediately write down that $\frac{z_1}{z_2}$ has modulus $r = 0.5$ and argument $\theta = \frac{\pi}{6}$ which is the principal argument.

b. We immediately write down that $\frac{z_1}{z_2}$ has modulus $r = \sqrt{\frac{3}{5}} = \frac{\sqrt{15}}{5}$ and argument $\theta = -\frac{\pi}{6}$ which is the principal argument.

c. We know that $r_1 = r_2 = \sqrt{3}$ and $\theta_1 = \theta_2 = 0.464$ and so the quotient has modulus 1 and argument $\theta = 0$ which is the principal argument. It should be of no surprise that this quotient is unity.

d. We know that $r_1 = \sqrt{104}$, $\theta_1 \approx 1.373$, $r_2 = \sqrt{2}$, and $\theta_2 = \frac{3\pi}{4}$ and so the quotient has modulus $r = \sqrt{52}$ and the principal argument is $\theta \approx -0.983$.

Note that the expressions (8.8) and (8.9) for multiplication and division of two complex numbers can be extended to any number of complex numbers in the obvious way. This is demonstrated in the following example.

EXAMPLE 8.11

If $z_1 = 4$, $z_2 = 3\mathrm{i}$, $z_3 = 4 + 4\mathrm{i}$, and $z_4 = -2 - 2\mathrm{i}$, write down the modulus and principal arguments of the following expressions.

a. $z_1z_2z_3$

b. $z_1z_2z_3z_4$

c. $\frac{z_1}{z_2z_4}$

d. $\frac{z_1z_2z_4}{z_3}$

Solution

We begin by determining the modulus and principal argument of each complex number.

$$\begin{aligned} r_1 &= 4 & \theta_1 &= 0 \\ r_2 &= 3 & \theta_2 &= \tfrac{\pi}{2} \\ r_3 &= 4\sqrt{2} & \theta_3 &= \tfrac{\pi}{4} \\ r_4 &= 2\sqrt{2} & \theta_4 &= -\tfrac{3\pi}{4} \end{aligned}$$

a. It should be clear that Eq. (8.8) can be used to determine $y = z_1z_2$ and used again to determine yz_3 which is the required product, $z_1z_2z_3$. The result of this would be that $|z_1z_2z_3| = r_1r_2r_3$ and $\arg(z_1z_2z_3) = \theta_1 + \theta_2 + \theta_3$ which can be determined directly as

$$|z_1z_2z_3| = 4 \times 3 \times 4\sqrt{2} = 48\sqrt{2}$$
$$\arg(z_1z_2z_3) = 0 + \frac{\pi}{2} + \frac{\pi}{4} = \frac{3\pi}{4} = \text{Arg}(z_1z_2z_3)$$

b. Using similar reasoning to part a., we arrive at

$$|z_1z_2z_4z_3| = r_1r_2r_3r_4 = 4 \times 3 \times 4\sqrt{2} \times 2\sqrt{2} = 192$$
$$\arg(z_1z_2z_3z_3) = \theta_1 + \theta_2 + \theta_3 + \theta_4$$
$$= 0 + \frac{\pi}{2} + \frac{\pi}{4} - \frac{3\pi}{4} = 0 = \text{Arg}(z_1z_2z_3z_4)$$

c. In this case, we can determine $y = z_2z_4$ from Eq. (8.8) and then $\frac{z_1}{y}$ from Eq. (8.9). The result of this would be that $\left|\frac{z_1}{z_2z_4}\right| = \frac{r_1}{r_2r_4}$ and $\arg\left(\frac{z_1}{z_2z_4}\right) = \theta_1 - \theta_2 - \theta_4$ which can be determined directly as

$$\left|\frac{z_1}{z_2z_4}\right| = \frac{4}{3 \times 2\sqrt{2}} = \frac{\sqrt{2}}{3}$$
$$\arg\left(\frac{z_1}{z_2z_4}\right) = 0 - \frac{\pi}{2} + \frac{3\pi}{4} = \frac{\pi}{4} = \text{Arg}\left(\frac{z_1}{z_2z_4}\right)$$

d. Using similar reasoning to part c., we arrive at

$$\left|\frac{z_1z_2z_4}{z_3}\right| = \frac{4 \times 3 \times 2\sqrt{2}}{4\sqrt{2}} = 6$$
$$\arg\left(\frac{z_1z_2z_4}{z_3}\right) = 0 + \frac{\pi}{2} + \frac{\pi}{4} + \frac{3\pi}{4} = \frac{3\pi}{2}$$
$$\Rightarrow \text{Arg}\left(\frac{z_1z_2z_4}{z_3}\right) = -\frac{\pi}{2}$$

8.5 A SIMPLIFIED POLAR FORM

8.5.1 Euler's formula

The polar form for a general complex number, as expressed in Eq. (8.5), is rather cumbersome to work with as it requires familiarity with trigonometric identities. We have, however, been able to derive some intuitive results for multiplication and division in the complex plane and these remain the main outcomes of the previous section. In this section, we look at a significantly simplified form of Eq. (8.5) that will allow us greater flexibility when working with complex numbers.

We begin by stating *Euler's formula*

$$e^{ix} \equiv \cos x + i \sin x$$

for $x \in \mathbb{R}$. This formula is stated here without proof and you are invited to work through a justification in Question 8.6. The particular case that $x = \pi$ leads to the identity that you may have seen previously

$$e^{i\pi} + 1 = 0$$

At a more pragmatic level, it should be immediately clear that Euler's formula can be used to rewrite the polar form (Eq. 8.5) as

$$a + bi = r(\cos\theta + i\sin\theta) = re^{i\theta} \tag{8.10}$$

This form has significant advantages over the previous polar form, not least as the rules for multiplication and division, Eqs. (8.8) and (8.9) and their extensions used in Example 8.11, are immediately obvious from the properties of exponential functions. In particular,

$$z_1 z_2 = r_1\,e^{i\theta_1} r_2\,e^{i\theta_2} = r_1 r_2\,e^{i(\theta_1+\theta_2)} \quad \text{and} \quad \frac{z_1}{z_2} = \frac{r_1\,e^{i\theta_1}}{r_2\,e^{i\theta_2}} = \frac{r_1}{r_2}e^{i(\theta_1-\theta_2)}$$

In addition, it enables one to determine powers of complex numbers without the need to work with cumbersome trigonometric identities. For example, z^n for $n \in \mathbb{R}$ is immediately expressed as

$$z^n = \left(re^{i\theta}\right)^n = r^n e^{in\theta} \tag{8.11}$$

This expression enables one to interpret the operation of raising a complex number to a real power in terms of rotation and stretching within the complex plane.

We begin by looking at integer powers. For example, to determine $(2+2i)^5$ we note that this can be expressed as z^5 where $z = 2 + 2i = 2\sqrt{2}e^{i\frac{\pi}{4}}$. Working in the simplified polar form, we then see that

$$\left(2\sqrt{2}e^{i\frac{\pi}{4}}\right)^5 = (2\sqrt{2})^5 e^{i\frac{5\pi}{4}}$$

In terms of the principle argument, this is expressed as $z^5 = 2\sqrt{2}e^{-i\frac{3\pi}{4}}$ and is easily converted to nonpolar form,

$$\begin{aligned}(2+2i)^5 &= \left(2\sqrt{2}\right)^5\left(\cos\left(-\frac{3\pi}{4}\right) + i\sin\left(-\frac{3\pi}{4}\right)\right)\\ &= -128 - 128i\end{aligned}$$

Of course it is possible to compute z^n directly from $(a+ib)^n$ for reasonably small values of integer n, but the advantages of using Euler's formula should be clear.

EXAMPLE 8.12

User Euler's formula to determine the following expressions in principal polar form.

a. $(5+5i)^5$
b. $(2+4i)^{12}$
c. $(-2+2i)^6$
d. $(-1-i)^{102}$

Solution

a. In polar form $\left(5\sqrt{2}e^{i\frac{\pi}{4}}\right)^5 = 12{,}500\sqrt{2}e^{i\frac{5\pi}{4}}$, which has principal form $12{,}500\sqrt{2}e^{-i\frac{3\pi}{4}}$.

b. $\left(2\sqrt{5}e^{0.3524\pi i}\right)^{12} = 64{,}000{,}000e^{12\times 0.3523\pi i}$, which has principal form $64{,}000{,}000\,e^{0.2290i}$.

c. $\left(2\sqrt{2}e^{i\frac{3\pi}{4}}\right)^6 = 512\sqrt{2}e^{i\frac{9\pi}{2}}$, which has principal form $512\,e^{i\frac{\pi}{2}} = 512i$.

d. $\left(\sqrt{2}e^{-i\frac{3\pi}{4}}\right)^{102} = 2^{51}e^{-i\frac{153\pi}{2}}$, which has principal form $2^{51}e^{-i\frac{\pi}{2}} = -2^{51}i$.

Consider now the operation of raising a complex number to a *fractional power*. For example, how can we determine z such that $z = (2+2i)^{1/2}$? Although it is tempting to simply use the polar form of $2+2i$ and apply Eq. (8.11) with $n = \frac{1}{2}$ to obtain a single value for z, this is not the full story. To see why, it is useful to express the above the expression as

$$z^2 - 2 - 2i = 0$$

This is a second-order polynomial and so has *two* roots in the complex plane, that is, $z = (2+2i)^{1/2}$ is expected to have two distinct values. By similar reasoning, one would expect $z = (1+6i)^{1/3}$ to have three distinct values, and $z = (5+2i)^{1/4}$ to have four.

In order to consider fractional powers, one must return to the distinction between the general argument and principal argument of a complex number: recall that the cyclic nature of the polar form means that any given complex number has infinitely many values for the argument, but only one principal argument with $\theta \in (-\pi, \pi]$. By raising to a fractional power, Eq. (8.11) shows that the consecutive arguments are "compressed" in angular terms such that multiple arguments can lie in the principal argument range, $(-\pi, \pi]$. To see this, we return to the motivating example $(2+2i)^{1/2}$. In polar form, this can be expressed as

$$(2+2i)^{1/2} = \left(2\sqrt{2}e^{i\left(\frac{\pi}{4}+2k\pi\right)}\right)^{1/2}$$

where $k \in \mathbb{Z}$ is introduced to account for all cyclical arguments of $2+2i$. Proceeding, we obtain

$$\left(2\sqrt{2}\right)^{1/2}\left(e^{i\left(\frac{\pi}{4}+2k\pi\right)}\right)^{1/2} = 2^{3/4}e^{i\left(\frac{\pi}{8}+k\pi\right)}$$

Note that $k = 0$ leads to a complex number with principal argument $\frac{\pi}{8}$ and $k = -1$ leads to an additional and distinct complex number with principal argument $-\frac{7\pi}{8}$; all other values of k lead to some cyclic version of these two distinct principals. The result is that we have obtained two distinct values for $(2+2i)^{1/2}$, as shown in Figure 8.8. The circle indicates that the distinct values have the same modulus and are evenly spaced around the complex plane. Reverting to nonpolar form, we have $(2+2i)^{1/2} \approx \pm(1.5538 + 0.6436i)$. The reader is invited to confirm that the square of each is $2+2i$.

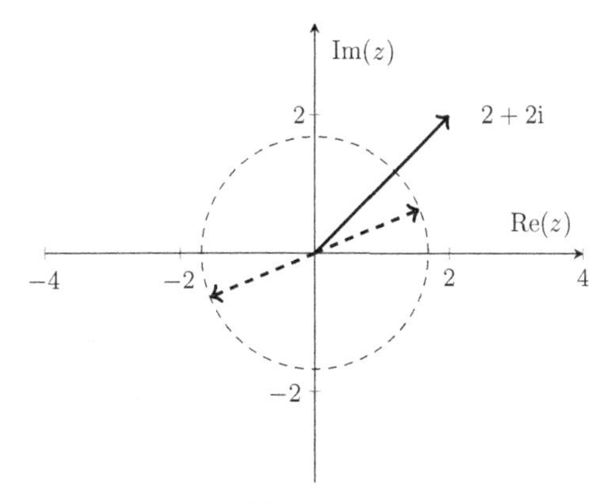

Figure 8.8 The two distinct values of $(2+2\mathrm{i})^{1/2}$ (dashed).

EXAMPLE 8.13

Determine all distinct complex values of $(3+\sqrt{3}\mathrm{i})^{1/n}$ for

a. $n = 3$

b. $n = 5$

Solution

We begin by expressing $z = 3+\sqrt{3}\mathrm{i}$ in its general polar form $2\sqrt{3}\mathrm{e}^{\mathrm{i}\left(\frac{\pi}{6}+2k\pi\right)}$ where $k \in \mathbb{Z}$.

a. For $n = 3$ we have

$$z^{1/3} = \left(2\sqrt{3}\right)^{1/3} \mathrm{e}^{\mathrm{i}\left(\frac{\pi}{18}+\frac{2k}{3}\pi\right)}$$

which leads to three distinct complex numbers when $k = -1, 0$, and 1.

$$z^{1/3} = \left(2\sqrt{3}\right)^{1/3} \mathrm{e}^{-\frac{11\pi}{18}\mathrm{i}}, \quad \left(2\sqrt{3}\right)^{1/3} \mathrm{e}^{\frac{\pi}{18}\mathrm{i}}, \quad \left(2\sqrt{3}\right)^{1/3} \mathrm{e}^{\frac{13\pi}{18}\mathrm{i}}$$

b. For $n = 5$ we have

$$z^{1/5} = \left(2\sqrt{3}\right)^{1/5} \mathrm{e}^{\mathrm{i}\left(\frac{\pi}{30}+\frac{2k}{5}\pi\right)}$$

which leads to five distinct complex numbers when $k = -3, -2, -1, 0$, and 1.

$$z^{1/5} = \left(2\sqrt{3}\right)^{1/5} \mathrm{e}^{-\frac{13\pi}{15}\mathrm{i}}, \quad \left(2\sqrt{3}\right)^{1/5} \mathrm{e}^{-\frac{7\pi}{15}\mathrm{i}}, \quad \left(2\sqrt{3}\right)^{1/5} \mathrm{e}^{-\frac{\pi}{15}\mathrm{i}},$$
$$\left(2\sqrt{3}\right)^{1/5} \mathrm{e}^{\frac{\pi}{3}\mathrm{i}}, \left(2\sqrt{3}\right)^{1/5} \mathrm{e}^{\frac{11\pi}{15}\mathrm{i}}$$

The reader is invited to confirm that, in each case, the resulting complex numbers are evenly spaced around a circle the complex plane.

EXAMPLE 8.14

Determine all values of the cube root of 1 in the complex plane.

Solution

We write 1 in general polar form as $1 = e^{i2k\pi}$ where $k \in \mathbb{Z}$. The cube root is then taken as

$$\sqrt[3]{1} = e^{i\frac{2k}{3}\pi}$$

which takes three distinct values for $k = -1, 0,$ and 1

$$\sqrt[3]{1} = e^{-i\frac{2}{3}\pi}, \quad e^{0}, \quad e^{i\frac{2}{3}\pi}$$

These can also be written as

$$\sqrt[3]{1} = -\frac{1}{2} - \frac{\sqrt{3}}{2}i, \quad 1, \quad -\frac{1}{2} + \frac{\sqrt{3}}{2}i$$

Recall that this problem was first considered in Example 8.6.

Let us now consider n to be a rational number, that is $n = \frac{p}{q}$, where $p, q \in \mathbb{Z}$ and $q \neq 0$. In this case, a complex number raised to this power can be manipulated to

$$z^{p/q} = (z^p)^{1/q}$$

Raising a complex number to a rational power $\frac{p}{q}$ can therefore be thought of as a two-stage process:

1. raising it to an integer power p
2. raising the resulting complex number to the fractional power $\frac{1}{q}$

Any complex number raised to a rational power $\frac{p}{q}$ is therefore expected to have q distinct values evenly arranged around a circle in the complex plane.

For example, $(2 + 2i)^{5/4}$ can be rewritten as

$$(2 + 2i)^{5/4} = \left((2 + 2i)^5\right)^{1/4} = (-128 - 128i)^{1/4} = \left(32\, e^{i\left(-\frac{3\pi}{4} + 2k\pi\right)}\right)^{1/4}$$

and we find four distinct values at $k = -1, 0, 1,$ and 2,

$$(2 + 2i)^{5/4} = 2\sqrt[4]{2}e^{\left(-\frac{11}{16}\pi i\right)}, \quad 2\sqrt[4]{2}e^{\left(-\frac{3}{16}\pi i\right)}, \quad 2\sqrt[4]{2}e^{\left(\frac{5}{16}\pi i\right)}, \quad 2\sqrt[4]{2}e^{\left(\frac{13}{16}\pi i\right)}$$

These are shown in Figure 8.9.

EXAMPLE 8.15

Plot all values of the following on the complex plane.

a. $i^{3/2}$

b. $(1 + i)^{2/3}$

c. $1^{1/5}$

d. $(-2)^{7/3}$

Solution

a. We first note that this can be rewritten as $\left(\mathrm{i}^3\right)^{1/2} = (-\mathrm{i})^{1/2}$. Working in the general polar form, this is expressed as $\left(\mathrm{e}^{\mathrm{i}\left(-\frac{\pi}{2}+2k\pi\right)}\right)^{1/2}$ leading to two distinct values $\mathrm{e}^{\frac{3\pi}{4}\mathrm{i}}$ and $\mathrm{e}^{-\frac{\pi}{4}\mathrm{i}}$. These can be written as $\frac{\sqrt{2}}{2}(-1+\mathrm{i})$ and $\frac{\sqrt{2}}{2}(1-\mathrm{i})$.

b. $(1+\mathrm{i})^{2/3} = \left(2\,\mathrm{e}^{\mathrm{i}\left(\frac{\pi}{2}+2k\pi\right)}\right)^{1/3}$ leads to three distinct values $\sqrt[3]{2}\mathrm{e}^{-\frac{\pi}{2}\mathrm{i}}$, $\sqrt[3]{2}\mathrm{e}^{\frac{\pi}{6}\mathrm{i}}$, and $\sqrt[3]{2}\mathrm{e}^{\frac{5\pi}{6}\mathrm{i}}$.

c. $1^{1/5} = \left(\mathrm{e}^{\mathrm{i}(0+2k\pi)}\right)^{1/5}$ leads to five distinct values $\mathrm{e}^{-\frac{4\pi}{5}\mathrm{i}}$, $\mathrm{e}^{-\frac{2\pi}{5}\mathrm{i}}$, 1, $\mathrm{e}^{\frac{2\pi}{5}\mathrm{i}}$, and $\mathrm{e}^{\frac{4\pi}{5}\mathrm{i}}$.

d. $(-2)^{7/3} = \left(128\,\mathrm{e}^{\mathrm{i}(\pi+2k\pi)}\right)^{1/3}$ leads the three distinct values $4\sqrt[3]{2}\mathrm{e}^{-\frac{\pi\mathrm{i}}{3}}$, $-4\sqrt[3]{2}$, and $4\sqrt[3]{2}\mathrm{e}^{\frac{\pi\mathrm{i}}{3}}$.

Figure 8.10 shows the resulting complex numbers in each case.

8.5.2 Connection to circular functions of real variables

In the particular case that $r = 1$, we can re-express Eq. (8.11) in terms of trigonometric functions to show that

$$(\cos\theta + \mathrm{i}\sin\theta)^n = \cos(n\theta) + \mathrm{i}\sin(n\theta) \tag{8.12}$$

This particular result is known as *De Moivre's formula*. As previously discussed, Eq. (8.11) takes multiple values when n is a fraction. However, De Moivre's formula is not capable of returning multiple values and so we must assume that De Moivre's formula does not in general hold for noninteger n.

Despite having origins in complex analysis, De Moivre's formula can be used to derive many of the trigonometric identities for real variables, such as those stated in Appendix A. This is demonstrated in the following examples.

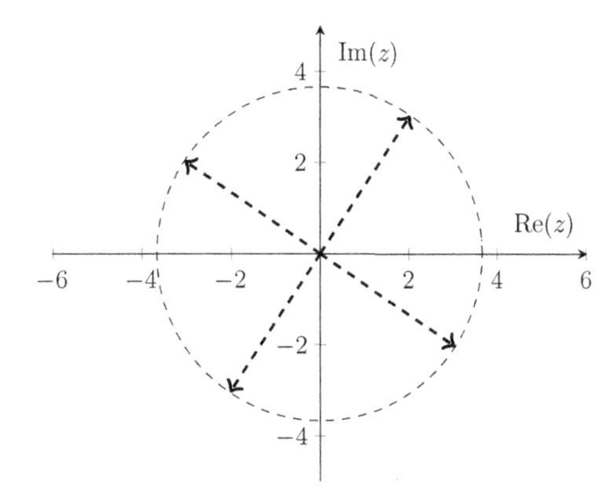

Figure 8.9 Four distinct values of $(2+2\mathrm{i})^{5/4}$.

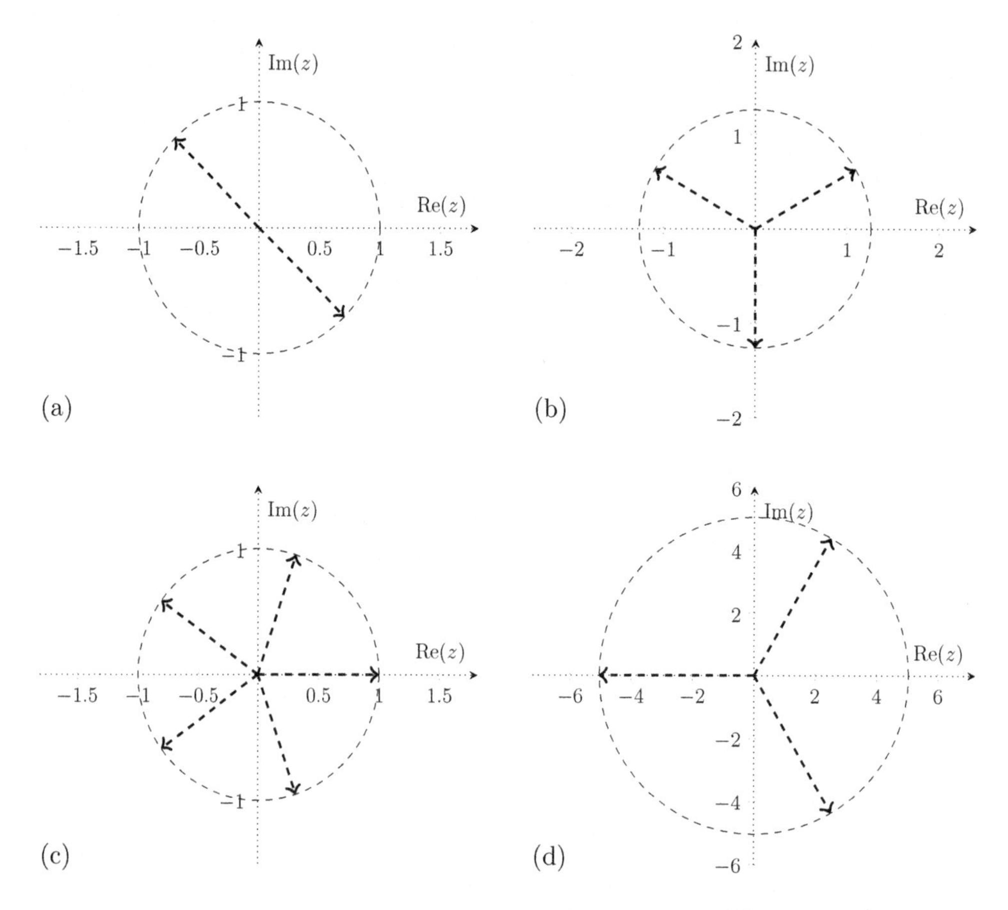

Figure 8.10 Argand diagrams for Example 8.15. (a) Two distinct values of $\mathrm{i}^{3/2}$. (b) Three distinct values of $(1+\mathrm{i})^{2/3}$. (c) Five distinct values of $1^{1/5}$. (d) Three distinct values of $(-2)^{7/3}$.

EXAMPLE 8.16

Use De Moivre's formula to prove the following double-angle identities

a. $\cos(2x) \equiv \cos^2 x - \sin^2 x$

b. $\sin(2x) \equiv 2\sin x \cos x$

Solution

Using Eq. (8.12) with $n = 2$, we have

$$(\cos x + \mathrm{i}\sin x)^2 = \cos(2x) + \mathrm{i}\sin(2x)$$

Expanding the left-hand side and equating real and imaginary parts we quickly see that

a. $\cos(2x) \equiv \cos^2 x - \sin^2 x$

b. $\sin(2x) \equiv 2\sin x \cos x$

as required.

EXAMPLE 8.17

Use De Moivre's formula to derive triple-angle identities for sin and cos functions.

Solution

Using Eq. (8.12) with $n = 3$, we have

$$(\cos x + \mathrm{i}\sin x)^3 = \cos(3x) + \mathrm{i}\sin(3x)$$

The left-hand side is expanded to give

$$\cos^3 x - 3\cos x\sin^2 x + \mathrm{i}\left(3\cos^2 x\sin x - \sin^3 x\right) = \cos(3x) + \mathrm{i}\sin(3x)$$

and equating real and imaginary parts leads to

$$\cos(3x) = \cos^3 x - 3\cos x\sin^2 x$$

$$\sin(3x) = 3\cos^2 x\sin x - \sin^3 x$$

These are the required triple-angle identities.

We finish this chapter by demonstrating that Euler's formula (Eq. 8.10) can be used to express $\sin x$ and $\cos x$ functions in terms of complex exponentials. The following expressions should be clear, owing to the odd and even properties of the trigonometric functions,

$$\mathrm{e}^{\mathrm{i}x} = \cos x + \mathrm{i}\sin x$$

$$\mathrm{e}^{-\mathrm{i}x} = \cos x - \mathrm{i}\sin x$$

With a little manipulation these can then be used to demonstrate that

$$\sin x = \frac{1}{2\mathrm{i}}\left(\mathrm{e}^{\mathrm{i}x} - \mathrm{e}^{-\mathrm{i}x}\right) \quad (8.13)$$

$$\cos x = \frac{1}{2}\left(\mathrm{e}^{\mathrm{i}x} + \mathrm{e}^{-\mathrm{i}x}\right) \quad (8.14)$$

While the benefits of expressing trigonometric functions of real variables in terms of complex exponentials may not be immediately clear, these expressions can also greatly simply the derivation of trigonometric identities. This is shown in the following example.

EXAMPLE 8.18

Use Eqs. (8.13) and (8.14) to prove the following identities for real variables.

a. $\sin^2 u = \frac{1}{2}(1 - \cos(2u))$

b. $\cos u\cos v = \frac{1}{2}(\cos(u + v) + \cos(u - v))$

Solution

a. Using Eq. (8.13), we proceed as

$$\begin{aligned}\sin^2 u &= \left(\frac{1}{2\mathrm{i}}\left(\mathrm{e}^{\mathrm{i}u} - \mathrm{e}^{-\mathrm{i}u}\right)\right)^2 \\ &= \frac{1}{2}\left(\frac{2 - \mathrm{e}^{2\mathrm{i}u} - \mathrm{e}^{-2\mathrm{i}u}}{2}\right) \\ &= \frac{1}{2}\left(1 - \cos(2u)\right)\end{aligned}$$

b. Using Eq. (8.14), we proceed as

$$\begin{aligned}\cos u \cos v &= \frac{1}{4}\left(\mathrm{e}^{\mathrm{i}u} + \mathrm{e}^{-\mathrm{i}u}\right)\left(\mathrm{e}^{\mathrm{i}v} + \mathrm{e}^{-\mathrm{i}v}\right) \\ &= \frac{1}{2}\left(\frac{\mathrm{e}^{\mathrm{i}(u+v)} + \mathrm{e}^{-\mathrm{i}(u+v)} + \mathrm{e}^{\mathrm{i}(u-v)} + \mathrm{e}^{-\mathrm{i}(u-v)}}{2}\right) \\ &= \frac{1}{2}\left(\cos(u+v) + \cos(u-v)\right)\end{aligned}$$

8.6 COMPLEX NUMBERS ON YOUR COMPUTER

As one might expect, Wolfram Alpha is not restricted to real numbers and it can be used to manipulate and work with complex numbers in a very intuitive way.

Wolfram Alpha automatically understands the input `i` as $\sqrt{-1}$ and so complex numbers are very easy to input. For example, one can simply use `1+2i` to input $1 + 2\mathrm{i}$. With this in mind, it is straightforward to work with complex numbers. For example, the result of the product $(1 + 2\mathrm{i}) \times (-2 + 6\mathrm{i})$ is obtained with the command

```
(1+2i)(-2+6i)
```

The engine returns the result in Cartesian form, $-14 + 2\mathrm{i}$, and also in polar form, $r \approx 14.1421$ and $\theta \approx 171.87°$. Note that the argument is expressed in degrees rather than radians and we can use the further command

```
171.87 degrees to radians
```

to find that this three radians.

As a further example, note that the five distinct values of $(-1 - 2\mathrm{i})^{2/5}$ are obtained with the command

```
(-1-2i)^(2/5)
```

The engine returns a range of information about the calculation and you need to scroll down to see the five values given in both Cartesian and simplified polar form.

The principle polar form of a complex number can also be obtained. For example, the polar form of $-1 - \mathrm{i}$ is found with the command

```
polar form (-1-i)
```

and Wolfram Alpha returns the modulus $r \approx 1.41421$ and the angle $\theta = -135°$. Again, the argument is returned in degrees and we use the command

```
-135 degrees to radians
```

to find that this $-\frac{3\pi}{4}$ radians. As an alternative to converting between degrees and radians, note that, for example, the command `argument(-1-i)` returns $-3\pi/4$ which is of course in radians. Furthermore, the command `modulus (-1-i)` can be used to obtain r and the polar form is obtained from these two results.

Now that we are aware of complex numbers, we are able to fully understand the output of, for example, the command

```
roots z^2-2z+2
```

In particular, Wolfram Alpha returns that the roots are $z = 1 \pm \mathrm{i}$. Which of course reflects that this quadratic polynomial has negative discriminant and so has no real roots.

On a related note, it is useful to be aware of the command `real roots`. For example, the cubic polynomial $z^3 - 3z^2 + 4z - 2$ must have three roots and the command

```
roots z^3-3z^2+4z-2
```

indicates that these are $z = 1$ and $z = 1 \pm \mathrm{i}$. However, if we were wishing to work entirely with real numbers, the command

```
real roots z^3-3z^2+4z-2
```

simply returns that $z = 1$. The equivalent command to determine only the complex roots is

```
complex roots z^3-3z^2+4z-2
```

EXAMPLE 8.19

Use Wolfram Alpha to perform the following.

a. Determine the three values of $(\pi \mathrm{i})^{2/3}$.

b. Determine the polar form of $\frac{(1+2\mathrm{i})^7}{(1-2\mathrm{i})^3}$.

c. Determine all roots of the cubic polynomial $g(x) = x^3 - 1$

Solution

a. We use the command

```
(pi*i)^(2/3)
```

and obtain $\pi^{2/3}\mathrm{e}^{\mathrm{i}\frac{\pi}{3}}$, $\pi^{2/3}\mathrm{e}^{-\mathrm{i}\frac{\pi}{3}}$, and $-\pi^{2/3}$.

b. We use the command

```
(1+2i)^7/(1-2i)^3
```

and directly obtain that $r = 25$ and $\theta = -85.6505°$. The argument is easily converted to -1.495 radians and we have the polar form $25\,e^{-1.495i}$. Alternatively the commands

```
argument((1+2i)^7/(1-2i)^3)
```

and

```
modulus((1+2i)^7/(1-2i)^3)
```

also return the required values of r and θ.

c. We use the command

```
roots x^3-1
```

and, as expected, find three roots in the complex plane. These are $x = 1$ and $-\frac{1}{2} \pm \frac{\sqrt{3}}{2}i$.

8.7 QUESTIONS

The following questions are intended to test your knowledge of the material discussed in this chapter. Full solutions are available in Chapter 8 Solutions of Part III. You should use an algebraic approach unless otherwise stated.

Question 8.1. Classify the following as real, imaginary, or complex. Plot each z_i on a single Argand diagram.

a. $z_1 = 3$

b. $z_2 = 2 + 4i$

c. $z_3 = 2i$

d. $z_4 = -1 + 3i$

e. $z_5 = -2 - 4i$

Question 8.2. Obtain the simplified polar form of each z_i in Question 8.1.

Question 8.3. Determine the following using the complex numbers defined in Question 8.1.

a. $z_1 + z_2 + z_3 + z_4 - z_5$

b. $z_2 z_3$

c. $\frac{z_1}{z_2}$

d. $z_2^2 z_3$

e. $\frac{z_1 z_4 + z_4}{z_2 z_3}$

Question 8.4. Determine all roots of the following polynomials and express each root in simplified polar form.

a. $f(x) = x^2 - 4$

b. $g(y) = y^3 + 3y^2 - 6y - 8$

c. $h(z) = (z^2 + 1)(z^2 + 2z + 2)$
d. $m(k) = k^5 + 32$

Question 8.5. If $z = 1 + \mathrm{i}$, determine the following in simplified polar form of the following.
a. zz^*
b. z^{10}
c. $\frac{1}{z}$
d. $\frac{1}{z^5}$

Question 8.6. Use the Maclaurin series of the exponential function to prove Euler's formula (Eq. 8.10).

Question 8.7. Derive expressions in terms of $\sin(x)$ and $\cos(x)$ for the following when $x \in \mathbb{R}$.
a. $\sin(4x)$
b. $\cos(4x)$

Question 8.8. Prove the following trigonometric identities for $u, v \in \mathbb{R}$.
a. $\cos^2(u) = \frac{1+\cos(2u)}{2}$
b. $\cos^3(u) = \frac{1}{4}(3\cos(u) + \cos(3u))$

CHAPTER 9

Probability Theory

Contents

Prerequisite knowledge	Intended learning outcomes

• Chapter 1	• Define and use the following terms
	  ◦ permutation and combination
	  ◦ sample space
	  ◦ event
	  ◦ probability of an event
	  ◦ independent and dependent events
	  ◦ conditional probability
	  ◦ partition of the sample space
	• Calculate the probability of events that are
	  ◦ independent
	  ◦ dependent
	  ◦ conditional
	• Know when and how to use
	  ◦ the total probability theorem
	  ◦ Bayes' theorem

This chapter presents an introduction to probability theory. Probability is of course at the heart of modern financial and actuarial mathematics and much of your future studies in these areas will necessarily rely on probability theory. It is therefore crucial that you have a solid understanding of the basics, and this is the aim of this chapter. You are likely to have studied probability in some form before, but you are encouraged to study this chapter in its entirety to ensure that there are no holes in your fundamental knowledge. While the presentation continues to avoid the formalities of theorems and proofs, the material in this chapter is presented in a reasonably formal manner that may, at first, not look familiar from your previous studies. This more formal approach is intended to prepare you for higher studies in probability theory where material is often presented in a very daunting way.

We begin with a description of the basic concepts before considering conditional probability and Bayes' Law.

9.1 FUNDAMENTAL CONCEPTS

The *probability* of an event occurring is intuitively understood to be the likelihood or chance of it occurring. In the very simplest cases, the probability of a particular event A occurring from an experiment is obtained from the number of ways that A can occur divided by the total number of possible outcomes. For example, the probability of obtaining ⚅ from a single roll of a fair[1] dice is obtained from the following observations,

- the possible outcomes are ⚀, ⚁, ⚂, ⚃, ⚄, and ⚅, that is there are six possible outcomes,
- the number of ways of obtaining ⚅ from a single roll is 1.

Therefore, probability of obtaining ⚅ from a single roll is $\frac{1}{6}$.

EXAMPLE 9.1

Calculate the probability of the following outcomes from a single roll of a fair dice.

a. ⚂

b. Either ⚀ or ⚁

c. Any one of ⚀, ⚁, ⚂, ⚃, ⚄, or ⚅

Solution

In each case, the total number of possible outcomes is six.

a. The total number of ways of obtaining ⚂ is one. The required probability is then $\frac{1}{6}$.

b. The total number of ways of obtaining either ⚀ or ⚁ is two. The required probability is then $\frac{2}{6} = \frac{1}{3}$.

c. The total number of ways of obtaining any one of ⚀, ⚁, ⚂, ⚃, ⚄, or ⚅ is six. The required probability is therefore $\frac{6}{6} = 1$.

---

[1] that is, an *unbiased*, six-sided dice.

EXAMPLE 9.2

Calculate the probability of obtaining the following total scores from simultaneously rolling *two* fair dice.

a. 1
b. 12
c. 6
d. 7

Solution

We need to determine the set of all possible outcomes. These are listed in Table 9.1 and we see that there are 36.

a. A total of 1 cannot be obtained. That is, the number of ways of obtaining a total of 1 is 0. The required probability is therefore 0.

b. A total of 12 can be obtained from only ⚅⚅. That is, there is only one way. The required probability is therefore $\frac{1}{36}$.

c. A total of 6 can be obtained from any of ⚀⚄, ⚁⚃, ⚂⚂, ⚃⚁, or ⚄⚀. That is, there are six ways. The required probability is therefore $\frac{6}{36} = \frac{1}{6}$.

d. A total of 7 can be obtained from any of ⚀⚅, ⚁⚄, ⚂⚃, ⚃⚂, ⚄⚁ or ⚅⚀. That is, there are six ways. The required probability is therefore $\frac{6}{36} = \frac{1}{6}$.

Table 9.1 Possible outcomes of simultaneously rolling two fair dice

Outcome	Total	Outcome	Total
⚀⚀	2	⚁⚀	3
⚀⚁	3	⚁⚁	4
⚀⚂	4	⚁⚂	5
⚀⚃	5	⚁⚃	6
⚀⚄	6	⚁⚄	7
⚀⚅	7	⚁⚅	8
⚂⚀	4	⚃⚀	5
⚂⚁	5	⚃⚁	6
⚂⚂	6	⚃⚂	7
⚂⚃	7	⚃⚃	8
⚂⚄	8	⚃⚄	9
⚂⚅	9	⚃⚅	10
⚄⚀	6	⚅⚀	7
⚄⚁	7	⚅⚁	8
⚄⚂	8	⚅⚂	9
⚄⚃	9	⚅⚃	10
⚄⚄	10	⚅⚄	11
⚄⚅	11	⚅⚅	12

EXAMPLE 9.3

Consider a bag containing a number of balls colored either red, blue, green, or yellow, denoted (R), (B), (G), or (Y), respectively. In particular, there are 2 × (R), 1 × (B), 1 × (G), and 1 × (Y). Calculate the probability that the draw of a single ball will be the following.

a. (Y)

b. (R)

c. either (R) or (B)

d. a black ball (Bl)

Solution

In each case, the total number of possible outcomes is five. That is, one can draw either (R), (R), (B), (G), or (Y).

a. The total number of ways of drawing (Y) is one. The required probability is then $\frac{1}{5}$.

b. The total number of ways of drawing (R) is two. The required probability is then $\frac{2}{5}$.

c. The total number of ways of drawing either (R) or (B) is three. The required probability is then $\frac{3}{5}$.

d. The total number of ways of drawing (Bl) is zero. The required probability is then 0.

It should be clear from the above examples that the probability of an outcome is 1 if that outcome is *certain*, and 0 if it that outcome is *impossible*. For example, in Example 9.1(c), the probability of obtaining either ⚀, ⚁, ⚂, ⚃, ⚄, or ⚅ from a single roll of a dice is 1; that is, one is certain to obtain one of those results. Furthermore, the probability of drawing a black ball in Example 9.3(d) is 0 because the bag only contains red, blue, green, and yellow balls.

We intuitively see that probabilities are real numbers on the interval $[0, 1]$. Probabilities are typically stated as either fractions, as in the previous examples, or decimals.

Now let us now consider *multiple* experiments. For example, rolling a dice twice and asking about the probability they we obtain ⚂ on each roll, or pulling two balls from a bag and asking about the probability that we obtain (B) and (Y). In order to properly consider multiple experiments, it is necessary to distinguish between *independent* and *dependent* events. Rolling dice and pulling balls from a bag are good examples for illustrating the difference.

First, let us consider rolling a fair dice. The probability that we obtain ⚂ on the first roll is $\frac{1}{6}$. It should be intuitively clear that the outcome of the second roll has no bearing on the outcome of the first roll and the probability of again obtaining ⚂ is $\frac{1}{6}$. The two rolls are said to be *independent*. Note that identical reasoning applies to simultaneously rolling two dice and asking about the probability that they both show ⚄.

Now consider drawing two balls from the bag described in Example 9.3. There are two variations of this experiment,

1. draw a ball, note its color and return it to the bag before drawing another,

2. draw a ball, note its color but do not return it to the bag before drawing another.

Since variation 1 involves replacement of the first ball, the outcome of the second draw has no bearing on the result of the first. The two draws in variation 1 are therefore *independent*.

Variation 2, however, is different. Since the first ball is not replaced, the outcome of the second draw is dependent on the outcome of the first. For example, if (Y) is drawn first from a bad containing (R)(R)(G)(Y)(B), it is impossible to draw it for a second time. This is explored in the following example.

EXAMPLE 9.4

Using the bag of balls in Example 9.3, calculate the probability of obtaining the following outcomes from two draws. State whether the draws are independent or dependent.

a. **1st:** (G) is drawn then replaced,
2nd: (Y) is drawn.

b. **1st:** (G) is drawn, but not replaced,
2nd: (Y) is drawn.

Solution

a. **1st:** There are five possible outcomes, only one is (G). The required probability is then $\frac{1}{5}$.
2nd: There are again five possible outcomes, only one is (Y). The required probability is then $\frac{1}{5}$.
The two draws are independent.

b. **1st:** There are five possible outcomes, only one is (G). The required probability is then $\frac{1}{5}$.
2nd: There are now four possible outcomes, only one is (Y). The required probability is then $\frac{1}{4}$.
The two draws are dependent.

The independence or otherwise of multiple draws is a crucial consideration when calculating the "overall" probability of drawing a particular set of sequential outcomes. We return these ideas in Section 9.4.

9.2 COMBINATIONS AND PERMUTATIONS

It is useful at this stage to briefly review combinations and permutations which you may recall from your previous studies.

We begin with a set of three differently colored balls: (R), (G), and (B). In how many ways can these three different balls be arranged? To answer this we might write down all possible arrangements as follows,

and quickly see that there are six possible arrangements. In general, we define a *permutation* to be an *ordered* arrangement of a number of items from a given set of distinct items. It should then be clear that we have identified all six permutations of the three balls from the set of three distinct balls. In contrast, we define a *combination* as an *unordered* set of a number of items from a given set of distinct items. It should then be clear that there is only one combination of the three balls from the set of three different balls.

Consider now a set of four differently colored balls: (R)(G)(B)(Y). We might ask how many possible permutations and combinations of three balls there are from this set of four. The 24 permutations are stated in Table 9.2 and it should be clear that there are four distinct combinations

Table 9.2 The 24 possible permutations picking three from four colored balls

RGB	BRG
RBG	BGR
RGY	BRY
RYG	BYR
RBY	BGY
RYB	BYG
GRB	YRB
GBR	YBR
GRY	YRG
GYR	YGR
GBY	YBG
GYB	YGB

It is common practice to denote the number of permutations of r items from a set of n distinct items by ${}_nP_r$. Similarly, the number of combinations of r items from a set of n distinct items is often denoted ${}_nC_r$. Some experimentation should convince you that the following expressions are true,

$$ {}_nP_r = \frac{n!}{(n-r)!} \tag{9.1} $$

$$ {}_nC_r = \frac{n!}{r!(n-r)!} \tag{9.2} $$

Recall that the symbol "!" denotes the *factorial* operation.

EXAMPLE 9.5

Calculate the number of possible permutations and combinations of the following.

a. Arranging three balls from a total of three distinct balls.

b. Arranging three balls from a total of four distinct balls.

Solution

We use Eqs. (9.1) and (9.2) in each case.

a. We have $n = 3$ and $r = 3$, therefore ${}_3P_3 = \frac{3!}{0!} = 6$ and ${}_3C_3 = \frac{3!}{3!0!} = 1$.

b. We have $n = 4$ and $r = 3$, therefore ${}_4P_3 = \frac{4!}{1!} = 24$ and ${}_4C_3 = \frac{4!}{3!1!} = 4$.

These values are equal to those stated previously. Recall the convention that $0! = 1$.

EXAMPLE 9.6

Consider a jar of 20 *distinct* marbles. How many ways are there are arranging a random draw of three marbles if

a. the order does not matter?

b. the order does matter?

Solution

The 20 marbles are distinct and so a random draw of 3 marbles will not involve repetitions. Equations (9.1) and (9.2) can therefore be used.

a. Since the order does not matter we require ${}_{20}C_3 = \frac{20!}{3!17!} = 1140$.

b. Since the order does matter we require ${}_{20}P_3 = \frac{20!}{17!} = 6840$.

Permutations and combinations can be extremely useful in straightforward probability calculations that involve large numbers of possible outcomes.

EXAMPLE 9.7

Consider a bag containing 10 balls numbered ⓪-⑨. Balls are drawn at random without replacement.

a. Calculate the probability that a random selection of three balls is ①②③ in any order.

b. Calculate the probability that balls ①②③ are drawn in that order.
c. Calculate the probability that all 10 balls are drawn in numerical order.

Solution

a. The draw ①②③ in any order represents one possible combination from a total of ${}_{10}C_3 = \frac{10!}{3!7!} = 120$. The probability is then $\frac{1}{120}$.
Alternatively one could work in terms of permutations. There are a total of ${}_{10}P_3 = \frac{10!}{7!} = 720$ permutations and the outcome of interest can occur in six ways

The probability is then $\frac{6}{720} = \frac{1}{120}$.
b. The draw ①②③ in that order is one permutation from a total of 720. The probability is then $\frac{1}{720}$.
c. The draw ⓪①②③④⑤⑥⑦⑧⑨ is one permutation from a total of ${}_{10}P_{10} = \frac{10!}{0!} = 3{,}628{,}800$. The probability is then $\frac{1}{3{,}628{,}800}$.

EXAMPLE 9.8

A winning lottery ticket is one that correctly matches n balls drawn in any order from the total of 49 distinct balls.
a. Calculate the probability that a single lottery ticket will win if $n = 6$.
b. Calculate n such that a single lottery ticket has more than a 0.01% probability of winning.

Solution

Since the order does not matter, a single lottery ticket represents one from a total of ${}_{49}C_n$ different possible outcomes.
a. If $n = 6$, the number of possible outcomes is $\frac{49!}{6!43!} = 13{,}983{,}816$. The probability of a single combination occurring is then $\frac{1}{13{,}983{,}816}$.
b. We require n such that

$$\frac{n!(49-n)!}{49!} > 0.0001$$

Trial and error shows that this is true when $n = 2$.

9.3 INTRODUCTORY FORMAL PROBABILITY THEORY

9.3.1 Terminology

We now turn to an introduction to formal probability theory which necessarily involves some new terminology. The *sample space* refers to the list of all possible outcomes of an experiment. Each possible outcome is represented by a single point in the sample space, irrespective of the number of ways that outcome can occur. For example, the sample space associated with the experiment of rolling a single fair dice is formed from the outcomes ⚀, ⚁, ⚂, ⚃, ⚄, and ⚅. Furthermore, the sample space of the bag of balls in Example 9.3 is Ⓡ, Ⓑ, Ⓖ, and Ⓖ. Note that the red ball does not appear twice, even though there are two red balls in the bag.

Mathematically, the sample space is a *set* often denoted by Ω. Recall that set theory was discussed in Chapter 1. The sample spaces in the motivating examples are then written formally as

$$\Omega_1 = \{⚀, ⚁, ⚂, ⚃, ⚄, ⚅\}$$

and

$$\Omega_2 = \{Ⓡ, Ⓑ, Ⓖ, Ⓨ\}$$

The discussion in this chapter will be entirely concerned with *discrete* sample spaces. That is, we will consider experiments that can result in one of any individually stated number of outcomes. However, it is also possible to have *continuous* sample spaces. For example, if we were looking at the lifespan of male hamsters, the outcome of the experiment of recording the time from birth until death will result in a positive real number[2] and in this case $\Omega = \mathbb{R}^+$.

EXAMPLE 9.9

State the sample space associated with the following experiments.

a. Tossing a coin.

b. Drawing a single ball from a bag containing balls ⒷⒷⒼⓎ.

c. Drawing two balls without replacement from the bag in part b. (assume order does not matter).

d. The total score obtained from rolling two fair dice.

Solution

a. $\Omega = \{\text{tails, heads}\}$

b. $\Omega = \{Ⓑ, Ⓖ, Ⓨ\}$

c. $\Omega = \{ⒷⒼ, ⒷⓎ, ⒼⓎ, ⒷⒷ\}$

d. $\Omega = \{2, 3, 4, 5, 6, 7, 8, 9, 10, 11, 12\}$

[2] You might argue that because time is actually subdivided into discrete seconds and no hamster is likely to live beyond 3 years, say, the continuum of positive real numbers is an inappropriate sample space. However, given the sheer number of seconds involved in a hamster's lifetime, it is practical to consider lifetimes measured in continuous years. All possible ages would then be represented in the continuous sample space, although the probability of observing a lifetime beyond a certain value, say 3 years, will be zero.

An *event* is defined as a collection of outcomes of an experiment that we might wish to associate a probability to. A simple example is the event that we obtain ⚄ from a single roll of a fair dice. We could denote this event by A and use set notation to define it as $A = \{$⚄$\}$. Events do not necessarily have to be a single outcome. For example, the event that we obtain an odd numbered result from a single roll is $B = \{$⚀, ⚂, ⚄$\}$.

Consider now the particular event $C = \{$⚀, ⚁, ⚂, ⚃, ⚄, or ⚅$\}$, which is interpreted as being the event that we obtain one of either ⚀, ⚁, ⚂, ⚃, ⚄, or ⚅. In this case,

$$C = \Omega$$

which means that C is a *certain event*. Technically the empty set $\emptyset$ is also a valid event and refers to an *impossible event*.

It is sensible to understand events as being any subset of the sample space, an empty set, or the sample space itself. There are some complications in this statement around sample spaces that are *uncountably infinite*, but these considerations are beyond the scope of this book.

EXAMPLE 9.10

Use set notation to express the following events associated with rolling a single dice.

a. A is the event that we obtain ⚅.
b. B is the event that we obtain an even score.
c. C is the event that we obtain a score less than 3.
d. D is the event that event A does not occur.
e. E is the event that a negative score is obtained.

Solution

a. $A = \{$⚅$\}$
b. $B = \{$⚁, ⚃, ⚅$\}$
c. $C = \{$⚀, ⚁$\}$
d. $D = \{$⚀, ⚁, ⚂, ⚃, ⚄$\}$
e. $E = \emptyset$

9.3.2 Use of set theory

You will be familiar with set *unions* and *intersections* from Chapter 1. These operations are very important within probability theory and they enable one to work with combinations of two or more basic events. For example, using the events defined in Example 9.10, a further event can be defined by the occurrence of *either* event A *or* event C and this is obtained from the union of A and C. In particular,

$$A \cup C = \{⚅\} \cup \{⚀, ⚁\} = \{⚀, ⚁, ⚅\}$$

Furthermore, the event that *both* event A *and* B occur defines the further event obtained from the intersection of A and B,

$$A \cap B = \{\text{⚅}\} \cap \{\text{⚁}, \text{⚂}, \text{⚅}\} = \{\text{⚅}\}$$

As a further example, the event that *both* event A *and* event C occur defines a further event,

$$A \cap C = \{\text{⚅}\} \cap \{\text{⚀}, \text{⚁}\} = \emptyset$$

The resulting empty set is interpreted as meaning that this particular intersection is impossible. That is, the outcome of a single roll cannot both score 6 and be less than 3. We would say that events A and C are pairwise *mutually exclusive* because the occurrence of one excludes the possibility of the other. Indeed this example demonstrates the mathematical definition of *pairwise mutual exclusivity*,

$$A \cap C = \emptyset$$

In addition to unions and intersections, the *absolute complement* operation is useful in probability theory. In fact these can be used to form events that are defined as being *not* another event, that is, are *complementary events*. For example, the complementary event of A is A^c and this is simply the event that A does not happen. An example of this was seen in Example 9.10(c) where $D = A^c$.

EXAMPLE 9.11

Obtain and interpret the events formed from the following operations on the events defined in Example 9.10.

a. B^c
b. $C \cup D$
c. $A \cap B$
d. $B \cap C^c$

Solution

a. $B^c = \{\text{⚁}, \text{⚃}, \text{⚅}\}^c = \{\text{⚀}, \text{⚂}, \text{⚄}\}$. This is event that the outcome is *not* even, this is, is odd.
b. $C \cup D = \{\text{⚀}, \text{⚁}\} \cup \{\text{⚀}, \text{⚁}, \text{⚂}, \text{⚃}, \text{⚄}\} = \{\text{⚀}, \text{⚁}, \text{⚂}, \text{⚃}, \text{⚄}\} = D$. This is the event that the score is either less than 3 *or* is not 6, which is the same as the event of not scoring 6.
c. $A \cap B = \{\text{⚅}\} \cap \{\text{⚁}, \text{⚃}, \text{⚅}\} = \{\text{⚅}\} = A$. This is the event that the outcome is both ⚅ *and* even, which is the same as obtaining ⚅.
d. $B \cap C^c = \{\text{⚁}, \text{⚃}, \text{⚅}\} \cap \{\text{⚀}, \text{⚁}\}^c = \{\text{⚁}, \text{⚃}, \text{⚅}\} \cap \{\text{⚂}, \text{⚃}, \text{⚄}, \text{⚅}\} = \{\text{⚃}, \text{⚅}\}$. This is the event that the outcome is even and not less than 3, which is the same as scoring either ⚃ or ⚅.

We are now able to move to a discussion of probabilities in a more formal way. Crucially, probabilities will be associated with events. That is, the sample space from any particular experiment will lead to the collection possible events $\{A_n\}$ and from these one can ask what is the probability of the experiment resulting in the particular event A_n.

More formally, for each event $A_n \subseteq \Omega$, we can associate a real number $P(A_n)$ that is called the *probability* of A_n and satisfies the following conditions,

$$P(A_n) \in [0, 1] \tag{9.3}$$

$$P(\Omega) = 1 \tag{9.4}$$

$$P(\emptyset) = 0 \tag{9.5}$$

These expressions are interpreted as saying that the probability of an event is a real number between 0 and 1, the probability of a certain event is 1, and the probability of an impossible event is 0.

Recall that two events A_i and A_j are said to be mutually exclusive if $A_i \cap A_j = \emptyset$ for $i \neq j$ and one intuitively understands mutual exclusivity as meaning that event A_i and A_j cannot be simultaneous outcomes of the same experiment. In terms of probabilities, this would clearly mean that $P(A_i \cap A_j) = 0$, which follows directly from Eq. (9.5). We would also have that

$$P(A_i \cup A_j) = P(A_i) + P(A_j) \tag{9.6}$$

If a particular collection of events $\{A_n\}$ are all pairwise mutually exclusive, Eq. (9.6) can be extended to give

$$P(A_1 \cup A_2 \cup A_3 \cup \cdots \cup A_k) = \sum_{j=1}^{k} P(A_j) \tag{9.7}$$

Equations (9.3)–(9.7) can be considered the fundamental properties of probabilities and can be used to derive many further properties. In particular, for two general events $A, B \subset \Omega$, it is possible to prove that

$$P(A) = 1 - P(A^c) \tag{9.8}$$

$$\text{if } A \subset B \text{ then } P(A) \leq P(B) \tag{9.9}$$

$$P(A \cup B) = P(A) + P(B) - P(A \cap B) \tag{9.10}$$

Equation (9.10) is often called the *addition rule* and is consistent with Eq. (9.7) when A and B are mutually exclusive ($A \cap B = \emptyset$). The addition rule enables one to compute the probabilities of either one event *or* another occurring.

EXAMPLE 9.12

Use set notation and Eqs. (9.3)–(9.6) to prove Eq. (9.8).

Solution

It is clear that $A_1 \cup A_1^c = \Omega$ and $A_1 \cap A_1^c = \emptyset$. We therefore have

$$P(\Omega) = P(A \cup A^c)$$
$$\Rightarrow 1 = P(A \cup A^c) \text{ from Eq. (9.4)}$$
$$\Rightarrow 1 = P(A) + P(A^c) \text{ from Eq. (9.6)}$$
$$\Rightarrow P(A) = 1 - P(A^c)$$

The proofs of Eqs. (9.9) and (9.10) are somewhat more involved and not considered here.

EXAMPLE 9.13

Demonstrate Eqs. (9.8)–(9.10) using examples from the roll of a single dice.

Solution

We define a number of events as follows

$A_1 = \{$⚀$\}$ $A_2 = \{$⚁, ⚂, ⚃, ⚄, ⚅$\}$ $A_3 = \{$⚁, ⚂, ⚃, ⚄$\}$
$A_4 = \{$⚀, ⚁, ⚂$\}$ $A_5 = \{$⚀, ⚁, ⚂, ⚃, ⚄$\}$ $A_6 = \{$⚁, ⚂$\}$

Since the event of rolling any particular score is pairwise mutually exclusive with any other score, for example $\{$⚀$\} \cap \{$⚁$\} = \emptyset$, we can use Eq. (9.7) to write that

$$P(A_1) = \tfrac{1}{6} \;\; P(A_2) = \tfrac{5}{6} \;\; A_3 = \tfrac{2}{3}$$
$$P(A_4) = \tfrac{1}{2} \;\; P(A_5) = \tfrac{5}{6} \;\; P(A_6) = \tfrac{1}{3}$$

Note that $A_2 = A_1^c$, that is the event that we obtain any score other than ⚀. Using the probabilities stated above, we can indeed confirm that

$$P(A_1) + P(A_2) = P(A_1) + P(A_1^c) = 1$$

which is a demonstration of Eq. (9.8).
It is clear that $A_1 \subset A_4$ and $P(A_1) < P(A_4)$. Furthermore, $A_6 \subset A_4 \subset A_5$ and $P(A_6) < P(A_4) < P(A_5)$. These are all demonstrations of Eq. (9.9).
Consider now the events $A_3 \cup A_4 = \{$⚀, ⚁, ⚂, ⚃, ⚄$\} = A_5$ and $A_3 \cap A_4 = \{$⚁, ⚂$\} = A_6$. The following observation then confirms Eq. (9.10).

$$P(A_3) + P(A_4) - P(A_6) = \frac{2}{3} + \frac{1}{2} - \frac{1}{3} = \frac{5}{6} = P(A_5)$$

EXAMPLE 9.14

A bag contains 1 × Ⓨ, 3 × Ⓑ, 2 × Ⓖ, and 4 × Ⓡ balls. If a single ball is drawn at random from the bag, calculate the probability of each of the follow events.

a. A_1 is the event that either Ⓡ or Ⓑ are drawn.

b. A_2 is the event that either Ⓨ, Ⓑ, or Ⓖ is drawn.

c. A_3 is the event that Ⓡ is not drawn.

d. A_4 is the event that neither Ⓨ nor Ⓖ is drawn.

Solution

We denote the sample space by $\Omega = \{(\mathrm{Y}),(\mathrm{B}),(\mathrm{G}),(\mathrm{R})\}$ and we define the events A_{Y}, A_{B}, A_{G}, and A_{R} to be drawing a yellow, blue, green, and red ball, respectively. That is

$$A_{\mathrm{Y}} = \{(\mathrm{Y})\} \; A_{\mathrm{B}} = \{(\mathrm{B})\} \; A_{\mathrm{G}} = \{(\mathrm{G})\} \; A_{\mathrm{R}} = \{(\mathrm{R})\}$$

It is clear that these basic events are all pairwise mutually exclusive, and

$$P(A_{\mathrm{Y}}) = \frac{1}{10} \quad P(A_{\mathrm{B}}) = \frac{3}{10} \quad P(A_{\mathrm{G}}) = \frac{1}{5} \quad P(A_{\mathrm{R}}) = \frac{2}{5}$$

a. $P(A_1) = P(A_{\mathrm{R}} \cup A_{\mathrm{B}}) = P(A_{\mathrm{R}}) + P(A_{\mathrm{B}}) = \frac{1}{5} + \frac{3}{10} = \frac{1}{2}$

b. $P(A_2) = P(A_{\mathrm{Y}} \cup A_{\mathrm{B}} \cup A_{\mathrm{G}}) = P(A_{\mathrm{Y}}) + P(A_{\mathrm{B}}) + P(A_{\mathrm{G}}) = \frac{1}{10} + \frac{3}{10} + \frac{1}{5} = \frac{3}{5}$

c. $P(A_3) = P(A_{\mathrm{R}}^c) = 1 - P(A_{\mathrm{R}}) = 1 - \frac{1}{5} = \frac{4}{5}$

d. This question is slightly more subtle to formulate using the formal notation. In particular, the event is more sensibly expressed as that the outcome is not (Y) *and* not (G). This is then immediately translated in mathematics as

$$P(A_4) = P(A_{\mathrm{Y}}^c \cap A_{\mathrm{G}}^c)$$

We note that $A_{\mathrm{Y}}^c = \{(\mathrm{B}),(\mathrm{G}),(\mathrm{R})\}$ and $A_{\mathrm{G}}^c = \{(\mathrm{Y}),(\mathrm{B}),(\mathrm{R})\}$ and so $A_{\mathrm{Y}}^c \cap A_{\mathrm{G}}^c = \{(\mathrm{B}),(\mathrm{R})\} = A_{\mathrm{B}} \cup A_{\mathrm{R}}$, which is such that $P(A_{\mathrm{Y}}^c \cap P_{\mathrm{G}}^c) = P(A_{\mathrm{B}} \cup A_{\mathrm{R}}) = P(A_{\mathrm{B}}) + P(A_{\mathrm{R}}) = \frac{3}{10} + \frac{2}{5} = \frac{7}{10}$.

EXAMPLE 9.15

A single card is drawn from a standard deck of 52 playing cards. Calculate the probability of each of the following events.

a. A_1 is the event that the card is either a king or a heart.
b. A_2 is the event that the card is either a queen or a jack.
c. A_3 is the event that the card is either a red or black.
d. A_4 is the event that card is either a red or an ace.

Solution

We begin by defining a number of particular events. In particular, A_{K}, A_{Q}, A_{J}, and A_{A} are the events that the card is a king, queen, jack, or ace, respectively. In addition, A_{B}, A_{R}, and A_{H} are the events that the card is red, black, or a heart, respectively. It is clear that,

$$P(A_{\mathrm{K}}) = P(A_{\mathrm{Q}}) = P(A_{\mathrm{J}}) = P(A_{\mathrm{A}}) = \frac{4}{52} = \frac{1}{13}$$

and

$$P(A_{\mathrm{R}}) = P(A_{\mathrm{B}}) = \frac{26}{52} = \frac{1}{2} \text{ and } P(A_{\mathrm{H}}) = \frac{13}{52} = \frac{1}{4}$$

a. $A_1 = A_K \cup A_H$. The events A_K and A_H are not mutually exclusive and so

$$P(A_1) = P(A_K \cup A_H) = P(A_K) + P(A_H) - P(A_K \cap A_H)$$

The event $A_K \cap A_H$ refers to drawing the unique king of hearts and so $P(A_K \cap A_H) = \frac{1}{52}$. Therefore, $P(A_1) = \frac{1}{13} + \frac{1}{4} - \frac{1}{52} = \frac{4}{13}$.

b. $A_2 = A_Q \cup A_J$. The events A_Q and A_J are mutually exclusive and so $P(A_2) = P(A_Q \cup A_J) = P(A_Q) + P(A_J) = \frac{1}{13} + \frac{1}{13} = \frac{2}{13}$.

c. $A_3 = A_R \cup A_B$. The events A_R and A_B are mutually exclusive and so $P(A_3) = P(A_R \cup A_B) = P(A_R) + P(A_B) = \frac{1}{2} + \frac{1}{2} = 1$. Note that this reflects that $A_R \cup A_B = \Omega$.

d. $A_4 = A_R \cup A_A$. The events A_R and A_A are not mutually exclusive and so $P(A_4) = P(A_R \cup A_A) = P(A_R) + P(A_A) - P(A_R \cap A_A)$. The event $A_R \cap A_A$ refers to drawing one of the two red aces and so $P(A_R \cap A_A) = \frac{2}{52} = \frac{1}{26}$. Therefore, $P(A_4) = \frac{1}{2} + \frac{1}{13} - \frac{1}{26} = \frac{7}{13}$.

9.4 CONDITIONAL PROBABILITIES

9.4.1 Dependent events

In this section, we return the discussion of independent and dependent probabilities and develop a formal framework for *conditional probabilities.*

Let A and B be two events associated with some experiment. We denote the *conditional probability* of the event B given that A has already occurred to be $P(B|A)$. One can read this expression as "the probability of B such that A." This conditional probability is such that

$$P(B|A) = \frac{P(A \cap B)}{P(A)} \tag{9.11}$$

which of course assumes that A is not an impossible event, that is $P(A) \neq 0$.

We motivate this with a simple example of 20 red balls and 80 blue balls in a bag. We define events A_1 and A_2 such that

$$A_1 = \left\{\text{first ball drawn is } \textcircled{R}\right\} \quad A_2 = \left\{\text{second ball drawn is } \textcircled{R}\right\}$$

Clearly $P(A_2)$ depends on whether the first ball drawn is replaced before the second is drawn. In the case that the first draw is replaced, events A_1 and A_2 are independent and $P(A_1) = P(A_2) = \frac{1}{5}$.

In that case that the first draw is not replaced, event A_2 is dependent on A_1. It is clear that $P(A_1) = \frac{1}{5}$, but one needs to know the outcome of the first draw in order to determine $P(A_2)$. In particular, if A_1 did occur there are $19 \times \textcircled{R}$ from 99 remaining which gives $P(A_2|A_1) = \frac{19}{99}$. If A_1 did not occur there are $20 \times \textcircled{R}$ from 99 remaining which gives $P(A_2|A_1^c) = \frac{20}{99}$.

In effect, computing the general conditional probability $P(B|A)$ is equivalent to computing $P(B)$ from a reduced sample space equal to A, rather than the entire sample space Ω. This goes some way to justifying Eq. (9.11).

Equation (9.11) can be arranged to give the *multiplication rule*

$$P(A \cap B) = P(B|A) \times P(A) \tag{9.12}$$

The multiplication rule enables one to compute the probability of one event *and* another occurring.[3] Recall that the addition rule (Eq. 9.10) enables one to calculate the probability of one event *or* another occurring.

If events A and B are independent, it is clear that $P(B|A) = P(B)$ and the multiplication rule simplifies to

$$P(A \cap B) = P(B) \times P(A)$$

That is, the probability of independent events A and B occurring is simply the product of the probabilities of each event occurring. Indeed this expression is taken to be the mathematical definition of independent events.

The multiplication rule can be extended to any number of general dependent events. For example,

$$\begin{aligned} P(A \cap B \cap C) &= P(C|A \cap B) \times P(A \cap B) \\ &= P(C|A \cap B) \times P(B|A) \times P(A) \end{aligned}$$

Which reduces to the following if the events are all independent.

$$P(A \cap B \cap C) = P(A) \times P(B) \times P(C)$$

EXAMPLE 9.16

A bag contains four colored balls (R), (G), (B), and (Y). Balls are drawn at random and replaced before the next draw. Calculate the probabilities of the following events.

a. $A_{\mathrm{R}} = \{(\mathrm{R})\}$, the event that (R) is drawn.

b. $A_{\mathrm{RR}} = \{(\mathrm{R})(\mathrm{R})\}$, the event that (R) is drawn twice.

c. $A_{\mathrm{RRB}} = \{(\mathrm{R})(\mathrm{R})(\mathrm{B})\}$, the event that the balls are drawn in order (R)(R)(B).

Solution

Since the balls are replaced, each draw is independent of the next.

a. $P(A_{\mathrm{R}}) = \frac{1}{4}$

b. $P(A_{\mathrm{RR}}) = P(A_{\mathrm{R}}|A_{\mathrm{R}}) = P(A_{\mathrm{R}})\,P(A_{\mathrm{R}}) = \frac{1}{16}$

c. $P(A_{\mathrm{RRB}}) = P(A_{\mathrm{B}}|A_{\mathrm{RR}}) = P(A_{\mathrm{B}})\,P(A_{\mathrm{RR}}) = P(A_{\mathrm{B}})\,P(A_{\mathrm{R}})\,P(A_{\mathrm{R}}) = \frac{1}{64}$, where A_{B} is the event that (B) is drawn.

---

[3] Since $A \cap B \equiv B \cap A$, we can also have $P(A \cap B) = P(A|B) \times P(B)$.

EXAMPLE 9.17

Two fair dice, one white and one black, are rolled. Calculate the probability of the following events.

a. $A_{\mathrm{a}} =$ {⚄⚅}
b. $A_{\mathrm{b}} =$ {⚂⚃}
c. $A_{\mathrm{c}} =$ {⚄⚅, ⚂⚃}

Solution

We use the notation $A_{\mathrm{w}1} =$ {⚀}, $A_{\mathrm{b}1} =$ {⚀}, $A_{\mathrm{w}2} =$ {⚀}, $A_{\mathrm{b}2} =$ {⚁}, etc. Note that the two dice are independent.

a. $P(A_{\mathrm{a}}) = P(A_{\mathrm{b}6}|A_{\mathrm{w}5}) = P(A_{\mathrm{b}6})P(A_{\mathrm{w}5}) = \frac{1}{6} \times \frac{1}{6} = \frac{1}{36}$
b. $P(A_{\mathrm{b}}) = P(A_{\mathrm{b}4}|A_{\mathrm{w}3}) = P(A_{\mathrm{b}4})P(A_{\mathrm{w}3}) = \frac{1}{6} \times \frac{1}{6} = \frac{1}{36}$
c. $P(A_{\mathrm{c}}) = P(A_{\mathrm{a}} \cup A_{\mathrm{b}}) = P(A_{\mathrm{a}}) + P(A_{\mathrm{b}}) = \frac{1}{18}$. Note that we have used the addition rule for the two mutually exclusive events A_{a} and A_{b}.

EXAMPLE 9.18

Three cards are drawn from a standard pack of playing cards without replacement. What is the probability that none of the cards are from the suit ♣.

Solution

We define the events $A_i =$ {the ith card is not ♣} for $i = 1, 2, 3$. It is clear that the events are dependent. The required probability is then

$$P(A_1 \cap A_2 \cap A_3) = P(A_1) \times P(A_2|A_1) \times P(A_3|A_1 \cap A_2)$$

Since there are 13 cards in each suit, $P(A_1) = \frac{39}{52}$. If the first card is not a ♣, then there are 51 cards available for the second draw, of which 38 are not ♣. Therefore $P(A_2|A_1) = \frac{38}{51}$. If neither the first or the second card was ♣, there are 50 cards available for the third draw, of which 37 are not ♣. Therefore, $P(A_3|A_1 \cap A_2) = \frac{37}{50}$. The required probability is then

$$P(A_1 \cap A_2 \cap A_3) = \frac{39}{52} \times \frac{38}{51} \times \frac{37}{50} = \frac{703}{1700} \approx 0.41$$

9.4.2 The use of partitions

We now introduce the concept of a *partition* before continuing with our discussion of dependent events. One would say that events B_k for $k = 1, \ldots, n$ are a *partition of the sample space* Ω if

1. $B_i \cap B_j = \emptyset$ for any $i \neq j$
2. $B_1 \cup B_2 \cup \cdots \cup B_n = \Omega$
3. $P(B_k) > 0$ for any k

A partition can therefore be thought of as a collection of possible, mutually exclusive events whose union is the entire sample space. That is, any possible result of an experiment must be included in one and only one of the events in a partition.

A partition of a sample space is not necessarily unique and there are typically many different partitions of the same sample space. For example, rolling a single dice has sample space $\Omega = \{\text{⚀}, \text{⚁}, \text{⚂}, \text{⚃}, \text{⚄}, \text{⚅}\}$ and the most obvious partition of this is the set of events

$$B_1 = \{\text{⚀}\} \quad B_2 = \{\text{⚁}\} \quad B_3 = \{\text{⚂}\} \quad B_4 = \{\text{⚃}\} \quad B_5 = \{\text{⚄}\} \quad B_6 = \{\text{⚅}\}$$

However, the following clearly also satisfy the conditions stated above and are perfectly valid partitions

$$B_1' = \{\text{⚀}, \text{⚁}\} \quad B_2' = \{\text{⚂}, \text{⚃}\} \quad B_3' = \{\text{⚄}, \text{⚅}\}$$

and also

$$B_1'' = \{\text{⚀}\} \quad B_2'' = (B_1'')^c$$

Note that these example partitions of this particular sample space are by no means exhaustive.

EXAMPLE 9.19

Determine why the following are *not* valid partitions of the sample space of rolling a standard *white* dice.

a. $B_1''' = \{\text{⚀}, \text{⚁}\} \quad B_2''' = \{\text{⚁}, \text{⚂}\} \quad B_3''' = \{\text{⚃}, \text{⚄}, \text{⚅}\}$
b. $B_1'''' = \{\text{⚀}, \text{⚂}\} \quad B_2'''' = \{\text{⚁}, \text{⚃}, \text{⚅}\}$
c. $B_1''''' = \{\text{⚀}, \text{⚁}, \text{⚂}, \text{⚃}, \text{⚄}, \text{⚅}\} \quad B_2''''' = \{\text{■}\}$

Solution

a. We note that $B_1''' \cap B_2''' = \{\text{⚁}\}$ and the first condition fails.
b. We note that $\{\text{⚄}\} \notin B_1'''' \cup B_2''''$ and the second condition fails.
c. We note that $P(B_2''''') = 0$ and the third condition fails.

If the events $\{B_k\}$ for $k = 1, \ldots, n$ form a partition of Ω, then the probability of an event $A \in \Omega$ can be expressed as

$$\begin{aligned} P(A) &= P(B_1 \cap A) + P(B_2 \cap A) + \cdots + P(B_n \cap A) \\ &= P(B_1)P(A|B_1) + P(B_2)P(A|B_2) + \cdots + P(B_n)P(A|B_n) \end{aligned} \tag{9.13}$$

Equation (9.13) is often referred to as the *total probability theorem*. An important use of this theorem is in computing the probability of event A for which the conditional probabilities $P(A|B_i)$ are already known or are easy to compute themselves. In practice, the crucial part of the application of the total probability theorem is the choice of the partition. This is explored in the following example.

EXAMPLE 9.20

What is the probability that the second card drawn from a standard pack of playing cards is a king?

Solution

We denote this event as A and use the simple partition of the sample space,

$$B_1 = \{\text{first card is a king}\} \quad B_2 = \{\text{first card is not a king}\} = B_1^c$$

This is such that $P(B_1) = \frac{1}{13}$ and $P(B_2) = \frac{12}{13}$. The required probability $P(A)$ can then be expressed in terms of this partition as

$$P(A) = P(B_1)P(A|B_1) + P(B_2)P(A|B_2)$$

We note that $P(A|B_1) = \frac{3}{51}$ and $P(A|B_2) = \frac{4}{51}$, and so

$$P(A) = \frac{1}{13} \times \frac{3}{51} + \frac{12}{13} \times \frac{4}{51} = \frac{1}{13}$$

EXAMPLE 9.21

Consider two bags of balls. Bag 1 contains 2 × Ⓖ and 8 × Ⓨ and bag 2 contains 4 × Ⓖ and 6 × Ⓨ. We now roll a fair dice. If the dice shows an even score we draw from bag 1, otherwise we draw from bag 2. Calculate the probability of drawing Ⓖ from a single experiment.

Solution

We use the partition $B_1 = \{\text{⚁}, \text{⚃}, \text{⚅}\}$ and $B_2 = \{\text{⚀}, \text{⚂}, \text{⚄}\}$ and use Eq. (9.13) to express the required probability as

$$\begin{aligned} P\left(\left\{\text{Ⓖ}\right\}\right) &= P(\{\text{⚁}, \text{⚃}, \text{⚅}\}) \times P\left(\text{Ⓖ} | \{\text{⚁}, \text{⚃}, \text{⚅}\}\right) + P(\{\text{⚀}, \text{⚂}, \text{⚄}\}) \times P\left(\text{Ⓖ} | \{\text{⚀}, \text{⚂}, \text{⚄}\}\right) \\ &= \frac{1}{2} \times \frac{2}{10} + \frac{1}{2} \times \frac{4}{10} \\ &= \frac{3}{10} \end{aligned}$$

We can use the total law of probability on the denominator and the multiplication rule on the numerator to rewrite Eq. (9.11) as follows.

$$P(B_k|A) = \frac{P(A|B_k)\,P(B_k)}{\sum_{j=1}^{n} P(A|B_j)\,P(B_j)} \tag{9.14}$$

This expression is often called *Bayes' Law* or *Bayes' Theorem* and enables one to compute the probability of a particular event B_k, given that event A has occurred. In order to do so, one needs to form a partition $\{B_k\}$ for $k = 1, \ldots, n$ for the sample space Ω.

For example, consider the results of a test for a particular medical condition that 5% of the population is known to suffer from. If the test is known to be 99% accurate for both positive and negative results, how can we compute the probability that someone who tests positive actually has the condition?

Let $Po = \{\text{positive test}\}$ and $C = \{\text{has medical condition}\}$. The information stated can then be written as

$$\begin{aligned} P(C) &= 0.05 \\ P(C^c) &= 0.95 \\ P(Po|C) &= 0.99 \\ P(Po^c|C^c) &= 0.99 \end{aligned}$$

and we require $P(C|Po)$. With the information we have, the right-hand side of Eq. (9.14) implies that the events C and C^c form a suitable partition for applying Bayes' Law in the form

$$P(C|Po) = \frac{P(Po|C)P(C)}{P(Po|C)P(C) + P(Po|C^c)P(C^c)}$$

It should be clear that $P(Po^c|C^c) = 0.99 \Rightarrow P(Po|C^c) = 0.01$ and so

$$P(C|Po) = \frac{0.99 \times 0.05}{0.99 \times 0.05 + 0.01 \times 0.95} \approx 0.839$$

That is, around 84% of those that test positive actually have the condition.

EXAMPLE 9.22

Consider a test for a particular condition that 20% of the population is known to have. Both positive and negative results from the test are known to be 95% accurate. Compute the probability of

a. a "false positive" test result

b. a "false negative" test result

Solution

If $Po = \{\text{positive test}\}$ and $C = \{\text{has medical condition}\}$, we have

$$\begin{aligned} P(C) &= 0.20 \\ P(C^c) &= 0.80 \\ P(Po|C) &= 0.95 \Rightarrow P(Po^c|C) = 0.05 \\ P(Po^c|C^c) &= 0.95 \Rightarrow P(Po|C^c) = 0.05 \end{aligned}$$

a. We require $P(C^c|Po)$, that is the probability that someone with a positive test does not actually have the condition. Using Bayes' Law with the partition C and C^c,

$$P(C^c|Po) = \frac{P(Po|C^c)P(C^c)}{P(Po|C)P(C) + P(Po|C^c)P(C^c)}$$

$$= \frac{0.05 \times 0.80}{0.95 \times 0.20 + 0.05 \times 0.80} \approx 0.174$$

That is, around 17% of those tested positive will not have the condition.

b. We require $P(C|Po^c)$, that is the probability that someone with a negative test result does actually have the condition. Using Bayes' Law with the partition C and C^c,

$$P(C|Po^c) = \frac{P(Po^c|C)P(C)}{P(Po^c|C)P(C) + P(Po^c|C^c)P(C^c)}$$

$$= \frac{0.05 \times 0.20}{0.05 \times 0.20 + 0.95 \times 0.80} \approx 0.013$$

That is, around 1% of those tested negative will actually have the condition.

EXAMPLE 9.23

An insurance company sells a particular motor policy only to the inhabitants of the two English cities, Birmingham and Leicester. Sales of the policy have been successful with 10% and 25% of the populations of Birmingham and Leicester, respectively, having polices with the company. If the population of Birmingham is three times larger than that of Leicester, what fraction of policyholders live in Birmingham?

Solution

We consider the combined populations of the two cities and define events $B =$ {lives in Birmingham} and $L =$ {lives in Leicester}, such that $P(B) = 0.75$ and $P(L) = 0.25$. Furthermore, if $Po =$ {has policy}, then $P(Po|B) = 0.10$ and $P(Po|L) = 0.25$. We require $P(B|Po)$. Using Bayes' Law with partition B and L,

$$P(B|Po) = \frac{P(Po|B)P(B)}{P(Po|B)P(B) + P(Po|L)P(L)}$$

$$= \frac{0.10 \times 0.75}{0.10 \times 0.75 + 0.25 \times 0.25} = 54\%$$

That is, 54% of policyholders live in Birmingham and 46% live in Leicester.

9.5 PROBABILITY ON YOUR COMPUTER

Wolfram Alpha provides an extremely intuitive means of investigating the probabilities of events within the standard scenarios we have used in this chapter: tossing coins, rolling dice, and drawing cards. For example, rather than asking it specific questions about particular events arising from, say, 12 coin tosses, the command

```
12 coin tosses
```

returns the probabilities of a number of different events. For example, the command returns that the probability of all heads is 0.02441%, and the probability of at least one head is 99.89%. Similarly, the command

```
15 dice
```

returns the probabilities of some interesting events arising from rolling 15 fair dice. For example, the probability that all faces will be shown in 64.42%, and the probability that all dice will show different faces is 0. This second result is of course obvious as there are only six possible faces on each dice. Furthermore, the probability of the particular event of holding three kings and one queen in a random hand of n cards from a standard deck is obtained from

```
probability 3 kings 1 queen
```

The engine returns the probabilities for various n. For example, if we have a random hand of nine cards, $n = 9$, and the probability of holding three kings and one queen is given as 0.4721%.

More specific questions can, of course, be asked of the engine. For example, the probability that 14 heads result from 30 tosses of a fair coin is obtained as approximately 13.54% from the command

```
probability 30 coin tosses 14 heads
```

The probabilities of related events are also returned in the same output. For example, the probability of more than 14 heads from 30 tosses is given as 57.22%.

The command `conditional probabilities` opens an interface that you may also find useful. In particular, you are invited to input the probabilities $P(A)$, $P(B)$, and $P(A \cap B)$ and the engine returns the conditional probabilities $P(A|B)$ and $P(B|A)$.

As you might expect, the command

```
combinations 4 from 50
```

returns the value of ${}_{50}C_4$ to be 230,300. Similarly, the command

```
permutations 4 from 50
```

returns that ${}_{50}P_4 = 5{,}527{,}200$.

While these facilities may have limited use in practice, they are very useful for generating questions for practicing your knowledge of probability. The reader is invited to explore how Wolfram Alpha could be used in this way.

EXAMPLE 9.24

Use Wolfram Alpha to determine the probabilities of the following events.

a. A dice is rolled five times, giving a different face each time.

b. A random draw of four playing cards will be two jacks, a king, and a queen.

c. Ten coins are tossed leading to five heads and five tails.

Solution

a. Each roll is an independent event and so five rolls of the same dice is exactly the same as rolling five dice simultaneously. We use the command `5 dice` and see that the probability that all dice are different is 0.09259, that is 9.26%.

b. We use the command `probability 2 jacks 1 king 1 queen` and see that the required probability is 3.546×10^{-4}.

c. We use the command `10 coins` and see that the probability of five heads and five tails is 24.61%.

9.6 ACTUARIAL APPLICATION: MORTALITY

Some examples of the use of probability theory in insurance and health care have already been discussed in this chapter. While these are both typical actuarial applications, we take the opportunity here to introduce some ideas from the additional and fundamental actuarial topic of *mortality*. Mortality is, of course, concerned with questions of an individual's life and death. Associating probabilities to the event of an individual's death within some time period is crucial in the traditional areas of actuarial practice: *life assurance* and *pensions*. In particular, a life assurance contract pays out when a death occurs, and a pension contract continues to pay until a death occurs. In both situations, an understanding of the probabilities of death at each age is required for estimating the expected present value of the liability to the insurer or pension provider.

Clearly a death event is a movement in one direction: an individual who is alive at some point in time can either be alive or dead at some later time. However, an individual who is now dead must remain dead for all later times. It is therefore sensible to formulate the mathematics of mortality in terms of the future status (alive or dead) of a current life. Standard actuarial practice is to use the notation

$$ {}_tp_x \text{ and } {}_tq_x $$

where ${}_tp_x$ denotes the probability that a life aged exactly x *survives* a further t years, and ${}_tq_x$ denotes the probability that the same life *dies* within the next t years. That is, both quantities are concerned with the future status of a life currently aged exactly x. Note that ${}_tp_x$ is often referred to as a *survival probability*, for obvious reasons. In the particular case that $t = 1$, the general notation is simplified such that

$$ {}_1p_x \equiv p_x \text{ and } {}_tq_x \equiv q_x $$

For example, p_{50} denotes the probability that a life, alive on his 50th birthday, will survive to his 51st birthday. Similarly, q_{50} denotes the probability that a life, alive on his 50th birthday, will die before his 51st birthday.

EXAMPLE 9.25

Use standard notation to express the following probabilities.

a. The probability that an individual alive on his 50th birthday will survive a further 20 years.

b. The probability that an individual alive on her 60th birthday will survive to her 90th birthday.

c. The probability that an individual currently aged 90 exact will die within the next year.

d. The probability that an individual alive on his 37th birthday will die before claiming a pension at age 65.

e. The probability that a life assurance contract will pay out during the next 2 years if the insured is currently aged 46.5 exact.

Solution

a. The probability that a life aged 50 exact will survive a further 20 years is denoted ${}_{20}p_{50}$.

b. The probability that a life aged 60 exact will survive a further 30 years is denoted ${}_{30}p_{60}$.

c. The probability that a life currently aged 90 exact will die within 1 year is denoted q_{90}.

c. The probability that a life aged 37 exact will die within the next 28 years is denoted ${}_{28}q_{37}$.

d. The contract pays out on death. The probability that a life aged 46.5 exact will die within 2 years is denoted ${}_{2}q_{46.5}$.

Given that the future life status of an individual life can only be alive or dead, the quantities ${}_{t}p_{x}$ and ${}_{t}q_{x}$ must sum to one,

$$ {}_{t}p_{x} + {}_{t}q_{x} = 1 $$

The implication of this statement is that we can work in terms of either ${}_{t}p_{x}$ or ${}_{t}q_{x}$ and convert to the other as required. Survival probabilities have some very useful properties that arise from that the transition from alive to dead is in one direction only. In particular, it should be clear that the probability that a life aged x exact survives a further t years is equal to the product of the probabilities of surviving each separate year, or indeed any number of sub-intervals over the future time period $[0, t]$. We might then write statements such as, for example,

$$ {}_{10}p_{x} = {}_{5}p_{x} \cdot {}_{2}p_{x+5} \cdot {}_{3}p_{x+7} = {}_{1}p_{x} \cdot {}_{9}p_{x+1} $$

In some sense, you might consider these and similar statements as a "principle of consistency" for survival probabilities (recall that this concept was discussed in Chapter 3 for compound interest problems). That is, in order to survive an extended period of time, the life must survive every possible arrangement of sub-intervals of this period.

EXAMPLE 9.26

Interpret the following expressions and indicate whether they are true or false.

a. ${}_{10}p_x = {}_5p_x \cdot {}_2p_{x+5} \cdot {}_3p_{x+7}$
b. ${}_{10}q_x \neq {}_5q_x \cdot {}_5q_{x+5}$

Solution

a. The probability that an individual aged x survives a further 10 years is equal to the probability that he survives the first 5 years, and then the next 2 years, and then the final 3 years. Note that the age of the life is correctly accounted for at the start of each time period. This is a true statement.
b. The probability that an individual aged x dies within the next 10 years is NOT equal to the probability that he dies within the first 5 years multiplied by the probability that an individual aged $x + 5$ dies within a further 5 years. This (negative) statement is true, that is, it is NOT possible to express death probabilities in this way.

It is very useful for actuaries to be able to determine survival probabilities over future time periods. For example, if a life is currently aged 30 exact, what is the probability that he dies at some time between the ages of 40 and 45 exact? It should be clear that, in order for the death to occur between ages 40 and 45, the life must first survive between ages 30 and 40. Using standard notation, the required probability is expressed as

$$ {}_{10}p_{30} \cdot {}_5q_{40} $$

EXAMPLE 9.27

Express the following probabilities in terms of standard notation.

a. The probability that a life currently aged 27 exact will die while aged 90.
b. The probability that a new-born will die while aged 100.

Solution

a. For a life currently aged 27 exact to die while aged 90, he must survive from age 27 to 90 and then die within 1 year of his 90th birthday. This probability is expressed as ${}_{63}p_{27} \cdot q_{90}$.
b. For a new-born to die while aged 100, she must survive from age 0 to 100 and then die within 1 year of her 100th birthday. This probability is expressed as ${}_{100}p_0 \cdot q_{100}$.

EXAMPLE 9.28

Using appropriate 1-year survival probabilities, construct an expression for the probability that a life aged 30 exact dies within 5 years.

Solution

For a life to die between ages 30 and 35, he must either die between ages 30 and 31, or survive to 31 and die between 31 and 32, etc. Each of these is an independent event and so we can add the probabilities. That is,

$$_5q_{30} = q_{30} + p_{30}.q_{31} + {}_2p_{30}.q_{32} + {}_3p_{30}.q_{33} + {}_4p_{30}.q_{34}$$

However, given that we are required to express the probability in terms of 1-year survival probabilities, we must use $q_x = 1 - p_x$ and expressions of the form ${}_2p_x = {}_1p_x.{}_1p_{x+1}$, for example. This leads to,

$$\begin{aligned} _5q_{30} = &(1 - p_{30}) + p_{30}.(1 - p_{31}) + p_{30}.p_{31}.(1 - p_{32}) + p_{30}.p_{31}.p_{32}.(1 - p_{33}) \\ &+ p_{30}.p_{31}.p_{32}.p_{33}.(1 - p_{34}) \end{aligned}$$

The question of how to obtain ${}_tp_x$ still remains. There are a number of ways to do this, but, inevitably, most rely on *empirical data*. That is, most methods for determining survival properties rely on the observation of *homogeneous populations* of individuals. The need for homogeneity should be clear from knowing that many factors influence an individual's mortality; for example, age, sex, health, and lifestyle all influence the probability that a life will survive another year. Accurate mortality predictions must therefore be based on data collected from observations of identical individuals.

Under the *life-table model*, mortality probabilities are computed from tabulated numbers of survivors at each age. A life table considers l_0 individuals born simultaneously and tracks the number of survivors l_x at each exact age, $x = 0, 1, 2, \ldots, \omega$, where ω is some limiting age, the survival beyond which is assumed to be zero.[4] Such long-term observations result in a table that is then assumed to apply to all other members of the particular homogeneous group observed. An example of a life table is illustrated in Table 9.3. Note that both l_x and d_x are reported, where $d_x = l_x - l_{x+1}$ is the *number of deaths* while aged x.

The following probabilities should be intuitively clear,

$${}_tp_x = \frac{l_{x+t}}{l_x} \qquad q_x = \frac{d_x}{l_x} \qquad {}_tq_x = \frac{l_x - l_{x+t}}{l_x}$$

For example, the probability that an individual aged 40 exact will survive at least 10 years is obtained from the number of lives aged 50 as a proportion of the number of lives aged 40. Using Table 9.3, we have

$${}_{10}p_{40} = \frac{l_{50}}{l_{40}} = \frac{93{,}925}{96{,}500} \approx 0.9733$$

Furthermore, the probability that an individual aged 60 exact will die before his 65th birthday is given by

$${}_5q_{60} = \frac{l_{60} - l_{65}}{l_{60}} = \frac{86{,}714 - 79{,}293}{86{,}714} \approx 0.0856$$

[4] Typically ω is taken to be some age between 100 and 120.

Table 9.3 Illustrative life table

x	l_x	d_x	x	l_x	d_x	x	l_x	d_x	x	l_x	d_x
0	100,000	814	21	98,413	85	42	96,155	194	63	82,701	1625
1	99,186	62	22	98,328	87	43	95,961	210	64	81,076	1783
2	99,124	38	23	98,241	87	44	95,751	230	65	79,293	1940
3	99,086	30	24	98,154	87	45	95,521	255	66	77,353	2097
4	99,056	24	25	98,067	84	46	95,266	283	67	75,256	2255
5	99,032	22	26	97,983	83	47	94,983	315	68	73,001	2403
6	99,010	20	27	97,900	83	48	94,668	352	69	70,598	2543
7	98,990	18	28	97,817	85	49	94,316	391	70	68,055	2674
8	98,972	19	29	97,732	87	50	93,925	436	71	65,381	2819
9	98,953	18	30	97,645	89	51	93,489	485	72	62,562	2969
10	98,935	18	31	97,556	91	52	93,004	537	73	59,593	3109
11	98,917	18	32	97,465	95	53	92,467	594	74	56,484	3218
12	98,899	19	33	97,370	97	54	91,873	656	75	53,266	3301
13	98,880	23	34	97,273	103	55	91,217	727	76	49,965	3386
14	98,857	29	35	97,170	113	56	90,490	806	77	46,579	3455
15	98,828	39	36	97,057	124	57	89,684	892	78	43,124	3494
16	98,789	52	37	96,933	133	58	88,792	987	79	39,630	3502
17	98,737	74	38	96,800	145	59	87,805	1091	80	36,128	3474
18	98,663	86	39	96,655	155	60	86,714	1207	81	32,654	3400
19	98,577	81	40	96,500	166	61	85,507	1334	82	29,254	3300
20	98,496	83	41	96,334	179	62	84,173	1472	⋮	⋮	⋮

Notes: The table is actually based on the mortality of the male population of Britain in the early 1990s and is an extract of the *English Life Tables No. 15* published by the Stationary Office.

EXAMPLE 9.29

Calculate the following using the data in Table 9.3.

a. The total number of deaths between ages 60 and 65.
b. The probability that a life aged 56 exact will survive to age 80 exact.
c. The probability that a life currently aged 10 exact will die between his 50th and 55th birthdays.
d. The probability that a new-born will die aged 81.

Solution

a. The total number of deaths between $x = 60$ and 65 is given by $l_{60} - l_{65} = 7421$. Note that this is also given by $d_{60} + d_{61} + d_{62} + d_{63} + d_{64}$.
b. This is straightforward survival probability given by

$$_{24}p_{56} = \frac{l_{80}}{l_{56}} = \frac{36{,}128}{90{,}490} \approx 0.3992$$

c. This requires the life to survive from age 10 to 50 and then die between 50 and 55. The probability is given by

$$ {}_{40}p_{10}.{}_{5}q_{50} = \frac{l_{50}}{l_{10}}.\frac{l_{50}-l_{55}}{l_{50}} = \frac{l_{50}-l_{55}}{l_{10}} \approx 0.0274 $$

d. This requires the life to survive from age 0 to 81 and then die within that year. The probability is given by

$$ {}_{81}p_{0}.q_{81} = \frac{l_{81}}{l_{0}}.\frac{l_{81}-l_{82}}{l_{81}} = \frac{l_{81}-l_{82}}{l_{0}} = 0.034 $$

EXAMPLE 9.30

Use Table 9.3 to confirm the expression for ${}_{5}q_{30}$ determined in Example 9.28.

Solution

We are required to confirm that

$$ \begin{aligned} {}_{5}q_{30} = {} & (1-p_{30}) + p_{30}.(1-p_{31}) + p_{30}.p_{31}.(1-p_{32}) + p_{30}.p_{31}.p_{32}.(1-p_{33}) \\ & + p_{30}.p_{31}.p_{32}.p_{33}.(1-p_{34}) \end{aligned} $$

The left-hand side is given by ${}_{5}q_{30} = \frac{l_{30}-l_{35}}{l_{30}} \approx 0.00486$. The right-hand side requires each p_x which are obtained as

$$ p_{30} = 0.99909 \quad p_{31} = 0.99907 \quad p_{32} = 0.99903 \quad p_{33} = 0.99903 \quad p_{34} = 0.99894 $$

and these lead to 0.00486, as required.

EXAMPLE 9.31

Calculate the following probabilities using Table 9.3.

a. The probability that an individual currently aged 44 exact dies while aged either 50 or 51.

b. The probability that an individual currently aged 37 exact dies while aged either 67 or 77.

Solution

a. The probability that an individual currently aged 44 exact dies either while aged 50 or aged 51 is given by

$$ {}_{6}p_{44}.{}_{2}q_{50} = \frac{l_{50}}{l_{44}}.\frac{l_{50}-l_{52}}{l_{50}} = .\frac{l_{50}-l_{52}}{l_{44}} \approx 0.0096 $$

b. In this case, we have to consider the probability of two independent events. Either the life (currently aged 37) dies aged 67 or he dies aged 77. The probability is therefore given by

$$ {}_{30}p_{37}.q_{67} + {}_{40}p_{37}.q_{77} = \frac{l_{67}-l_{68}}{l_{30}} + \frac{l_{77}-l_{78}}{l_{30}} \approx 0.0585 $$

9.7 QUESTIONS

The following questions are intended to test your knowledge of the material discussed in this chapter. Full solutions are available in Chapter 9 Solutions of Part IV. You should use an algebraic approach unless otherwise stated.

Question 9.1. Twenty otherwise identical balls are individually numbered 1-20 and placed inside a bag. Calculate the probabilities of the following events.

a. A random draw of a single ball produces a ball numbered as a multiple of both 3 and 4.

b. A random draw of a single ball produces a ball numbered as either a multiple of 3 or 4.

c. A random draw of five balls produces the particular balls (1)(2)(3)(4)(5) in any order.

d. A random draw of three balls produces the particular balls (2)(4)(6) in that order.

Question 9.2. A production line in a factory produces 50 units per week and, on average, 10% of these units are expected to have some defect. Calculate the probability that a random selection of two units will both be faulty in the following situations.

a. The first unit is replaced before the second is selected.

b. The first unit is not replace before the second is selected.

Question 9.3. The math department of a university has been moved temporarily to another building while their building is being modernized. Each professor is to be allocated an office at random. If the probability that a randomly chosen office has a blackboard is 0.6, the probability that it has a computer terminal is 0.7, and the probability that it has neither is 0.1, what is the probability that a professor is allocated an office with at least one item missing from the office?

Question 9.4. A bag contains five balls colored red (R), green (G), blue (B), yellow (Y), and black (K). State which of the following are valid partitions of the sample space.

a. $A_1 = \{(R), (G)\}$, $A_2 = \{(B)\}$, $A_3 = \{(K)\}$

b. $B_1 = \{(R), (G), (B), (K)\}$, $B_2 = \{(Y)\}$

c. $C_1 = \{(R), (G)\}$, $C_2 = \{(K)\}$, $C_3 = \{(R), (B), (Y)\}$

d. $D_1 = \{(R), (G)\}$, $D_2 = \{(B), (P), (K)\}$, $D_3 = \{(Y)\}$

Question 9.5. A new medical test has recently been approved to test for a rare genetic condition. The test is known to be 99% accurate. You ask that your doctor perform this test on you and the result comes back positive, that is, the test identifies you as having the condition. Calculate the probability that you do not actually have the condition in the following situations.

a. The condition is reasonably common and about 1% of the population is expected to have it.

b. The condition is rare and only 0.1% of the population is expected to have it.

Question 9.6. Determine how common the condition in Question 9.5 must be for you to be 90% sure that a positive test result means that you have the condition. You should still assume that the test is 99% accurate.

Question 9.7. A single dice is rolled repeatedly until it results in ⚅. Calculate the probability that this occurs on roll n, where n is any even number, in each of the following situations.

a. The dice is fair.

b. The dice is biased such that $P(⚅) = 0.3$.

c. The dice is biased such that $P(⚅) = 1$.

Question 9.8. Express the following in standard actuarial notation.

a. The probability that a life aged 50 exact will survive to age 56 exact.

b. The probability that a life aged 40 exact will die aged either 90, 91, or 92.

c. The probability that a new-born will die either aged 60, 61, or 78.

Question 9.9. The 1-year survival probability of a particular species of animal is known to be given by

$$p_x = \frac{0.9}{(1+x)^2}$$

where $x = 0, 1, 2, \ldots$ is age.

a. Calculate the probability that an animal aged 5 exact will survive another year.

b. Calculate the probability that a new-born animal will die within the first 3 years of life.

Question 9.10. A young man receives a new car and an unusual financial contract as a gift for his 16th birthday. If the car is worth \$11,000 and the contract promises to pay \$10,000 if he survives to his 20th birthday, estimate the financial value of the gift. You should assume that,

- his 1-year survival function is given by $p_x = 0.8 + 0.006 \times (x - 20)^2$ for ages $x = 16, 17, \ldots, 19$,
- if paid, the \$10,000 would be received on his 20th birthday,
- the effective rate of interest is 5% per annum.

CHAPTER 10

Introductory Linear Algebra

Contents

Prerequisite knowledge	**Intended learning outcomes**
• Chapter 1	• Define and recognize $m \times n$ matrices • Identify whether given matrices can be added, subtracted and multiplied, and, where possible, perform these operations • Obtain the transpose of a matrix and recall the general properties of transposed matrices • Define, recognize and work with square matrices, including the particular cases I_n and 0_n • State the conditions for a matrix to be invertible and, where possible, obtain the inverse in simple cases using the adjoint method • Solve simple matrix equations • Use matrix algebra to solve systems of simultaneous equations

In this chapter, we are concerned with introductory *linear algebra*. Linear algebra is a vast area of mathematics and we are necessarily limited in what we can cover here. In particular, we will limit our scope to *matrix algebra*. After defining what is meant by a *matrix*, we discuss the fundamental operations of matrix addition and multiplication and introduce an analogue of division. While we begin by discussing matrices as objects in their own right, our motivation will be focussed towards understanding techniques for the solution of systems of linear simultaneous equations; and these are considered at the end of the chapter.

10.1 BASIC MATRIX ALGEBRA

In terms of the scope of this book, a *matrix* will be defined to be a rectangular (or square) *array* of real numbers. For example,

$$A = \begin{bmatrix} 5 & -2 \\ 1 & 7 \\ 2 & 8 \end{bmatrix}, \quad B = \begin{bmatrix} 1 & -2 & 8 \\ 7 & 5 & -1 \end{bmatrix}, \quad C = \begin{bmatrix} -1 & 1 & 2 \end{bmatrix}, \quad D = \begin{bmatrix} 7 \\ 9 \\ -10 \end{bmatrix} \tag{10.1}$$

are all examples of *matrices* (the plural of the singular "matrix"). Matrices can be classified by the dimensions of the array, that is, the number of *rows* and *columns* stated in that order. For example, matrix A in Eq. (10.1) is a 3×2 matrix, whereas B is a 2×3 matrix. In general, one speaks of an $m \times n$ matrix that is understood to refer to m rows and n columns. Matrices C and D in Eq. (10.1) are therefore 1×3 and 3×1 matrices, respectively. For obvious reasons, C is also an example of a *row matrix* and D an example of a *column matrix*.

The individual numbers that constitute a matrix are called its *entries*. A particular entry is identified by its location within the matrix, that is, the row and column position that it appears, numbered from the top left corner. For example, referring to Eq. (10.1), the (3,2)-entry of A is 8 and the (2,1)-entry of B is 7. Matrices are typically denoted by capital letters and their entries by lowercase letters.

The general notation $A = [a_{ij}]$ for an $m \times n$ matrix is particularly useful to define matrix A by its particular entries a_{ij}. The square brackets should be interpreted as meaning that matrix A is formed from the ordered collection of individual entries a_{ij}, each at row i and column j, where $i = 1, \ldots, m$ and $j = 1, \ldots, n$. The notation $A = [a_{ij}]$ is then understood as meaning

$$A = \begin{bmatrix} a_{11} & a_{12} & \cdots & a_{1n} \\ a_{21} & a_{22} & \cdots & a_{2n} \\ \vdots & \vdots & \vdots & \vdots \\ a_{m1} & a_{m2} & \cdots & a_{mn} \end{bmatrix}$$

For example, the 1×3 matrix G formed from $g_{11} = 1$, $g_{12} = 3$, and $g_{13} = 13$ is

$$G = [g_{ij}] = \begin{bmatrix} 1 & 3 & 13 \end{bmatrix}$$

EXAMPLE 10.1

State the dimensions of the following matrices in $m \times n$ format. In each case, state any three separate entries using "a_{ij}" notation.

a. $A = \begin{bmatrix} 1 & 4 \\ 2 & 5 \\ 3 & 6 \end{bmatrix}$

b. $B = \begin{bmatrix} -2 & -4 & -3 \\ 1 & 2 & 4 \\ 3 & 2 & 1 \end{bmatrix}$

c. $C = \begin{bmatrix} -2 & -1 & 3 & 8 & 2 \end{bmatrix}$

c. $D = \begin{bmatrix} 1 & 3 & 2 \\ 2 & 3 & 1 \\ 3 & 1 & 2 \\ 0 & 2 & 1 \\ 0 & 0 & 7 \end{bmatrix}$

Solution

a. 3×2 matrix with particular entries including $a_{11} = 1$, $a_{12} = 4$, and $a_{21} = 2$.
b. 3×3 matrix with particular entries including $b_{22} = 2$, $b_{23} = 4$, and $b_{32} = 2$.
c. 1×5 matrix with particular entries including $c_{12} = -1$, $c_{13} = 3$, and $c_{15} = 2$.
d. 5×3 matrix with particular entries including $d_{42} = 2$, $d_{22} = 3$, and $d_{53} = 7$.

For two matrices to be equal it should be clear that all of their corresponding entries must be equal; this necessarily means that their dimensions are identical too. That is, if A is an $m \times n$ matrix and B an $p \times q$ matrix, the statement $A = B$ must mean necessarily that $m = p$, $n = q$ and $a_{ij} = b_{ij}$ for all $i = 1, \ldots, m$ and $j = 1, \ldots, n$.

EXAMPLE 10.2

State, with reasons, whether the following pairs of matrices are equal:

a. $A = \begin{bmatrix} 4 & 4 \\ 3 & 1 \\ 1 & -1 \end{bmatrix}, \quad B = \begin{bmatrix} 4 & 4 \\ 3 & 1 \\ 1 & -1 \end{bmatrix}$

b. $C = \begin{bmatrix} 1 & 2 & 1 \\ 2 & 3 & 4 \end{bmatrix}, \quad D = \begin{bmatrix} 1 & 2 \\ 2 & 3 \\ 1 & 4 \end{bmatrix}$

c. $E = \begin{bmatrix} 1 & 2 & 1 \\ 2 & 5 & 4 \end{bmatrix}, \quad F = \begin{bmatrix} 1 & 2 & 1 \\ 2 & 6 & 4 \end{bmatrix}$

Solution

a. These matrices have identical dimensions and $a_{ij} = b_{ij}$ for all i, j. Therefore $A = B$.
b. These matrices have different dimensions, C is 2×3 and D is 3×2. Therefore $C \neq D$.
c. These matrices have identical dimensions. However, $e_{22} \neq f_{22}$ and so $E \neq D$.

As with individual numbers, which we now refer to as *scalars* to distinguish them from matrices, it is possible to add two or more matrices. However, matrix addition is only defined for matrices of the same dimension. For example, if A, B, and C are 2×3, 2×3, and 2×2 matrices, respectively, it is possible to determine $A + B$, but not $A + C$ or $B + C$. The matrix resulting from the addition of A and B is in fact determined by

$$A + B = [a_{ij} + b_{ij}] \tag{10.2}$$

That is, one simply adds the corresponding entries. This definition is extended to the sum of three or more equally-sized matrices in the obvious way.

Now, let A, B, and C denote matrices of the same dimension such that the addition of any two or all the three is defined. Equation (10.2) shows that the general properties of matrix addition follow directly from the familiar properties of scalar addition. In particular, matrix addition is *commutative* and *associative*. These two properties are symbolically represented as

$$A + B = B + A \tag{10.3}$$

$$(A + B) + C = A + (B + C) \tag{10.4}$$

Similarly, the obvious definition of matrix subtraction is true. That is,

$$A - B = A + (-B) \tag{10.5}$$

EXAMPLE 10.3

You are given the following matrices:

$$D = \begin{bmatrix} 1 & -2 \\ 0 & 1 \\ 0 & 2 \\ 1 & 2 \end{bmatrix}, \quad E = \begin{bmatrix} 1 & 4 & 4 \\ 4 & 2 & 2 \\ 1 & 2 & 3 \end{bmatrix}, \quad F = \begin{bmatrix} 2 & 2 \\ 1 & 1 \\ 1 & -3 \\ 3 & -1 \end{bmatrix} \tag{10.6}$$

Wherever possible, obtain the following:

a. $D + E$
b. $D + F$
c. $E + F$
d. $D - F$

Solution

a. D and E have different dimensions and so computing $D + E$ is not possible.

b. D and F have the same dimensions and it is possible to compute $D + F$. Using Eq. (10.2), this is

$$D + F = \begin{bmatrix} 1 & -2 \\ 0 & 1 \\ 0 & 2 \\ 1 & 2 \end{bmatrix} + \begin{bmatrix} 2 & 2 \\ 1 & 1 \\ 1 & -3 \\ 3 & -1 \end{bmatrix} = \begin{bmatrix} 1+2 & -2+2 \\ 0+1 & 1+1 \\ 0+1 & 2-3 \\ 1+3 & 2-1 \end{bmatrix} = \begin{bmatrix} 3 & 0 \\ 1 & 2 \\ 1 & -1 \\ 4 & 1 \end{bmatrix}$$

c. E and F have different dimensions and so computing $E + F$ is not possible.

d. D and F have the same dimensions and it is possible to compute $D - F$. Using Eq. (10.5), this is

$$D - F = \begin{bmatrix} 1 & -2 \\ 0 & 1 \\ 0 & 2 \\ 1 & 2 \end{bmatrix} - \begin{bmatrix} 2 & 2 \\ 1 & 1 \\ 1 & -3 \\ 3 & -1 \end{bmatrix} = \begin{bmatrix} 1-2 & -2-2 \\ 0-1 & 1-1 \\ 0-1 & 2+3 \\ 1-3 & 2+1 \end{bmatrix} = \begin{bmatrix} -1 & -4 \\ -1 & 0 \\ -1 & 5 \\ -2 & 3 \end{bmatrix}$$

The term $-B$ in Eq. (10.5) is of course interpreted as $-1 \times B$ and this raises the important concept of *scalar multiplication* of matrices.

Scalar multiplication follows in the obvious way and is distinct from *matrix multiplication*, which we consider in Section 10.2. In particular, for any scalar $k \in \mathbb{R}$ (or indeed $\mathbb{C}$) we define the scalar multiplication of a matrix A by k as

$$kA = [ka_{ij}]$$

For example,

$$2E = 2\begin{bmatrix} 1 & 4 & 4 \\ 4 & 2 & 2 \\ 1 & 2 & 3 \end{bmatrix} = \begin{bmatrix} 2\times 1 & 2\times 4 & 2\times 4 \\ 2\times 4 & 2\times 2 & 2\times 2 \\ 2\times 1 & 2\times 2 & 2\times 3 \end{bmatrix} = \begin{bmatrix} 2 & 8 & 8 \\ 8 & 4 & 4 \\ 2 & 4 & 6 \end{bmatrix}$$

It is also possible to define the *scalar division* of a matrix by any non-zero scalar. This follows in the obvious way

$$\frac{1}{k}A = \left[\frac{a_{ij}}{k}\right]$$

Clearly the result of multiplying (or dividing) an $m \times n$ matrix by the scalar k is another $m \times n$ matrix.

The idea of scalar multiplication is easily extended to prove the obvious results for two scalars k and l,

$$k(A + B) = kA + kB \tag{10.7}$$

$$(k + l)A = kA + lA \tag{10.8}$$

EXAMPLE 10.4

Use "a_{ij}" notation to prove the results given in Eqs. (10.3), (10.4), and (10.7).

Solution

These are easily proved using "a_{ij}" notation and exploiting the properties of scalar addition and multiplication at the element level. In particular,

$$A + B = [a_{ij} + b_{ij}] = [b_{ij} + a_{ij}] = B + A$$

$$(A+B)+C=[(a_{ij}+b_{ij})+c_{ij}]=[a_{ij}+(b_{ij}+c_{ij})]=A+(B+C)$$
$$k(A+B)=k[a_{ij}+b_{ij}]=[ka_{ij}+kb_{ij}]=kA+kB$$

as required. Similar reasoning can be used to prove Eq. (10.8).

EXAMPLE 10.5

If $k=2$, $l=3$, $A=\begin{bmatrix}1 & 2\\ 2 & 1\end{bmatrix}$ and $B=\begin{bmatrix}0 & 1\\ 3 & 2\end{bmatrix}$, compute the following expressions:

a. $(k+l)A$

b. $kA+lB$

c. $\frac{1}{k}A$

d. $2lA-4kB$

Solution

a. $(2+3)\times\begin{bmatrix}1 & 2\\ 2 & 1\end{bmatrix}=\begin{bmatrix}5 & 10\\ 10 & 5\end{bmatrix}$

b. $2\times\begin{bmatrix}1 & 2\\ 2 & 1\end{bmatrix}+3\times\begin{bmatrix}0 & 1\\ 3 & 2\end{bmatrix}=\begin{bmatrix}2 & 4\\ 4 & 2\end{bmatrix}+\begin{bmatrix}0 & 3\\ 9 & 6\end{bmatrix}=\begin{bmatrix}2 & 7\\ 13 & 8\end{bmatrix}$

c. $\frac{1}{2}\times\begin{bmatrix}1 & 2\\ 2 & 1\end{bmatrix}=\begin{bmatrix}\frac{1}{2} & 1\\ 1 & \frac{1}{2}\end{bmatrix}$

d. $2\times 3\times\begin{bmatrix}1 & 2\\ 2 & 1\end{bmatrix}-4\times 2\times\begin{bmatrix}0 & 1\\ 3 & 2\end{bmatrix}=\begin{bmatrix}6 & 12\\ 12 & 6\end{bmatrix}+\begin{bmatrix}0 & -8\\ -24 & -16\end{bmatrix}=\begin{bmatrix}6 & 4\\ -12 & -10\end{bmatrix}$

The notion of matrix subtraction leads to the concept of a *zero matrix*. Much like the scalar computation $2-2$, for example, necessitates the need for a zero scalar, the matrix computation $E-E$, for example, necessitates the need for a zero matrix. In particular,

$$E-E=\begin{bmatrix}1 & 4 & 4\\ 4 & 2 & 2\\ 1 & 2 & 3\end{bmatrix}-\begin{bmatrix}1 & 4 & 4\\ 4 & 2 & 2\\ 1 & 2 & 3\end{bmatrix}=\begin{bmatrix}0 & 0 & 0\\ 0 & 0 & 0\\ 0 & 0 & 0\end{bmatrix}$$

A zero matrix is, therefore, a matrix of any dimension that has every entry equal to zero. It should be clear that a zero matrix also arises from the scalar multiplication of any matrix by the zero scalar, $k=0$.

To this point our discussion has been limited to manipulating matrices held in a fixed array. For example, both scalar multiplication and matrix addition begins and results with $m\times n$ matrices. However, there is a fundamental operation that enables one to transform an $m\times n$ matrix to a particular $n\times m$ matrix; this operation is called *transposition*. The transposition of matrix A leads to the *transpose* of A, denoted A^{T}. Put precisely,

$$\text{if } A=[a_{ij}] \text{ then } A^{\mathrm{T}}=[a'_{ij}]=[a_{ji}]$$

That is, the rows and columns of A become the columns and rows, respectively, of A^T. For example,

$$A = \begin{bmatrix} 10 & 2 & 5 & 6 \end{bmatrix} \Rightarrow A^T = \begin{bmatrix} 10 \\ 2 \\ 5 \\ 6 \end{bmatrix}$$

$$B = \begin{bmatrix} 1 & 2 \\ 3 & 4 \\ 5 & 6 \end{bmatrix} \Rightarrow B^T = \begin{bmatrix} 1 & 3 & 5 \\ 2 & 4 & 6 \end{bmatrix}$$

$$C = \begin{bmatrix} 1 & 0 & 1 \\ 0 & 2 & -4 \\ 1 & -4 & 3 \end{bmatrix} \Rightarrow C^T = \begin{bmatrix} 1 & 0 & 1 \\ 0 & 2 & -4 \\ 1 & -4 & 3 \end{bmatrix}$$

This final example is an interesting one as $C^T = C$, which indicates a particular symmetry of matrix C. Indeed a *symmetric* matrix is defined as one that is equal to its transpose. Clearly all symmetric matrices must be *square*, that is, they must have dimensions $m \times m$. The properties of transposed matrices are explored in the following examples and we will discuss square matrices in Section 10.3.

EXAMPLE 10.6

Use "a_{ij}" notation to prove the following properties of transposition. In each case, verify the property with a simple example using 2×3 matrices.

a. $(A^T)^T = A$

b. $(kA)^T = kA^T$

c. $(A + B)^T = A^T + B^T$, with A and B having identical dimensions

Solution

a. Since $A = a_{ij}$, $A^T = [a'_{ij}] = [a_{ji}]$ and so $(A^T)^T = [a'_{ji}] = [a_{ij}] = A$, as required.

For example, if $A = \begin{bmatrix} 1 & 2 & 3 \\ 4 & 5 & 6 \end{bmatrix}$, $(A^T)^T = \begin{bmatrix} 1 & 4 \\ 2 & 5 \\ 3 & 6 \end{bmatrix}^T = \begin{bmatrix} 1 & 2 & 3 \\ 4 & 5 & 6 \end{bmatrix} = A.$

b. $(kA)^T = [ka'_{ij}] = [ka_{ji}] = k[a_{ji}] = kA^T$.

For example, if $k = 2$,

$$(2A)^T = \begin{bmatrix} 2 & 4 & 6 \\ 8 & 10 & 12 \end{bmatrix}^T = \begin{bmatrix} 2 & 8 \\ 4 & 10 \\ 6 & 12 \end{bmatrix} = 2\begin{bmatrix} 1 & 4 \\ 2 & 5 \\ 3 & 6 \end{bmatrix} = 2\begin{bmatrix} 1 & 2 & 3 \\ 4 & 5 & 6 \end{bmatrix}^T = 2A^T$$

c. $(A + B)^T = [a_{ij} + b_{ij}]^T = [a'_{ij} + b'_{ij}] = [a_{ji} + b_{ji}] = [a_{ji}] + [b_{ji}] = A^T + B^T$.

For example, if $B = \begin{bmatrix} 2 & 0 & 1 \\ 1 & -2 & 3 \end{bmatrix}$ and A is taken from part a,

$$(A+B)^{\mathrm{T}} = \begin{bmatrix} 3 & 2 & 4 \\ 5 & 3 & 9 \end{bmatrix}^{\mathrm{T}} = \begin{bmatrix} 3 & 5 \\ 2 & 3 \\ 4 & 9 \end{bmatrix} = \begin{bmatrix} 1 & 4 \\ 2 & 5 \\ 3 & 6 \end{bmatrix} + \begin{bmatrix} 2 & 1 \\ 0 & -2 \\ 1 & 3 \end{bmatrix}$$

$$= \begin{bmatrix} 1 & 2 & 3 \\ 4 & 5 & 6 \end{bmatrix}^{\mathrm{T}} + \begin{bmatrix} 2 & 0 & 1 \\ 1 & -2 & 3 \end{bmatrix}^{\mathrm{T}} = A^{\mathrm{T}} + B^{\mathrm{T}}$$

EXAMPLE 10.7

Prove that if C and D are symmetric matrices of equal dimensions, then $C+D$ is also symmetric. Demonstrate this property with a simple example using 4×4 matrices.

Solution

The symmetry property for a single matrix is $C^{\mathrm{T}} = C$. We are, therefore, required to prove that $(C+D)^{\mathrm{T}} = C+D$ and this follows as below.

$$(C+D)^{\mathrm{T}} = C^{\mathrm{T}} + D^{\mathrm{T}} \text{ (from property c. in Example 10.6)}$$

$$= C + D \text{ (since } C \text{ and } D \text{ are individually symmetric)}$$

For example, working with $C = \begin{bmatrix} 1 & 1 & 0 & 1 \\ 1 & 2 & 3 & -1 \\ 0 & 3 & -1 & 7 \\ 1 & -1 & 7 & 3 \end{bmatrix}$ and $D = \begin{bmatrix} 0 & 2 & 1 & 4 \\ 2 & -3 & -2 & -1 \\ 1 & -2 & 1 & 0 \\ 4 & -1 & 0 & 5 \end{bmatrix}$

which are both individually symmetric,

$$(C+D)^{\mathrm{T}} = \begin{bmatrix} 1 & 3 & 1 & 5 \\ 3 & -1 & 1 & -2 \\ 1 & 1 & 0 & 7 \\ 5 & -2 & 7 & 8 \end{bmatrix}^{\mathrm{T}} = \begin{bmatrix} 1 & 3 & 1 & 5 \\ 3 & -1 & 1 & -2 \\ 1 & 1 & 0 & 7 \\ 5 & -2 & 7 & 8 \end{bmatrix} = C + D$$

10.2 MATRIX MULTIPLICATION

We now turn to matrix multiplication, that is, the multiplication of one matrix by another. Unlike the scalar multiplication of a single matrix, the multiplication of two matrices may not be intuitive to you. The first thing to note is that not all pairs of matrices can be multiplied. In particular, if A is an $m \times n$ matrix and B a $p \times q$ matrix, the product $A \times B$ is only defined if $n = p$. The reason for this will be clear from the following discussion.

Let us first consider the multiplication of a row matrix by a column matrix. For example, if $A = \begin{bmatrix} 1 & 2 & 3 \end{bmatrix}$, a 1×3 matrix, and $B = \begin{bmatrix} 4 \\ 5 \\ 6 \end{bmatrix}$, a 3×1 matrix, then the product AB is defined, but BA is not. The product AB is in fact calculated as

$$AB = \begin{bmatrix} 1 & 2 & 3 \end{bmatrix} \begin{bmatrix} 4 \\ 5 \\ 6 \end{bmatrix} = 1 \times 4 + 2 \times 5 + 3 \times 6 = 32$$

which is a 1×1 matrix (a scalar). In terms of the dimensions of the matrices involved,

$$\underbrace{(1 \times 3)}_{A} \times \underbrace{(3 \times 1)}_{B} \rightarrow \underbrace{1 \times 1}_{AB}$$

You might be aware from your previous studies that this product is identical to the *dot product* of the associated three-dimensional vectors. More formally, we state that the product of a general row matrix A and general column matrix B is given by

$$AB = \sum_{k=1}^{n} a_{1k} b_{k1} \tag{10.9}$$

In this example $n = 3$ and

$$AB = \sum_{k=1}^{3} a_{1k} b_{k1} = a_{11} b_{11} + a_{12} b_{21} + a_{13} b_{31}$$

However, the expression extends to any $n \in \mathbb{N}^{+}$.

EXAMPLE 10.8

Calculate all possible products of two matrices that exist from

$$C = \begin{bmatrix} 0 & 1 & 5 & 3 \end{bmatrix}, \quad D = \begin{bmatrix} 1 \\ 4 \\ -1 \\ 2 \end{bmatrix}, \quad E = \begin{bmatrix} -3 \\ 2 \end{bmatrix}, \quad F = \begin{bmatrix} 0 & 1 \end{bmatrix}$$

Solution

The dimensions of the matrices are such that only CD and FE are defined. Using Eq. (10.9), we have

$$CD = \begin{bmatrix} 0 & 1 & 5 & 3 \end{bmatrix} \begin{bmatrix} 1 \\ 4 \\ -1 \\ 2 \end{bmatrix} = 0 \times 1 + 1 \times 4 + 5 \times (-1) + 3 \times 2 = 5$$

$$FE = \begin{bmatrix} 0 & 1 \end{bmatrix} \begin{bmatrix} -3 \\ 2 \end{bmatrix} = 0 \times (-3) + 1 \times 2 = 2$$

The product of row and column matrices, as described here, is actually the fundamental operation for calculating the products of more general matrices. In particular, Eq. (10.9) is generalized to

$$(ab)_{ij} = \sum_{k=1}^{n} a_{ik} b_{kj} \tag{10.10}$$

In other words, this can be interpreted as

the ijth entry in the matrix formed from the product AB is given by the product of row matrix i from A with column matrix j from B.

Equation (10.10) is valid only if A is an $m \times n$ matrix and B an $n \times p$ matrix; the resulting matrix AB is then of size $m \times p$. That is,

$$\underbrace{(m \times n)}_{A} \times \underbrace{(n \times p)}_{B} \rightarrow \underbrace{m \times p}_{AB}$$

For example, let us consider the following 2×4 and 4×3 matrices:

$$A = \begin{bmatrix} 0 & -1 & 1 & 2 \\ -3 & 2 & 2 & 3 \end{bmatrix}, \quad B = \begin{bmatrix} 1 & 2 & 0 \\ 4 & 3 & 1 \\ -1 & 1 & 2 \\ 2 & -2 & 4 \end{bmatrix}$$

It should be clear that the product AB is defined and will be a 2×3 matrix. However, the product BA is not defined. The matrix resulting from AB is computed from the repeated application of Eq. (10.10) to determine all entries. For example, the (1,1)th entry of AB can be obtained from the "dot" product of row 1 of A with column 1 of B,

$$\begin{bmatrix} 0 & -1 & 1 & 2 \end{bmatrix} \begin{bmatrix} 1 \\ 4 \\ -1 \\ 2 \end{bmatrix} = -1$$

Similarly, the (2,3)th entry of AB, for example, can be obtained from the "dot" product of row 2 of A with column 3 of B,

$$\begin{bmatrix} -3 & 2 & 2 & 3 \end{bmatrix} \begin{bmatrix} 0 \\ 1 \\ 2 \\ 4 \end{bmatrix} = 18$$

The process is repeated for all entries in AB and the reader is invited to confirm the final result

$$AB = \begin{bmatrix} -1 & -6 & 9 \\ 9 & -4 & 18 \end{bmatrix}$$

EXAMPLE 10.9

If

$$C = \begin{bmatrix} 1 & 2 & 3 & 4 \\ 5 & 6 & 7 & 8 \end{bmatrix} \quad \text{and} \quad D = \begin{bmatrix} 1 & 2 \\ 3 & 4 \\ 5 & 6 \\ 7 & 8 \end{bmatrix}$$

compute the products CD and DC.

Solution

It is clear that matrix C is 2×4 and matrix D is 4×2. Both products CD and DC therefore exist and will have dimensions 2×2 and 4×4, respectively. In particular,

$$CD = \begin{bmatrix} 1 & 2 & 3 & 4 \\ 5 & 6 & 7 & 8 \end{bmatrix} \begin{bmatrix} 1 & 2 \\ 3 & 4 \\ 5 & 6 \\ 7 & 8 \end{bmatrix} = \begin{bmatrix} 50 & 60 \\ 104 & 140 \end{bmatrix}$$

and

$$DC = \begin{bmatrix} 1 & 2 \\ 3 & 4 \\ 5 & 6 \\ 7 & 8 \end{bmatrix} \begin{bmatrix} 1 & 2 & 3 & 4 \\ 5 & 6 & 7 & 8 \end{bmatrix} = \begin{bmatrix} 11 & 14 & 17 & 20 \\ 23 & 30 & 37 & 44 \\ 35 & 46 & 57 & 68 \\ 47 & 62 & 77 & 92 \end{bmatrix}$$

Example 10.9 demonstrates a crucial difference between matrix multiplication and scalar multiplication: even if both products are defined, in general, $AB \neq BA$. That is, matrix multiplication does not *commute*. We must therefore talk about "pre" and "post" multiplication. For example, matrix B can be pre-multiplied by A to give AB, or post-multiplied by A to give BA.

For an additional appropriately sized matrix C and scalar k, it should be clear that

$$ABC = (AB)C = A(BC) \tag{10.11}$$

$$kAB = k(AB) = (kA)B = A(kB) \tag{10.12}$$

Although matrix multiplication is not commutative, it is still distributive. That is,

$$A(B + C) = AB + AC \tag{10.13}$$

$$(A + B)C = AC + BC \tag{10.14}$$

The reader is invited to confirm these properties with example matrices of appropriate dimension.

EXAMPLE 10.10

Use "a_{ij}" notation to prove that for appropriately sized matrices

$$(AB)^{\mathrm{T}} = B^{\mathrm{T}} A^{\mathrm{T}} \tag{10.15}$$

Demonstrate this with a simple example of rectangular matrices.

Solution

If $A = [a_{ij}]$ and $B = [b_{ij}]$ are $m \times n$ and $n \times p$ matrices, respectively, then $A^{\mathrm{T}} = [a'_{ij}] = [a_{ji}]$ and $B^{\mathrm{T}} = [b'_{ij}] = [b_{ji}]$ are $n \times m$ and $p \times n$, respectively. Using Eq. (10.10), the product $B^{\mathrm{T}} A^{\mathrm{T}}$ is

$$\sum_{k=1}^{m} b'_{ik} a'_{kj} = \sum_{k=1}^{m} b_{ki} a_{jk} = \sum_{k=1}^{m} a_{jk} b_{ki}$$

This relies on the fact that $b_{ki} a_{jk} = a_{jk} b_{ki}$, that is, the scalar entries commute. The expression defines the jith entry of AB, that is the ijth entry of $(AB)^{\mathrm{T}}$, as required. For example, if

$$A = \begin{bmatrix} 1 & 2 \\ 3 & 4 \\ 5 & 6 \end{bmatrix} \Rightarrow A^{\mathrm{T}} = \begin{bmatrix} 1 & 3 & 5 \\ 2 & 4 & 6 \end{bmatrix}$$

and

$$B = \begin{bmatrix} 1 & -2 & -3 & 4 \\ 3 & -2 & 1 & 2 \end{bmatrix} \Rightarrow B^{\mathrm{T}} = \begin{bmatrix} 1 & 3 \\ -2 & -2 \\ -3 & 1 \\ 4 & 2 \end{bmatrix}$$

Then

$$(AB)^{\mathrm{T}} = \begin{bmatrix} 7 & -6 & -1 & 8 \\ 15 & -14 & -5 & 20 \\ 23 & -22 & -9 & 32 \end{bmatrix}^{\mathrm{T}} = \begin{bmatrix} 7 & 15 & 23 \\ -6 & -14 & -22 \\ -1 & -5 & -9 \\ 8 & 20 & 32 \end{bmatrix}$$

and

$$B^{\mathrm{T}} A^{\mathrm{T}} = \begin{bmatrix} 1 & 3 \\ -2 & -2 \\ -3 & 1 \\ 4 & 2 \end{bmatrix} \begin{bmatrix} 1 & 3 & 5 \\ 2 & 4 & 6 \end{bmatrix} = \begin{bmatrix} 7 & 15 & 23 \\ -6 & -14 & -22 \\ -1 & -5 & -9 \\ 8 & 20 & 32 \end{bmatrix}$$

Note that the property demonstrated in Eq. (10.15) is readily extended to the product of any number of appropriately sized matrices. For example,

$$(ABCDE)^{\mathrm{T}} = E^{\mathrm{T}} D^{\mathrm{T}} C^{\mathrm{T}} B^{\mathrm{T}} A^{\mathrm{T}}$$

The important point to note is that the transpose "reverses" the order of the product.

10.3 SQUARE MATRICES

10.3.1 Properties

We now turn our attention to *square matrices*, that is, matrices of dimension $m \times m$ for some m. Square matrices are an interesting subset of general rectangular matrices. For example, if A and B are both $m \times m$, it should be clear that AB, BA, A^2 and B^2 all exist and also have dimension $m \times m$. This "preservation of dimension" under multiplication is a key property.

We had previously introduced the zero matrix as being any matrix with zero in every entry. The notation 0_m will now be used to denote the special case of a zero matrix of dimension $m \times m$, that is a *square zero matrix*. For example,

$$0_2 = \begin{bmatrix} 0 & 0 \\ 0 & 0 \end{bmatrix} \quad \text{and} \quad 0_5 = \begin{bmatrix} 0 & 0 & 0 & 0 & 0 \\ 0 & 0 & 0 & 0 & 0 \\ 0 & 0 & 0 & 0 & 0 \\ 0 & 0 & 0 & 0 & 0 \\ 0 & 0 & 0 & 0 & 0 \end{bmatrix}$$

It should be clear that 0_m is such that

$$A0_m = 0_m A = 0_m \tag{10.16}$$

for all A with dimension $m \times m$. It is, therefore, seen that pre- and post-multiplication by 0_m leads to 0_m. The matrices A and 0_m therefore do commute, as one might intuitively expect.

Furthermore, we can define the square matrix equivalent of unity. This is denoted by I_m and consists of unit entries on the *main diagonal* and 0 elsewhere. For example,

$$I_2 = \begin{bmatrix} 1 & 0 \\ 0 & 1 \end{bmatrix} \quad \text{and} \quad I_5 = \begin{bmatrix} 1 & 0 & 0 & 0 & 0 \\ 0 & 1 & 0 & 0 & 0 \\ 0 & 0 & 1 & 0 & 0 \\ 0 & 0 & 0 & 1 & 0 \\ 0 & 0 & 0 & 0 & 1 \end{bmatrix}$$

It should be clear from Eq. (10.10) that I_m commutes with any $m \times m$ matrix A and is such that

$$AI_m = I_m A = A \tag{10.17}$$

A trivial property of both 0_m and I_m is that they are symmetric, that is $0_m^{\mathrm{T}} = 0_m$ and $I_m^{\mathrm{T}} = I_m$.

10.3.2 Inverse matrices

We have so far defined matrix addition, subtraction, and multiplication, but we are yet to define an equivalent for the division of matrices. This is considered here.

In terms of scalar arithmetic, it is clear that dividing by a non-zero number is equivalent to multiplying by the *reciprocal* of that number. For example,

$$\frac{7}{2} \equiv 2^{-1} \times 7$$

Indeed standard algebraic convention is such that multiplying by 2^{-1} is universally understood as meaning "divide by 2."

Similarly, dividing by matrix A should be considered as equivalent to multiplying by A^{-1}, where A^{-1} is referred to as the *inverse* of A. As we shall see, only square matrices can have an inverse but squareness alone is not sufficient to ensure that a matrix is *invertible* (i.e., has an inverse).

For scalars, it is clear that $k^{-1} \times k = k \times k^{-1} = 1$ for $k \neq 0$. Similarly, the inverse of square matrix A is such that

$$A^{-1}A = AA^{-1} = I_m \tag{10.18}$$

Indeed matrix A is said to be invertible if it is square and A^{-1} exists such that Eq. (10.18) is satisfied. This expression, therefore, acts as the definition of A^{-1}.

It should be immediately clear from Eq. (10.18) that I_m^{-1} is the inverse of I_m, and 0_m is non-invertible.

EXAMPLE 10.11

Demonstrate that $A = \begin{bmatrix} \frac{1}{2} & \frac{1}{2} \\ \frac{1}{2} & -\frac{1}{2} \end{bmatrix}$ has inverse $A^{-1} = \begin{bmatrix} 1 & 1 \\ 1 & -1 \end{bmatrix}$.

Solution

To show that A^{-1} is the inverse of A, we need to confirm that Eq. (10.18) is satisfied. In particular, we require $AA^{-1} = I_2$ and $A^{-1}A = I_2$. Indeed

$$\begin{bmatrix} \frac{1}{2} & \frac{1}{2} \\ \frac{1}{2} & -\frac{1}{2} \end{bmatrix} \begin{bmatrix} 1 & 1 \\ 1 & -1 \end{bmatrix} = \begin{bmatrix} 1 & 0 \\ 0 & 1 \end{bmatrix} = I_2$$

$$\begin{bmatrix} 1 & 1 \\ 1 & -1 \end{bmatrix} \begin{bmatrix} \frac{1}{2} & \frac{1}{2} \\ \frac{1}{2} & -\frac{1}{2} \end{bmatrix} = \begin{bmatrix} 1 & 0 \\ 0 & 1 \end{bmatrix} = I_2$$

as required.

Determining the inverse of a given invertible matrix is a difficult process that will be considered later. We begin by familiarizing ourselves with some fundamental properties of inverse matrices.

The first property is a direct analogue of scalar reciprocals and is stated without proof,

$$\left(A^{-1}\right)^{-1} = A \tag{10.19}$$

This property is particularly useful when solving matrix equations that involve inverse matrices, as discussed in Section 10.5. Transposes and products of invertible matrices have particular properties that may also prove useful in practice. In particular, if A is an invertible matrix then A^{T} is also invertible and equal to $\left(A^{-1}\right)^{\mathrm{T}}$. By definition, this means that

$$A^{\mathrm{T}}\left(A^{-1}\right)^{\mathrm{T}} = \left(A^{-1}\right)^{\mathrm{T}} A^{\mathrm{T}} = I_m \tag{10.20}$$

which is simply an application of Eq. (10.18). The proof of this stems from Eq. (10.15) for the transposition of products. In particular,

$$A^{\mathrm{T}}(A^{-1})^{\mathrm{T}} = (A^{-1}A)^{\mathrm{T}} = I_m^{\mathrm{T}} = I_m$$

Similarly, we see that $(A^{-1})^{\mathrm{T}}A^{\mathrm{T}} = (AA^{-1})^{\mathrm{T}} = I_m^{\mathrm{T}} = I_m$. Furthermore, it should be clear that

$$\left(A^{-1}\right)^{\mathrm{T}} = \left(A^{\mathrm{T}}\right)^{-1} \tag{10.21}$$

EXAMPLE 10.12

Use matrix $A = \begin{bmatrix} \frac{1}{2} & \frac{1}{2} \\ \frac{1}{2} & -\frac{1}{2} \end{bmatrix}$ to confirm Eq. (10.20).

Solution

The inverse of A is known from Example 10.11, and is such that

$$\left(A^{-1}\right)^{\mathrm{T}} = \begin{bmatrix} 1 & 1 \\ 1 & -1 \end{bmatrix}$$

Equation (10.20) is then confirmed by

$$A^{\mathrm{T}}\left(A^{-1}\right)^{\mathrm{T}} = \begin{bmatrix} \frac{1}{2} & \frac{1}{2} \\ \frac{1}{2} & -\frac{1}{2} \end{bmatrix}\begin{bmatrix} 1 & 1 \\ 1 & -1 \end{bmatrix} = \begin{bmatrix} 1 & 0 \\ 0 & 1 \end{bmatrix}$$

$$\left(A^{-1}\right)^{\mathrm{T}} A^{\mathrm{T}} = \begin{bmatrix} 1 & 1 \\ 1 & -1 \end{bmatrix}\begin{bmatrix} \frac{1}{2} & \frac{1}{2} \\ \frac{1}{2} & -\frac{1}{2} \end{bmatrix} = \begin{bmatrix} 1 & 0 \\ 0 & 1 \end{bmatrix}$$

If A, B, and C are invertible matrices of identical dimension, the inverse of any product is obtained from the product of the individual inverses in reverse order. For example,

$$\begin{aligned} (ABC)^{-1} &= C^{-1}B^{-1}A^{-1} \\ (BAC)^{-1} &= C^{-1}A^{-1}B^{-1} \end{aligned}$$

$$(ACA)^{-1} = A^{-1}C^{-1}A^{-1}$$

This can be extended to the products of any number of invertible matrices. Comparing these results to Eq. (10.15), for example, demonstrates some parallels between the properties of transposed and inverted matrix products.

EXAMPLE 10.13

Demonstrate that $(ABC)^{-1} = C^{-1}B^{-1}A^{-1}$.

Solution

This can be demonstrated easily using Eq. (10.18). In particular,

$$ABC \times C^{-1}B^{-1}A^{-1} = AB\left(CC^{-1}\right)B^{-1}A^{-1} = A\left(BB^{-1}\right) = AA^{-1} = I_m$$

$$C^{-1}B^{-1}A^{-1} \times ABC = C^{-1}B^{-1}\left(A^{-1}A\right)BC = C^{-1}\left(B^{-1}B\right)C = C^{-1}C = I_m$$

Successive steps in each case rely on the fact that $BB^{-1} = B^{-1}B = I_m$, for example, and that multiplying by I_m is the matrix equivalent of multiplying by unity.

EXAMPLE 10.14

If P, Q, R, S, T, U, and V are invertible matrices of identical dimension, write down the following in terms of products of individual inverse matrices.

a. $(PQ)^{-1}$
b. $(PQRST)^{-1}$
c. $(VSPQRTVQ)^{-1}$
d. $\left((PQR)^{\mathrm{T}}\right)^{-1}$

Solution

a. $(PQ)^{-1} = Q^{-1}P^{-1}$
b. $(PQRST)^{-1} = T^{-1}S^{-1}R^{-1}Q^{-1}P^{-1}$
c. $(VSPQRTVQ)^{-1} = Q^{-1}V^{-1}T^{-1}R^{-1}Q^{-1}P^{-1}S^{-1}V^{-1}$
d. $\left((PQR)^{\mathrm{T}}\right)^{-1} = \left(R^{\mathrm{T}}Q^{\mathrm{T}}P^{\mathrm{T}}\right)^{-1} = (P^{\mathrm{T}})^{-1}(Q^{\mathrm{T}})^{-1}(R^{\mathrm{T}})^{-1} = (P^{-1})^{\mathrm{T}}(Q^{-1})^{\mathrm{T}}(R^{-1})^{\mathrm{T}}$ from Eq. (10.21).

10.3.3 The adjoint method for inversion

You should now be comfortable with the properties of invertible matrices. The question now is how can one determine whether a matrix is invertible and, if it is, how to determine the inverse?

We begin with the comment that the inverse of an invertible matrix is unique. That is, an invertible matrix has only one inverse. This is of course also true of scalar reciprocals, for example, it is clear that 2^{-1} equals 0.5 and no other value. The proof of uniqueness is given in the following example.

EXAMPLE 10.15

Use Eq. (10.18) to demonstrate the uniqueness of A^{-1} for an invertible $m \times m$ matrix A.

Solution

We assume that A is invertible and denote two distinct inverses of A by C and D. The aim is to prove that $C \equiv D$. Since C and D are both assumed to be inverses of A, we have $CA = I_m$ and $DA = I_m$ by Eq. (10.18). The proof proceeds by exploiting these expressions,

$$\begin{aligned} D &= I_m D && \text{by the definition of } I_m \\ &= (CA)D && \text{since } CA = I_m \\ &= C(AD) && \text{from Eq. (10.11)} \\ &= CI_m && \text{since } AD = I_m \\ &= C && \text{by the definition of } I_m \end{aligned}$$

which is as required.

There are a number of methods available for computing the inverse of an invertible matrix and we focus on the *adjoint method* in this book. Other methods include the *Gauss-Jordon method* or simply using some commercial software. We return to the use of Wolfram Alpha for matrix algebra at the end of this chapter. As we will see, manual methods for inverting matrices are very labor intensive. In practice, one would always revert to a computer, but it is important to understand how matrices can be inverted by hand. We begin with small matrices and gradually increase their size.

For a 2×2 matrix it is easily shown that

$$A = \begin{bmatrix} a_{11} & a_{12} \\ a_{21} & a_{22} \end{bmatrix} \Rightarrow A^{-1} = \frac{1}{a_{11}a_{22} - a_{12}a_{21}} \begin{bmatrix} a_{22} & -a_{12} \\ -a_{21} & a_{11} \end{bmatrix} \tag{10.22}$$

The inverse is therefore only defined if $a_{11}a_{22} - a_{12}a_{21} \neq 0$ and this restriction gives a method by which one can distinguish invertible from non-invertible 2×2 matrices. In particular, if $a_{11}a_{22} = a_{12}a_{22}$, then the 2×2 matrix is *not* invertible.

EXAMPLE 10.16

Demonstrate that Eq. (10.22) is true for an invertible 2×2 matrix.

Solution

We are required to confirm that the inverse matrix in Eq. (10.22) satisfies the conditions in Eq. (10.18). Since the matrix is invertible, we have $a_{11}a_{22} - a_{12}a_{21} \neq 0$ and so

$$\begin{aligned} AA^{-1} &= \begin{bmatrix} a_{11} & a_{12} \\ a_{21} & a_{22} \end{bmatrix} \times \frac{1}{a_{11}a_{22} - a_{12}a_{21}} \begin{bmatrix} a_{22} & -a_{12} \\ -a_{21} & a_{11} \end{bmatrix} \\ &= \frac{1}{a_{11}a_{22} - a_{12}a_{21}} \begin{bmatrix} a_{11} & a_{12} \\ a_{21} & a_{22} \end{bmatrix} \begin{bmatrix} a_{22} & -a_{12} \\ -a_{21} & a_{11} \end{bmatrix} \end{aligned}$$

$$= \frac{1}{a_{11}a_{22} - a_{12}a_{21}} \begin{bmatrix} a_{11}a_{22} - a_{12}a_{21} & 0 \\ 0 & a_{11}a_{22} - a_{12}a_{21} \end{bmatrix}$$
$$= I_2$$

The reader is invited to confirm that $A^{-1}A = I$ is also satisfied.

The quantity $a_{11}a_{22} - a_{12}a_{21}$ is known as the *determinant* of the 2×2 matrix A, denoted algebraically by $\det(A)$. It is often convenient to use the matrix notation for a determinant, indicated by a vertical line either side of the array as follows:

$$\det(A) = \begin{vmatrix} a_{11} & a_{12} \\ a_{21} & a_{22} \end{vmatrix} = a_{11}a_{22} - a_{12}a_{21} \tag{10.23}$$

Although we will not consider the wider uses of determinants at length here, they are actually very important values associated with square matrices and have wide application in linear algebra.

The matrix in Eq. (10.22) is called the *adjoint* of the 2×2 matrix A, denoted $\text{adj}(A)$. That is,

$$\text{adj}(A) = \begin{bmatrix} a_{22} & -a_{12} \\ -a_{21} & a_{11} \end{bmatrix}$$

Equation (10.22) can then be expressed as

$$A^{-1} = \frac{1}{\det(A)}\text{adj}(A) \tag{10.24}$$

and is often referred to as the *adjoint method* for inverting matrix A. Note that this method is also sometimes referred to as the *adjugate method.*

The adjoint method is in fact applicable to all invertible $m \times m$ matrices. Furthermore, it leads to the general rule that a general $m \times m$ matrix, A, is invertible if and only if $\det(A) \neq 0$. Clearly the adjoint method changes the problem of inverting a general square matrix to determining its determinant and adjoint form. We now consider these aspects for general m.

Consider a general $m \times m$ matrix, $A = [a_{ij}]$. As we will see, both $\det(A)$ and $\text{adj}(A)$ are defined in terms of the *cofactor matrix* of A. The ijth entry of the cofactor matrix $C(A)$ is denoted $c_{ij}(A)$ and defined to be

$$c_{ij}(A) = (-1)^{i+j} m_{ij}(A) \tag{10.25}$$

where $m_{ij}(A)$ is the ijth element of the *minor* of A, $M(A)$. Each element of the minor is obtained from the determinant of the $(m-1) \times (m-1)$ matrix formed by removing the ith row and jth column of A. Cofactors and minors are best illustrated with an example.

EXAMPLE 10.17

You are given that

$$B = \begin{bmatrix} 1 & 2 & 3 \\ 4 & 5 & 6 \\ 7 & 8 & 9 \end{bmatrix}$$

Calculate the $(1, 1)$- and $(3, 2)$-elements of the minor and cofactor matrices, $M(B)$ and $C(B)$.

Solution

We have

$$m_{11}(A) = \begin{vmatrix} 5 & 6 \\ 8 & 9 \end{vmatrix} = -3, \quad m_{32}(A) = \begin{vmatrix} 1 & 3 \\ 4 & 6 \end{vmatrix} = -6$$

and so, by Eq. (10.25),

$$c_{11}(A) = (-1)^2 m_{11}(A) = -3, \quad c_{32}(A) = (-1)^5 m_{32}(A) = 6$$

The determinant of A is calculated from its cofactor matrix $M(A)$ using a *Laplace expansion*. For example, the Laplace expansion *along the first column* of A is obtained by summing the product of each entry of the first column of A with its associated entry in the first column of the cofactor matrix $C(A)$,

$$\det(A) = \sum_{k=1}^{m} a_{k1} C_{k1} = a_{11}c_{11}(A) + a_{21}c_{21}(A) + \cdots + a_{m1}c_{m1}(A) \tag{10.26}$$

In fact, the determinant can actually be calculated by a Laplace expansion taken along *any* row or column. This result is stated without proof and the intention is that the reader concentrates on the practical use of Laplace expansions.

EXAMPLE 10.18

Express the Laplace expansion for the determinant of a general $m \times m$ matrix A along

a. the 2nd column

b. the $(m - 1)$th column

c. the 4th row

Solution

a. $\det(A) = a_{12}c_{12}(A) + a_{22}c_{22}(A) + \cdots + a_{m2}c_{m2}(A)$

b. $\det(A) = a_{1,m-1}c_{1,m-1}(A) + a_{2,m-1}c_{2,m-1}(A) + \cdots + a_{m,m-1}c_{m,m-1}(A)$

c. $\det(A) = a_{41}c_{41}(A) + a_{42}c_{42}(A) + \cdots + a_{4m}c_{4m}(A)$

Consider the matrix

$$A = \begin{bmatrix} 3 & 0 & 2 \\ 2 & 0 & -2 \\ 0 & 1 & 1 \end{bmatrix} \tag{10.27}$$

It should be clear that not all minors and cofactors are needed to compute det(A), only those along the particular row or column that the Laplace expansion is taken along. However, by way of illustration, we calculate all nine minors here. These are found to be

$$m_{11}(A) = \begin{vmatrix} 0 & -2 \\ 1 & 1 \end{vmatrix} = 2, \quad m_{12}(A) = \begin{vmatrix} 2 & -2 \\ 0 & 1 \end{vmatrix} = 2, \quad m_{13}(A) = \begin{vmatrix} 2 & 0 \\ 0 & 1 \end{vmatrix} = 2$$

$$m_{21}(A) = \begin{vmatrix} 0 & 2 \\ 1 & 1 \end{vmatrix} = -2, \quad m_{22}(A) = \begin{vmatrix} 3 & 2 \\ 0 & 1 \end{vmatrix} = 3, \quad m_{23}(A) = \begin{vmatrix} 3 & 0 \\ 0 & 1 \end{vmatrix} = 3$$

$$m_{31}(A) = \begin{vmatrix} 0 & 2 \\ 0 & -2 \end{vmatrix} = 0, \quad m_{32}(A) = \begin{vmatrix} 3 & 2 \\ 2 & -2 \end{vmatrix} = -10, \quad m_{33}(A) = \begin{vmatrix} 3 & 0 \\ 2 & 0 \end{vmatrix} = 0$$

Equation (10.25) can then be used to obtain the nine cofactors

$$c_{11}(A) = 2, \; c_{12}(A) = -2, \; c_{13}(A) = 2$$

$$c_{21}(A) = 2, \; c_{22}(A) = 3, \quad c_{23}(A) = -3$$

$$c_{31}(A) = 0, \; c_{32}(A) = 10, \; c_{33}(A) = 0$$

One could opt to take the Laplace expansion along the first column of A, as in Eq. (10.26), and this would lead to

$$\begin{aligned} \det(A) &= a_{11}c_{11}(A) + a_{21}c_{21}(A) + a_{31}c_{31}(A) \\ &= 3 \times 2 + 2 \times 2 + 0 \times 0 \\ &= 10 \end{aligned}$$

Alternatively, note that the process is greatly simplified if we were to calculate the Laplace expansion along the 2nd column. This is because there is only one non-zero entry in that column. In this case,

$$\det(A) = a_{32}c_{32}(A) = 1 \times 10 = 10$$

It should be clear that the effort required to calculate the determinant of a matrix can be reduced greatly by determining the Laplace expansion along a carefully chosen row or column.

EXAMPLE 10.19

Confirm that the following matrix is invertible.

$$B = \begin{bmatrix} 3 & 0 & 2 \\ 2 & 0 & 0 \\ 1 & 1 & 1 \end{bmatrix}$$

Solution

The matrix is invertible if $\det(B) \neq 0$. We note that the Laplace expansion along row 2 is an appropriate choice. In this case,

$$\det(B) = b_{21}c_{21}(B)$$

where

$$c_{21}(B) = (-1)^3 m_{21}(B) = -1 \times \begin{vmatrix} 0 & 2 \\ 1 & 1 \end{vmatrix} = 2$$

Therefore, $\det(B) = 2 \times 2 = 4$ and B is invertible.

Note that an expansion along the 2nd column would also be a sensible choice. In this case,

$$\det(B) = b_{32}c_{32}(B) = 1 \times (-1)^5 \times \begin{vmatrix} 3 & 2 \\ 2 & 0 \end{vmatrix} = 4$$

The reader is invited to compare these answers to that obtained along any other row or column.

We now have a method for calculating the determinant of a square matrix, from which one can determine whether the matrix is invertible. If the matrix is found to be invertible, Eq. (10.24) requires the adjoint (or adjugate) matrix to determine the inverse. In fact the adjoint of matrix A is also given in terms of the cofactors of A. In particular,

$$\mathrm{adj}(A) = C(A)^{\mathrm{T}} = \left[c_{ij}(A)\right]^{\mathrm{T}} \tag{10.28}$$

This expression is stated without justification and should be taken to give the definition of the adjoint matrix. Returning to the example given in Eq. (10.27), we can use the nine cofactors previously computed to write down that

$$\mathrm{adj}(A) = \begin{bmatrix} 2 & -2 & 2 \\ 2 & 3 & -3 \\ 0 & 10 & 0 \end{bmatrix}^{\mathrm{T}} = \begin{bmatrix} 2 & 2 & 0 \\ -2 & 3 & 10 \\ 2 & -3 & 0 \end{bmatrix}$$

The adjoint method then gives that

$$A^{-1} = \frac{1}{10}\begin{bmatrix} 2 & 2 & 0 \\ -2 & 3 & 10 \\ 2 & -3 & 0 \end{bmatrix}$$

The reader is invited to confirm that this is indeed the inverse of A by checking the two conditions in Eq. (10.18).

EXAMPLE 10.20

Compute the inverse of matrix B given in Example 10.19.

Solution

We know that $\det(B) = 4 \neq 0$ and so the inverse does exist. The adjoint form is calculated from the nine minors

$$\begin{aligned} &m_{11}(B) = 0, \quad m_{12}(B) = 2, \quad m_{13}(B) = 2 \\ &m_{21}(B) = -2, \; m_{22}(B) = 1, \quad m_{23}(B) = 3 \\ &m_{31}(B) = 0, \quad m_{32}(B) = -4, \; m_{33}(B) = 0 \end{aligned}$$

which lead to the cofactors

$$\begin{aligned} &c_{11}(B) = 0, \; c_{12}(B) = -2, \; c_{13}(B) = 2 \\ &c_{21}(B) = 2, \; c_{22}(B) = 1, \quad c_{23}(B) = -3 \\ &c_{31}(B) = 0, \; c_{32}(B) = 4, \quad c_{33}(B) = 0 \end{aligned}$$

We, therefore, see that

$$\text{adj}(B) = \begin{bmatrix} 0 & -2 & 2 \\ 2 & 1 & -3 \\ 0 & 4 & 0 \end{bmatrix}^{T} = \begin{bmatrix} 0 & 2 & 0 \\ -2 & 1 & 4 \\ 2 & -3 & 0 \end{bmatrix}$$

and the inverse is the obtained as

$$B^{-1} = \frac{1}{4}\begin{bmatrix} 0 & 2 & 0 \\ -2 & 1 & 4 \\ 2 & -3 & 0 \end{bmatrix}$$

We have seen that the cofactors of $m \times m$ matrix A are obtained from the minors, which are in turn obtained from the determinants of interim $(m-1) \times (m-1)$ matrices formed by removing particular rows and columns from A. This process is straightforward for $m = 3$ as the interim determinants are easily obtained from Eq. (10.23). However, for $m = 4$, the interim determinants are more difficult to obtain. In fact, one needs to apply the full Laplace expansion method to obtain the determinant of each interim 3×3 matrix. This reasoning can be extended to any matrix with $m > 3$ and it should be clear that more and more layers of interim matrices and determinants are needed as

m increases. The implication of this is that the computation of determinants and adjoints is extremely time consuming for large m.

We demonstrate this by calculating just one element of cofactor matrix, c_{23}, for the following 4×4 matrix

$$D = \begin{bmatrix} 3 & 0 & 2 & 1 \\ 2 & 1 & 0 & 1 \\ 1 & 1 & 1 & 1 \\ 3 & 0 & 1 & 2 \end{bmatrix}$$

We have

$$c_{23}(D) = -m_{23}(D)$$

where

$$m_{23}(D) = \begin{vmatrix} 3 & 0 & 1 \\ 1 & 1 & 1 \\ 3 & 0 & 2 \end{vmatrix}$$

In this case, the interim determinant is obtained most efficiently using a Laplace expansion along the second column. That is,

$$m_{23}(D) = 1 \times (-1)^2 \times \begin{vmatrix} 3 & 1 \\ 3 & 2 \end{vmatrix} = 1 \times (3 \times 2 - 3 \times 1) = 3$$

and so $c_{23}(D) = -3$. This process must then be repeated to a total of 16 times to obtain the full cofactor matrix of D

$$C(D) = [c_{ij}(D)] = \begin{bmatrix} 2 & 0 & 2 & -4 \\ 3 & 3 & -3 & -3 \\ -3 & 3 & 3 & 3 \\ -1 & -3 & -1 & 5 \end{bmatrix} \tag{10.29}$$

The reader is invited to spend the time confirming this cofactor matrix by hand.

EXAMPLE 10.21

Use the cofactor matrix stated in Eq. (10.29) to obtain D^{-1}.

Solution

We immediately write down

$$\text{adj}(D) = [C_{ij}(D)]^{\text{T}} = \begin{bmatrix} 2 & 3 & -3 & -1 \\ 0 & 3 & 3 & -3 \\ 2 & -3 & 3 & -1 \\ -4 & -3 & 3 & 5 \end{bmatrix}$$

The determinant det(D) is obtained by using the Laplace expansion along the third column of D as

$$\begin{aligned}\det(D) &= 2\times\begin{vmatrix}2&2&1\\1&1&1\\3&0&2\end{vmatrix}+1\times\begin{vmatrix}3&0&1\\2&1&1\\3&0&2\end{vmatrix}-1\times\begin{vmatrix}3&0&1\\2&1&1\\1&1&1\end{vmatrix}\\&=2\times 2+1\times 3-1\times 1\\&=6\end{aligned}$$

where considerable effort was required to obtained the interim 3×3 determinants. Equation (10.18) therefore leads to

$$D^{-1}=\frac{1}{6}\begin{bmatrix}2&3&-3&-1\\0&3&3&-3\\2&-3&3&-1\\-4&-3&3&5\end{bmatrix}$$

EXAMPLE 10.22

Determine $c_{23}(E)$ where E is the following 5×5 matrix:

$$E=\begin{bmatrix}3&0&2&1&1\\2&1&0&1&2\\1&1&1&1&2\\3&0&1&2&1\\1&0&1&2&1\end{bmatrix}$$

Solution

By definition,

$$c_{23}(E)=(-1)^5 m_{23}(E)=-\begin{vmatrix}3&0&1&1\\1&1&1&2\\3&0&2&1\\1&0&2&1\end{vmatrix}=-|M|$$

This interim determinant, which we denote $|M|$, is a non-trivial computation in itself. Using a Laplace expansion along the second column of M, we have

$$|M|=\begin{vmatrix}3&0&1&1\\1&1&1&2\\3&0&2&1\\1&0&2&1\end{vmatrix}=1\times c_{22}(M)$$

with

$$c_{22}(M)=(-1)^4\begin{vmatrix}3&1&1\\3&2&1\\1&2&1\end{vmatrix}=3\times(2-2)-1\times(3-1)+1\times(6-2)=2$$

Note that a further Laplace expansion along the first row of this second interim matrix has been used. We have $|M| = 2$ and so $c_{23}(E) = -2$.

This process can be repeated to a total of 25 times to obtain the full cofactor matrix. After considerable effort we would find that

$$C(E) = \begin{bmatrix} 0 & -6 & 0 & -2 & 4 \\ 0 & -8 & -2 & -2 & 6 \\ 0 & 10 & 2 & 2 & -6 \\ 1 & 11 & 1 & 3 & -8 \\ -1 & -9 & -1 & -1 & 6 \end{bmatrix}$$

EXAMPLE 10.23

Determine E^{-1} for E defined in Example 10.22.

Solution

The reader is invited to confirm that $\det(E) = 2 \neq 0$, and so E^{-1} does exist. Using the cofactor matrix stated in Example 10.22, we can immediately write down that

$$\text{adj}(E) = \begin{bmatrix} 0 & 0 & 0 & 1 & -1 \\ -6 & -8 & 10 & 11 & -9 \\ 0 & -2 & 2 & 1 & -1 \\ -2 & -2 & 2 & 3 & -1 \\ 4 & 6 & -6 & -8 & 6 \end{bmatrix}$$

and Eq. (10.18) leads to

$$E^{-1} = \frac{1}{2}\begin{bmatrix} 0 & 0 & 0 & 1 & -1 \\ -6 & -8 & 10 & 11 & -9 \\ 0 & -2 & 2 & 1 & -1 \\ -2 & -2 & 2 & 3 & -1 \\ 4 & 6 & -6 & -8 & 6 \end{bmatrix}$$

It should be clear that, while the adjoint method for inverting matrices is a simple procedure to state, it is extremely labor intensive in practice for larger matrices. This is also true of the alternative methods not discussed here. In practice one would use computational software to invert matrices for $m > 2$.

10.4 SOLVING MATRIX EQUATIONS

In this section, we return again to general (non-square) $m \times n$ matrices.

It is clear how one can solve the "scalar" equation $2 + 2x = 4$ for x. But how would our strategy change when attempting to solve the following *matrix equation* for A?

$$\begin{bmatrix} 1 & 2 & 3 \\ -1 & 2 & 4 \end{bmatrix} + 2A = \begin{bmatrix} 5 & 0 & 2 \\ 3 & -6 & 10 \end{bmatrix} \tag{10.30}$$

It should be immediately clear that A must be a 2×3 matrix. At this stage one might opt to consider the expression in terms of the unknown entries. That is, rewrite Eq. (10.30) in terms of a_{ij} as

$$\begin{bmatrix} 1 + 2a_{11} & 2 + 2a_{12} & 3 + 2a_{13} \\ -1 + 2a_{21} & 2 + 2a_{22} & 4 + 2a_{23} \end{bmatrix} = \begin{bmatrix} 5 & 0 & 2 \\ 3 & -6 & 10 \end{bmatrix}$$

This represents six independent equations, one for each a_{ij}, that are easily solved to give the solution

$$A = \begin{bmatrix} 2 & -1 & -0.5 \\ 2 & -2 & 3 \end{bmatrix}$$

Alternatively, one could work with Eq. (10.30) at the matrix level. Using the standard matrix manipulations this would proceed as

$$\begin{aligned} 2A &= \begin{bmatrix} 5 & 0 & 2 \\ 3 & -6 & 10 \end{bmatrix} - \begin{bmatrix} 1 & 2 & 3 \\ -1 & 2 & 4 \end{bmatrix} \\ \Rightarrow A &= \frac{1}{2} \begin{bmatrix} 4 & -2 & -1 \\ 4 & -4 & 6 \end{bmatrix} = \begin{bmatrix} 2 & -1 & -0.5 \\ 2 & -2 & 3 \end{bmatrix} \end{aligned}$$

It is valid to work either in the level of entries or in the matrix level, and both are, of course, equivalent. However, working at the matrix level is often more efficient unless matrix multiplication is involved. We return to equations involving multiplication later in this section.

EXAMPLE 10.24

Determine matrix C such that

$$2C - \begin{bmatrix} 1 & 2 \\ -1 & 2 \end{bmatrix} = \begin{bmatrix} 10 & 4 \\ -5 & 5 \end{bmatrix} - 2C$$

Solution

We opt to work in terms of matrix manipulations. The equation is easily manipulated to give

$$4C = \begin{bmatrix} 11 & 6 \\ -6 & 7 \end{bmatrix}$$

and so is solved by

$$C = \begin{bmatrix} \frac{11}{4} & \frac{3}{2} \\ -\frac{3}{2} & \frac{7}{4} \end{bmatrix}$$

EXAMPLE 10.25

Solve the following expression for A, where A, B, and C are matrices of equal dimension.

$$2A + B + 4C = -5A + 3B + 0.5C$$

Solution

We manipulate the expression at the matrix level as if it were a simple scalar expression. In particular,

$$\begin{aligned} 2A + 5A &= -B + 3B - 4C + 0.5C \\ \Rightarrow A &= \frac{1}{7}(2B - 3.5C) \end{aligned}$$

Matrix equations can be further complicated with the transpose operation. For example, the very simple equation

$$A^{\mathrm{T}} = B$$

where A and B are appropriately sized matrices, is solved for A as follows

$$\begin{aligned} \left(A^{\mathrm{T}}\right)^{\mathrm{T}} &= B^{\mathrm{T}} \\ \Rightarrow A &= B^{\mathrm{T}} \end{aligned}$$

The following example considers a more complicated equation.

EXAMPLE 10.26

Find the matrix A that solves the equation

$$\left(A^{\mathrm{T}} - \begin{bmatrix} 0 & 2 \\ 4 & -1 \end{bmatrix}\right)^{\mathrm{T}} = 2A - \begin{bmatrix} 1 & 0 \\ 0 & 1 \end{bmatrix}$$

Solution

In this particular case a sensible strategy would be to begin by applying the property $(C + D)^{\mathrm{T}} = C^{\mathrm{T}} + D^{\mathrm{T}}$ to the left hand side. In particular, we proceed as

$$\begin{aligned} \left(A^{\mathrm{T}} - \begin{bmatrix} 0 & 2 \\ 4 & -1 \end{bmatrix}\right)^{\mathrm{T}} &= 2A - \begin{bmatrix} 1 & 0 \\ 0 & 1 \end{bmatrix} \\ (A^{\mathrm{T}})^{\mathrm{T}} - \begin{bmatrix} 0 & 2 \\ 4 & -1 \end{bmatrix}^{\mathrm{T}} &= 2A - \begin{bmatrix} 1 & 0 \\ 0 & 1 \end{bmatrix} \\ A - \begin{bmatrix} 0 & 4 \\ 2 & -1 \end{bmatrix} &= 2A - \begin{bmatrix} 1 & 0 \\ 0 & 1 \end{bmatrix} \\ \Rightarrow A &= \begin{bmatrix} 1 & -4 \\ -2 & 2 \end{bmatrix} \end{aligned}$$

To this point we have only considered linear matrix equations, that is, equations that *do not* involve matrix products. As we know from Section 10.2, matrix multiplication is only possible for appropriately sized matrices and this necessarily limits the equations that can exist. In the particular case of square matrices one is free to manipulate complicated algebraic expressions involving the product of two or more matrices without regard to whether the products actually exist.

For example, if A, B, and C are square matrices of equal dimension, then the expression

$$A(5B - C) + (A - 3BC)C + 2CB(I_m - 3AB)$$

can be manipulated as

$$\begin{aligned} &5AB - AC + AC - 3BC^2 + 2CB - 6CBAB \\ \Rightarrow &5AB - 3BC^2 + 2CB - 6CBAB \end{aligned}$$

Note that since matrix multiplication does not commute, the order in each product must be maintained under algebraic manipulations. In particular, one must resist the urge to collect repeated terms as squares or cubes. For example, in general, the term $6CBAB \neq 6CB^2A \neq 6AB^2C$. The non-commuting property is crucial when solving such matrix expressions. This is illustrated in the following example.

EXAMPLE 10.27

Find the matrices that solve the following equations.

a. $2A + A\begin{bmatrix} 0 & 0 \\ 0 & 3 \end{bmatrix} = \begin{bmatrix} 1 & 2 \\ 1 & 1 \end{bmatrix}$

b. $2B + \begin{bmatrix} 0 & 0 \\ 0 & 3 \end{bmatrix} B = \begin{bmatrix} 1 & 2 \\ 1 & 1 \end{bmatrix}$

Solution

Now that multiplication is involved, it is necessary to work at the level of entries.

a. We expand the equation as

$$\begin{aligned} 2\begin{bmatrix} a_{11} & a_{12} \\ a_{21} & a_{22} \end{bmatrix} + \begin{bmatrix} a_{11} & a_{12} \\ a_{21} & a_{22} \end{bmatrix}\begin{bmatrix} 0 & 0 \\ 0 & 3 \end{bmatrix} &= \begin{bmatrix} 1 & 2 \\ 1 & 1 \end{bmatrix} \\ \Rightarrow \begin{bmatrix} 2a_{11} & 2a_{12} \\ 2a_{21} & 2a_{22} \end{bmatrix} + \begin{bmatrix} 0 & 3a_{12} \\ 0 & 3a_{22} \end{bmatrix} &= \begin{bmatrix} 1 & 2 \\ 1 & 1 \end{bmatrix} \\ \Rightarrow \begin{bmatrix} 2a_{11} & 5a_{12} \\ 2a_{21} & 5a_{22} \end{bmatrix} &= \begin{bmatrix} 1 & 2 \\ 1 & 1 \end{bmatrix} \\ \Rightarrow A &= \begin{bmatrix} \frac{1}{2} & \frac{2}{5} \\ \frac{1}{2} & \frac{1}{5} \end{bmatrix} \end{aligned}$$

b. In a similar way,

$$2\begin{bmatrix} b_{11} & b_{12} \\ b_{21} & b_{22} \end{bmatrix} + \begin{bmatrix} 0 & 0 \\ 0 & 3 \end{bmatrix}\begin{bmatrix} b_{11} & b_{12} \\ b_{21} & b_{22} \end{bmatrix} = \begin{bmatrix} 1 & 2 \\ 1 & 1 \end{bmatrix}$$

$$\Rightarrow \begin{bmatrix} 2b_{11} & 2b_{12} \\ 2b_{21} & 2b_{22} \end{bmatrix} + \begin{bmatrix} 0 & 0 \\ 3b_{21} & 3b_{22} \end{bmatrix} = \begin{bmatrix} 1 & 2 \\ 1 & 1 \end{bmatrix}$$

$$\Rightarrow \begin{bmatrix} 2b_{11} & 2b_{12} \\ 5b_{21} & 5b_{22} \end{bmatrix} = \begin{bmatrix} 1 & 2 \\ 1 & 1 \end{bmatrix}$$

$$\Rightarrow B = \begin{bmatrix} \frac{1}{2} & 1 \\ \frac{1}{5} & \frac{1}{5} \end{bmatrix}$$

We note that the $A \neq B$ which stems from the non-commuting property of matrices.

If we continue to be restricted to square matrices, it is possible to conceive of matrix equations involving inversions. For example,

$$A^{-1} + I_2 = \begin{bmatrix} 2 & 1 \\ 2 & -1 \end{bmatrix}$$

can be manipulated to give

$$A^{-1} = \begin{bmatrix} 2 & 1 \\ 2 & -1 \end{bmatrix} - \begin{bmatrix} 1 & 0 \\ 0 & 1 \end{bmatrix} = \begin{bmatrix} 1 & 1 \\ 2 & -2 \end{bmatrix}$$

and so, by Eq. (10.19),

$$A = \begin{bmatrix} 1 & 1 \\ 2 & -2 \end{bmatrix}^{-1} = \frac{1}{4}\begin{bmatrix} 2 & 1 \\ 2 & -1 \end{bmatrix}$$

Matrix multiplications, inversions and transpositions can be combined to give yet more complicated equations that one should be able to solve.

EXAMPLE 10.28

Solve the following matrix equations for B.

a. $(B^{-1})^{\mathrm{T}} = \begin{bmatrix} 0 & 1 \\ 2 & 3 \end{bmatrix}$

b. $\left(2B^{-1} - I_2\right)^{\mathrm{T}} = \begin{bmatrix} 1 & 0 \\ -1 & 2 \end{bmatrix}$

c. $\left(\begin{bmatrix} 1 & 0 \\ 1 & 1 \end{bmatrix} B\right)^{-1} = \begin{bmatrix} 2 & 1 \\ 1 & 2 \end{bmatrix}$

d. $\left(B\begin{bmatrix} 1 & 0 \\ 1 & 1 \end{bmatrix}\right)^{-1} = \begin{bmatrix} 2 & 1 \\ 1 & 2 \end{bmatrix}$

Solution

a. $(B^{-1})^{\mathrm{T}} = \begin{bmatrix} 0 & 1 \\ 2 & 3 \end{bmatrix} \Rightarrow B^{-1} = \begin{bmatrix} 0 & 2 \\ 1 & 3 \end{bmatrix}$. Therefore, $B = \begin{bmatrix} 0 & 2 \\ 1 & 3 \end{bmatrix}^{-1} = \frac{1}{2}\begin{bmatrix} -3 & 2 \\ 1 & 0 \end{bmatrix}$

b. $\left(2B^{-1} - I_2\right)^{\mathrm{T}} = \begin{bmatrix} 1 & 0 \\ -1 & 2 \end{bmatrix} \Rightarrow B^{-1} = \frac{1}{2}\begin{bmatrix} 2 & -1 \\ 0 & 3 \end{bmatrix}$. Therefore, $B = \frac{1}{3}\begin{bmatrix} 3 & 1 \\ 0 & 2 \end{bmatrix}$

c. We begin by manipulating the equation to give

$$\begin{bmatrix} 1 & 0 \\ 1 & 1 \end{bmatrix} B = \begin{bmatrix} 2 & 1 \\ 1 & 2 \end{bmatrix}^{-1}$$

$$\Rightarrow \begin{bmatrix} 1 & 0 \\ 1 & 1 \end{bmatrix} B = \frac{1}{3}\begin{bmatrix} 2 & -1 \\ -1 & 2 \end{bmatrix}$$

At this stage it should be clear that we need to pre-multiply by sides of this expression by $\begin{bmatrix} 1 & 0 \\ 1 & 1 \end{bmatrix}^{-1}$ to give

$$\begin{bmatrix} 1 & 0 \\ 1 & 1 \end{bmatrix}^{-1} \begin{bmatrix} 1 & 0 \\ 1 & 1 \end{bmatrix} B = \frac{1}{3}\begin{bmatrix} 1 & 0 \\ 1 & 1 \end{bmatrix}^{-1} \begin{bmatrix} 2 & -1 \\ -1 & 2 \end{bmatrix}$$

$$\Rightarrow B = \frac{1}{3}\begin{bmatrix} 1 & 0 \\ -1 & 1 \end{bmatrix}\begin{bmatrix} 2 & -1 \\ -1 & 2 \end{bmatrix}$$

$$\Rightarrow B = \frac{1}{3}\begin{bmatrix} 2 & -1 \\ -3 & 3 \end{bmatrix}$$

d. We begin as in part c. and obtain

$$B\begin{bmatrix} 1 & 0 \\ 1 & 1 \end{bmatrix} = \frac{1}{3}\begin{bmatrix} 2 & -1 \\ -1 & 2 \end{bmatrix}$$

Now it is necessary to post-multiply both sides of this expression by $\begin{bmatrix} 1 & 0 \\ 1 & 1 \end{bmatrix}^{-1}$, which yields

$$B\begin{bmatrix} 1 & 0 \\ 1 & 1 \end{bmatrix}\begin{bmatrix} 1 & 0 \\ 1 & 1 \end{bmatrix}^{-1} = \frac{1}{3}\begin{bmatrix} 2 & -1 \\ -1 & 2 \end{bmatrix}\begin{bmatrix} 1 & 0 \\ 1 & 1 \end{bmatrix}^{-1}$$

$$\Rightarrow B = \frac{1}{3}\begin{bmatrix} 2 & -1 \\ -1 & 2 \end{bmatrix}\begin{bmatrix} 1 & 0 \\ -1 & 1 \end{bmatrix}$$

$$\Rightarrow B = \frac{1}{3}\begin{bmatrix} 3 & -1 \\ -3 & 2 \end{bmatrix}$$

10.5 SOLVING SYSTEMS OF LINEAR SIMULTANEOUS EQUATIONS

10.5.1 Matrix inversion method

You may already be familiar with linear simultaneous equations. That is, systems of interlinked linear equations that must to be solved simultaneously to yield a solution. For example, the system

$$\left.\begin{array}{l} 2x + y = 7 \\ 3x - y = 8 \end{array}\right\}$$

represents a system of two linear simultaneous equations for two unknowns, x and y. This system is easily solved by, for example, adding the equations and eliminating the ys to obtain

$$5x = 15 \Rightarrow x = 3$$

This value can then be used in the first equation, for example, to show that $y = 7 - 2 \times 3 = 1$. This system, therefore, solved by $x = 3$ and $y = 1$.

Simultaneous equations can exist for any number of unknowns and, as long as the number of unknowns is equal to the number of *independent* equations, a unique collection of values for the independent variables can be found that solve the system. The method of eliminating variables between equations can become very cumbersome as the number of variables and equations increases, and a more efficient method is required. We end this chapter with a description of such a method that is based on matrix algebra.

Consider the following system of three simultaneous equations for three unknowns:

$$\left.\begin{array}{l} 2x + y - z = 1 \\ -x + 2y + z = 6 \\ 3x + y + z = 8 \end{array}\right\}$$

It should be clear that the equations are independent of each other and so the system is expected to have a unique set of values for x, y, and z that solve the system. The reader is invited to begin by using an elimination method to confirm that the solution is $x = 1$, $y = 2$, and $z = 3$.

As an alternative, we begin by using our knowledge of matrix multiplication to rewrite the system as

$$\begin{bmatrix} 2 & 1 & -1 \\ -1 & 2 & 1 \\ 3 & 1 & 1 \end{bmatrix} \begin{bmatrix} x \\ y \\ z \end{bmatrix} = \begin{bmatrix} 1 \\ 6 \\ 8 \end{bmatrix}$$

This system can then be considered a matrix problem of the form

$$AX = B$$

Here A is the square *coefficient matrix* for the system of equations, X is the column matrix of unknown variables, and B is a column matrix of constants. The system of simultaneous equations is then solved by determining X. This is done by pre-multiplying both sides by A^{-1} to obtain

$$A^{-1}AX = A^{-1}B$$

$$\Rightarrow X = A^{-1}B \tag{10.31}$$

Our example system is then solved as

$$\begin{bmatrix} x \\ y \\ z \end{bmatrix} = \begin{bmatrix} 2 & 1 & -1 \\ -1 & 2 & 1 \\ 3 & 1 & 1 \end{bmatrix}^{-1} \begin{bmatrix} 1 \\ 6 \\ 8 \end{bmatrix}$$

and the problem is reduced to one of obtaining the inverse of the coefficient matrix. Using the adjoint method, we see that

$$\begin{bmatrix} 2 & 1 & -1 \\ -1 & 2 & 1 \\ 3 & 1 & 1 \end{bmatrix}^{-1} = \frac{1}{13}\begin{bmatrix} 1 & -2 & 3 \\ 4 & 5 & -1 \\ -7 & 1 & 5 \end{bmatrix}$$

and so

$$\begin{bmatrix} x \\ y \\ z \end{bmatrix} = \frac{1}{13}\begin{bmatrix} 1 & -2 & 3 \\ 4 & 5 & -1 \\ -7 & 1 & 5 \end{bmatrix}\begin{bmatrix} 1 \\ 6 \\ 8 \end{bmatrix} = \begin{bmatrix} 1 \\ 2 \\ 3 \end{bmatrix}$$

This confirms our previous solution.

Consider now the system

$$\begin{cases} 2x + y = 7 \\ 4x + 2y = 14 \end{cases}$$

It should be immediately clear that the two equations are not *independent* as the second is twice the first. This system is, therefore, solved by *any* pair of values such that $y = 7 - 2x$, that is, it does not have a unique solution. But how does this affect the use of our matrix inversion method?

The system is rewritten as

$$\begin{bmatrix} 2 & 1 \\ 4 & 2 \end{bmatrix}\begin{bmatrix} x \\ y \end{bmatrix} = \begin{bmatrix} 7 \\ 14 \end{bmatrix}$$

and so

$$\begin{bmatrix} x \\ y \end{bmatrix} = \begin{bmatrix} 2 & 1 \\ 4 & 2 \end{bmatrix}^{-1}\begin{bmatrix} 7 \\ 14 \end{bmatrix}$$

and we are required to invert the coefficient matrix. However, we immediately note that the determinant of this is 0 and so the matrix cannot be inverted. The matrix inversion method cannot therefore be used here which reflects that no unique solution exists.

EXAMPLE 10.29

Solve the following systems of simultaneous equations:

a.

$$\left.\begin{array}{l} 3a+2c+d=10 \\ 2a+b+d=1 \\ a+b+c+d=12 \\ 3a+c+2s=-1 \end{array}\right\}$$

b.

$$\left.\begin{array}{l} 3a+2c+2+e=2 \\ 2a+b+d+2e=4 \\ a+b+c+d+2e=10 \\ 3a+c+2d+e=14 \\ a+c+2d+e=4 \end{array}\right\}$$

Solution

a. The system can be written as

$$\begin{bmatrix} 3 & 0 & 2 & 1 \\ 2 & 1 & 0 & 1 \\ 1 & 1 & 1 & 1 \\ 3 & 0 & 1 & 2 \end{bmatrix}\begin{bmatrix} a \\ b \\ c \\ d \end{bmatrix}=\begin{bmatrix} 10 \\ 1 \\ 12 \\ -1 \end{bmatrix}$$

Note that the coefficient matrix was actually inverted in Example 10.21. The system is then rewritten as

$$\begin{bmatrix} a \\ b \\ c \\ d \end{bmatrix}=\begin{bmatrix} 3 & 0 & 2 & 1 \\ 2 & 1 & 0 & 1 \\ 1 & 1 & 1 & 1 \\ 3 & 0 & 1 & 2 \end{bmatrix}^{-1}\begin{bmatrix} 10 \\ 1 \\ 12 \\ -1 \end{bmatrix}=\frac{1}{6}\begin{bmatrix} 2 & 3 & -3 & -1 \\ 0 & 3 & 3 & -3 \\ 2 & -3 & 3 & -1 \\ -4 & -3 & 3 & 5 \end{bmatrix}\begin{bmatrix} 10 \\ 1 \\ 12 \\ -1 \end{bmatrix}$$

and some further effort leads to

$$\begin{bmatrix} a \\ b \\ c \\ d \end{bmatrix}=\begin{bmatrix} -2 \\ 7 \\ 9 \\ -2 \end{bmatrix}$$

This is the unique solution of the system.

b. The system can be written as

$$\begin{bmatrix} 3 & 0 & 2 & 1 & 1 \\ 2 & 1 & 0 & 1 & 2 \\ 1 & 1 & 1 & 1 & 2 \\ 3 & 0 & 1 & 2 & 1 \\ 1 & 0 & 1 & 2 & 1 \end{bmatrix} \begin{bmatrix} a \\ b \\ c \\ d \\ e \end{bmatrix} = \begin{bmatrix} 2 \\ 4 \\ 10 \\ 14 \\ 4 \end{bmatrix}$$

Note that the coefficient matrix was inverted in Example 10.22. The system is then rewritten as

$$\begin{bmatrix} a \\ b \\ c \\ d \\ e \end{bmatrix} = \begin{bmatrix} 3 & 0 & 2 & 1 & 1 \\ 2 & 1 & 0 & 1 & 2 \\ 1 & 1 & 1 & 1 & 2 \\ 3 & 0 & 1 & 2 & 1 \\ 1 & 0 & 1 & 2 & 1 \end{bmatrix}^{-1} \begin{bmatrix} 2 \\ 4 \\ 10 \\ 14 \\ 4 \end{bmatrix} = \frac{1}{2} \begin{bmatrix} 0 & 0 & 0 & 1 & -1 \\ -6 & -8 & 10 & 11 & -9 \\ 0 & -2 & 2 & 1 & -1 \\ -2 & -2 & 2 & 3 & -1 \\ 4 & 6 & -6 & -8 & 6 \end{bmatrix} \begin{bmatrix} 2 \\ 4 \\ 10 \\ 14 \\ 4 \end{bmatrix}$$

and some further effort leads to

$$\begin{bmatrix} a \\ b \\ c \\ d \\ e \end{bmatrix} = \begin{bmatrix} 5 \\ 87 \\ 11 \\ 23 \\ -58 \end{bmatrix}$$

This is the unique solution of the system.

10.5.2 Cramer's rule

We have seen that it is possible to use matrix inversion techniques to solve systems of linear simultaneous equations. The technique is necessarily computationally intensive and generates the entire set of unknown variables. We now discuss *Cramer's rule* which enables one to determine any particular unknown variable at a time and, perhaps more importantly, does not require one to invert the coefficient matrix.

We begin by rewriting Eq. (10.31) at the level of matrix entries. In particular, using Eqs. (10.24) and (10.28), we have

$$\begin{bmatrix} x_1 \\ x_2 \\ \vdots \\ x_m \end{bmatrix} = \frac{1}{\det(A)} \begin{bmatrix} c_{11}(A) & c_{12}(A) & \dots & c_{1m}(A) \\ c_{21}(A) & c_{22}(A) & \dots & c_{2m}(A) \\ \vdots & \vdots & \vdots & \vdots \\ c_{m1}(A) & c_{m2}(A) & \dots & c_{mm}(A) \end{bmatrix}^{\mathrm{T}} \begin{bmatrix} b_1 \\ b_2 \\ \vdots \\ b_m \end{bmatrix}$$

$$= \frac{1}{\det(A)} \begin{bmatrix} c_{11}(A) & c_{21}(A) & \dots & c_{m1}(A) \\ c_{12}(A) & c_{22}(A) & \dots & c_{m2}(A) \\ \vdots & \vdots & \vdots & \vdots \\ c_{1m}(A) & c_{2m}(A) & \dots & c_{mm}(A) \end{bmatrix} \begin{bmatrix} b_1 \\ b_2 \\ \vdots \\ b_m \end{bmatrix}$$

The variable x_1, for example, is therefore obtained from

$$x_1 = \frac{1}{\det(A)} (b_1 c_{11}(A) + b_2 c_{21}(A) + \dots + b_m c_{m1}(A))$$

Comparing this expression to Eq. (10.26), it should be clear that $b_1 c_{11}(A) + b_2 c_{21}(A) + \dots + b_m c_{m1}(A)$ can be identified with the determinant of a new matrix A_1, obtained by replacing the first column of the coefficient matrix A with the column matrix B. That is

$$b_1 c_{11}(A) + b_2 c_{21}(A) + \dots + b_m c_{m1}(A) = \begin{vmatrix} b_1 & a_{12} & \dots & a_{1m} \\ b_2 & a_{22} & \dots & a_{2m} \\ \vdots & \vdots & \vdots & \vdots \\ b_m & a_{m2} & \dots & a_{mm} \end{vmatrix}$$

The unknown variable x_1 is, therefore, seen to be determined by

$$x_1 = \frac{\det(A_1)}{\det(A)}$$

This result is easily generalized to give Cramer's rule

$$x_i = \frac{\det(A_i)}{\det(A)} \tag{10.32}$$

where x_i is the ith unknown variable and A_i is the matrix resulting from replacing the ith column with the column matrix B.

EXAMPLE 10.30

Use Cramer's rule to determine the values of b and c that solve the system in Example 10.29(a).

Solution

Using Eq. (10.32), we have

$$b = \frac{\det(D_2)}{\det(D)}, \quad c = \frac{\det(D_3)}{\det(D)}$$

where

$$D = \begin{bmatrix} 3 & 0 & 2 & 1 \\ 2 & 1 & 0 & 1 \\ 1 & 1 & 1 & 1 \\ 3 & 0 & 1 & 2 \end{bmatrix} \Rightarrow \det(D) = 6$$

$$D_2 = \begin{bmatrix} 3 & 10 & 2 & 1 \\ 2 & 1 & 0 & 1 \\ 1 & 12 & 1 & 1 \\ 3 & -1 & 1 & 2 \end{bmatrix} \Rightarrow \det(D_2) = 42$$

$$D_3 = \begin{bmatrix} 3 & 0 & 10 & 1 \\ 2 & 1 & 1 & 1 \\ 1 & 1 & 12 & 1 \\ 3 & 0 & -1 & 2 \end{bmatrix} \Rightarrow \det(D_3) = 54$$

We, therefore, see that $b = 42/6 = 7$ and $c = 54/6 = 9$, consistent with the results of Example 10.29(a). The reader is invited to confirmed the determinants using appropriate Laplace expansions.

10.6 MATRIX ALGEBRA ON YOUR COMPUTER

Wolfram Alpha can be used to perform the matrix operations discussed in this chapter. We will work with the following matrices throughout our discussion:

$$A = \begin{bmatrix} 2 & 2 & -1 \\ 2 & 1 & 1 \\ 1 & 1 & 3 \end{bmatrix} \text{ and } B = \begin{bmatrix} 0 & 4 & 3 \\ 1 & 2 & -2 \\ 1 & 2 & 1 \end{bmatrix}$$

Of course the first hurdle to using Wolfram Alpha for matrix algebra is how to input a matrix, but this is easily done. For example, the command for defining matrix A is

```
{{2,2,-1},{2,1,1},{1,1,3}}
```

That is, we define the matrix in terms of its rows. Similarly, matrix B is input as

```
{{0,4,3},{1,2,-2},{1,2,1}}
```

Now that we know how to input a matrix, simple calculations follow in an intuitive way. For example, the sum $A + B$ is computed from the command

```
{{2,2,-1},{2,1,1},{1,1,3}}+{{0,4,3},{1,2,-2},{1,2,1}}
```

The sum $10A - 3B$ is computed from

```
10{{2,2,-1},{2,1,1},{1,1,3}}-3{{0,4,3},{1,2,-2},{1,2,1}}
```

and the product AB is computed from

```
{{2,2,-1},{2,1,1},{1,1,3}}*{{0,4,3},{1,2,-2},{1,2,1}}
```

Note that in each case Wolfram Alpha gives its interpretation of the input so that we can confirm that we have correctly communicated our request before relying on the answer.

Wolfram Alpha can also be used to determine the transpose, determinant and inverse of any given matrix. The commands are again very intuitive. For example, A^{T} is obtained from the command

```
{{2,2,-1},{2,1,1},{1,1,3}}^T
```

det(A) is obtained from

```
det {{2,2,-1},{2,1,1},{1,1,3}}
```

and A^{-1} is obtained from

```
inv {{2,2,-1},{2,1,1},{1,1,3}}
```

For the reasons made clear in this chapter, the `inv` command is a particularly important use of the engine. However, aside from our need to determine inverse matrices in an efficient way, our main use of Wolfram Alpha throughout this book has always been as an educational tool. It is therefore useful to be aware of the commands `minor` and `cofactor`. While these are clearly redundant if we are intending to compute directly the inverses, they do provide a means of checking the necessary stages in the pen-and-paper methods discussed in this chapter. For example, the minor $M(A)$ is obtained from

```
minor {{2,2,-1},{2,1,1},{1,1,3}}
```

and the cofactor $C(A)$ is obtained from

```
cofactor {{2,2,-1},{2,1,1},{1,1,3}}
```

Furthermore, the adjoint (a.k.a. adjugate) matrix adj(A) is obtained with the command `adjugate`. In particular

```
adjugate {{2,2,-1},{2,1,1},{1,1,3}}
```

Some of our interest in matrix algebra was motivated by the solution of systems of linear simultaneous equations. It is probably of no surprise that Wolfram Alpha can be used to solve such systems directly. Again, the ability to check your own pen-and-paper solutions is a very useful educational tools. For example, the solution to the system

$$\left.\begin{aligned} x + 2y &= 0 \\ 4x - 2y &= 3 \end{aligned}\right\}$$

can be obtained from the command

```
solve x+2y=0,4x-2y=3
```

Of course there a many other commands that Wolfram Alpha understands and the reader is invited to explore how the engine can be used for matrix algebra.

EXAMPLE 10.31

You are given that

$$A = \begin{bmatrix} 1 & 2 & 1 \\ 1 & 3 & 2 \\ 2 & 2 & 3 \end{bmatrix}, \quad B = \begin{bmatrix} 1 & 0 & 1 \\ 4 & 3 & 2 \\ 0 & 1 & 2 \end{bmatrix}$$

Use Wolfram Alpha to perform the following.

a. Determine AB and BA.

b. Compute $\det(A)$ and $\det(B)$.

c. Compute A^{-1} directly.

d. Compute A^{-1} from the cofactor of A.

Solution

a. We use the command

```
{{1,2,1},{1,3,2},{2,2,3}}*{{1,0,1},{4,3,2},{0,1,2}}
```

and obtain that

$$AB = \begin{bmatrix} 9 & 7 & 7 \\ 13 & 11 & 11 \\ 10 & 9 & 12 \end{bmatrix}$$

Similarly, we use the command

```
{{1,0,1},{4,3,2},{0,1,2}}*{{1,2,1},{1,3,2},{2,2,3}}
```

and obtain that

$$BA = \begin{bmatrix} 3 & 4 & 4 \\ 11 & 21 & 16 \\ 5 & 7 & 8 \end{bmatrix}$$

b. We use the command

```
det {{1,2,1},{1,3,2},{2,2,3}}
```

and determine that $\det(A) = 3$. Similarly, the command

```
det {{1,0,1},{4,3,2},{0,1,2}}
```

leads to $\det(B) = 8$.

c. We use the command

```
inv {{1,2,1},{1,3,2},{2,2,3}}
```

and determine that

$$A^{-1} = \frac{1}{3}\begin{bmatrix} 5 & -4 & 1 \\ 1 & 1 & -1 \\ -4 & 2 & 1 \end{bmatrix}$$

d. A^{-1} can be obtained using the adjoint method as

$$A^{-1} = \frac{1}{\det(A)}\text{adj(A)}$$

Where $\text{adj}(A) = C(A)^{\text{T}}$ with

$$C(A) = \begin{bmatrix} 5 & 1 & -4 \\ -4 & 1 & 2 \\ 1 & -1 & 1 \end{bmatrix}$$

as obtained from the command

```
cofactors {{1,2,1},{1,3,2},{2,2,3}}
```

We therefore find that

$$A^{-1} = \frac{1}{3}\begin{bmatrix} 5 & -4 & 1 \\ 1 & 1 & -1 \\ -4 & 2 & 1 \end{bmatrix}$$

Note that the adjoint/adjugate matrix, $\text{adj}(A)$, can be confirmed with the command

```
adjugate {{1,2,1},{1,3,2},{2,2,3}}
```

10.7 ACTUARIAL APPLICATION: MARKOV CHAINS

We end this chapter with a brief discussion of an application of matrices to the insurance industry. The following discussion assumes some familiarity with basic probability theory and the reader is invited to review Chapter 9 if the terminology proves difficult.

Consider a random process that returns one of a finite number of possibilities at each discrete time step $t = 0, 1, 2, \ldots$. That is, a process that returns some realization, X_t, of a random variable at each t. While this wording may appear unfamiliar, it is sufficient to understand that we are simply defining a process that, as time moves on in steps, returns one of a finite number of possible results, each with a particular probability. For example, we might be repeatedly rolling a dice. In this case, each discrete time step t is considered as a new roll and X_t is the result of that roll. X_t therefore returns one outcome from the discrete state space {⚀, ⚁, ⚂, ⚃, ⚄, ⚅}. One particular run of the process for 9 time steps might return, for example,

⚄, ⚁, ⚀, ⚂, ⚃, ⚃, ⚅, ⚃, ⚀

That is, $X_1 =$ ⚄, $X_2 =$ ⚁, $\ldots, X_9 =$ ⚀.

A general random process is said to possess the *Markov property* if the future development of the process depends only on the *current* state. We can express this property mathematically in terms of conditional probabilities,

$$P(X_t = x | X_0 = x_0, X_1 = x_1, \ldots, X_{t-1} = x_{t-1}) = P(X_t = x | X_{t-1} = x_{t-1}) \quad (10.33)$$

This expression is interpreted as stating that, given the history of the random process from time 0 to time $t-1$, the probability that the event at time t returns some particular value depends only on the most recent result at time $t-1$. We might summarize the

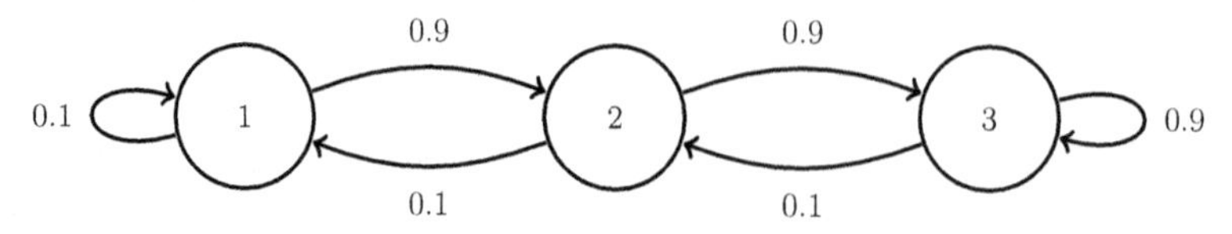

Figure 10.1 Illustration of a no-claims discount policy with three states of discount.

Markov property with the simple phrase "history does not matter." A process that has the Markov property is often called a *Markov process* or a *Markov chain*.

Returning to the dice-rolling process, the probability that the 10th roll is, say, ⚅ is not affected by the previous rolls and so it is a Markov process. This of course follows from the independence of each roll and is such that Eq. (10.33) can be simplified yet further,

$$
\begin{aligned}
&P\left(X_{10} = \text{⚅} | X_1 = \text{⚄}, X_1 = \text{⚁}, \ldots, X_9 = \text{⚀}\right) \\
&= P\left(X_{10} = \text{⚅} | X_9 = \text{⚀}\right) = P\left(X_{10} = \text{⚅}\right) = \frac{1}{6}
\end{aligned}
$$

You may be familiar with *no-claims discount policies* for motor insurance. Under these policies the insurance premium for a policyholder is reduced if no claims have been made in the previous year. A simple example of such a policy has three "states" of discount,

- State 1: 0% discount
- State 2: 10% discount
- State 3: 50% discount

A policyholder moves up a state (or remains in State 3) as a result of making no claims in the previous year, and moves down a state (or remains in State 1) as a result of making one or more claims in the previous year. Making a claim is a random event and the state that a policyholder finds himself in at time t is $X_t \in \{1, 2, 3\}$. We therefore see that the states form a discrete state space for a discrete-time random process. If the probability of making no claims in any particular year is 0.9, the probability of making at least one claim is 0.1. That is, the probability of moving up a state (or remaining in State 3) is 0.9, and the probability of moving down a state (or remaining in State 1) is 0.1. This process is illustrated in Figure 10.1 and we can confirm that it is a Markov process.

EXAMPLE 10.32

You are given that a particular policyholder's 14-year discount history is

$$1, 2, 1, 2, 3, 2, 3, 3, 2, 1, 1, 1, 2, 3$$

where the number refers to a particular discount state, as defined above. Determine the probability that the policyholder will be in the following states at $t = 15$.

a. State 1

b. State 2

c. State 3

Is the process Markov?

Solution

The policyholder is able to move down only one state from his current state, $X_{14} = 3$, or remain in that state. The probabilities therefore follow.

a. $P(X_{15} = 1|X_1 = 1, X_2 = 2, \ldots, X_{14} = 3) = P(X_{15} = 1|X_{14} = 3) = 0$
b. $P(X_{15} = 2|X_1 = 1, X_2 = 2, \ldots, X_{14} = 3) = P(X_{15} = 2|X_{14} = 3) = 0.1$
c. $P(X_{15} = 3|X_1 = 1, X_2 = 2, \ldots, X_{14} = 3) = P(X_{15} = 3|X_{14} = 3) = 0.9$

We see that only the current state, X_{14}, matters and the so process is Markov.

Since this no-claims discount process is a Markov process, it is possible to define a 1-step *transition matrix*, $P = [p_{ij}]$. Each element p_{ij} is the probability that the policyholder will move to State j at time $t+1$ given that he is in State i at time t. In our motivating example, this matrix is given by

$$P = \begin{bmatrix} 0.1 & 0.9 & 0 \\ 0.1 & 0 & 0.9 \\ 0 & 0.1 & 0.9 \end{bmatrix} \tag{10.34}$$

Note that the 3rd row is formed from the probabilities calculated in Example 10.32; the other rows can be calculated in a similar way. The matrix P can be thought of as containing all the information about a single step of the random process: it indicates which transitions are impossible ($p_{ij} = 0$) and gives the probability of those that are possible ($p_{ij} \neq 0$)

One-step transition matrices are very useful for determining the distribution of possible events after *multiple* time steps. For example, the transition matrix that applies after two time steps, that is from time t to time $t+2$, is given by

$$P^{(2)} = P \times P = \begin{bmatrix} 0.1 & 0.9 & 0 \\ 0.1 & 0 & 0.9 \\ 0 & 0.1 & 0.9 \end{bmatrix} \begin{bmatrix} 0.1 & 0.9 & 0 \\ 0.1 & 0 & 0.9 \\ 0 & 0.1 & 0.9 \end{bmatrix} = \begin{bmatrix} 0.1 & 0.09 & 0.81 \\ 0.01 & 0.18 & 0.81 \\ 0.01 & 0.09 & 0.9 \end{bmatrix}$$

We see that the probability that a policyholder starts in State 1 and ends in State 1 after two time steps is given by element $p_{11}^{(2)} = 0.1$. Furthermore, the probability that a policyholder starts in State 2 and ends in State 3 after two time steps is $p_{23}^{(2)} = 0.81$.

EXAMPLE 10.33

Confirm that $p_{11}^{(2)} = 0.1$ by considering all possible transitions over two time steps.

Solution

Consider a policyholder initially in State 1. After one time step, the policyholder will be in State 1 with probability 0.1 or in State 2 with probability 0.9. For the second time step, the policyholder can end in State 1 in two possible ways:

1. start in State 1 and remain in State 1 with probability 0.1,

2. start in State 2 move to State 1 with probability 0.1.
The probability that the chain State 1 → State 1 → State 1 occurs is $0.1 \times 0.1 = 0.01$.
The probability that the chain State 1 → State 2 → State 1 occurs is $0.9 \times 0.1 = 0.09$.
These two events are independent and so the probability that either happens is given by $0.01 + 0.09 = 0.1 = p_{11}^{(2)}$, as required. Similar reasoning can be used to confirm each element of $P^{(2)}$.

The two-stage transition matrix $P^{(2)}$ for a general Markov process is given by

$$P^{(2)} = P \times P = \begin{bmatrix} p_{11} & p_{12} & p_{13} \\ p_{21} & p_{22} & p_{23} \\ p_{31} & p_{32} & p_{33} \end{bmatrix} \begin{bmatrix} p_{11} & p_{12} & p_{13} \\ p_{21} & p_{22} & p_{23} \\ p_{31} & p_{32} & p_{33} \end{bmatrix}$$

and the particular element $p_{11}^{(2)}$ is then

$$p_{11}^{(2)} = p_{11}p_{11} + p_{12}p_{21} + p_{13}p_{31}$$

Given that each p_{ij} represents the probability of an independent transition, this expression can be interpreted as stating that the probability of starting and ending in State 1 is given by

- the probability of the chain State 1 → State 1 → State 1, plus
- the probability of the chain State 1 → State 2 → State 1, plus
- the probability of the chain State 1 → State 3 → State 1.

A little work should convince you that it is possible to generalize the above expressions to give that the n-step transition matrix

$$P^{(n)} \equiv P^n$$

This relationship stems directly from the properties of matrix multiplication, as demonstrated above for $n = 2$.

EXAMPLE 10.34

Obtain an expression for $p_{32}^{(2)}$ in terms of elements of the 1-step transition matrix P for a general Markov process. Interpret your expression.

Solution

The element $p_{32}^{(2)}$ in the matrix $P^{(2)} = P^2$ arises from the vector product of row 3 of P with column 2 of P. That is,

$$p_{32}^{(2)} = p_{31}p_{12} + p_{32}p_{22} + p_{33}p_{32}$$

This states that the probability of starting in State 2 and ending in State 3 after two time steps is given by

- the probability of the chain State 3 → State 1 → State 2, plus
- the probability of the chain State 3 → State 2 → State 2, plus
- the probability of the chain State 3 → State 3 → State 2.

These are all possible routes between State 3 and State 2 over two time steps.

EXAMPLE 10.35

Obtain an expression for $p_{11}^{(3)}$ in terms of elements of the 1-step transition matrix P for a general Markov process. Interpret this expression.

Solution

We are required to consider the $p_{11}^{(3)}$ element of $P^{(3)} = P^3$. Note that $P^3 = P^2 \times P$ with

$$P^2 = \begin{bmatrix} p_{11}p_{11}+p_{12}p_{21}+p_{13}p_{31} & p_{11}p_{12}+p_{12}p_{22}+p_{13}p_{32} & p_{11}p_{13}+p_{12}p_{23}+p_{13}p_{33} \\ p_{21}p_{11}+p_{22}p_{21}+p_{23}p_{31} & p_{21}p_{12}+p_{22}p_{22}+p_{23}p_{32} & p_{21}p_{13}+p_{22}p_{23}+p_{23}p_{33} \\ p_{31}p_{11}+p_{32}p_{21}+p_{33}p_{31} & p_{31}p_{12}+p_{32}p_{22}+p_{33}p_{32} & p_{31}p_{13}+p_{32}p_{23}+p_{33}p_{33} \end{bmatrix}$$

Therefore, the element $p_{11}^{(3)}$ is given by

$$\begin{aligned} p_{11}^{(3)} &= (p_{11}p_{11}+p_{12}p_{21}+p_{13}p_{31})p_{11} \\ &\quad +(p_{11}p_{12}+p_{12}p_{22}+p_{13}p_{32})p_{21}+(p_{11}p_{13}+p_{12}p_{23}+p_{13}p_{33})p_{31} \end{aligned}$$

that is

$$\begin{aligned} p_{11}^{(3)} &= p_{11}p_{11}p_{11}+p_{12}p_{21}p_{11}+p_{13}p_{31}p_{11}+p_{11}p_{12}p_{21} \\ &\quad +p_{12}p_{22}p_{21}+p_{13}p_{32}p_{21}+p_{11}p_{13}p_{31}+p_{12}p_{23}p_{31}+p_{13}p_{33}p_{31} \end{aligned}$$

Each term gives the probability of moving along a particular chain that begins in State 1 and ends in State 1. The expression is therefore summing over all possible routes from State 1 to State 1 in 3 time steps, as we would expect.

EXAMPLE 10.36

Use the transition matrix in Eq. (10.34) to determine the probabilities of all possible transitions within the no-claims discount system over 4 years.

Solution

It is clear that we require $P^{(4)} = P^4$. After some work, we determine that

$$P^4 = \begin{bmatrix} 0.1 & 0.9 & 0 \\ 0.1 & 0 & 0.9 \\ 0 & 0.1 & 0.9 \end{bmatrix}^4 = \begin{bmatrix} 0.019 & 0.0981 & 0.8829 \\ 0.0109 & 0.1062 & 0.8829 \\ 0.0109 & 0.0981 & 0.891 \end{bmatrix}$$

We see that it is possible to move between all states and, for example, the probability of starting in State 1 and ending in State 3 after 4 time steps is $p_{13}^{(4)} = 0.8829$.

EXAMPLE 10.37

An insurance company has 1000 policyholders distributed over the three states of the no-claims discount process with transition matrix given by Eq. (10.34). The initial distribution is,

- 100 in State 1
- 600 in State 2
- 300 in State 3

Determine the distribution of policyholders 4 years later.

Solution

The relevant transition matrix is that obtained in Example 10.36,

$$P^{(4)} = P^4 = \begin{bmatrix} 0.019 & 0.0981 & 0.8829 \\ 0.0109 & 0.1062 & 0.8829 \\ 0.0109 & 0.0981 & 0.891 \end{bmatrix}$$

and we are interested in the action of this matrix on the initial distribution. It should be clear that the total number of policyholders that end in State 1 is given by

$$100p_{11}^{(4)} + 600p_{21}^{(4)} + 300p_{31}^{(4)}$$

Similarly, the total number that end in States 2 and 3 are, respectively, given by

$$100p_{12}^{(4)} + 600p_{22}^{(4)} + 300p_{32}^{(4)} \text{ and } 100p_{13}^{(4)} + 600p_{23}^{(4)} + 300p_{33}^{(4)}$$

In fact, these quantities arise from the matrix multiplication

$$\begin{bmatrix} 100 & 600 & 300 \end{bmatrix} \begin{bmatrix} 0.019 & 0.0981 & 0.8829 \\ 0.0109 & 0.1062 & 0.8829 \\ 0.0109 & 0.0981 & 0.891 \end{bmatrix} = \begin{bmatrix} 11.71 & 102.96 & 885.33 \end{bmatrix}$$

That is, after 4 years, we would expect 12 policyholders in State 1, 103 in State 2 and 885 in State 3.

EXAMPLE 10.38

Determine whether there is a long-term, stable distribution of the 1000 policyholders in Example 10.37.

Solution

We are required to determine a particular distribution of policyholders between the three states that does *not* change over time. We use n_i to denote the number of policyholders in State i, such that $n_1 + n_2 + n_3 = 1000$, and note that these must be such that

$$\begin{bmatrix} n_1 & n_2 & n_3 \end{bmatrix} \begin{bmatrix} 0.1 & 0.9 & 0 \\ 0.1 & 0 & 0.9 \\ 0 & 0.1 & 0.9 \end{bmatrix} = \begin{bmatrix} n_1 & n_2 & n_3 \end{bmatrix}$$

That is, the distribution does not change any over one time step. This leads to four simultaneous equations

$$\begin{cases} 0.1n_1 + 0.1n_2 = n_1 \\ 0.9n_1 + 0.1n_3 = n_2 \\ 0.9n_2 + 0.9n_3 = n_3 \\ n_1 + n_2 + n_3 = 1000 \end{cases}$$

that can be solved to give

$$n_1 = \frac{1000}{91}, \quad n_2 = \frac{9000}{91}, \quad n_3 = \frac{81,000}{91}$$

This is interpreted as giving 11 policyholders in State 1, 99 in State 2 and 890 in State 3. Note that it is common practice to refer to this long-term, stable distribution as the *stationary* distribution. This reflects that, once the distribution is reached, it does not change with time. It is possible to show that every Markov chain (with a finite state space) has at least one stationary distribution.

10.8 QUESTIONS

The following questions are intended to test your knowledge of the concepts discussed in this chapter. Full solutions are available in Chapter 10 Solutions of Part III. You should use an algebraic approach unless otherwise stated.

Question 10.1. If A, B, and C are matrices of the same dimension, simplify

$$4(A + 4B + C) + 3(2B - A) - 4(5(2B + A - C) - 2(A + 2B - C))$$

Question 10.2. A matrix is said to be *skew symmetric* if $S^{\mathrm{T}} = -S$. Let A be *any* square matrix.

a. Show that $A - A^{\mathrm{T}}$ is skew symmetric.

b. Determine matrices C and D such that $A = C + D$ and C is symmetric and D is skew symmetric.

c. Demonstrate that all diagonal entries on a general skew symmetric matrix S are zero.

Question 10.3. Determine A such that $A^2 = 0_n$ in each of the following situations:

a. A is a general 2×2 matrix.

b. A is a symmetric 2×2 matrix.

Question 10.4. Where possible, determine the unknown matrix that solves the following matrix equations.

a.

$$2(B^{\mathrm{T}} - 2I_2)^{-1} = \begin{bmatrix} 0 & 2 \\ 1 & -1 \end{bmatrix}$$

b.

$$\frac{1}{3}(B^{\mathrm{T}} - 2I_2)^{-1} = \begin{bmatrix} 0 & 2 \\ 0 & -1 \end{bmatrix}$$

Question 10.5. If

$$A = \begin{bmatrix} 1 & 2 & 1 & 2 \\ 1 & 0 & 1 & 2 \\ 2 & 1 & 2 & 1 \\ 0 & 0 & 1 & 2 \end{bmatrix}$$

determine A^{-1}. Confirm that your result is indeed the inverse.

Question 10.6. Solve the following system of linear simultaneous equations using matrix methods.

$$\begin{cases} x + 2y + z + a = 4 \\ x - 2y + 3z = 2 \\ 2x + y - a = -3 \\ a + z = 2 \end{cases}$$

Question 10.7. Use Cramer's rule to confirm your answer to Question 10.6.

Question 10.8. A no-claims discount model has four states and associated 1-step transition probabilities as shown in Figure 10.2. Calculate the probability that a policyholder initially in State 1 will be in State 3 after 5 years.

Question 10.9. You are given that the states in Question 10.8 correspond to

- State 1: 0% discount
- State 2: 10% discount
- State 3: 40% discount
- State 4: 70% discount

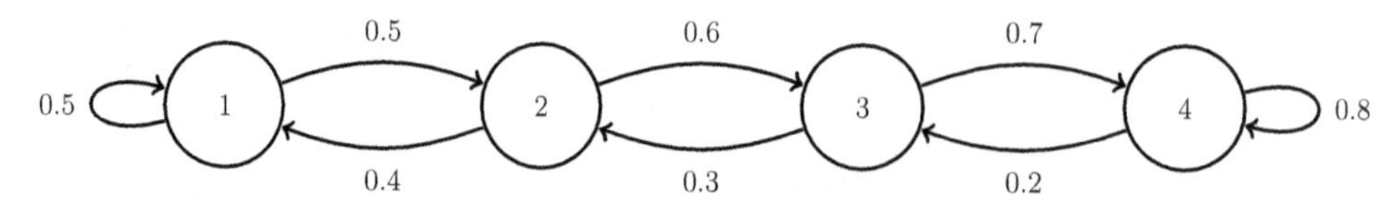

Figure 10.2 No-claims discount model and transition probabilities for Question 10.8.

If the undiscounted premium is \$600 per annum, calculate the *steady* annual revenue the policy is likely to generate from 27,000 policyholders.

Question 10.10. A frog moves home to a narrow strip of pond along which there are 5 evenly spaced lily pads. From each lily pad, the frog has an equal chance of jumping to the pad on either side until it reaches the pad at either end. Here the frog will sit indefinitely to feed off the abundant supply of flies now accessible to it.

a. If the frog begins on the middle lily pad, calculate the probability that it has not eaten within 3 jumps.

b. Estimate the probability that a frog beginning on a random lily pad will never eat.

CHAPTER 11

Implicit Functions and ODEs

Contents

Prerequisite knowledge	Intended learning outcomes

• Chapter 1
• Chapter 2
• Chapter 3
• Chapter 4
• Chapter 6
• Chapter 7
	• Define, recognize, and work with implicit functions
	• Recognize and perform implicit derivatives
	• Recognize and define an ordinary differential equation and boundary value problem/initial value problem
	• Classify an ODE in terms of
	  ◦ order
	  ◦ linearity
	  ◦ constant or variable coefficients
	• Select and use the standard methods for solving first-order problems,
	  ◦ separable
	  ◦ integrating factor
	  ◦ exact

In this chapter, we extend yet further the broad concepts of differential and integral calculus considered in the earlier chapters of this book. We remain with functions of a single variable, for example, $f(x)$, and introduce the important concepts of *implicit functions* and *ordinary differential equations*. In particular, we define what ordinary differential equations (ODEs) are and discuss the basic strategies for their algebraic solution in some simple classes.

11.1 IMPLICIT FUNCTIONS

The discussion of differential calculus presented so far in this book has been concerned almost entirely with functions of the form $y = f(x)$. For example, given the function

$$y(x) = x^2 + \sin x$$

we are able to immediately write down its derivative as

$$\frac{dy}{dx} = 2x + \cos x$$

As a further example, given

$$y - \sin x = 2 + yx \tag{11.1}$$

we can determine an expression for $y'(x)$ by first manipulating the expression to make $y(x)$ the subject and then performing the derivative of $y(x)$ in the usual manner. That is,

$$y(x) = \frac{2 + \sin x}{1 - x} \Rightarrow \frac{dy}{dx} = \frac{\sin x + (1 - x)\cos x + 2}{(1 - x)^2} \tag{11.2}$$

In both of these examples, we differentiated an expression for $y(x)$. This process is therefore known as *explicit differentiation.*

In contrast, consider now the expression

$$y^2 + (2x - 1)y = 4 \tag{11.3}$$

This is not easily manipulated into the form $y = f(x)$, rather it defines the function $y = f(x)$ *implicitly*. We would therefore say that $y = f(x)$ is an *implicit function* defined by Eq. (11.3). If we understand that y is a function of x we can take the derivative of Eq. (11.3) term by term with respect to x. Repeated use of the product rule discussed in Chapter 3 proceeds as follows,

$$\frac{d}{dx}y^2 + \frac{d}{dx}(2x - 1)y = \frac{d}{dx}4$$

$$\Rightarrow 2y\frac{dy}{dx} + 2y + (2x - 1)\frac{dy}{dx} = 0$$

This is an example of *implicit differentiation*, that is, we have obtained an implicit expression for $y'(x)$ by taking the derivative of an implicit function for $y(x)$. In some cases, but by no means all, it is then possible to rearrange the expression resulting from the implicit differentiation to form an explicit expression for $y'(x)$. In this example, we find that

$$\frac{dy}{dx} = \frac{-2y}{2(y + x) - 1}$$

Of course it is possible to take the implicit derivative of any expression. For example, taking the derivative of Eq. (11.1) leads directly to

$$\frac{\mathrm{d}}{\mathrm{d}x}y - \frac{\mathrm{d}}{\mathrm{d}x}(yx) - \frac{\mathrm{d}}{\mathrm{d}x}\sin x = \frac{\mathrm{d}}{\mathrm{d}x}2$$

$$\frac{\mathrm{d}y}{\mathrm{d}x} - x\frac{\mathrm{d}y}{\mathrm{d}x} - y - \cos x = 0$$

$$\Rightarrow \frac{\mathrm{d}y}{\mathrm{d}x} = \frac{y + \cos x}{1 - x}$$

which can easily be shown to be equal to Eq. (11.2) after substitution of $y = \frac{2+\sin x}{1-x}$. Explicit and implicit derivatives are therefore seen to be equivalent.

EXAMPLE 11.1

Use implicit derivatives to determine an explicit expression for $y'(x)$ in the following cases.

a. $y^2 + x^2 = 4$
b. $y = \cos(3x + 7y)$
c. $\frac{y}{x^2} + \frac{x}{y^4} = 3y^2$
d. $\mathrm{e}^{xy} = 2x^2y^3$

Solution

a. We assume $y = y(x)$ and take derivatives with respect to x.

$$\frac{\mathrm{d}}{\mathrm{d}x}y^2 + \frac{\mathrm{d}}{\mathrm{d}x}x^2 = \frac{\mathrm{d}}{\mathrm{d}x}4$$

$$2y\frac{\mathrm{d}y}{\mathrm{d}x} + 2x = 0$$

$$\Rightarrow \frac{\mathrm{d}y}{\mathrm{d}x} = -\frac{x}{y}$$

b. We assume $y = y(x)$ and take derivatives with respect to x.

$$\frac{\mathrm{d}}{\mathrm{d}x}y = \frac{\mathrm{d}}{\mathrm{d}x}\cos(3x + 7y)$$

$$\frac{\mathrm{d}y}{\mathrm{d}x} = -\sin(3x + 7y)\left(3 + 7\frac{\mathrm{d}y}{\mathrm{d}x}\right)$$

$$\Rightarrow \frac{\mathrm{d}y}{\mathrm{d}x} = \frac{-3\sin(3x + 7y)}{1 + 7\sin(3x + 7y)}$$

c. We assume $y = y(x)$ and take derivatives with respect to x.

$$\frac{\mathrm{d}}{\mathrm{d}x}(yx^{-2} + xy^{-4}) = \frac{\mathrm{d}}{\mathrm{d}x}(3y^2)$$

$$x^{-2}\frac{\mathrm{d}y}{\mathrm{d}x} - 2x^{-3}y - 4y^{-5}x\frac{\mathrm{d}y}{\mathrm{d}x} + y^{-4} = 6y\frac{\mathrm{d}y}{\mathrm{d}x}$$

$$\Rightarrow \frac{dy}{dx} = \frac{y(x^3 - 2y^5)}{4x^4 + 6x^3y^6 - xy^5}$$

d. We assume $y = y(x)$ and take derivatives with respect to x.

$$\frac{d}{dx}\,e^{xy} = \frac{d}{dx}(2x^2y^3)$$

$$e^{xy}\left(y + x\frac{dy}{dx}\right) = 2\left(2xy^3 + 3x^2y^2\frac{dy}{dx}\right)$$

$$\Rightarrow \frac{dy}{dx} = \frac{y\left(e^{xy} - 4xy^2\right)}{x\left(6xy^2 - e^{xy}\right)}$$

For location (x, y) that satisfies an implicit expression for y, the value of $y'(x)$ can be interpreted as the *gradient of the tangent to a curve* at that point. For example, you may recognize the expression in Example 11.1a., $x^2 + y^2 = 4$, as the equation of a circle of radius 2 centered at $(x, y) = (0, 0)$; this is illustrated in Figure 11.1. The point $(\sqrt{2}, \sqrt{2})$ satisfies $x^2 + y^2 = 4$ and so is on the circle, and the gradient of the tangent to the circle at that point is then given by the expression obtained in Example 11.1a. evaluated at $(\sqrt{2}, \sqrt{2})$. That is,

$$\frac{dy}{dx} = -\left.\frac{x}{y}\right|_{(\sqrt{2},\sqrt{2})} = -1$$

It is then possible to determine an expression for the tangent line at this point by fitting a straight line of gradient -1 through $(\sqrt{2}, \sqrt{2})$. In particular, the equation of the tangent with this gradient is $y = mx + c$. Here, $m = -1$ is the gradient and $c = 2\sqrt{2}$ is the y-intercept. The process of fitting this straight line was discussed in Chapter 3.

Obtaining the tangent line to a curve defined by an implicit expression is a useful application of implicit differentiation.

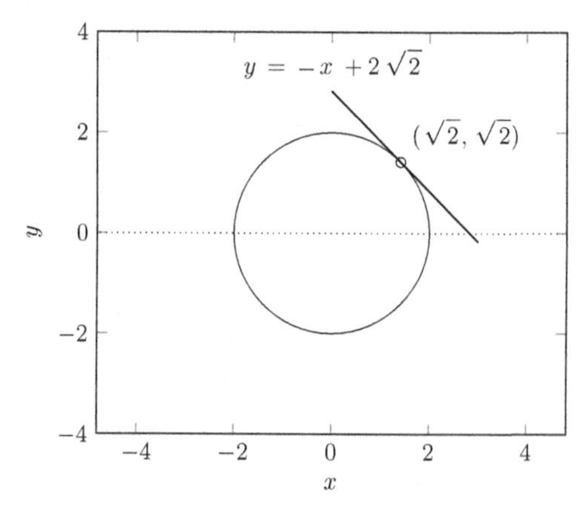

Figure 11.1 $x^2 + y^2 = 4$ and the tangent line at $(\sqrt{2}, \sqrt{2})$.

EXAMPLE 11.2

Confirm, in each case, that the point stated is on the curve defined by the implicit expression. Determine the equation of the tangent at that point.

a. $y^2 + x^2 = 4$ at $(x, y) = (0, 2)$

b. $y = \cos(3x + 7y)$ at $(x, y) = (2, 0.2315)$

Solution

Note that the expressions are identical to those in Example 11.1a. and b.

a. We note that $2^2 + 0^2 = 4$ and so confirm that the point $(0, 2)$ is on the curve. Using the expression stated in Example 11.1a., the gradient of the tangent at this point is 0. The equation of the tangent to the curve at $(0, 2)$ is then found to be $y = 2$.

b. We note that $\cos(3 \times 2 + 7 \times 0.2315) \approx 0.2315$, as required for the point $(2, 0.2315)$ to be on the curve. Using the expression stated in Example 11.1b., the gradient of the tangent at this point is -0.3737. The equation of the tangent at $(2, 0.2315)$ is then found to be $y = 0.9789 - 0.3737x$.

11.2 ORDINARY DIFFERENTIAL EQUATIONS

In Section 11.1, we saw how implicit expressions involving the derivatives of $y(x)$ could result from the differentiation of implicit functions. Such expressions are in fact examples of a class of equation called *ordinary differential equations* or *ODEs*.[1] Typically, however, ODEs occur naturally in the mathematical description of physical and financial systems and the task is to determine the function $y(x)$ that solves them. This is of course the reverse situation to the previous section where the ODE arose from our attempt to obtain $y'(x)$ from an expression involving $y(x)$.

For example, a simple financial model might assume the following statement

the rate of increase of a country's GDP at some time is directly proportional to its current GDP

If we use $f(t)$ to denote the GDP as a function of time t, this statement is easily expressed mathematically as the ODE

$$\frac{\mathrm{d}}{\mathrm{d}t} f(t) = kf(t) \tag{11.4}$$

where $k > 0$ is the constant of proportionality. Given the value of a country's GDP at a particular time, say $f(t_0) = G$, we might wish to solve Eq. (11.4) and determine the resulting model for the future time evolution of the GDP $f(t)$ for $t > t_0$.

[1] The "ordinary" indicates that the expression consists of derivatives of a function of a single independent variable.

EXAMPLE 11.3

Demonstrate that the function $f(t) = G\,\mathrm{e}^{k(t-t_0)}$ satisfies Eq. (11.4) and the associated condition $f(t_0) = G$.

Solution

We see that $f(t_0) = G\,\mathrm{e}^{k(t_0-t_0)} = G$ and so the condition is satisfied. Furthermore, we note that $f'(t) = kG\,\mathrm{e}^{k(t-t_0)} = kf(t)$, as required. Under this model, the country's GDP is expected to grow exponentially with time.

Before we consider strategies for the algebraic solution of simple classes of ODEs, we discuss some aspects of their description and classification that will prove useful.

In general, an ODE can be of any form consistent with the expression

$$F\left(x, y, y', y'', \ldots, y^{(n)}\right) = 0 \tag{11.5}$$

That is, one can consider an ODE to be an implicit function of the dependent variable y, its derivatives, and the independent variable x. A particular ODE is then classified by its basic algebraic form, particularly how y and its derivatives appear when expressed in the form of Eq. (11.5).

- The value of n determines the *order* of the ODE, that is, the highest order derivative of $y(x)$ that appears. For example, all expressions involving derivatives in Section 11.1 and Eq. (11.4) consist of first-order derivatives, $n = 1$, and are first-order ODEs. In contrast, the expressions

$$\frac{\mathrm{d}^2y}{\mathrm{d}x^2} + \frac{\mathrm{d}y}{\mathrm{d}x} + y = 0 \quad \text{and} \quad \frac{\mathrm{d}^3y}{\mathrm{d}x^3} + 4y = 2 \tag{11.6}$$

  are second- and third-order ODEs, respectively.
- An ODE can be *linear* or *nonlinear*, as determined by the appearance of products of the dependent variable and its derivatives. For example, the ODEs in Eqs. (11.4) and (11.6) are linear, whereas

$$\frac{\mathrm{d}y}{\mathrm{d}x} + y^2 - 2 = 0 \quad \text{and} \quad \frac{\mathrm{d}^3y}{\mathrm{d}x^3} + y\frac{\mathrm{d}^2y}{\mathrm{d}x^2} + 5\frac{\mathrm{d}y}{\mathrm{d}x} = 0 \tag{11.7}$$

  are both nonlinear owing to the y^2 and $y\frac{\mathrm{d}^2y}{\mathrm{d}x^2}$ terms, respectively.
- A further important property of ODEs is whether they have *constant* or *variable coefficients*. All the ODEs in Eqs. (11.4) and (11.6) have the dependent variable and its derivatives multiplied by constants only; these ODEs are therefore said to have *constant coefficients*. In contrast,

$$\frac{d^3y}{dx^3} + 4x^2y = 2 \quad \text{and} \quad \left(\frac{dy}{dx}\right)^2 + xy - 2 = 0$$

involve coefficients that are functions of x and so have variable coefficients.

EXAMPLE 11.4

Classify the following ODEs in terms of their order, linearity, and coefficients.

a. $h'' + 5h - 5 = 0$

b. $gg''' + xg = \sin x$

c. $\left(\frac{dy}{dx}\right)^2 - 4y = 2$

d. $xf' + xf = x$

Solution

a. Second order, linear, with constant coefficients.

b. Third order, nonlinear, with variable coefficients.

c. First order, nonlinear, with constant coefficients.

d. We note that the ODE is not in its basic form and divide through by x to obtain $f' + f = 1$, this of course assumes that $x \neq 0$. This is then classified as first order, linear, with constant coefficients.

Now that we are able to classify any given ODE, the next natural question is how can one solve it. That is, given $F(x, y, y', y'', \ldots, y^{(n)}) = 0$ how does one obtain $y(x)$?

While the solution of Eq. (11.4) was relatively straightforward and can be obtained from recalling the properties of exponential functions, there is no single approach to obtaining the solution to any given ODE. This is due to the complexity that ODEs can possess; for example, the behavior of an ODE is very dependent on its order, (non)linearity, and coefficients. As with the integrals introduced in Chapter 6, there are a number of strategies that can be incorporated to solve an ODE and one needs to pick the correct approach. As a further parallel to integration, it is often not possible to obtain an analytical solution to an ODE and numerical methods would be needed to make progress. The study of ODEs is a huge topic in mathematics and much of it is beyond the scope of this book. Here, we will consider only ODEs that permit an algebraic solution and basic strategies for their solution.

We will be concerned with *boundary value problems* in all that follows. That is, each ODE will be presented together with a set of *boundary conditions*. A solution will then be sought that satisfies both the ODE and the conditions. The solution of Eq. (11.4) subject to boundary condition $f(t_0) = G$ is an example of a boundary value problem. Note that if the conditions are given as at some initial time from which the solution must evolve, the problem can also be referred to as an *initial value problem*. The GDP example is therefore also an example of an initial value problem.

The importance of the boundary conditions to the solution of an ODE is demonstrated in the following example.

EXAMPLE 11.5

The GDP of two different countries is thought to be modeled by Eq. (11.4).

a. If, at time t_0, the GDPs of countries A and B are G_A and G_B, respectively, write down a boundary value problem in each case. You should assume the value of k is the same for each country.

b. Use the result of Example 11.3 to write down a solution for each boundary value problem and comment on your answers if $G_A \neq G_B$.

Solution

a. Let $f_A(t)$ and $f_B(t)$ denote the GDP of countries A and B at time $t > t_0$, respectively. The two boundary value problems are then,

A: $f_A'(t) = kf_A(t)$ such that $f_A(t_0) = G_A$

B: $f_B'(t) = kf_B(t)$ such that $f_B(t_0) = G_B$

b. From the results of Example 11.3, it should be clear that the two boundary value problems have solutions

$$f_A(t) = G_A e^{k(t-t_0)} \quad \text{and} \quad f_A(t) = G_B e^{k(t-t_0)}$$

We see that the two GDPs will evolve differently if $G_A \neq G_B$. That is, the different boundary conditions have led to different solutions to the same ODE.

All boundary value problems that we consider here will be *well posed*, that is they permit only a single, unique solution. Well-posed boundary value problems are clearly of significant practical importance in financial and physical modeling.

11.3 ALGEBRAIC SOLUTION OF FIRST-ORDER BOUNDARY VALUE PROBLEMS

We now turn to solution strategies for first-order boundary value problems. It is necessary to restrict the discussion to first-order ODEs so as to limit the discussion to manageable mathematics, consistent with the scope of this introductory text.

As we now know, a first-order ODE is an implicit expression that includes a single derivative of the dependent variable. In some sense, one would therefore expect a single indefinite integral to be required to solve it, leading to a single arbitrary constant. It should then be clear that a well-posed first-order boundary value problems consist of a first-order ODE and a *single* boundary condition. Indeed, an nth order ODE requires n boundary conditions for it to be well posed, although this is not necessarily a sufficient condition for well-posedness.

11.3.1 Separable ODEs

We begin with a solution strategy that can be applied to boundary value problems involving *separable* ODEs. A separable ODE is one that can be written in the form

$$F(y)\frac{\mathrm{d}y}{\mathrm{d}x} = G(x) \tag{11.8}$$

For example, the following ODEs are separable according this definition

$$\frac{\mathrm{d}y}{\mathrm{d}x} = x^2, \qquad y\frac{\mathrm{d}y}{\mathrm{d}x} = x^2, \qquad \frac{\mathrm{d}y}{\mathrm{d}x} = \frac{y}{\sin x} \tag{11.9}$$

The first two expressions are already in the form of Eq. (11.8) and some manipulation of the third expression is required to confirm that it is separable. In particular,

$$\frac{\mathrm{d}y}{\mathrm{d}x} = \frac{y}{\sin x} \Rightarrow \frac{1}{y}\frac{\mathrm{d}y}{\mathrm{d}x} = \frac{1}{\sin x}$$

EXAMPLE 11.6

Determine whether the following ODEs are separable.

a. $g(x) + g'(x) = x^2$

b. $\mathrm{e}^x g'(x) = \mathrm{e}^{-g(x)} x$

c. $\frac{y(x)}{\sin x}\frac{\mathrm{d}y(x)}{\mathrm{d}x} = \pi$

d. $\mathrm{e}^y \frac{\mathrm{d}z(y)}{\mathrm{d}y} = \frac{z(y)+1}{y-2} + y^2$

Solution

a. $g(x) + g'(x) = x^2$ is not separable.

b. $\mathrm{e}^x g'(x) = \mathrm{e}^{-g(x)} x$ is separable and can be expressed as $\mathrm{e}^{g(x)} g'(x) = x\,\mathrm{e}^{-x}$.

c. $\frac{y(x)}{\sin x}\frac{\mathrm{d}}{\mathrm{d}x}y(x) = \pi$ is separable and can be expressed as $y(x)\frac{\mathrm{d}}{\mathrm{d}x}y(x) = \pi \sin x$.

d. $\mathrm{e}^y \frac{\mathrm{d}z(y)}{\mathrm{d}y} = \frac{z(y)+1}{y-2} + y^2$ is not separable.

The advantage of separable ODEs is that they can often be integrated directly. In particular,

$$F(y)\frac{\mathrm{d}y}{\mathrm{d}x} = G(x)$$

$$\Rightarrow \int F(y)\frac{\mathrm{d}y}{\mathrm{d}x}\,\mathrm{d}x = \int G(x)\,\mathrm{d}x$$

A simple change of the integration variable from x to y on the left-hand side leads to

$$\int F(y)\,\mathrm{d}y = \int G(x)\,\mathrm{d}x$$

Performing both integrals would lead to an expression for $y(x)$, as required. For example, the first-order ODE of Eq. (11.9) can be solved as follows,

$$\frac{\mathrm{d}y}{\mathrm{d}x} = x^2$$

$$\Rightarrow \int 1\,\mathrm{d}y = \int x^2\,\mathrm{d}x$$

$$\Rightarrow y(x) = \frac{1}{3}x^3 + c$$

Here, c is a combined arbitrary constant resulting from the two indefinite integrals. A particular value for c can be obtained from the boundary condition of the boundary value problem. For example, let us suppose that the solution is such that $y(1) = \frac{1}{3}$ then

$$\frac{1}{3} + c = \frac{1}{3} \Rightarrow c = 0$$

leading to the solution $y(x) = \frac{1}{3}x^3$.

An alternative approach to imposing the boundary condition in separable ODEs is to incorporate it into the limits of the two definite integrals. For example, an alternative solution to the same problem is as follows.

$$\frac{dy}{dx} = x^2$$

$$\int_{\frac{1}{3}}^{y(x)} 1\, d\tilde{y} = \int_1^x \tilde{x}^2\, d\tilde{x}$$

$$y(x) - \frac{1}{3} = \frac{1}{3}(x^3 - 1^3)$$

$$\Rightarrow y(x) = \frac{1}{3}x^3$$

Note that we have used dummy variables $\tilde{y}$ and $\tilde{x}$ in the intermediate integrals to avoid confusion with the upper integration limits y and x.

EXAMPLE 11.7

Solve the following first-order separable boundary value problems.

a. $\frac{df}{dx} = 3f$ such that $f(0) = 1$.

b. $xy\frac{dy}{dx} = x^3 + 2$ such that $y(1) = \pi$.

c. $g'(x) = g(x)\sin(x)$ such that $g(0.5) = 3$

d. $e^{2x}\frac{dy}{dx} = e^{-y}$ such that $y(10) = 2$.

Solution

a. We express the problem in the form of Eq. (11.8) and integrate both sides,

$$\frac{df}{dx} = 3f$$

$$\int_1^{f(x)} \frac{1}{\tilde{f}}\, d\tilde{f} = \int_0^x 3\, d\tilde{x}$$

$$\ln f(x) - \ln 1 = 3x - 0$$

$$\Rightarrow f(x) = e^{3x}$$

Note that this demonstrates the solution process for Example 11.3.

b. We express the problem in the form of Eq. (11.8) and integrate both sides,

$$
\begin{aligned}
xy\frac{\mathrm{d}y}{\mathrm{d}x} &= x^3 + 2 \\
\int_{\pi}^{y(x)} \tilde{y}\,\mathrm{d}\tilde{y} &= \int_{1}^{x} \tilde{x}^2 + \frac{2}{\tilde{x}}\,\mathrm{d}\tilde{x} \\
\frac{1}{2}\left(y^2(x) - \pi^2\right) &= \frac{1}{3}\left(x^3 - 1\right) + 2\left(\ln(x) - \ln(1)\right) \\
\Rightarrow y^2(x) &= \frac{2}{3}\left(x^3 - 1\right) + 4\ln(x) + \pi^2 \\
\Rightarrow y(x) &= \sqrt[+]{\frac{2}{3}\left(x^3 - 1\right) + 4\ln(x) + \pi^2}
\end{aligned}
$$

Note that the positive square root has been taken to ensure that the boundary condition is satisfied.

c. We express the problem in the form of Eq. (11.8) and integrate both sides,

$$
\begin{aligned}
\frac{\mathrm{d}g}{\mathrm{d}x} &= g\sin x \\
\int_{3}^{g(x)} \frac{1}{g}\,\mathrm{d}g &= \int_{0.5}^{x} \sin x\,\mathrm{d}x \\
\ln g(x) - \ln 3 &= -\cos x + \cos 0.5 \\
\Rightarrow g(x) &= 3\,\mathrm{e}^{\cos 0.5 - \cos x}
\end{aligned}
$$

d. We express the problem in the form of Eq. (11.8) and integrate both sides,

$$
\begin{aligned}
\mathrm{e}^{2x}\frac{\mathrm{d}y}{\mathrm{d}x} &= \mathrm{e}^{-y} \\
\int_{2}^{y(x)} \mathrm{e}^{\tilde{y}}\mathrm{d}\tilde{y} &= \int_{10}^{x} \mathrm{e}^{-2\tilde{x}}\mathrm{d}\tilde{x} \\
\mathrm{e}^{y(x)} - \mathrm{e}^2 &= -\frac{1}{2}\left(\mathrm{e}^{-2x} - \mathrm{e}^{-20}\right) \\
\Rightarrow y(x) &= \ln\left(\mathrm{e}^2 + 0.5\,\mathrm{e}^{-20} - 0.5\,\mathrm{e}^{-2x}\right)
\end{aligned}
$$

The reader is invited to verify that the solution does indeed solve the boundary value problem in each case. Two checks are necessary that (1) the boundary condition is satisfied and (2) the expression for $y(x)$ satisfies the ODE.

11.3.2 Linear ODEs

Consider now the general class of linear first-order boundary value problems of the form

$$
\frac{\mathrm{d}y}{\mathrm{d}x} + p(x)y = q(x) \quad \text{such that} \quad y(x_0) = y_0. \tag{11.10}
$$

Here, $p(x)$ and $q(x)$ are either functions of x or constants. Although not separable, such ODEs can permit a solution using *integrating factors*. We define the integrating factor to be

$$I(x) = \exp\left(\int p(x)\,\mathrm{d}x\right) \tag{11.11}$$

with zero constant of integration in the exponent. Multiplying Eq. (11.10) by $I(x)$ leads to an expression that can be simplified using the product rule. In particular,

$$\begin{aligned} I(x)\frac{\mathrm{d}y}{\mathrm{d}x} + I(x)p(x)y &= I(x)q(x) \\ \Rightarrow \frac{\mathrm{d}}{\mathrm{d}x}\left(y(x)I(x)\right) &= I(x)q(x) \end{aligned} \tag{11.12}$$

Progress in solving the boundary value problem can then often be made by taking the indefinite integral of both sides of Eq. (11.12) with respect to x, leading to

$$\begin{aligned} \int \frac{\mathrm{d}}{\mathrm{d}x}(y(x)I(x))\,\mathrm{d}x &= \int I(x)q(x)\,\mathrm{d}x \\ \Rightarrow y(x) &= \frac{1}{I(x)}\int I(x)q(x)\,\mathrm{d}x + \frac{c}{I(x)} \end{aligned}$$

Note that the form of Eq. (11.11) is such that $I(x) \neq 0$. The arbitrary constant c, resulting from combining both constants of integration, can then be determined from the boundary condition $y(x_0) = y_0$.

EXAMPLE 11.8

Use an integrating factor to solve the boundary value problem

$$\frac{\mathrm{d}y}{\mathrm{d}x} + 3x^2 y = 0 \quad \text{such that} \quad y(0) = 2$$

Solution

Using Eq. (11.11), it should be clear that the appropriate integrating factor is

$$I(x) = \exp\left(\int 3x^2\,\mathrm{d}x\right) = \mathrm{e}^{x^3}$$

Multiplying the ODE through by $I(x)$ leads to

$$\begin{aligned} \mathrm{e}^{x^3}\frac{\mathrm{d}y}{\mathrm{d}x} + 3x^2\mathrm{e}^{x^3}y &= 0 \\ \Rightarrow \frac{\mathrm{d}}{\mathrm{d}x}\left(y\,\mathrm{e}^{x^3}\right) &= 0 \end{aligned}$$

This is then integrated to yield,

$$y = c\,\mathrm{e}^{-x^3}$$

The condition $y(0) = 2$ implies that $c = 2$ and so the solution to this particular boundary value problem is

$$y(x) = 2\,\mathrm{e}^{-x^3}$$

EXAMPLE 11.9

Solve the boundary value problem

$$x\frac{\mathrm{d}y}{\mathrm{d}x} + y = 4x\sin x \quad \text{such that} \quad y(0) = 0$$

Solution

You might note that this ODE is such that we can immediately use the product rule on the left-hand side

$$\begin{aligned} x\frac{\mathrm{d}y}{\mathrm{d}x} + y &= 4x\sin x \\ \frac{\mathrm{d}}{\mathrm{d}x}(xy) &= 4x\sin x \end{aligned}$$

Integration by parts can then be used to integrate this expression with respect x,

$$\begin{aligned} xy &= \int 4x\sin x\,\mathrm{d}x \\ \Rightarrow xy &= 4(\sin(x) - x\cos(x)) + c \end{aligned}$$

The boundary condition is satisfied when $c = 0$ and we find that the particular solution to the boundary value problem is

$$y(x) = 4\left(\frac{\sin(x)}{x} - \cos(x)\right)$$

Note that if we had not initially noticed the shortcut, we might have sought to transform the ODE to the standard form of Eq. (11.10). This is done by dividing through by x, leading to

$$\frac{\mathrm{d}y}{\mathrm{d}x} + \frac{y}{x} = 4\sin x$$

The integrating factor is then

$$I(x) = \exp\left(\int \frac{1}{x}\mathrm{d}x\right) = \mathrm{e}^{\ln x} = x$$

Multiplying through by x leads us back to the original expression. At this stage we would naturally be seeking to apply the product rule and so proceed as above.

11.3.3 Exact ODEs

The final class of ODEs that we consider here are called *exact ODEs*. Although they are ODEs, that is they are formed from derivatives of functions of a single variable, exact ODEs have a particular definition that involves the introduction of implicit functions which will be manipulated as if they have *two* variables. Multivariate calculus is considered in the following Chapter 12 and the reader may wish to return to this discussion after having read that. Alternatively, the following preliminary discussion of *partial derivatives* may be considered an introduction to Section 12.1 where the topic is discussed in detail.

An ODE of the form

$$P(x, y) + Q(x, y)\frac{dy}{dx} = 0 \tag{11.13}$$

is said to be *exact* if

$$\frac{\partial P}{\partial y} = \frac{\partial Q}{\partial x} \tag{11.14}$$

Let us first consider Eq. (11.13). The first thing to note is that the functions P and Q are implicit functions that involve both y and x. For example, the ODE

$$-\frac{y}{x^2} + \frac{1}{x}\frac{dy}{dx} = 0 \tag{11.15}$$

is of the required form with

$$P(x, y) = -\frac{y}{x^2} \quad \text{and} \quad Q(x, y) = \frac{1}{x}$$

Whether this ODE is exact is determined by Condition (11.14) which involves the unfamiliar notation "$\frac{\partial}{\partial x}$." At this stage it is sufficient to understand that $\frac{\partial Q}{\partial x}$ denotes the *partial derivative of Q with respect to x*. As we shall see in Section 12.1, this is the derivative of multivariate function $Q(x, y)$ taken with respect to x while treating y as a constant. Of course $y = y(x)$ but this is ignored in the partial derivative. Similarly, $\frac{\partial P}{\partial y}$ is the *partial derivative of P with respect to y*, which is the derivative of P with respect to y taken while holding x as constant. In our example,

$$\frac{\partial P}{\partial y} = \frac{\partial}{\partial y}\left(-\frac{y}{x^2}\right) = -\frac{1}{x^2} \quad \text{and} \quad \frac{\partial Q}{\partial x} = \frac{\partial}{\partial x}\left(\frac{1}{x}\right) = -\frac{1}{x^2}$$

we then note that in this particular example Condition (11.14) is satisfied and ODE (Eq. 11.15) is exact.[2]

[2] You may note that ODE (Eq. 11.15) is also separable and so amenable to the techniques used previously. In particular, the ODE can be rearranged to $\frac{1}{y}y' = \frac{1}{x} \Rightarrow \ln y = \ln x + c$ and so $y(x) = cx$.

EXAMPLE 11.10

Determine whether the following ODEs are exact.

a. $\frac{y}{x^2} + \frac{1}{x}\frac{dy}{dx} = 0$

b. $(f^2 \cos z - \sin z) + (2f \sin z + 2)f' = 0$

c. $2xg^2 + 4 + 2(x^2 g - 3)g' = 0$

d. $\frac{(2x+x^2y^3)}{(x^3y^2+4y^3)} + \frac{dy}{dx} = 0$

Solution

a. We can immediately write down that

$$P(x, y) = \frac{y}{x^2} \Rightarrow \frac{\partial P}{\partial y} = \frac{1}{x^2}$$

and

$$Q(x, y) = \frac{1}{x} \Rightarrow \frac{\partial Q}{\partial x} = -\frac{1}{x^2} \neq \frac{\partial P}{\partial y}$$

Condition (11.14) fails, and this ODE is not exact.

b. We can immediately write down that

$$P(z, f) = f^2 \cos z - \sin z \Rightarrow \frac{\partial P}{\partial f} = 2f \cos z$$

and

$$Q(z, f) = 2f \sin z + 2 \Rightarrow \frac{\partial Q}{\partial z} = 2f \cos z = \frac{\partial P}{\partial f}$$

This ODE is therefore exact.

c. We immediately write down that

$$P(x, g) = 2xg^2 + 4 \Rightarrow \frac{\partial P}{\partial g} = 4xg$$

and

$$Q(x, g) = 2(x^2 g - 3) \Rightarrow \frac{\partial Q}{\partial x} = 4xg = \frac{\partial P}{\partial g}$$

This ODE is therefore exact,

d. We begin by rearranging the ODE to a form consistent with Eq. (11.13),

$$(2x + x^2y^3) + (x^3y^2 + 4y^3)\frac{dy}{dx} = 0$$

From this we see that

$$P(x, y) = 2x + x^2y^3 \Rightarrow \frac{\partial P}{\partial y} = 3x^2y^2$$

and

$$Q(x, y) = x^3y^2 + 4y^3 \Rightarrow \frac{\partial Q}{\partial x} = 3x^2y^2 = \frac{\partial P}{\partial y}$$

The ODE is therefore exact.

The properties of exact ODEs are particularly useful when seeking an algebraic solution. To understand the solution process, consider a further and unknown multivariate function, $\Lambda(x, y)$, which is such that

$$\frac{\partial \Lambda}{\partial x} = P(x, y) \quad \text{and} \quad \frac{\partial \Lambda}{\partial y} = Q(x, y) \tag{11.16}$$

An ODE of the form given by Eq. (11.13) can therefore be rewritten as

$$\frac{\partial \Lambda}{\partial x} + \frac{\partial \Lambda}{\partial y}\frac{\mathrm{d}y}{\mathrm{d}x} = 0$$

and, if it the ODE is exact, that is Condition (11.14) is satisfied, the chain rule can be used to express this as

$$\frac{\mathrm{d}}{\mathrm{d}x}\Lambda(x, y(x)) = 0$$

This expression can then be integrated to yield an *implicit* solution of the ODE involving $y(x)$ and an arbitrary constant c,

$$\Lambda(x, y(x)) = c \tag{11.17}$$

Finding the solution of an exact ODE therefore reduces to finding the implicit function $\Lambda(y, x)$ from $P(x, y)$ and $Q(x, y)$. In some cases, it may then be possible to obtain an explicit expression for $y(x)$ from this implicit solution.

Let us return to our example ODE given by Eq. (11.15) which has been previously confirmed as exact. We have

$$\frac{\partial \Lambda}{\partial x} = P(x, y) = -\frac{y}{x^2} \quad \text{and} \quad \frac{\partial \Lambda}{\partial y} = Q(x, y) = \frac{1}{x}$$

Progress toward obtaining $\Lambda(x, y)$ can be made by first "reversing" the partial derivative with respect to x. That is, we can integrate $P(x, y)$ with respect to x while keeping y fixed,

$$\Lambda = \int -\frac{y}{x^2}\,\mathrm{d}x + c(y) = \frac{y}{x} + c(y)$$

The introduction of the arbitrary function $c(y)$ is necessary since we have assumed y to be fixed in the reversal of the partial derivative. Its use should be clear from the observation that

$$\frac{\partial \Lambda}{\partial x} = \frac{\partial}{\partial x}\left(\frac{y}{x} + c(y)\right) = -\frac{y}{x^2}$$

The form of the $c(y)$ is then found by taking the partial derivative of $\Lambda(x, y)$ with respect to y and equating this to $Q(x, y)$, as required by Eq. (11.16). That is,

$$\frac{\partial \Lambda}{\partial y} = \frac{\partial}{\partial y}\left(\frac{y}{x} + c(y)\right) = \frac{1}{x} + \frac{\mathrm{d}}{\mathrm{d}y}c(y)$$

Comparing this to $Q(x, y)$ means that, in this case,

$$\frac{\mathrm{d}}{\mathrm{d}y}c(y) = 0 \Rightarrow c(y) = \tilde{c}$$

with $\tilde{c}$ an arbitrary constant. We have therefore found that $\Lambda = \frac{y}{x} + \tilde{c}$ satisfies both expressions in Eq. (11.16) and this can be used in Eq. (11.17) to give

$$\frac{y}{x} + \tilde{c} = c$$

Finally, we can combine the two arbitrary constants and obtain the solution to the ODE as

$$y = cx$$

In practice, the value of c would be determined from a boundary condition. Note that the solution is identical to that obtained in Footnote 2 and the reader is invited to check that this is indeed a solution of the ODE.

In the above discussion, we opted to find $\Lambda(x, y)$ by first reversing the partial derivative with respect to x and then using the partial derivative with respect to y to determine the arbitrary function $c(y)$. However, it is also possible to begin by reversing the partial derivative with respect to y and use the partial derivative with respect to x to determine the arbitrary function $c(x)$. This alternative solution is detailed in the following example.

EXAMPLE 11.11

Find the solution to the boundary value problem given by the exact ODE Eq. (11.15) subject to the boundary condition $y(1) = 10$. You should begin by using an appropriate integral of $Q(x, y)$.

Solution

We have

$$\frac{\partial \Lambda}{\partial x} = P(x, y) = -\frac{y}{x^2} \quad \text{and} \quad \frac{\partial \Lambda}{\partial y} = Q(x, y) = \frac{1}{x}$$

and attempt reverse the y-derivative to obtain $Q(x, y)$. This is done by holding x as a constant and performing the integration with respect to y,

$$\Lambda = \int \frac{1}{x}\,\mathrm{d}y = \frac{y}{x} + c(x)$$

Here, $c(x)$ is an arbitrary function of x which is obtained from the following expression compared to $P(x, y)$.

$$\frac{\partial \Lambda}{\partial x} = \frac{\partial}{\partial x}\left(\frac{y}{x} + c(x)\right) = -\frac{y}{x^2} + \frac{\mathrm{d}}{\mathrm{d}x}c(x)$$

Therefore, $\frac{\mathrm{d}c}{\mathrm{d}x} = 0$ and $c(x) = \tilde{c}$. An implicit solution to the exact ODE is then obtained from Eq. (11.17) as

$$\frac{y}{x} = c \Rightarrow y(x) = cx$$

This general solution to the ODE is identical to that obtained previously. The particular solution to this boundary value problem is then found from the boundary condition as

$$y(x) = 10x$$

EXAMPLE 11.12

Solve the following boundary value problems formed from exact ODEs and a single boundary condition.

a. $(f^2 \cos z - \sin z) + (2f \sin z + 2)f' = 0$ such that $f(\pi) = 0$.

b. $2xg^2 + 4 + 2(x^2 g - 3)g' = 0$ such that $g(1) = 1$.

c. $\frac{(2x+x^2y^3)}{(x^3y^2+4y^3)} + \frac{\mathrm{d}y}{\mathrm{d}x} = 0$ such that $y(0) = 1$.

Solution

We note that the ODEs were demonstrated to be exact in Example 11.10.

a. We have

$$\frac{\partial \Lambda}{\partial z} = P(z,f) = \cos z f^2 - \sin z$$
$$\Rightarrow \Lambda = f^2 \sin z + \cos z + c(f)$$
$$\Rightarrow \frac{\partial \Lambda}{\partial f} = 2f \sin z + \frac{\mathrm{d}}{\mathrm{d}f}c(f)$$

Comparing this to $Q(z,f) = 2f \sin z + 2$ we find that

$$\frac{\mathrm{d}}{\mathrm{d}f}c(f) = 2$$
$$\Rightarrow c(f) = 2f + \tilde{c}$$

We therefore have $\Lambda(z,f) = f^2 \sin z + \cos z + 2f + \tilde{c}$ which, with Eq. (11.17) and combining the arbitrary constants, leads to the implicit solution

$$f^2 \sin z + \cos z + 2f = c$$

The boundary condition $f(\pi) = 0$ is such that $c = 0 - 1 + 0 = -1$ and so the implicit solution to the boundary value problem is

$$f^2(z) \sin z + 2f(z) + 1 + \cos z = 0$$

which is a quadratic in $f(z)$. Going further, we can obtain an explicit expression for the solution as

$$f(z) = \frac{-2 + \sqrt{4 - 4\sin z(1+\cos z)}}{2\sin z} = \frac{-1 + \sqrt{1 - \sin z(1+\cos z)}}{\sin z}$$

Note that the positive square root has been taken to ensure that the boundary condition is satisfied.

b. We have

$$\begin{aligned} \frac{\partial \Lambda}{\partial x} &= P(x,g) = 2xg^2 + 4 \\ \Rightarrow \Lambda &= x^2 g^2 + 4x + c(g) \\ \Rightarrow \frac{\partial \Lambda}{\partial g} &= 2gx^2 + \frac{\mathrm{d}}{\mathrm{d}g} c(g) \end{aligned}$$

Comparing this to $Q(x,g) = 2(x^2 g - 3)$ we find that

$$\begin{aligned} \frac{\mathrm{d}}{\mathrm{d}g} c(g) &= -6 \\ \Rightarrow c(g) &= -6g + \tilde{c} \end{aligned}$$

We therefore have $\Lambda(x,g) = g^2x^2 + 4x - 6g + \tilde{c}$ which, with Eq. (11.17) and combining the arbitrary constants, leads to the implicit solution

$$x^2 g^2(x) - 6g(x) + 4x = c$$

The boundary condition $g(1) = 1$ is such that $c = 1 - 6 + 4 = -1$ and so the implicit solution to the boundary value problem is

$$x^2 g^2(x) - 6g(x) + 1 + 4x = 0$$

which is a quadratic in $g(x)$. Going further, we can obtain an explicit expression for the solution as

$$g(x) = \frac{6 - \sqrt{36 - 4x^2(1+4x)}}{2x^2} = \frac{3 - \sqrt{9 - x^2(1+4x)}}{x^2}$$

Note that the negative square root has been taken to ensure that the boundary condition is satisfied.

c. Using the standard form of the ODE, as stated in the solution to Example 11.10d., we find that

$$\begin{aligned} \frac{\partial \Lambda}{\partial x} &= P(x,y) = 2x + x^2 y^3 \\ \Rightarrow \Lambda &= x^2 + \frac{x^3 y^3}{3} + c(y) \\ \Rightarrow \frac{\partial \Lambda}{\partial y} &= y^2 x^3 + \frac{\mathrm{d}}{\mathrm{d}y} c(y) \end{aligned}$$

Comparing this to $Q(x, y) = x^3y^2 + 4y^3$ we find that

$$\frac{\mathrm{d}}{\mathrm{d}y}c(y) = 4y^3$$
$$\Rightarrow c(y) = y^4 + \tilde{c}$$

We therefore have $\Lambda(x, y) = x^2 + \frac{x^3y^3}{3} + y^4 + \tilde{c}$ which, with Eq. (11.17) and combining the arbitrary constants, leads to the implicit solution

$$x^2 + \frac{x^3y^3(x)}{3} + y^4(x) = c$$

The boundary condition $y(0) = 1$ is such that $c = 0 + 0 + 1 = 1$ and so the implicit solution to the boundary value problem is

$$x^2 + \frac{x^3y^3(x)}{3} + y^4(x) - 1 = 0$$

which is a quartic in $y(x)$. We choose to leave this solution in implicit form.

The reader is invited to repeat each solution, beginning by reversing the other partial derivative.

11.4 IMPLICIT FUNCTIONS AND ODEs ON YOUR COMPUTER

Wolfram Alpha can be used to both aid your studies of the material in this chapter and also to quickly determine solutions to ODEs.

We begin with the implicit function $x^2 + 2xy^2 - 3x^3 = 0$. This can be expressed as an explicit function using, for example, the command

```
x^2+2x*y^2-3x^3=0 y=?
```

Wolfram Alpha returns that

$$y = \pm\frac{\sqrt{x}\sqrt{3x-1}}{\sqrt{2}}$$

In order to obtain the explicit derivative of this function with respect to x, we have to be careful. In particular, we need to include that $y = y(x)$, otherwise the engine will assume that it is some constant value. This is done by slightly rewriting the expression as $x^2 + 2xy(x)^2 - 3x^3$; it is then clear that y is a function of x. The command

```
d/dx (x^2+2x*y(x)^2-3x^3)=0
```

returns the result $-9x^2 + 4xyy' + 2y^2 + 2x = 0$. The engine goes further and actually solves this ODE as

$$y(x) = \pm\frac{\sqrt{c + (3x-1)x^2}}{\sqrt{2x}}$$

Consider now the particular ODE $y' + xy = 0$. It should be clear that this is separable and has a general solution $y = c\,e^{-\frac{x^2}{2}}$ for some arbitrary constant c. We can check this solution using Wolfram Alpha with the command

```
solve y'+xy=0
```

Note that in this case it has not been necessary to state that $y = y(x)$. This is because the prime is interpreted as being a derivative with respect to x and so y must be a function of x. The engine first confirms it has understood our input correctly, classifies the ODE as a first-order linear ODE, and returns the solution for y in terms of the arbitrate constant c_1.

But what if we want to impose the boundary condition that $y(1) = 4$, for example? In this case, the boundary value problem is intuitively input with the command

```
solve y'+xy=0 if y(1)=4
```

and we obtain that $y(x) = 4\,e^{\frac{1-x^2}{2}}$.

As a more complicated example, the solution of the initial value problem

$$\frac{df}{dt} - 2f + t^2 = 0 \quad \text{and} \quad f(0) = 10$$

is obtained from the command

```
solve f'-2f+t^2=0 if f(0)=10
```

Wolfram Alpha returns that this has solution $f(t) = \frac{1}{2}(2t^2 + 2t + 39\,e^{2t} + 1)$.

EXAMPLE 11.13

Use Wolfram Alpha to solve the following ODEs.

a. $f(x)f'(x) = f^2(x) - x$ such that $f(0) = 10$.

b. $g'(t) = 5t^2 + \sin t$ such that $g(10) = 1$.

Solution

a. We use the command,

```
solve f(x)*f'(x)=f(x)^2 - x if f(0)=1
```

and immediately obtain that

$$f(x) = \frac{1}{\sqrt{2}} \overset{+}{\sqrt{2x + e^{2x} + 1}}$$

b. We use the command,

```
solve g'(t)=5t^2+sin(t) if g(10)=1
```

and immediately obtain that

$$g(t) = \frac{5t^3}{3} - \cos(t) - \frac{4997}{3} + \cos(10)$$

11.5 QUESTIONS

The following questions are intended to test your knowledge of the material discussed in this chapter. Full solutions are available in Chapter 11 Solutions of Part III. You should use an algebraic approach unless otherwise stated.

Question 11.1. Determine all tangent lines of the following implicit function at $x = 1$.

$$x^2 + y^2 = 2 - xy$$

Question 11.2. Classify the following ODEs.

a. $\frac{df}{dx} + f = 0$

b. $g'' + xg = 2x$

c. $x^2h^2 + h^{(4)} = \sin x$

d. $\frac{d^2y}{dz^2} + y\cos z = e^y$

Question 11.3. Solve the following boundary value problems.

a. $y' - \frac{y^2-1}{x} = 0$ such that $y(5) = 0$

b. $\frac{dy}{dz} = \frac{3}{z^2\cos y}$ such that $y(10) = 0$

c. $g' = (5t + 1)g$ such that $g(0) = 1$

d. $k' = 5(ky + 2y + k + 2)$ such that $k(12) = 3$

Question 11.4. Solve the following boundary value problems.

a. $f' = x + \frac{f}{x}$ such that $f(1) = \pi$

b. $\frac{h'}{y} = 5 + 2h$ such that $h(-4) = 2$

c. $g' = 10 - 2g$ such that $g(0) = 0$

d. $y' + y = \sin(x)$ such that $y(\pi) = 1$

Question 11.5. Solve the following boundary value problems.

a. $(2xy + 1)y' = -(y^2 - 2x)$ such that $y(0) = 1$

b. $\frac{df}{dy} = \frac{\cos y - \cos f - f}{y - y\sin f}$ such that $f(\pi) = 1$

c. $(g + 1)e^z = (2g - e^z)g'$ such that $g \to 4$ as $z \to -\infty$

d. $\frac{1}{t}\frac{dz}{dt} = -2\frac{(1+z^2)\arctan(z)}{t^2}$ such that $z(1) = 1$

Question 11.6. While flying between San Francisco and Honolulu, a trainee actuary notices a new island 500 miles off the coast of California. He alerts the authorities and, soon later, the island is subject to a detailed study. Scientists discover that a new species of beetle inhabits the island and a sample of 50 beetles are returned to a research institute in San Diego. In the laboratory environment, the beetle population is observed to grow at a rate proportional to its population and such that, without any other influences, the population would quadruple every 5 weeks. On any given week, 20 insects are extracted from the population and dispatched around the world for further study, and 5 die of natural causes. Determine how long you might expect the beetle population to survive in San Diego.

Question 11.7. Solve the following *second-order* initial value problem

$$\frac{\mathrm{d}^2}{\mathrm{d}x^2}y + 3\frac{\mathrm{d}}{\mathrm{d}x}y - 10y = 0$$

such that $y'(0) = 1$ and $y(0) = 0$.
Hint. Factorize the ODE.

Question 11.8. Solve the following initial value problem

$$f^{(4)} - 10f''' + 35f'' - 50f' + 24y = 0$$

such that $f'''(0) = 0, f''(0) = -2, f'(0) = 1$, and $f(0) = 0$.
Hint. Attempt to generalize the process used in Question 11.7.

Question 11.9. In standard actuarial notation, the quantity ${}_tp_{50}$ denotes the probability that, on his 50th birthday, an individual will survive for a further t years. Under the *Weibull model* of mortality, ${}_tp_{50}$ is determined by

$$\frac{\mathrm{d}}{\mathrm{d}t}{}_tp_{50} = -\alpha\beta t^{\beta-1}\,{}_tp_{50}$$

where α and $\beta \neq 1$ are constants. You are given that the probability that the individual will survive to age 60 is 0.6, and to age 70 is 0.5.
a. Find the probability that the individual will die before his 90th birthday.
b. Determine, using this model, the maximum age the individual is 95% sure to reach.

Question 11.10. Under the *Gompertz model* of mortality, the quantity ${}_tp_{50}$ is determined by

$$\frac{\mathrm{d}}{\mathrm{d}t}{}_tp_{50} = -Bc^{50+t}\,{}_tp_{50}$$

where B and c are constants. Repeat Question 11.9 under this model.

CHAPTER 12

Multivariate Calculus

Contents

Prerequisite knowledge	**Intended learning outcomes**

• Chapter 1
• Chapter 2
• Chapter 3
• Chapter 4
• Chapter 6
• Chapter 7
• Chapter 11
	• Define and recognize partial derivatives
	• Explore the properties of multivariate functions, including
	  ◦ obtain the first- and higher-order partial derivatives
	  ◦ obtain mixed derivatives at higher orders
	  ◦ locate and classify turning points
	• Apply the method of Lagrange multipliers to solve constrained optimization problems
	• Perform double integrals of bivariate functions over domains that are
	  ◦ rectangular
	  ◦ nonrectangular

In this chapter, we consider the basic calculus of functions of two or more independent variables, that is the *calculus of multivariate functions*, or *multivariate calculus*. This book has so far focused on functions of a single independent variable, for example $y(x)$ or $f(z)$; however, most realistic models of physical and financial systems involve more than one independent variable. For example, the evolution of a country's GDP is likely to be determined by, for example, consumption, government spending, imports, and exports, to name only a few; a realistic model for a GDP must therefore be capable of being interrogated for changes in many independent variables. In addition, as we saw in the previous chapter with the discussion of exact ODEs, it is often very convenient with to work with mathematical tools that require some knowledge of multivariate calculus. We will see an additional example of such tools later in this chapter.

The description that follows will be concerned with the general function f of n independent variables $x_1, x_2, \ldots, x_n$. That is, we will work with

$$f(x_1, x_2, \ldots, x_n)$$

While it is possible to develop a formal analysis of such functions, for example, we could derive formal definitions of continuity and limits in n dimensions, our interest here is motivated by the practical use of multivariate calculus. A formal analysis is therefore beyond the scope of this book and we move directly to the definitions of partial derivatives and multiple integration. It is assumed that the reader has gained sufficient "intuition" from the study of *univariate calculus*, that is single-variable calculus, to be able to understand the arguments presented here.

12.1 PARTIAL DERIVATIVES AND THEIR USES

Recall that we can interpret a univariate function $g(x_1)$ as defining the height of a curve above the x_1-axis, as illustrated in Figure 12.1(a). The derivative $g'(x_1)$ then simply results in a function that gives the local gradient in the direction of increasing x_1 at each x_1. In Chapter 3, we discussed that

$$\frac{\mathrm{d}g}{\mathrm{d}x_1} = \lim_{\delta \to 0} \frac{g(x_1 + \delta) - g(x_1)}{\delta}$$

for $\delta \in \mathbb{R}$. The task now is to generalize this expression to the "derivative" of multivariate functions, beginning the *bivariate function* $f(x_1, x_2)$.

Extending the analogy that a univariate function represents the height above the x_1-axis, we might interpret $f(x_1, x_2)$ as defining the height of a *surface* above the x_1-x_2 plane. This is illustrated in Figure 12.1(b). However, at this stage, it is not clear what is

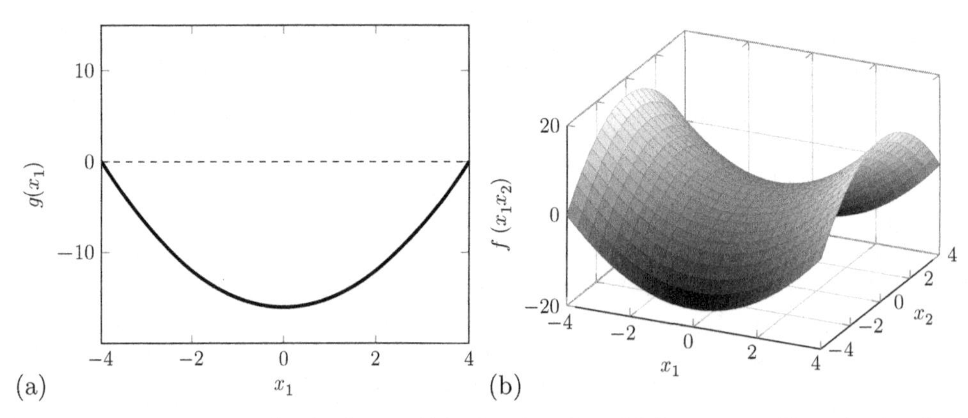

Figure 12.1 Interpretation of univariate and bivariate functions. (a) Curve created by univariate function $g(x_1) = x_1^2 - (-4)^2$. (b) Surface created by bivariate function $f(x_1, x_2) = x_1^2 - x_2^2$.

meant by the "derivative" of $f(x_1, x_2)$ because we have two independent directions that the gradient could be considered in. One gradient could be taken along the direction of increasing x_1 for some fixed x_2, and another could be taken along the direction of increasing x_2 for some fixed x_1; in general, these two derivatives would be different. In fact, these interpretations are the essence of *partial derivatives*.

We define the *partial derivative* of $f(x_1, x_2)$ with respect to x_1 as the derivative with respect to x_1 of the univariate curve formed by a "slice" through the surface at some *fixed* x_2. An example of such a slice is shown in Figure 12.1 where it should be clear that $g(x_1) \equiv f(x_1, -4)$. A further example, $f(x_1, -2)$, is shown by the bold line in Figure 12.2(a). In general, we would not specify a particular fixed value of x_2 and would report the partial derivative as a function of both x_1 and x_2. The partial derivative of the bivariate function $f(x_1, x_2)$ with respect to x_1 is then defined as

$$\frac{\partial}{\partial x_1} f(x_1, x_2) = \lim_{\delta \to 0} \frac{f(x_1 + \delta, x_2) - f(x_1, x_2)}{\delta}$$

Similarly, the partial derivative of $f(x_1, x_2)$ with respect to x_2 is the derivative with respect to x_2 of the curve formed from a slice through the surface at some fixed x_1. See, for example, Figure 12.2(b) which shows $f(-2, x_2)$ by a bold line. Mathematically, this is defined as

$$\frac{\partial}{\partial x_2} f(x_1, x_2) = \lim_{\delta \to 0} \frac{f(x_1, x_2 + \delta) - f(x_1, x_2)}{\delta}$$

Note that in each case the symbol ∂ replaces d to indicate that it is a partial derivative of a multivariate function.

Although the interpretation of multivariate functions in terms of "heights" of n-dimensional *hypersurfaces* is impossible to picture, the mathematical definition of a partial derivative can be easily generalized to functions of n variables. In particular, we understand the partial derivative of $f(x_1, x_2, \ldots, x_n)$ with respect to x_i as

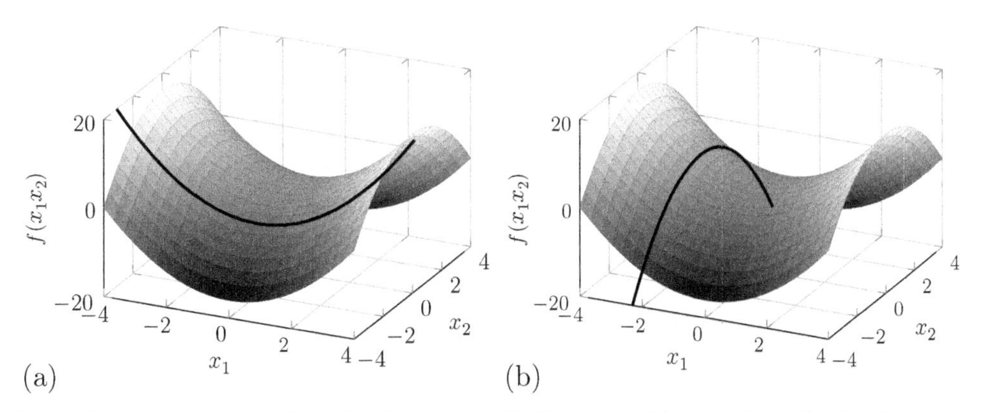

Figure 12.2 Interpretation of partial derivatives. (a) Variation of $f(x_1, x_2)$ in x_1 for fixed $x_2 = -2$. (b) Variation of $f(x_1, x_2)$ in x_2 for fixed $x_1 = -2$.

$$\frac{\partial}{\partial x_i} f(x_1, x_2, \ldots, x_n) = \lim_{\delta \to 0} \frac{f(x_1, x_2, \ldots, x_i + \delta, \ldots, x_n) - f(x_1, x_2, \ldots, x_n)}{\delta} \tag{12.1}$$

In practice, no new mathematics is involved in performing a partial derivative. As we have seen from the previous discussion, all other independent variables are treated as constants and we proceed exactly as we would for the derivative of a univariate function. For example, consider the bivariate function

$$f(x_1, x_2) = 4x_1 x_2^2 + 2$$

It should be clear that two different first-order partial derivatives are possible

$$\frac{\partial f}{\partial x_1} = \frac{\partial}{\partial x_1}(4x_1 x_2^2 + 2) = 4x_2^2$$

and

$$\frac{\partial f}{\partial x_2} = \frac{\partial}{\partial x_2}(4x_1 x_2^2 + 2) = 8x_1 x_2$$

A number of different but equivalent notations exist for the partial derivative of $f(x_1, \ldots, x_n)$ with respect to independent variable x_i. The common notations are

$$\frac{\partial f}{\partial x_i} \equiv \frac{\partial}{\partial x_i} f \equiv \partial_{x_i} f \equiv f_{x_i}$$

and these will be used interchangeably throughout this chapter. The crucial requirement of any notation for a partial derivative is that it makes clear the particular variable with which the partial derivative is being taken. For example, the notation for the derivative of a univariate function f' has no meaning if f is multivariate. On a related note, it should be clear from a comparison of Eqs. (12.1) and (3.4) that the partial derivative of a univariate function is simply the derivative of that function. That is,

$$\frac{\partial}{\partial x} f(x) \equiv \frac{\mathrm{d}}{\mathrm{d}x} f(x)$$

However, despite their equivalence, it is typical practice to reserve the use of ∂ for the derivatives of multivariate functions only.

EXAMPLE 12.1

You are given that

$$f(x_1, x_2) = \frac{x_1 \sin x_2}{x_1 + x_2} \quad \text{and} \quad g(x, y, z) = x^3 y^2 z \ln x$$

Obtain the following partial derivatives.

a. $\frac{\partial f}{\partial x_1}$

b. $\partial_{x_2} f$

c. g_x

d. $\partial_y g$

e. $\frac{\partial}{\partial z} g$

Solution

After some algebraic manipulation we obtain the following.

a. $\frac{\partial f}{\partial x_1} = \frac{x_2 \sin x_2}{(x_1+x_2)^2}$

b. $\partial_{x_2} f = \frac{x_1((x_1+x_2)\cos x_1 - \sin x_2)}{(x_1+x_2)^2}$

c. $g_x = x^2 y^2 z(3 \ln x + 1)$

d. $\partial_y g = 2x^3 yz \ln x$

e. $\frac{\partial}{\partial z} g = x^3 y^2 \ln x$

As demonstrated by Example 12.1, the partial derivative of a multivariate function, in general, results in another multivariate function. This is entirely analogous to the observation in Chapter 3 that the derivative of univariate function results in another function of that variable. With this is in mind it is easy to consider the concept of *higher-order partial derivatives*. For example, we can easily define the second-order partial derivative of $f(x_1, x_2)$ with respect to x_1 as the partial derivative of f_{x_1} with respect to x_1,

$$\frac{\partial^2 f}{\partial x_1^2} = \frac{\partial}{\partial x_1}\left(\frac{\partial f}{\partial x_1}\right)$$

The notation for such second-order and indeed every higher-order partial derivatives follows as one might expect. For example, the second partial derivative of $f(x_1, x_2)$ with respect to x_1 can be denoted as,

$$\frac{\partial^2 f}{\partial x_i^2} \equiv \frac{\partial^2}{\partial x_i^2} f \equiv \partial_{x_i x_i} f \equiv f_{x_i x_i}$$

An interesting feature of multivariate functions is that, in general, it is possible to take the partial derivative of f_{x_i} with respect to x_j for $i \neq j$. That is, one can obtain *mixed* partial derivatives at higher order; note that these are sometimes also referred to *cross* partial derivatives. Mixed partial derivative are denoted in the obvious way. For example, at second order the following notations should be understood as being equivalent,

$$\frac{\partial}{\partial x_j}\left(\frac{\partial f}{\partial x_i}\right) \equiv \frac{\partial^2 f}{\partial x_i \partial x_j} \equiv \partial_{x_i x_j} f = f_{x_i x_j}$$

for all i, j. In the particular case that $i = j$, we obtain the second-order partial derivative with respect to x_i,

$$\frac{\partial^2}{\partial x_i \partial x_i} \equiv \frac{\partial^2 f}{\partial x_i^2}$$

As you might expect, mixed partial derivatives at yet higher order can be obtained and the following notation is again understood to be equivalent,

$$\frac{\partial}{\partial x_k}\left(\frac{\partial}{\partial x_j}\left(\frac{\partial f}{\partial x_i}\right)\right) \equiv \frac{\partial^3 f}{\partial x_i \partial x_j \partial x_k} \equiv \partial_{x_i x_j x_k} f \equiv f_{x_i x_j x_k}$$

for all i, j, k.

If the higher-order derivatives of a multivariate function are continuous with respect to all independent variables, the order of the mixed derivatives does not matter.[1] For example,

$$\frac{\partial^2 f}{\partial x_1 \partial x_2} \equiv \frac{\partial^2 f}{\partial x_2 \partial x_1}$$

and

$$\frac{\partial^3 f}{\partial x_1 \partial x_2 \partial x_3} \equiv \frac{\partial^3 f}{\partial x_1 \partial x_3 \partial x_2} \equiv \frac{\partial^3 f}{\partial x_3 \partial x_1 \partial x_2} \equiv \frac{\partial^3 f}{\partial x_3 \partial x_2 \partial x_1} \equiv \frac{\partial^3 f}{\partial x_2 \partial x_1 \partial x_3} \equiv \frac{\partial^3 f}{\partial x_2 \partial x_3 \partial x_1}$$

This is demonstrated in the following example.

EXAMPLE 12.2

You are given that $g(x_1, x_2, x_3, x_4) = x_1^2 x_2^4 \sin x_3 + x_3^5 \ln x_4$. Obtain the following mixed partial derivatives.

a. $g_{x_1 x_2}$
b. $g_{x_2 x_1}$
c. $g_{x_1 x_2 x_3}$
d. $g_{x_3 x_1 x_2}$
e. $g_{x_1 x_4 x_1 x_4}$
f. $g_{x_1 x_4 x_4 x_1}$

Solution

a. $g_{x_1 x_2} = \frac{\partial}{\partial x_2}(2x_1 x_2^4 \sin x_3) = 8x_1 x_2^3 \sin x_3$
b. $g_{x_2 x_1} = \frac{\partial}{\partial x_1}(4x_1^4 y^3 \sin x_3) = 8x_1 x_2^3 \sin x_3$
c. $g_{x_1 x_2 x_3} = \frac{\partial^2}{\partial x_2 \partial x_3}(2x_1 x_2^4 \sin x_3) = \frac{\partial}{\partial x_3}(8x_1 x_2^3 \sin x_3) = 8x_1 x_2^3 \cos x_3$
d. $g_{x_3 x_1 x_2} = \frac{\partial^2}{\partial x_1 \partial x_2}(x_1^2 x_2^4 \cos x_3 + 5x_3^4 \ln x_4) = \frac{\partial}{\partial x_2}(2x_1 x_2^4 \cos x_3) = 8x_1 x_2^3 \cos x_3$
e. $g_{x_1 x_4 x_1 x_4} = \frac{\partial^3}{\partial x_4 \partial x_1 \partial x_4}(2x_1 x_2^4 \sin x_3) = \frac{\partial^2}{\partial x_1 \partial x_4} 0 = 0$
f. $g_{x_1 x_4 x_4 x_1} = \frac{\partial^3}{\partial x_4 \partial x_4 \partial x_1}(2x_1 x_2^4 \sin x_3) = \frac{\partial^2}{\partial x_4 \partial x_1} 0 = 0$

The interpretation of higher-order derivatives of univariate functions is straightforward. For example, $\frac{d^2 g}{dx_1^2}$ is the simply the rate of change of g_{x_1} in the direction of increasing x_1; $g_{x_1 x_1} > 0$ indicates that the curve is concave upwards and $g_{x_1 x_1} < 0$ indicates that the curve is concave downwards. However, things are less clear for partial

[1] The rigorous definition of continuity in the context of multivariate functions is beyond the scope of this book. However, you may assume that the definition of continuity given in Chapter 3 for univariate functions can be extended to each independent variable x_i.

derivatives and indeed close to impossible to interpret for functions of many variables. In this case of the bivariate function $f(x_1, x_2)$, $f_{x_1x_2}$ can be interpreted as the rate of change of the gradient in the x_1-direction for increasing x_2. Also, given that $f_{x_1x_2} \equiv f_{x_2x_1}$, $f_{x_1x_2}$, can also be interpreted as the rate of change of the gradient in the x_2-direction for increasing x_1.

These considerations naturally lead to the analogy of critical points in bivariate functions which we consider in the next section.

12.2 CRITICAL POINTS OF BIVARIATE FUNCTIONS

Recall from Chapter 3 that the critical (or turning) points of a univariate function are located by finding the roots of $f'(x_1)$. These are then classified from either the sign of $f''(x_1)$ or by testing its behavior either side of the critical point. One might therefore expect that these ideas can be extended to bivariate functions of the form $f(x_1, x_2)$. However, as we shall see, things are more subtle. Figure 12.3 shows the possible critical

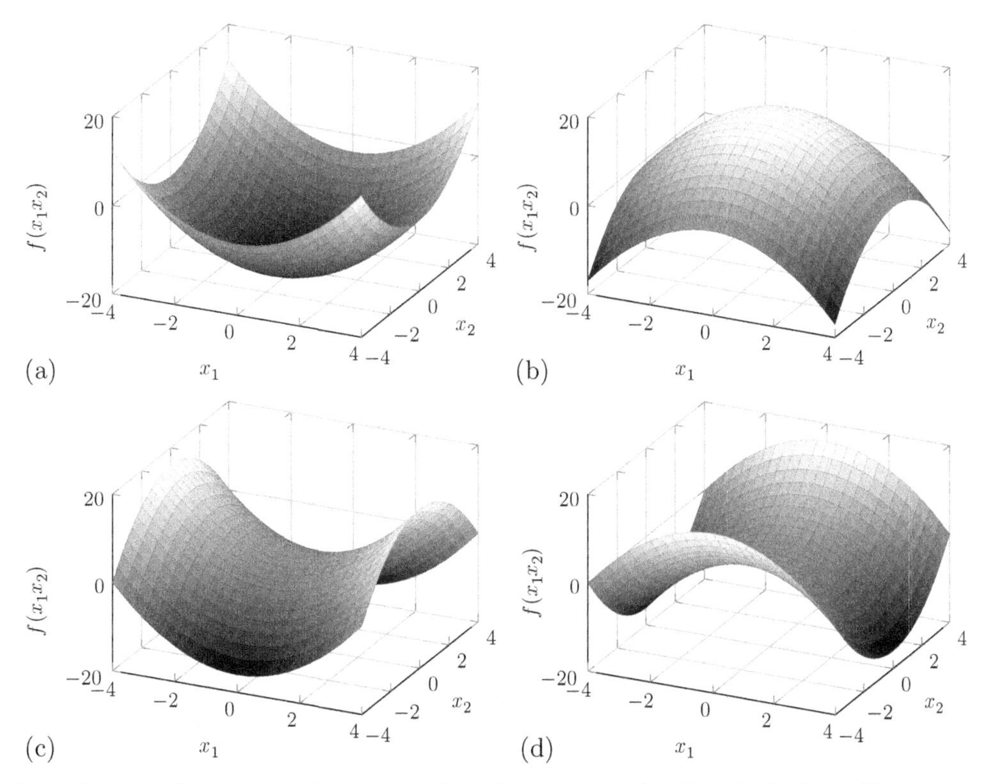

Figure 12.3 Local maxima, minima, and saddle point in bivariate functions. (a) Surface with minimum at (0, 0) created by $f(x_1, x_2) = -20 + x_1^2 + x_2^2$. (b) Surface with maximum at (0, 0) created by $f(x_1, x_2) = 15 - x_1^2 - x_2^2$. (c) Surface with saddle point at (0, 0) created by $f(x_1, x_2) = x_1^2 - x_2^2$. (d) Surface with saddle point at (0, 0) created by $f(x_1, x_2) = -x_1^2 + x_2^2$.

points in three dimensions, namely maxima, minima, and *saddle points*. Note that maxima and minima are collectively known as *extrema* for obvious reasons.

It should be clear that the minimum at $(0, 0)$ in Figure 12.3(a) is a minimum in both the x_1- and x_2-directions. Using our intuition from univariate calculus we might expect that the two curves formed from $f(x_1, 0)$ and $f(0, x_2)$ both have a minimum at $(0, 0)$, that is

$$\left.\frac{\partial f}{\partial x_1}\right|_{(0,0)} = 0 \quad \text{with} \quad \left.\frac{\partial^2 f}{\partial x_1^2}\right|_{(0,0)} > 0$$

and

$$\left.\frac{\partial f}{\partial x_2}\right|_{(0,0)} = 0 \quad \text{with} \quad \left.\frac{\partial^2 f}{\partial x_2^2}\right|_{(0,0)} > 0$$

Indeed, these can be confirmed from the functional form of the particular surface $f(x_1, x_2) = -20 + x_1^2 + x_2^2$. However, these conditions are not the full story. For the point $(0, 0)$ to be a minimum, we also require the product $f_{x_1x_1} f_{x_2x_2}$ to be greater than the product of the mixed partial derivatives $f_{x_1x_2} f_{x_2x_1} \equiv f_{x_1x_2}^2$. This is a technical consideration and the justification is beyond the scope of this book.

In general, the necessary conditions for a local minimum in a bivariate function at, say, (a, b) are

$$f_{x_1}(a, b) = f_{x_2}(a, b) = 0, \quad f_{x_1x_1}(a, b) > 0 \quad \text{and} \quad f_{x_1x_1}(a, b) f_{x_2x_2}(a, b) > f_{x_1x_2}^2(a, b) \tag{12.2}$$

Similar reasoning leads to the necessary conditions for a local maximum at (a, b) to be

$$f_{x_1}(a, b) = f_{x_2}(a, b) = 0, \quad f_{x_1x_1}(a, b) < 0 \quad \text{and} \quad f_{x_1x_1}(a, b) f_{x_2x_2}(a, b) > f_{x_1x_2}^2(a, b) \tag{12.3}$$

In both cases, the condition that $f_{x_1x_1}(a, b)$ and $f_{x_2x_2}(a, b)$ are of the *same* sign is contained in the final condition on the mixed derivatives.

The saddle point shown in Figure 12.3(c) consists of a minimum in the x_1-direction and a maximum in the x_2-direction and resembles a horse saddle. Note that a saddle with the reverse properties (maximum in x_1 and minimum in x_2) is also possible, as shown in Figure 12.3(d). The condition on both first partial derivatives is therefore the same as for maxima and minima; that is, there is a turning point in both directions. The condition on the second-order and mixed partial derivatives are, however, different. In fact, we classify the point (a, b) to be a saddle point if

$$f_{x_1}(a, b) = f_{x_2}(a, b) = 0 \quad \text{and} \quad f_{x_1x_1}(a, b) f_{x_2x_2}(a, b) < f_{x_1x_2}^2(a, b) \tag{12.4}$$

Note that $f_{x_1x_1}(a, b)$ and $f_{x_2x_2}(a, b)$ will have *different* signs at a saddle point.

The second derivative tests fail for extreme and saddle points if the analysis leads to $f_{x_1x_1}(a,b)f_{x_2x_2}(a,b) = f^2_{x_1x_2}(a,b)$. In this case, the test is deemed to have given no information and one should consider using other means to classify the behavior of the surface. For example, a computational plot of the surface could be generated.

EXAMPLE 12.3

Confirm the location of the three critical points illustrated in Figure 12.3 and use the second partial derivative test to classify them.

Solution

a. The surface is defined by $f(x_1,x_2) = -20 + x_1^2 + x_2^2$. The location of the critical point is (a,b) such that $f_{x_1} = f_{x_2} = 0$ at that point. We note that

$$f_{x_1} = 2x_1 \quad \text{and} \quad f_{x_2} = 2x_2$$

and so $x_1 = 0$ and $x_2 = 0$ at the critical point, that is $(a,b) = (0,0)$. The second derivatives here are $f_{x_1x_1} = 2 = f_{x_2x_2} > 0$ and so the critical point is potentially a minimum. Finally, we have $f_{x_1x_2} = 0$ and so $f_{x_1x_1}f_{x_2x_2} > f^2_{x_1x_2}$, therefore $(0,0)$ is a minimum by Eq. (12.2).

b. The surface is defined by $f(x_1,x_2) = 15 - x_1^2 - x_2^2$. We note that

$$f_{x_1} = -2x_1 \quad \text{and} \quad f_{x_2} = -2x_2$$

and so $x_1 = 0$ and $x_2 = 0$ at the critical point; that is, $(a,b) = (0,0)$. The second derivatives are $f_{x_1x_1} = -2 = f_{x_2x_2} < 0$ and so the critical point is potentially a maximum. Finally, we have $f_{x_1x_2} = 0$ and so $f_{x_1x_1}f_{x_2x_2} > f^2_{x_1x_2}$, therefore $(0,0)$ is a maximum by Eq. (12.3).

c. The surface is defined by $f(x_1,x_2) = x_1^2 - x_2^2$. We note that

$$f_{x_1} = 2x_1 \quad \text{and} \quad f_{x_2} = -2x_2$$

and so $x_1 = 0$ and $x_2 = 0$ at the critical point; that is, $(a,b) = (0,0)$. The second derivatives are $f_{x_1x_1} = 2$ and $f_{x_2x_2} = -2$ and so the critical point is potentially a saddle. Finally, we have $f_{x_1x_2} = 0$ and so $f_{x_1x_1}f_{x_2x_2} < f^2_{x_1x_2}$, therefore $(0,0)$ is a saddle point by Eq. (12.4).

d. The surface is defined by $f(x_1,x_2) = -x_1^2 + x_2^2$. We note that

$$f_{x_1} = -2x_1 \quad \text{and} \quad f_{x_2} = 2x_2$$

and so $x_1 = 0$ and $x_2 = 0$ at the critical point; that is, $(a,b) = (0,0)$. The second derivatives are $f_{x_1x_1} = -2$ and $f_{x_2x_2} = 2$ and so the critical point is potentially a saddle. Finally, we have $f_{x_1x_2} = 0$ and so $f_{x_1x_1}f_{x_2x_2} < f^2_{x_1x_2}$, therefore $(0,0)$ is a saddle point by Eq. (12.4).

EXAMPLE 12.4

Find and classify any critical points of the bivariate function $g(y, z) = 5y^2 - 8y - 2yz - 6z + 4z^2$.

Solution

We begin by finding the critical points by solving the simultaneous equations resulting from both first-order partial derivatives,

$$g_y = 10y - 8 - 2z = 0 \quad g_z = -2y - 6 + 8z = 0 \quad \Rightarrow (a, b) = (1, 1)$$

The second partial derivatives at this location are evaluated as

$$g_{yy} = 10 \quad g_{zz} = 8 \quad g_{yz} = -2$$

We see that $g_{yy} > 0$ and $g_{yy}g_{zz} > g_{yz}^2$, and Eq. (12.2) indicates that we have a minimum at $(1, 1)$. This is confirmed in Figure 12.4.

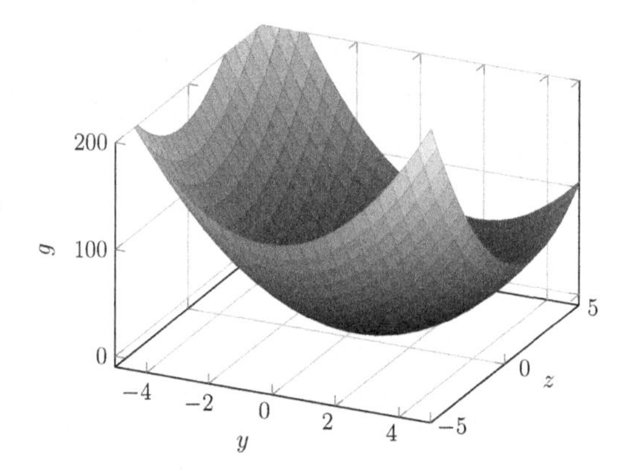

Figure 12.4 Figure showing the critical point of $g(y, z) = 5y^2 - 8y - 2yz - 6z + 4z^2$ at $(y, z) = (0, 0)$ for Example 12.4.

EXAMPLE 12.5

Find and classify any critical points of the bivariate function $h(x, y) = 2x^2 - 4xy + y^4 + 2$.

Solution

We begin by finding the critical points by solving the simultaneous equations resulting from both first-order partial derivatives,

$$h_x = 4x - 4y = 0 \quad h_y = -4x + 4y^3 = 0$$

We therefore see that $x = y$ and $4x = 4y^3$ and so $x = 0, \pm 1$. Potential critical points therefore exist at $(a, b) = (0, 0), (1, 1), (-1, -1)$. The second partial derivatives are

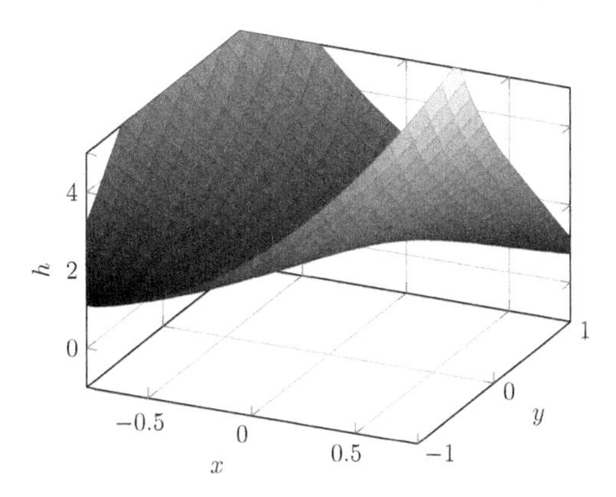

Figure 12.5 Figure showing the critical points of $h(x, y) = 2x^2 - 4xy + y^4 + 2$ for Example 12.5.

$$h_{xx} = 4 \quad h_{yy} = 12y^2 \quad h_{xy} = -4$$

and Eqs. (12.2)–(12.4) indicate that we have a saddle at $(0, 0)$ and minima at $\pm(1, 1)$. This is illustrated in Figure 12.5.

EXAMPLE 12.6

Find and classify any critical points of the bivariate function $k(x_1, x_2) = x_1^3 + x_2^3 - 2x_1 + 6x_2$.

Solution

We begin by finding the critical points by attempting to solve the simultaneous equations resulting from both first-order partial derivatives,

$$k_{x_1} = 3x_1^2 - 2 = 0 \quad k_{x_2} = 3x_2^2 + 6 = 0$$

The expression for k_{x_2} has no real solution and so no critical points exist.

12.3 THE METHOD OF LAGRANGE MULTIPLIERS

A very useful application of partial derivatives is in *constrained optimization*; this is a crucial concept in economics and finance. For example, if $f(x_1, x_2)$ gives the minimum return available from an investment of \$$x_1$ and \$$x_2$ in Asset 1 and Asset 2, respectively, we might be interested in maximizing this minimum return subject to a financial constraint on the total we have to invest. That is, we might want to maximize $f(x_1, x_2)$ subject to the constraint that $x_1 + x_2 = c$ for some constant c. In this particular example, we might simply solve the constraint for x_1 and maximize the *univariate* function $f(x_1, c - x_1)$. But what if we had a more complicated implicit expression for the constraint that we could not solve in practice?

In general, we can use the method of *Lagrange multipliers* to find the extreme value (maximum or minimum) of a general multivariate function $f(x_1, \ldots, x_n)$ subject to a number of constraints, each expressed in the form $g(x_1, \ldots, x_n) = c$. The method incorporates $f(x_1, \ldots, x_n)$ and each $g(x_1, \ldots, x_n) - c$ into a *Lagrange function*, L, in such a way that the extreme value of $f(x_1, \ldots, x_n)$ is obtained only when the constraints are satisfied.

We begin by restricting the discussion to the optimization of bivariate functions subject to a single constraint and return to more complicated examples at the end of this section. As we have previously discussed, the bivariate function $f(x_1, x_2)$ can be considered as defining a surface above the x_1-x_2 plane. Any unconstrained extrema of this surface can be found easily and classified using the second partial derivative test discussed in the previous section. However, imposing the constraint $g(x_1, x_2) = c$ effectively limits the domain on the x_1-x_2 plane and results in a particular subset of possible points on the surface of $f(x_1, x_2)$. The constrained extreme points are then the maxima or minima from within this subset of points.

By way of illustration, Figure 12.6(a) shows the unconstrained maximum of the surface defined by $f(x_1, x_2) = 20 - x_1^2 + x_2^2$ over the entire x_1-x_2 plane. Figures 12.6(b)-(d) then illustrate the interpretation of finding the maximum under various constraints. For example, Figure 12.6(c) shows the effect of imposing the constraint $g(x_1, x_2) - c \equiv x_1 + x_2 + 2$, that is $x_1 + x_2 = -2$. As we see, each constraint restricts the domain to the values of x_1 and x_2 along a straight line (indicated with the dashed line) and results in the subset of points on the surface directly above this line (indicated with the solid line). A constrained maximum is then a maximum value on this solid line.

Equivalent plots for various constrained minima of $f(x_1, x_2) = x_1^2 + x_2^2$ are shown in Figure 12.7. Note that the constraint $g(x_1, x_2) - c \equiv x_1 - x_2$, shown in Figures 12.6(d) and 12.7(d), lead to a constrained maximum and minimum that coincides with the unconstrained value; the other examples do not.

The "straight-line" constraints shown in both figures are particularly simple and we will see more complicated examples in what follows.

The Lagrange function for a general bivariate function $f(x_1, x_2)$ with a single constraint $g(x_1, x_2) = c$ is defined to be

$$L(x_1, x_2, \lambda) = f(x_1, x_2) - \lambda \times (g(x_1, x_2) - c)$$

Note that we have introduced a new independent variable λ and so the Lagrange function for a bivariate function with a single constraint has *three* variables. As discussed previously, the location of an *unconstrained* critical point of $f(x_1, x_2)$, (a, b), can be obtained from simultaneously solving

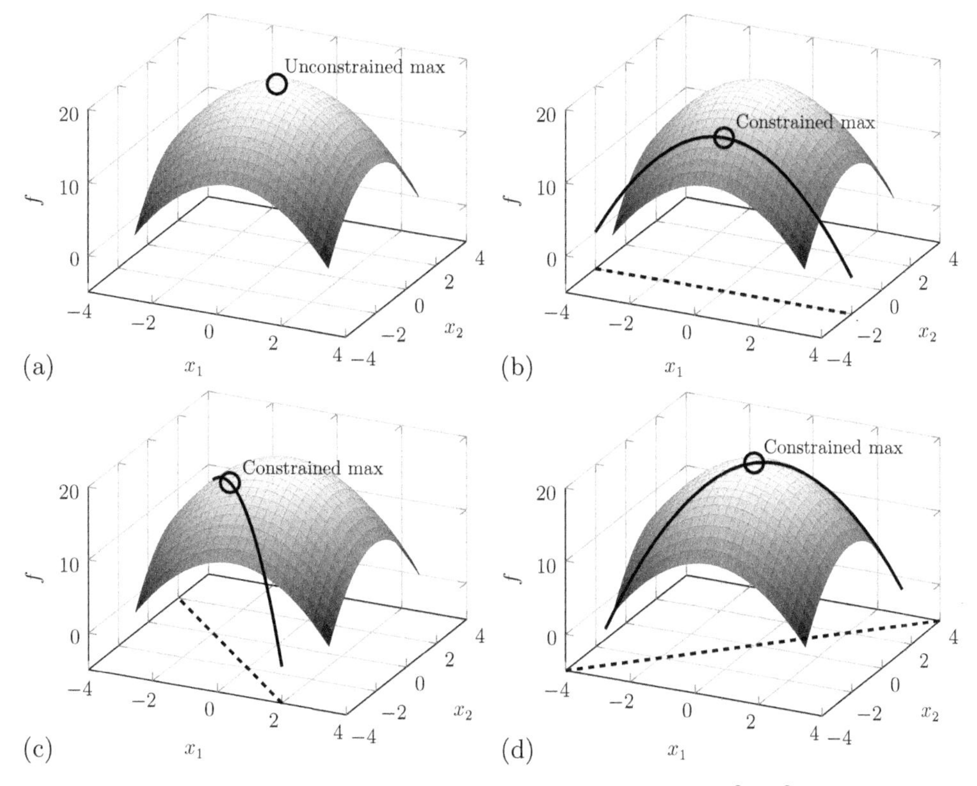

Figure 12.6 Constrained and unconstrained maxima of $f(x_1, x_2) = 20 - x_1^2 - x_2^2$. (a) Unconstrained maximum. (b) Constrained maximum subject to $g(x_1, x_2) - c = x_2 + 2$. (c) Constrained maximum subject to $g(x_1, x_2) - c = x_1 + x_2 + 2$. (d) Constrained maximum subject to $g(x_1, x_2) - c = x_1 - x_2$.

$$\frac{\partial f}{\partial x_1} = 0 \quad \text{and} \quad \frac{\partial f}{\partial x_2} = 0$$

The Lagrange multiplier method extends this requirement to finding the critical points of L, $(x_1 x_2, \lambda) = (a_0, b_0, \lambda_0)$, from the simultaneous solution of

$$\frac{\partial L}{\partial x_1} = f_{x_1} - \lambda g_{x_1} = 0, \quad \frac{\partial L}{\partial x_2} = f_{x_2} - \lambda g_{x_2} = 0, \quad \text{and} \quad \frac{\partial L}{\partial \lambda} = g(x_1, x_2) - c = 0 \tag{12.5}$$

Since the constraint is expected to hold at all points, including each extreme point, the additional term in the Lagrange function, $\lambda \times (g(x_1, x_2) - c)$, is zero and $L(x_1, x_2, \lambda) = f(x_1, x_2)$. Each resulting (a_0, b_0) is then an extreme point of $f(x_1, x_2)$ subject to $g(x_1, x_2) = c$, as required.

You will note that the Lagrange multiplier method as described here does not consider the second partial derivatives of either L or f and so gives no information about the type of constrained extrema found (either maxima or minima). The method

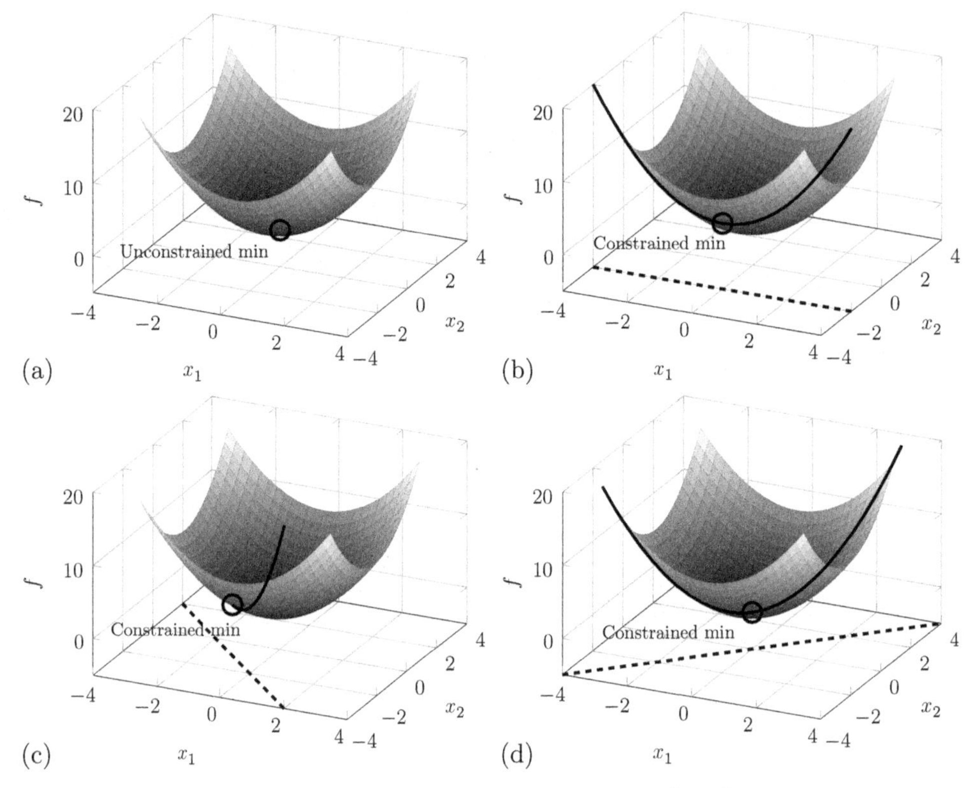

Figure 12.7 Constrained and unconstrained minima of $f(x_1, x_2) = x_1^2 + x_2^2$. (a) Unconstrained minimum. (b) Constrained minimum subject to $g(x_1, x_2) - c = x_2 + 2$. (c) Constrained minimum subject to $g(x_1, x_2) - c = x_1 + x_2 + 2$. (d) Constrained minimum of subject to $g(x_1, x_2) - c = x_1 - x_2$.

therefore requires the resulting extreme points to be classified manually. In particular, one should evaluate the original function $f(x_1, x_2)$ at each constrained extreme point as demonstrated in the following examples.

EXAMPLE 12.7

Maximize $x_1 + x_2$ subject to the constraint $x_1^2 + x_2^2 = 4$. Give a geometrical interpretation of this problem.

Solution

This can be solved with the Lagrange multiplier method. We have

$$f(x_1, x_2) = x_1 + x_2 \quad \text{and} \quad g(x_1, x_2) - c \equiv x_1^2 + x_2^2 - 4$$

and so form the Lagrange function

$$L(x_1, x_2, \lambda) = x_1 + x_2 + \lambda(x_1^2 + x_2^2 - 4)$$

We then simultaneously solve

$$L_{x_1} = 1 + 2\lambda x_1 = 0 \quad L_{x_2} = 1 + 2\lambda x_2 = 0 \quad L_\lambda = x_1^2 + x_2^2 - 4 = 0$$

that is, $\lambda = -\frac{1}{2x_1} = -\frac{1}{2x_2} \Rightarrow x_1 = x_2$ and so $x_1 = x_2 = \pm\sqrt{2}$. Evaluating $f(x_1, x_2)$ at these points shows that $f(\sqrt{2}, \sqrt{2}) > f(-\sqrt{2}, -\sqrt{2})$ and so $(-\sqrt{2}, -\sqrt{2})$ is a constrained minimum of $f(x_1, x_2)$ and $(\sqrt{2}, \sqrt{2})$ is the required constrained maximum.

The function $f = x_1 + x_2$ defines an upwardly sloping plane and the constraint $x_1^2 + x_2^2 = 4$ forms a circle of radius 2 centered at the original in the x_1-x_2 plane. We have therefore obtained the maximum value of the points sitting on the plane directly above the circle.

EXAMPLE 12.8

Working in units of \$1000, the minimum return from a portfolio consisting of an investment of x_1 and x_2 in Asset 1 and Asset 2, respectively, is thought to be given by

$$f(x_1, x_2) = 65 - 1.5x_1^2 - x_2^2$$

You decide to invest \$10,000 in the portfolio. Determine the optimal holding of each asset.

Solution

This is a constrained optimization problem that requires us to maximize $f(x_1, x_2)$ subject to $x_1 + x_2 = 10$. The Lagrange function is

$$L(x_1, x_2, \lambda) = 65 - 1.5x_1^2 - x_2^2 + \lambda(x_1 + x_2 - 10)$$

and, from Eq. (12.5), the critical points of L are obtained from simultaneously solving

$$L_{x_1} = -3x_1 + \lambda = 0 \quad L_{x_2} = -2x_2 + \lambda = 0 \quad L_\lambda = x_1 + x_2 - 10 = 0$$

This leads to $\lambda = 3x_1 = 2x_2$ and $x_1 + \frac{3}{2}x_1 = 10$, therefore $x_1 = 4$ and $x_2 = 6$. The shape of the surface defined by $f(x_1, x_2)$ is such that this extreme point is the required maximum. We therefore invest \$4000 in Asset 1 and \$6000 in Asset 2 and expect to receive a minimum return of $f(4, 6)$ which is \$5000.

The reader is invited to confirm that any other values of x_1 and x_2 that satisfy $x_1 + x_2 = 10$ give a lower value of f.

The method of Lagrange multipliers can be extended to finding extreme points of multivariate functions of n variables subject to $m < n$ constraints. In principle, no change to the method described is required in the optimization of an n-dimensional problem with a single constraint. Indeed, the Lagrange function is simply modified to

$$L(x_1, \ldots, x_n, \lambda) = f(x_1, \ldots, x_n) + \lambda \times (g(x_1, \ldots, x_n) - c)$$

and we find the values of $(x_1, \ldots, x_n, \lambda)$ that simultaneously solve the $n + 1$ equations

$$\{L_{x_i} = 0\} \quad \text{and} \quad L_\lambda = 0$$

for all $i = 1, \ldots, n$.

EXAMPLE 12.9

Determine and classify the extreme values of $x_1 + 3x_2 + 2x_3$ subject to the constraint $x_1^2 + x_2^2 + x_3^2 = 9$.

Solution

This is an optimization problem with three variables and one constraint. We have $f(x_1, x_2, x_3) = x_1 + 3x_2 + 2x_3$ and $g(x_1, x_2, x_3) - c \equiv x_1^2 + x_2^2 + x_3^2 - 9$ and so the Lagrange function is

$$L(x_1, x_2, x_3, \lambda) = x_1 + 3x_2 + 2x_3 + \lambda \times (x_1^2 + x_2^2 + x_3^2 - 9)$$

The conditions are therefore

$$L_{x_1} = 1 + 2\lambda x_1 = 0 \quad L_{x_2} = 3 + 2\lambda x_2 = 0 \quad L_{x_3} = 2 + 2\lambda x_3 = 0$$
$$L_\lambda = x_1^2 + x_2^2 + x_3^2 - 9 = 0$$

The first three conditions lead to $\lambda = -\frac{1}{2x_1} = -\frac{3}{2x_2} = -\frac{1}{x_3}$ and so $\lambda = \pm\frac{1}{3}\sqrt{\frac{7}{2}}$. These lead to extreme points at locations

$$\left(-\frac{3}{\sqrt{14}}, -\frac{9}{\sqrt{14}}, -3\sqrt{\frac{2}{7}}\right) \quad \text{and} \quad \left(\frac{3}{\sqrt{14}}, \frac{9}{\sqrt{14}}, 3\sqrt{\frac{2}{7}}\right)$$

Substituting these points back into $f(x_1, x_2, x_3)$ shows that these represent a minimum and a maximum, respectively.

EXAMPLE 12.10

Find the shortest distance between the origin and the plane defined by $x_1 + 2x_2 - 2x_3 = 9$.

Solution

While this does not at first appear to be a problem amenable to the method of Lagrange multipliers, it can be rewritten as an optimization problem. In particular, we are required to minimize $f(x_1, x_2, x_3)$ subject to $g(x_1, x_2, x_3) - c \equiv x_1 + 2x_2 - 2x_3 - 9$ where f is some measure of the distance from the origin to a point (x_1, x_2, x_3). The algebra is greatly simplified if we take f to be the *square* of the distance, that is $f(x_1, x_2, x_3) = x_1^2 + x_2^2 + x_3^2$. We then form the Lagrange function

$$L(x_1, x_2, x_3, \lambda) = x_1^2 + x_2^2 + x_2^2 + \lambda \times (x_1 + 2x_2 - 2x_3 - 9)$$

which leads to the conditions

$$L_{x_1} = 2x_1 + \lambda = 0 \quad L_{x_2} = 2(x_2 + \lambda) = 0 \quad L_{x_3} = 2(x_3 - \lambda) = 0$$
$$L_\lambda = x_1 + 2x_2 - 2x_3 - 9 = 0$$

The first three conditions lead to $\lambda = -2x_1 = -x_2 = x_3$ and the fourth is satisfied if $x_1 = 1$, $x_2 = 2$, and $x_3 = -2$. The point $(1, 2, -2)$ is therefore the required constrained extreme point. Note that since the plane can become infinitely far from the origin, we conclude that this point must be a minimum, in fact it is a distance $\sqrt{1^2 + 2^2 + 2^2} = 3$ away.

The extension of the method to m constraints is slightly more complicated but, after some thought, follows directly from our previous discussion. To state the problem more formally, we are required to find the extreme points of $f(x_1, \ldots, x_n)$ subject to the multiple constraints $g_j(x_1, \ldots, x_n) = c_j$ for $j = 1, \ldots, m$ where $m < n$. In this most general case of the method, the Lagrange function is written as

$$L(x_1, \ldots, x_n, \lambda_1, \ldots, \lambda_m) = f(x_1, \ldots, x_n) + \sum_{j=1}^{m} \lambda_j \times (g_j(x_1, \ldots, x_n) - c_j) \tag{12.6}$$

and we attempt to find the values of $(x_1, \ldots, x_n, \lambda_1, \ldots, \lambda_m)$ that simultaneously solve

$$\{L_{x_i} = 0\} \quad \text{and} \quad \{L_{\lambda_j} = 0\}$$

for $i = 1, \ldots, n$ and $j = 1, \ldots, m$. Some thought should convince you that it is impossible to impose m constraints on a multivariate function of n variables if $m \geq n$.

EXAMPLE 12.11

Find and classify the extreme points of $f(x_1, x_2, x_3) = x_1 + 4x_2 - 2x_3$ subject to the constraints that $2x_1 - x_2 - x_3 = 2$ and $x_1^2 + x_2^2 = 4$.

Solution

This problem is amenable to the method of Lagrange multipliers with two constraints. In particular, we form the Lagrange function Eq. (12.6) as

$$L(x_1, x_2, x_3, \lambda_1, \lambda_2) = x_1 + 4x_2 - 2x_3 + \lambda_1 \times (2x_1 - x_2 - x_3 - 2) + \lambda_2 \times (x_1^2 + x_2^2 - 4)$$

The conditions that we are required to solve arise from $L_{x_1} = 0$, $L_{x_2} = 0$, $L_{x_3} = 0$, $L_{\lambda_1} = 0$, and $L_{\lambda_2} = 0$. That is,

$$1 + 2\lambda_1 + 2\lambda_2 x_1 = 0, \quad 4 - \lambda_1 + 2\lambda_2 x_2 = 0, \quad -2 - \lambda_1 = 0, \quad 2x_1 - x_2 - x_3 = 2,$$
$$x_1^2 + x_2^2 = 4$$

Some manipulation leads to the solutions $(x_1, x_2, x_3) = \left(-\frac{2}{\sqrt{5}}, \frac{4}{\sqrt{5}}, -\frac{2}{5}(5 + 4\sqrt{5})\right)$ and $\left(\frac{2}{\sqrt{5}}, -\frac{4}{\sqrt{5}}, \frac{2}{5}(-5 + 4\sqrt{5})\right)$ which, upon substitution into $f(x_1, x_2, x_3)$, are easily seen to be maximum and minimum of the constrained problem, respectively.

EXAMPLE 12.12

Locate and classify the extremes of the function $f(x, y, z) = 2xyz$ subject to the constraints $x + y + z = 2$ and $2x + 2y + z = 1$.

Solution

This is Lagrange multiplier problem for $f(x, y, z)$ with two constraints, $g_1(x, y, z) - c_1 \equiv x + y + z - 2$ and $g_2(x, y, z) - c_2 \equiv 2x + 2y + z - 1$. We form the Lagrange function as

$$L = 2xyz + \lambda_1 \times (x + y + z - 2) + \lambda_2 \times (2x + 2y + z - 1)$$

which leads to conditions

$$2yz + \lambda_1 + 2\lambda_2 = 0, \quad 2xz + \lambda_1 + 2\lambda_2 = 0, \quad 2xy + \lambda_1 + \lambda_2 = 0, \quad x + y + z = 2,$$
$$2x + 2y + z = 1$$

Some manipulation leads to the solution point $(x, y, z) = (-0.5, -0.5, 3)$ where $f(-0.5, -0.5, 3) = 1.5$. Note that the two constraints are in fact planes and their intersection, that is when they are both satisfied, will be straight line. In fact, the line is given by $2 - x - y = 1 - 2x - 2y$ or $y = -1 - x$. The function $f(x, y, z) = 2xyz$ will take very large values as x increases or decreases along this line and so we expect the point $(-0.5, -0.5, 3)$ to be a constrained minimum.

12.4 BIVARIATE INTEGRAL CALCULUS

We end this chapter with a discussion of the integral calculus of bivariate functions, $f(x, y)$. The ideas presented here can be extended to the integral calculus of multivariate functions but a detailed discussion is beyond the scope of this introductory text. Note that in what follows we use x and y to denote the two independent variables; this simplifies the discussion and is in contrast to the notation x_1 and x_2 that has been used elsewhere in this chapter.

Recall from Chapter 7 that the definite integral of a univariate function can be interpreted as the *area* under the curve defined by that function. Extending this analogy would mean that, in some sense, we should consider the integral of a bivariate function as the "area" under a surface. However, the surface extends in both the x- and y-directions and so encloses a three-dimensional *volume*. The question now is how to obtain that volume for a function $f(x, y)$ over some rectangular domain $x \in [a, b]$ and $y \in [c, d]$. The following discussion is a generalization of that given in Section 7.1 and the reader is invited to review that material before continuing. It will be assumed that the function is continuous in both the x- and y-directions throughout this domain.

Figure 12.8 illustrates a rectangular domain on the x-y plane ($z = 0$). The volume of interest, V, is then that directly above this rectangle and bounded by the surface

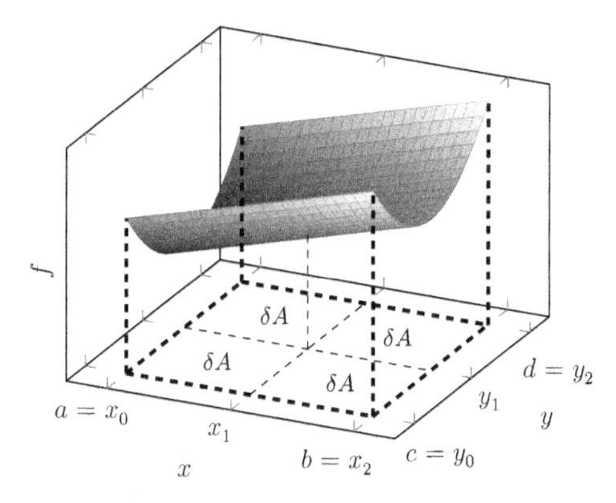

Figure 12.8 The volume under a surface defined by $f(x_1, x_2)$.

of $f(x, y)$. We begin by discretizing the x and y domains into n and m subintervals of regular width δx and δy, respectively. The discretization is such that $a = x_0 < x_1 < \cdots < x_n = b$ and $c = y_0 < y_1 < \cdots < y_m = d$, and so divides the rectangular domain into $n \times m$ smaller rectangles, each of area $\delta x \delta y = \delta A$. This is illustrated in the particular case that $n = m = 2$. The total volume V can then be considered as consisting of $n \times m$ individual volume elements, each enclosed between an area element on the x-y plane and the surface directly above. Although the area elements have been established as having identical area, the height of the surface changes with x and y and so the volume elements are not necessarily equal. If $\delta V_{i,j}$ denotes the particular volume element with a base that has its upper right corner at x_i and y_j, the total volume is given by

$$V = \sum_{j=1}^{m} \sum_{i=1}^{n} \delta V_{i,j}$$

The problem is now one of obtaining each $\delta V_{i,j}$. To do this we pick some point, (x_i^*, y_j^*), within each rectangular element which could be the midpoint, for example, and approximate the height of the surface above the element by $f(x_i^*, y_j^*)$. Each element is then approximated by $\delta V_{i,j} \approx f(x_i^*, y_j^*)\delta A$ with an error that is dependent on the local curvature of the surface above (x_i^*, y_j^*). The approximation will of course improve as $\delta x, \delta y \to 0$. The total volume is then approximated as

$$V \approx \sum_{j=1}^{m} \sum_{i=1}^{n} f(x_i^*, y_j^*)\delta A = \sum_{j=1}^{m} \sum_{i=1}^{n} f(x_i^*, y_j^*)\delta x \delta y$$

and the error is completely removed in the limit that $\delta x, \delta y \to 0$, that is, $n, m \to \infty$. We therefore have

$$V = \lim_{n,m \to \infty} \sum_{j=1}^{m} \sum_{i=1}^{n} f(x_i^*, y_j^*)\delta x \delta y$$

Comparing this to Eq. (7.3), it is sensible to use this double summation as the definition of the *double integral* over a rectangular domain $[a, b] \times [c, d]$, that is

$$V = \lim_{n,m\to\infty} \sum_{j=1}^{m}\sum_{i=1}^{n} f(x_i^*, y_j^*)\delta x \delta y = \int_{y=c}^{d}\int_{x=a}^{b} f(x, y)\, dx\, dy \tag{12.7}$$

A volume above the x-y plane ($z = 0$) is returned as positive, while a volume below the plane is negative. This is entirely analogous to the areas above and below the x-axis resulting from the definite integrals of univariate functions considered in Chapter 7.

Integrals of this type of often referred to as *iterated integrals* which reflects the approach that we take to evaluate them. To explore this further we first note that the process by which we derived Eq. (12.7) implies that the order of the summations is irrelevant. We could therefore write the double integral in two different but equivalent ways,

$$\int_{y=c}^{d}\left(\int_{x=a}^{b} f(x, y)\, dx\right) dy \equiv \int_{x=a}^{b}\left(\int_{y=c}^{d} f(x, y)\, dy\right) dx \tag{12.8}$$

The common interpretation of the left-hand side of this expression is that one should first perform the "inner" partial integral which is with respect to x while holding y constant. This would result in some function of y, $\bar{h}(y)$, which would then be integrated by the "outer" integral. That is,

$$V = \int_{y=c}^{d}\left(\int_{x=a}^{b} f(x, y)\, dx\right) dy = \int_{y=c}^{d} \bar{h}(y)\, dy$$

In contrast, one would interpret the right-hand side of Eq. (12.8) as meaning that one should first perform the inner partial integral with respect to y while holding x constant. This would result in some function of x, $\tilde{h}(x)$, which is then integrated by the outer integral,

$$V = \int_{x=a}^{b}\left(\int_{y=c}^{d} f(x, y)\, dy\right) dx = \int_{x=a}^{b} \tilde{h}(x)\, dx$$

In both cases, we are using an iterative process to evaluate the integral. The equivalence of the two approaches for smooth surfaces is demonstrated in Example 12.13.

Note that we have so far been careful to indicate whether each pair of limits is in the x- or y-direction. This is not actually necessary if we understand that the inner integral is paired with the inner distance element. That is, the integral

$$\int_{2}^{5}\int_{-1}^{2} (x^2 + y^2)\, dx\, dy$$

is interpreted as meaning

$$\int_{y=2}^{5} \int_{x=-1}^{2} (x^2 + y^2) \, dx \, dy$$

Whereas the integral

$$\int_{2}^{5} \int_{-1}^{2} (x^2 + y^2) \, dy \, dx$$

is interpreted as

$$\int_{x=2}^{5} \int_{y=-1}^{2} (x^2 + y^2) \, dy \, dx$$

EXAMPLE 12.13

Perform the following integrals in the order indicated by the brackets.

a. $I_1 = \int_0^2 \left(\int_1^3 x^2 + y^2 \, dx \right) \, dy$

b. $I_2 = \int_1^3 \left(\int_0^2 x^2 + y^2 \, dy \right) \, dx$

Solution

a. The inner integral gives

$$\int_1^3 x^2 + y^2 \, dx = \left[\frac{x^3}{3} + y^2 x \right]_{x=1}^{3} = \frac{26}{3} + 2y^2$$

and we complete the evaluation by performing the outer integral

$$\int_0^2 \frac{26}{3} + 2y^2 \, dy = \left[\frac{26}{3} y + \frac{2}{3} y^3 \right]_{y=0}^{2} = \frac{68}{3}$$

We therefore have $I_1 = \frac{68}{3}$.

b. The inner integral gives

$$\int_0^2 x^2 + y^2 d = \left[yx^2 + \frac{y^3}{3} \right]_{y=0}^{2} = 2x^2 + \frac{8}{3}$$

and we complete the evaluation by performing the outer integral

$$\int_1^3 2x^2 + \frac{8}{3} \, dy = \left[\frac{2x^3}{3} + \frac{8x}{3} \right]_{y=1}^{3} = \frac{68}{3}$$

We therefore have $I_2 = \frac{68}{3} = I_1$.

The order in which we perform the iterated integral in Example 12.13 was seen not to matter and this is always the case for well-behaved, continuous surfaces over rectangular domains. However, it may not be true for surfaces that are not continuous over the domain, or indeed continuous surfaces over nonrectangular domains. Discontinuous surfaces are beyond the scope of this book and we consider nonrectangular domains later in this section.

EXAMPLE 12.14

Evaluate the following iterative integrals.

a. $\int_1^2 \int_{-1}^2 y\,\mathrm{e}^x\,\mathrm{d}y\,\mathrm{d}x$

b. $\int_0^{\pi} \int_{-2\pi}^{2\pi} \cos(x)\sin(y)\,\mathrm{d}x\,\mathrm{d}y$

c. $\int_9^{10} \int_1^{\mathrm{e}} \left(x^2 + \frac{1}{y}\right)\,\mathrm{d}y\,\mathrm{d}x$

d. $\int_0^1 \int_{-5}^5 \left(x^2y^3 - 2y\cos x\right)\,\mathrm{d}x\,\mathrm{d}y$

Solution

Since, in each case, the integrand is clearly a continuous surface, the order in which we perform the integrals does not matter. We opt to perform each integral in the order naturally indicated by the question.

a. The inner integral is

$$\int_{-1}^{2} y\,\mathrm{e}^x\,\mathrm{d}y = \frac{3}{2}\mathrm{e}^x$$

which leads to

$$\int_1^2 \int_{-1}^2 y\,\mathrm{e}^x\,\mathrm{d}y\,\mathrm{d}x = \int_1^2 \frac{3}{2}\,\mathrm{e}^x\,\mathrm{d}x = \frac{3}{2}\left(\mathrm{e}^2 - \mathrm{e}^1\right) \approx 7.006$$

b. The inner integral is

$$\int_{-2\pi}^{2\pi} \cos(x)\sin(y)\,\mathrm{d}x = \sin y\,[\sin(2\pi) - \sin(2\pi)] = 0$$

which leads to

$$\int_0^{\pi} \int_{-2\pi}^{2\pi} \cos(x)\sin(y)\,\mathrm{d}x\,\mathrm{d}y = 0$$

c. The inner integral is

$$\int_1^{\mathrm{e}} \left(x^2 + \frac{1}{y}\right)\,\mathrm{d}y = \left[x^2y + \ln y\right]_{y=1}^{\mathrm{e}} = (\mathrm{e} - 1)x^2 + 1$$

which leads to

$$\int_9^{10} \int_1^{\mathrm{e}} \left(x^2 + \frac{1}{y}\right)\,\mathrm{d}y\,\mathrm{d}x = \int_9^{10} ((\mathrm{e} - 1)x^2 + 1)\mathrm{d}x = \frac{1}{3}\,(271\,\mathrm{e} - 268) \approx 156.218$$

d. The inner integral is

$$\int_{-5}^{5} \left(x^2y^3 - 2y\cos x\right)\,\mathrm{d}x = \left[\frac{x^3y^3}{3} - 2y\sin x\right]_{x=-5}^{5} = \frac{250y^3}{3} - 4y\sin 5$$

which leads to

$$\int_0^1 \int_{-5}^5 \left(x^2 y^3 - 2y\cos x\right)\, \mathrm{d}x\, \mathrm{d}y = \int_0^1 \frac{250y^3}{3} - 4y \sin 5\, \mathrm{d}y = \frac{125}{6} - 2\sin 5 \approx 22.751$$

The reader is invited to perform the integrals in the reverse order in each case.

Note the following useful observation for evaluating the iterative integral of a *separable* integrand over a rectangular domain,

$$\int_a^b \int_c^d f(x)g(y)\, \mathrm{d}y\, \mathrm{d}x = \int_a^b f(x)\, \mathrm{d}x \int_c^d g(y)\, \mathrm{d}y \tag{12.9}$$

That is, one can evaluate the bivariate integral of a separable integrand as the product of two univariate integrands. This approach often simplifies the evaluation of double integrals.

EXAMPLE 12.15

Where possible, apply Eq. (12.9) to the integrals in Example 12.14.

Solution

Only parts (a) and (b) are separable. In these cases, we write

$$\int_1^2 \int_{-1}^2 y\,\mathrm{e}^x\, \mathrm{d}y\, \mathrm{d}x = \left(\int_{-1}^2 y\, \mathrm{d}y\right) \times \left(\int_1^2 \mathrm{e}^x\, \mathrm{d}x\right) = \frac{3}{2}(\mathrm{e}^2 - \mathrm{e})$$

$$\int_0^\pi \int_{-2\pi}^{2\pi} \cos(x)\sin(y)\, \mathrm{d}x\, \mathrm{d}y = \left(\int_{-2\pi}^{2\pi} \cos(x)\, \mathrm{d}x\right) \times \left(\int_0^\pi \sin(y)\, \mathrm{d}y\right) = 0$$

These agree with the evaluations in Example 12.14.

In practice, one may be required to integrate a bivariate function over a domain that is not rectangular. Before we consider how to evaluate such integrals, it is necessary to understand how we might denote them. Of course the issue here is how to express the limits. It is often convenient to write the integral and limits separately. In particular, a more general form of iterative integral Eq. (12.7) is

$$V = \iint_D f(x, y)\, \mathrm{d}x\, \mathrm{d}y \tag{12.10}$$

where D would be defined in an additional expression. For example, one can use this general notation to denote the following integral over a rectangular domain as

$$\int_0^1 \int_{-5}^5 (x^2y^3 - 2y)\, \mathrm{d}x\, \mathrm{d}y = \iint_D (x^2y^3 - 2y)\, \mathrm{d}x\, \mathrm{d}y \quad \text{for } D = \left\{x, y | x \in [-5, 5], y \in [0, 1]\right\}$$

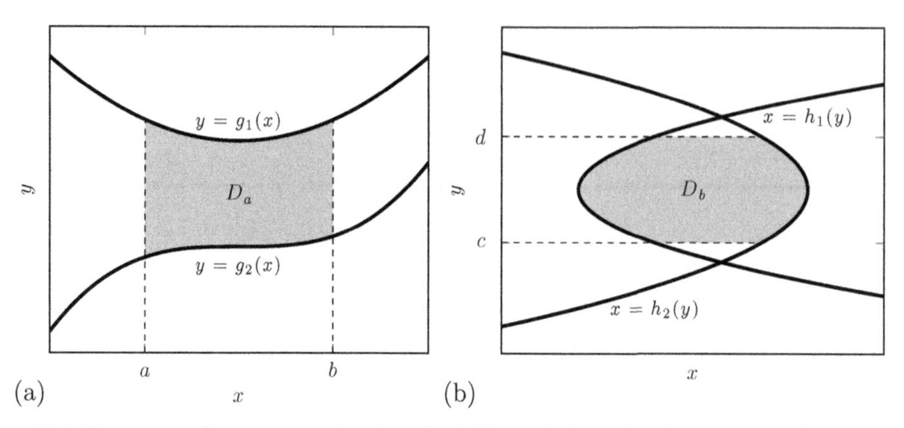

Figure 12.9 Two types of nonrectangular domain. (a) Domain bounded between two functions of x. (b) Domain bounded between two functions of y.

Figure 12.9 illustrates two general types of nonrectangular domains in the x-y plane that we might wish to integrate over. In Figure 12.9(a), we see the domain D_a bounded by $x \in [a, b]$ in the horizontal direction and $g_1(x)$ and $g_2(x)$ in the vertical direction. In contrast, Figure 12.9(b) illustrates the domain D_b bounded by $y \in [c, d]$ in the vertical direction and $h_1(y)$ and $h_2(y)$ in the horizontal direction. Integrals of some $f(x, y)$ over these domains could then be expressed as

$$V_a = \iint_{D_a} f(x, y)\,\mathrm{d}x\,\mathrm{d}y \quad \text{where } D_a = \left\{x, y | x \in [a, b], y \in [g_1(x), g_2(x)]\right\} \tag{12.11}$$

and

$$V_b = \iint_{D_b} f(x, y)\,\mathrm{d}x\,\mathrm{d}y \quad \text{where } D_b = \left\{x, y | x \in [h_1(y), h_2(y)], y \in [c, d]\right\} \tag{12.12}$$

EXAMPLE 12.16

Using the form of Eq. (12.10), express the integral of $f(x, y) = xy$ over the domains illustrated in Figure 12.10.

Solution

a. $\iint_{D_a} xy\,\mathrm{d}x\,\mathrm{d}y$ where $D_a = \left\{x, y | x \in [0, 1], y \in [-x - 2, 2x + 1]\right\}$

b. $\iint_{D_b} xy\,\mathrm{d}x\,\mathrm{d}y$ where $D_b = \left\{x, y | x \in [\mathrm{e}^y, \mathrm{e}^y + 5], y \in [1, 2]\right\}$

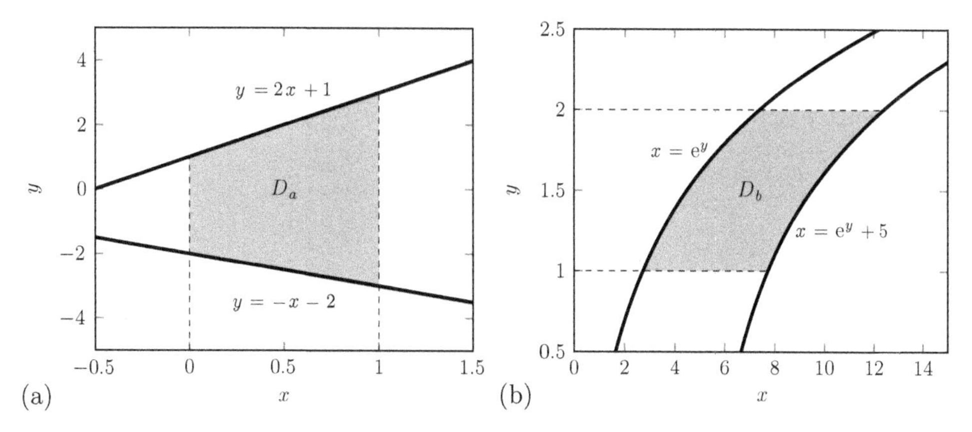

Figure 12.10 Domains for Example 12.16. (a) Domain D_a. (b) Domain D_b.

Consider an iterative integral over domain D_a as shown in Figure 12.9(a), that is Eq. (12.11). We note that the limits in the y-direction are functions of x. It is therefore necessary to insist that the integral with respect to y is performed before we consider the outer integral with respect to x. We are therefore forced to write V_a as

$$V_a = \iint_{D_a} f(x, y)\, \mathrm{d}x\, \mathrm{d}y = \int_a^b \int_{g_1(x)}^{g_2(x)} f(x, y)\, \mathrm{d}y\, \mathrm{d}x$$

As with rectangular domains, the inner integral results in some function of x, but now the x-dependence arises from both $f(x, y)$ and $g_{1,2}(x)$. No new mathematics is required to perform the outer integral, although the integrand could well be a complicated expression.

Consider now an integral over D_b. In this case, the limits in the x-direction are functions of y and it is necessary to perform this integral first. We are then forced to write V_b as

$$V_b = \iint_{D_a} f(x, y)\, \mathrm{d}x\, \mathrm{d}y = \int_c^d \int_{h_1(y)}^{h_2(y)} f(x, y)\, \mathrm{d}x\, \mathrm{d}y$$

It should be clear from this discussion that, unlike with rectangular domains, the order of integration does matter for iterative integrals over general domains.

EXAMPLE 12.17

Evaluate the integrals in Example 12.16.

Solution

a. We begin by writing the integral as

$$\iint_{D_a} xy\,\mathrm{d}x\,\mathrm{d}y = \int_0^1 \int_{-x-2}^{2x+1} xy\,\mathrm{d}y\,\mathrm{d}x$$

The inner integral is obtained as

$$\int_{-x-2}^{2x+1} xy\,\mathrm{d}y = \left[\frac{xy^2}{2}\right]_{y=-x-2}^{2x+1} = \frac{x}{2}\left((2x+1)^2 - (-x-2)^2\right) = \frac{3}{2}x(x^2-1)$$

which leads to

$$\iint_{D_a} xy\,\mathrm{d}x\,\mathrm{d}y = \int_0^1 \frac{3}{2}x(x^2-1)\,\mathrm{d}x = -0.375$$

b. We begin by writing the integral as

$$\iint_{D_b} xy\,\mathrm{d}x\,\mathrm{d}y = \int_1^2 \int_{\mathrm{e}^y}^{\mathrm{e}^{2y}+1} xy\,\mathrm{d}x\,\mathrm{d}y$$

The inner integral is obtained as

$$\int_{\mathrm{e}^y}^{\mathrm{e}^{2y}+1} xy\,\mathrm{d}x = \frac{y}{2}[x^2]_{\mathrm{e}^y}^{\mathrm{e}^{2y}+1} = \frac{y}{2}\left(\mathrm{e}^{4y} + \mathrm{e}^{2y} + 1\right)$$

which leads to an outer integral

$$\int_1^2 \frac{y}{2}\left(\mathrm{e}^{4y} + \mathrm{e}^{2y} + 1\right)\,\mathrm{d}y$$

Integration by parts is required to evaluate this, and, after some manipulation, we obtain

$$\begin{aligned}\iint_{D_b} xy\,\mathrm{d}x\,\mathrm{d}y &= \left[\frac{1}{32}\left(8y^2 + \mathrm{e}^{4y}(4y-1) + \mathrm{e}^{2y}(8y-4)\right)\right] \\ &= \frac{1}{32}\left(7\,\mathrm{e}^8 + 9\,\mathrm{e}^4 - 4\,\mathrm{e}^2 + 24\right) \approx 667.267\end{aligned}$$

EXAMPLE 12.18

Evaluate the following integrals.

a. $V_1 = \iint\limits_{D_1} (x_1^2 - 2)\,\mathrm{d}x_1\,\mathrm{d}x_2$ where $D_1 = \{x_1, x_2 | x_1 \in [-1, 1], x_2 \in [-1, \sin x_1]\}$

b. $V_2 = \iint\limits_{D_2} \mathrm{e}^{\frac{2x_1}{x_2}}\,\mathrm{d}x_1\,\mathrm{d}x_2$ where $D_2 = \left\{x_1, x_2 | x_1 \in [x_2, x_2^3], x_2 \in [0, 1]\right\}$

Solution

a. We proceed by forming the iterative integral

$$V_1 = \int_{-1}^{1} \int_{-1}^{\sin x_1} (x_1^2 - 2)\,\mathrm{d}x_2\,\mathrm{d}x_1$$

The inner integral is evaluated as

$$\int_{-1}^{\sin x_1} (x_1^2 - 2)\,\mathrm{d}x_2 = \left[(x_1^2 - 2)x_2\right]_{x_2=-1}^{\sin x_1} = (x_1^2 - 2)(\sin x_1 + 1)$$

This leads to an outer integral that is evaluated using integration by parts,

$$V_1 = \int_{-1}^{1} (x_1^2 - 2)(\sin x_1 + 1)\,\mathrm{d}x = \left[\frac{x^3}{3} - (x^2 - 2)\cos x - 2x(1 - \sin x) + 2\cos x\right]_{-1}^{1}$$
$$= -\frac{10}{3}$$

b. We proceed by forming the iterative integral

$$V_2 = \int_0^1 \int_{x_2}^{x_2^3} \mathrm{e}^{\frac{2x_1}{x_2}}\,\mathrm{d}x_1\,\mathrm{d}x_2$$

The inner integral is evaluated as

$$\int_{x_2}^{x_2^3} \mathrm{e}^{\frac{2x_1}{x_2}}\,\mathrm{d}x_1 = \left[\frac{x_2}{2}\,\mathrm{e}^{\frac{2x_1}{x_2}}\right]_{x_1=x_2}^{x_2^3} = \frac{x_2}{2}\left(\mathrm{e}^{2x_2^2} - \mathrm{e}^2\right)$$

which leads to an outer integral

$$\int_0^1 \frac{x_2}{2}\left(\mathrm{e}^{2x_2^2} - \mathrm{e}^2\right)\mathrm{d}x_2 = \int_0^1 \frac{1}{2}x_2\,\mathrm{e}^{2x_2^2}\,\mathrm{d}x_2 - \int_0^1 \frac{1}{2}x_2\,\mathrm{e}^2\,\mathrm{d}x_2$$

The first term is evaluated by a change of variable $u = 2x_2^2$, and we obtain

$$V_2 = \int_0^1 \frac{1}{2}x_2\,\mathrm{e}^{2x_2^2}\,\mathrm{d}x_2 - \int_0^1 \frac{1}{2}x_2\,\mathrm{e}^2\,\mathrm{d}x_2 = \frac{1}{4}\left[\frac{1}{2}\mathrm{e}^{2x_2^2} - \mathrm{e}^2 x_2^2\right]_0^1 = -\frac{1}{8}\left(1 + \mathrm{e}^2\right)$$
$$\approx -1.0486$$

EXAMPLE 12.19

Evaluate

$$V_3 = \iint\limits_{D_3} (x_1^2 - 2)\,\mathrm{d}x_1\,\mathrm{d}x_2$$

where D_3 is a triangular region on the x_1-x_2 plane with vertices at $(-1, 0)$, $(1, 0)$, and $(0, 4)$.

Solution

The triangular domain is illustrated in Figure 12.11. Straight lines can be fitted between the vertices, as is also illustrated. It is clear that the domain is of the type $D_3 = \{x_1, x_2 | x_1 \in [-1, 1], x_2 \in [g_1(x_1), g_2(x_1)]\}$, however, the region is defined by three expressions for $g(x_1)$. It is therefore necessary to divide the domain into two regions D_{3a} and D_{3b}. In particular,

$$D_{3a} = \{x_1, x_2 | x_1 \in [-1, 0], x_2 \in [0, 4x_1 + 4]\}$$

and

$$D_{3b} = \{x_1, x_2 | x_1 \in [0, 1], x_2 \in [0, -4x_1 + 4]\}$$

We then divide the required integration into two as

$$V_3 = \underbrace{\iint\limits_{D_{3a}} (x_1^2 - 2)\,\mathrm{d}x_1\,\mathrm{d}x_2}_{V_{3a}} + \underbrace{\iint\limits_{D_{3b}} (x_1^2 - 2)\,\mathrm{d}x_1\,\mathrm{d}x_2}_{V_{3b}}$$

and evaluate each part separately.

$$V_{3a} = \int_{-1}^{0} \left(\int_{0}^{4x_1+4} (x_1^2 - 2)\,\mathrm{d}x_2 \right) \mathrm{d}x_1$$

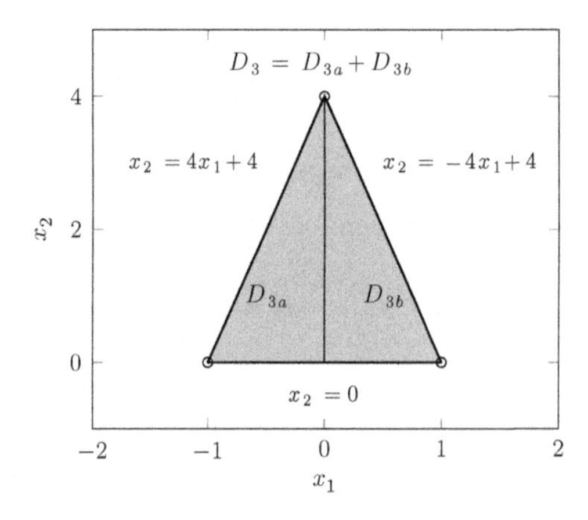

Figure 12.11 The domain of integration for Example 12.19.

$$= \int_{-1}^{0} \left(4(x_1+1)(x_1^2-2)\right) \mathrm{d}x_1$$

$$= 4\left[\frac{x^4}{4} + \frac{x^3}{3} - x^2 - 2x\right]_{-1}^{0} = -\frac{11}{3}$$

$$V_{3b} = \int_0^1 \left(\int_0^{-4x_1+4} (x_1^2-2)\,\mathrm{d}x_2\right) \mathrm{d}x_1$$

$$= \int_0^1 \left(-4(x_1-1)(x_1^2-2)\right) \mathrm{d}x_1$$

$$= -4\left[\frac{x^4}{4} - \frac{x^3}{3} - x^2 + 2x\right]_0^1 = -\frac{11}{3}$$

We therefore have $V_3 = -\frac{22}{3}$.

12.5 MULTIVARIATE CALCULUS ON YOUR COMPUTER

In essence, all we have done in this chapter is generalize the univariate calculus discussed in earlier chapters. While this generalization has led to many complicated issues for algebraic work, the commands required for performing multivariate calculus on Wolfram Alpha are modified only very slightly.

Beginning with partial derivatives, there are, in fact, no changes to the commands. Recall that in Chapter 3 we saw that the command

```
d/dx (x^2-y^2-xy)
```

was understand as

$$\frac{\mathrm{d}}{\mathrm{d}x}\left(x^2 - y^2 - xy\right)$$

for some constant number y. This operation is of course the same as taking the *partial derivative*

$$\frac{\partial}{\partial x}\left(x^2 - y^2 - xy\right)$$

where x and y are now both independent variables. That is, to take a partial derivative with respect to x, the variable y is treated as if it were a constant.

Similarly, the partial derivative with respect to y is obtained from the command

```
d/dy (x^2-y^2-xy)
```

Higher-order and mixed derivatives follow in the obvious way. For example,

$$\frac{\partial^2}{\partial y^2}(x^2 - y^2 - xy) \quad \text{and} \quad \frac{\partial^2}{\partial x \partial y}(x^2 - y^2 - xy)$$

are obtained from

```
d^2/dy^2 (x^2-y^2-xy)
```

and

```
d^2/dydx (x^2-y^2-xy)
```

The command for obtaining and classifying any stationary points of multivariate functions is exactly are for univariate functions. For example, the command

```
stationary points (x^2-y^2-xy)
```

can be used to determine that $f(x, y) = x^2 - y^2 - xy$ has a saddle point at $(x, y) = (0, 0)$.

The command for constrained optimization is also intuitive. For example, the result of Example 12.11 can be confirmed from the command

```
optimize x+4y-2z such that 2x-y-z=2 and x^2+y^2=4
```

That is, these commands return any maxima and minima of the problem. Alternatively, if we were particularly interested in the maximum of the constrained problem then the following command is appropriate.

```
maximize x+4y-2z such that 2x-y-z=2 and x^2+y^2=4
```

Similarly, the minimum of the same constrained problem is obtained from

```
minimize x+4y-2z such that 2x-y-z=2 and x^2+y^2=4
```

Let us know turn our attention to bivariate integration. The definite integral

$$\int_0^4 \int_{-3}^5 (x^2 - y^2 - xy)\, dx\, dy = -32$$

can be confirmed from the command

```
int int x^2-y^2-xy dx dy between x=-3 and 5 and between y=0 and 4
```

Furthermore, the double integral over a nonrectangular domain

$$\int_0^4 \int_{-3+y}^{5+y^2} (x^2 - y^2 - xy)\, dx\, dy = \frac{50,472}{35}$$

is confirmed from the command

```
int int x^2-y^2-xy dx dy between x=-3+y and 5+y^2 and between y=0 and 4
```

The reader is, however, reminded to be careful with regard the order of integration over nonrectangular regions. For example, the command

```
int int x^2-y^2-xy dy dx between x=-3+y and 5+y^2 and between y=0 and 4
```

is interpreted as an inner integral with respect to y between $y = 0$ and 4 and an outer integral with respect to x between $-3 + y$ and $5 + y^2$. The result is therefore returned as

$$\frac{4}{3}\left(y^6 + 12y^4 - y^3 + 41y^2 - 29y - 24\right)$$

which is possibly not what you intended.

EXAMPLE 12.20

Use Wolfram Alpha to answer the following questions.

a. Locate and classify stationary points of the function $f(x_1, x_2) = x_1^4x_2^3 + x_1^2 + 2x_2^2 - 2x_1x_2$.

b. Find the location of the minimum value of $g(x, y) = 5x^2 + 10xy$ such that x and y sit on a circle of radius 3 and centered at $(x, y) = (3, 1)$ in the x-y plane.

c. Find the location of the maximum value of $h(x, y, z) = x(y^2 + 3z)$ such that $x + y + z = 4$ and $xy = 4$.

d. Evaluate the definite integral

$$\iint_D (2y\cos(x) + x\sin(y))\,\mathrm{d}x\,\mathrm{d}y \quad \text{where } D = \left\{x, y | x \in [-\pi, \pi/2], y \in [\pi, 2\pi]\right\}$$

e. Evaluate the definite integral

$$\iint_D (xy + 2x - 3y)\,\mathrm{d}x\,\mathrm{d}y \quad \text{where } D = \left\{x, y | x \in [3, 4], y \in [2 - x, 2 + x]\right\}$$

Solution

a. Note that it is easier to use variables x and y in place of x_1 and x_2, respectively. We use the command

```
stationary points of x^4y^3-2xy+x^2+2y^2
```

and find that there is a minimum at $(x, y) = (x_1, x_2) = (0, 0)$ and saddle points at $(x, y) = (x_1, x_2) \approx (-0.8762, -0.5941)$ and $(x, y) = (x_1, x_2) \approx (1.3564, -0.7501)$.

b. This is a constrained optimization problem on $5x^2 + 10xy$ such that $(x - 3)^2 + (y - 1)^2 = 9$. We use the command

```
minimize 5x^2+10xy such that (x-3)^2+(y-1)^2=9
```

and obtain that the problem has a global minimum at $(x, y) \approx (2.42018, -1.94343)$ where $g \approx -17.7483$.

c. This is a constrained optimization problem on $x(y^2 + 3z)$ such that $x + y + z = 4$ and $xy = 4$. We use the command

```
maximize x(y+3z) such that x+y+z=4 and xy=4
```

and obtain that the problem has a global maximum at $(x, y) = (2, 2, 0)$ where $h = 4$.

d. This is an integral over a simple rectangular domain. We use the command

```
int int (2y*cos(x)+x*sin(y)) dx dy between x=-pi and pi/2 and between
y=pi and 2*pi
```

and obtain the value $\frac{15\pi^2}{4}$.

e. This is an integral over a nonrectangular domain and we need to perform the integral with respect to y first. That is, it must be expressed as

$$\int_3^4 \int_{2-x}^{2+x} (xy + 2x - 3y)\, dy\, dx$$

We therefore use the command

```
int int (xy+2x-3y) dy dx between x=3 and 4 and between y=2-x and 2+x
```

and obtain the result $\frac{170}{3}$.

12.6 QUESTIONS

The following questions are intended to test your knowledge of the material discussed in this chapter. Full solutions are available in Chapter 12 Solutions of Part III. You should use an algebraic approach unless otherwise stated.

Question 12.1. You are given the multivariate functions

$$f(x, y, x) = x^2yz + \sin(xy + x) \quad g(x, y) = e^{x^2+\cos(y)}$$

Determine the following partial derivatives.

a. $\frac{\partial^3 f}{\partial x \partial y \partial z}$

b. $\frac{\partial^3 g}{\partial x^2 \partial y}$

c. $\partial_{zz}(fg)$

Question 12.2. Where they exist, determine, and classify all stationary points of the following bivariate functions.

a. $f(x, y) = 5x - 2x^2 + 10y - y^2$

b. $g(x, z) = (1 + z^2)\, e^{2-x}$

c. $h(x, z) = \frac{x+z}{x^2+2z^2+6}$

d. $k(a, b) = \sin^2(a) + b^2$

Question 12.3. Solve the following constrained optimization problems, where possible.

a. Maximize $x^2 + 2y^2$ such that $x^2 + y^2 = 4$.

b. Minimize $x + 2y^2$ such that $x + y^2 = 1$.

c. Minimize $2x^2 + y^2 - z^2$ such that $3xy - 4z^2 = 0$.

d. Maximize $-y + z - x + 6$ such that $x - y - z = 2$ and $x^2 + y^2 = 1$.

Question 12.4. Demonstrate that the function $f(x, y, z) = x^2$ has only one critical point on the surface $x^2 + y^2 = z$. Classify the critical point.

Question 12.5. Maximize the volume of a rectangular box with surface area 2 m^2 and height 0.1 m.

Question 12.6. Evaluate the following definite integrals over the rectangular domains stated.

a. $\iint_{D_a} 2xy \,dx\,dy$ for $D_a = \{x, y | x \in [-1, 2], y \in [2, 4]\}$.

b. $\iint_{D_b} (x^2 - z^2 + 10) \,dx\,dz$ for $D_b = \{x, z | x \in [0, 10], z \in [-1, 1]\}$.

c. $\iint_{D_c} 0.1p\cos(q) \,dp\,dq$ for $D_c = \{p, q | p \in [-2, -1], q \in [0, \pi]\}$.

d. $\iint_{D_d} e^{x+y} \sin(x + y) \,dx\,dy$ for $D_d = \{x, y | x \in [1, 2], y \in [1, 2]\}$.

Hint. Attempt to express the integrand in a separable form.

Question 12.7. Evaluate the following definite integrals over the domains stated.

a. $\iint_{D_a} 2xy \,dx\,dy$ for $D_a = \{x, y | x \in [-1, 2], y \in [-x, x]\}$.

b. $\iint_{D_b} (x^2 - z^2 + 10) \,dx\,dz$ for $D_b = \{x, z | x \in [2z, 4z], z \in [1, 2]\}$.

c. $\iint_{D_c} 0.1p\sin(q) \,dp\,dq$ for $D_c = \{p, q | p \in [-q, q^2], q \in [-\pi, \pi]\}$.

d. $\iint_{D_d} e^{x/y} \,dx\,dy$ for $D_d = \{x, y | x \in [y, y^2], y \in [4, 5]\}$.

Question 12.8. An insurance company provides buildings and contents insurance to its policyholders. We use X to denote the random variable for annual claims under buildings insurance and Y to denote annual claims under contents insurance. Working in units of \$10M, X and Y are known to have a joint density function of the form,

$$p(x, y) = \begin{cases} \frac{1}{a}(x - y + 5) & \text{for } 0 \le x \le 5,\ 0 \le y \le 7 \\ 0 & \text{otherwise} \end{cases}$$

where a is an unknown constant.

a. Determine the value of a.

b. Calculate the probability that the total claims exceeds \$60M.

Hint. As we will see in Chapter 13, a *joint density function* $p(x, y)$ is such that the probability that two *continuous random variables* X and Y take values $X \in [x_1, x_2]$ and $Y \in [y_1, y_2]$ is given by

$$\int_{x_1}^{x_2} \int_{y_1}^{y_2} p(x, y) \,dy\,dx$$

Question 12.9. You are aware that your happiness in the early afternoon depends on what you have eaten for lunch. After many years of research, you are reasonably sure that your happiness, H, is a function of the number of sandwiches, s, and chocolate bars, c, consumed and is given by

$$H(s,c) = s^{1/2}c^{2/5}$$

If each sandwich costs \$3.50, each chocolate bar costs \$1.25, and you budget for \$10 each lunch, determine the optimal purchase to maximize your afternoon happiness. You should assume that it is possible to purchase any fraction of a sandwich or chocolate bar.

Question 12.10. A factory manufactures three products and has an annual profit given by

$$P(x,y,z) = 6x + 8y + 4z - 0.4x^2 - y^2 - z^2$$

where x, y, and z represent the number of units manufactured (in units of 10,000) of products X, Y, and Z, respectively, in the year. In order to meet the demands of its current contracts, the factory is required to manufacture a combined total of 120,000 units of X and Y.

a. Determine the optimal values of x, y, and z.

b. Determine how the maximum profit could change if the factory restricts orders such that the total combined units of X and Y is 100,000.

CHAPTER 13

Introductory Numerical Methods

Contents

Prerequisite knowledge	**Intended learning outcomes**

• Chapter 1
• Chapter 2
• Chapter 3
• Chapter 4
• Chapter 5
• Chapter 6
• Chapter 7
• Chapter 11
• Chapter 12
	• Recognize, select, and implement numerical methods for root finding:
	  ◦ interval (bisection and regula falsi) methods
	  ◦ gradient (Newton-Raphson and secant) methods
	• Recognize, select, and implement numerical methods for performing the numerical differentiation of univariate data:
	  ◦ elementary approach
	  ◦ central-difference approaches of $O(h^2)$ and $O(h^4)$
	• Recognize, select, and implement numerical methods for univariate integration:
	  ◦ trapezoidal approach
	  ◦ Simpson's approach
	• Determine an estimate of the error bounds in each numerical method

As explained throughout this book, it is often impossible to perform many of the computations that we may see using algebraic, that is "pen-and-paper," techniques.

For example, we may not be able to determine the roots of a given function or compute a definite integral. This chapter is concerned with numerical approaches to such problems. In particular, we will consider elementary numerical approaches for root finding, evaluating derivatives and computing definite integrals. Note that we will restrict ourselves to univariate functions, although it is often possible to generalize the methods discussed to multivariate functions.

With the continued rapid growth in affordable computer power, the design and implementation of numerical methods is a huge growth area relevant to many aspects of mathematics. You should not therefore begin this chapter thinking that computers can only be used for root finding and basic calculus. Indeed, we have seen that Wolfram Alpha, for example, has numerical algorithms for the linear algebra problems of Chapter 10, the simulation of random experiments related to Chapter 9 and the solution of ODEs in Chapter 11, to name just a few areas. Our motivation for studying numerical methods here is toward understanding some of the concepts that sit behind commercial packages. The study of the use of numerical methods beyond root finding and calculus is outside of the scope of this introductory text but it is hoped that this chapter gives an insight into the usefulness of computers in all aspects of applied mathematics.

A thorough discussion of the implementation of numerical methods should include reference to computer codes. However, rather than assuming familiarity with any particular language or indeed spending time teaching the basics of a particular language, we opt for a presentation, where relevant, in terms of *pseudocodes*. That is, we state the algorithms in English rather than in some compliable computing language. Those familiar with a particular language, for example, MATLAB, Fortran or C, may wish to attempt to implement the algorithms discussed.

In nearly all cases, the calculations described in the examples presented in this chapter can be implemented in Excel and confirmed with Wolfram Alpha and you are invited to repeat them in these ways. We will also find that some of the numerical methods discussed here are also useful for pen-and-paper approximations; these will be flagged along the way.

13.1 ROOT FINDING

We had previously reverted to using Excel's Goalseek facility or Wolfram Alpha's `roots` command to determine the roots of complicated functions. While using these standard functions of Excel and Wolfram Alpha are very convenient, the fact that they sit behind commercial software limits their use in practice. That is, we cannot incorporate these into larger computational projects that we may be coding. Furthermore, the default accuracy settings of Goalseek may not be sufficient for your needs or it may struggle to solve highly nonlinear functions. For these reasons it is important that you have an understanding of the basic strategies of numerical methods for root finding, such as those that are used behind the scenes in Excel and Wolfram Alpha. This motivates the first section of this chapter.

Recall from Chapter 2 that a real root, x_r, of general real-valued function $f(x)$ is such that $f(x_r) = 0$. We will remain in the real plane throughout this chapter, that is $x, f(x) \in \mathbb{R}$. In practice, we will assume that we have some rough idea of where this root is, perhaps from a quick sketch of the function or some information from the context of the problem, and that the function has continuous derivatives near to this root. This information, together with other properties of the function, can then be used to compute a sequence of increasingly accurate approximations of the root via some numerical method. In general, $f(x)$ may have more than one real root but it is assumed that we are seeking one particular root at a time.

13.1.1 Interval methods

We begin with a discussion of *interval methods* for approximating the roots of functions. Interval methods are very simple and require only knowledge of an initial interval within which a single root is known to exist. Interval methods then proceed to use repeated evaluations of the function to iteratively narrow the interval that contains the root until a desired level of accuracy is reached.

The bisection method

Consider $f(x)$ to be a function continuous on the interval $[a, b]$ that contains a single root x_r, that is $x_r \in [a, b]$. Since a and b are either side of the root, it should be clear that

$$f(a)f(b) < 0$$

which simply means that $f(a)$ and $f(b)$ have opposite sign. Two situations satisfying this criteria can exist, depending on the gradient of the function close to the root, as shown in Figure 13.1. In practice, interval methods do not use the local gradient and the procedure would proceed identically in either case.

The first iteration of the *bisection method* consists of bisecting the original interval $[a, b]$ into two equal subintervals $[a, c]$ and $[c, b]$, such that $c = \frac{1}{2}(a + b)$, and keeping the half over which $f(x)$ changes sign. This is determined by evaluations of $f(c)$ and $f(b)$. In particular,

$$f(b)f(c) > 0 \Rightarrow x_r \in [a, c] \quad \text{and} \quad f(b)f(c) < 0 \Rightarrow x_r \in [b, c]$$

The iteration therefore results in a narrower interval within which x_r is contained, and this can be used as the input into the next iteration of the process. Repeated iteration of the bisection method generates increasingly narrow intervals that contain the root. Note that this method cannot be used to determine *repeated roots* where the function does not cross the x-axis.

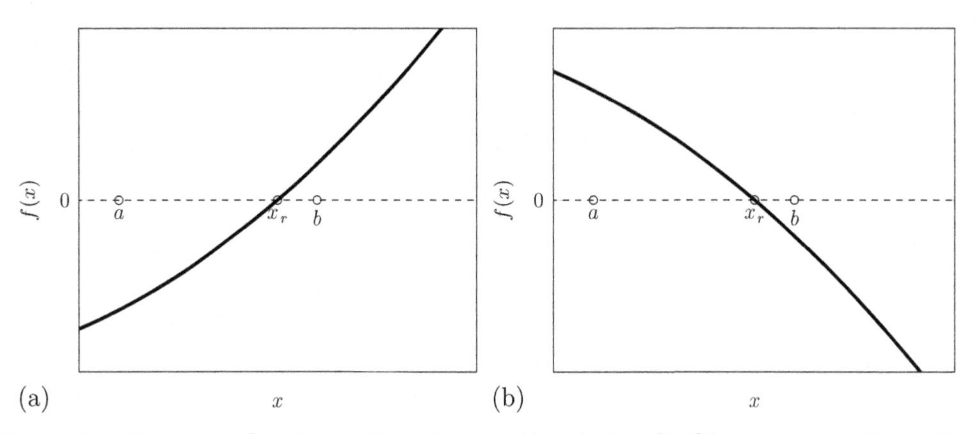

Figure 13.1 Continuous functions with root $x_r \in [a, b]$ such that $f(a)f(b) < 0$. (a) Locally positive gradient with $f(a) < 0$ and $f(b) > 0$. (b) Locally negative gradient with $f(a) > 0$ and $f(b) < 0$.

In general, we denote the interval bounds used in the nth iteration of the process by a_n and b_n for $n \in \mathbb{N}^+$. These are such that $x_r \in [a_n, b_n]$, and the midpoint, c_n, can be taken to be the nth approximation of the root. That is, $x_r \simeq c_n$.

The indexing is such that we begin at $n = 1$ with $a_1 = a$ and $b_1 = b$. The *error* in the nth estimate is commonly defined to be $\epsilon_n = b_n - c_n$, that is half of the interval's width. An acceptable error tolerance, ϵ, would be stated in advance and this defines the *termination criteria* for the iterative process: the process stops at iteration n such that $\epsilon_n \leq \epsilon$. We discuss the convergence of the bisection method toward the actual root later in this section, but first illustrate the practical use of the method with an example.

EXAMPLE 13.1

Use the bisection method to estimate all roots of the following function with an error tolerance of $\epsilon = 0.002$,

$$g(x) = x^3 + x - 2$$

Solution

It may be immediately clear to you that $g(x)$ has a *single root* at $x_r = 1$, but let us assume that a rough sketch has led us to suspect only that the root is within the interval $[0, 1.5]$, as shown in Figure 13.2(a). This is confirmed with the observation that

$$g(0) \times g(1.5) < 0$$

The interval $[0, 1.5]$ therefore defines the starting interval $[a_1, b_1] = [a, b]$ with midpoint $c_1 = \frac{1.5+0}{2} = 0.75$. The first estimate $x_r \simeq c_1$ has error $1.5 - 0.75 = 0.75 > \epsilon$ and we must move to $n = 2$.

Note that

$$g(c_1) \times g(b_1) = g(0.75) \times g(1.5) < 0$$

Table 13.1 Bisection method for Example 13.1

n	a_n	b_n	$x_r \simeq c_n$	ϵ_n	$g(a_n)$	$g(b_n)$	$g(c_n)$
1	0.00000	1.50000	0.75000	0.75000	−2.00000	2.87500	−0.82813
2	0.75000	1.50000	1.12500	0.37500	−0.82813	2.87500	0.54883
3	0.75000	1.12500	0.93750	0.18750	−0.82813	0.54883	−0.23853
4	0.93750	1.12500	1.03125	0.09375	−0.23853	0.54883	0.12796
5	0.93750	1.03125	0.98438	0.04688	−0.23853	0.12796	−0.06177
6	0.98438	1.03125	1.00781	0.02344	−0.06177	0.12796	0.03143
7	0.98438	1.00781	0.99609	0.01172	−0.06177	0.03143	−0.01558
8	0.99609	1.00781	1.00195	0.00586	−0.01558	0.03143	0.00782
9	0.99609	1.00195	0.99902	0.00293	−0.01558	0.00782	−0.00390
10	0.99902	1.00195	1.00049	0.00146			0.00196

which means that x_r lies within the upper half of the interval, that is $x_r \in [0.75, 1.5]$ which defines our new $[a_2, b_2]$. The midpoint of this second interval is $c_2 = 1.125$ and the estimate $x_r \simeq c_2$ has error $b_2 - c_2 = 0.375 > \epsilon$. We must therefore proceed to $n = 3$.

Note that

$$g(c_2) \times g(b_2) = g(1.125) \times g(1.5) > 0$$

We therefore see that $x_r \in [0.75, 1.125] = [a_3, b_3]$ with midpoint $c_3 = 0.9375$. The estimate $x_r \simeq c_3$ has error $1.125 - 0.9375 > \epsilon$ and we proceed again.

The process continues to generate $[a_n, b_n]$ as shown in Table 13.1 until the estimate $x_r \simeq c_{10}$ is obtained within the desired error tolerance. Note that the result $x_r \simeq 1.00049$ is reasonably close to the true value of $x_r = 1$ and is such that $g(c_{10}) = 0.00196 \approx 0$. Of course, a more accurate approximation can be obtained with yet further iterations. The first four iterations of the process are illustrated graphically in Figure 13.2.

Given a starting interval $[a, b]$ such that $f(a) \times f(b) < 0$ and a required error tolerance ϵ, the algorithm for the bisection method can be written as follows.

ALGORITHM FOR THE BISECTION METHOD

1. $n = 1$, $a_n = a$, and $b_n = b$
2. $c_n = \frac{1}{2}(a_n + b_n)$
3. if $b_n - c_n \leq \epsilon$ accept $x_r \simeq c_n$ and end, otherwise
4. if $f(b_n) \times f(c_n) < 0$ then $a_{n+1} = c_n$ and $b_{n+1} = b_n$, otherwise $a_{n+1} = a_n$ and $b_{n+1} = c_n$
5. $n = n + 1$ and go to 2.

While we have seen this algorithm to be somewhat cumbersome to implement by hand, it should be clear that a computer can be programmed to run it very rapidly.

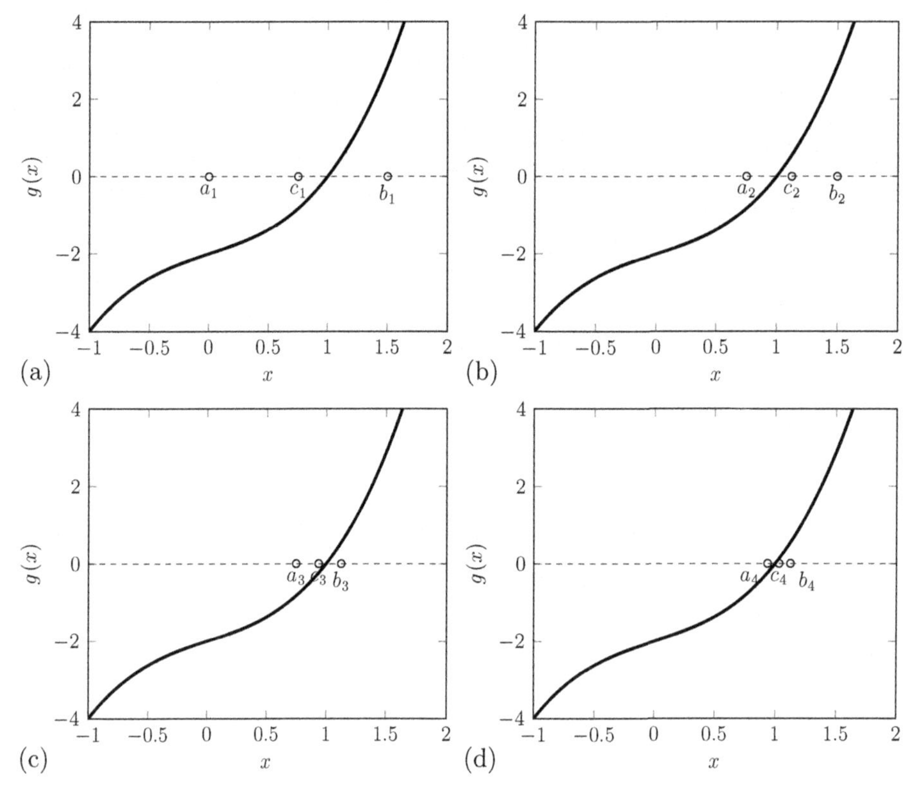

Figure 13.2 An illustration of the first four iterations of the bisection method for Example 13.1. (a) $n = 1$. (b) $n = 2$. (c) $n = 3$. (d) $n = 4$.

EXAMPLE 13.2

Use the bisection method to obtain the roots required of the following functions to an error tolerance of $\epsilon = 0.01$.

a. The single real positive root of $f(x) = x^2 + 2.011x - 5.04795$.
b. The single real negative root of $f(x)$ in part a.
c. The single positive root of $g(x) = \sin(x^2) - x + 1$.

Solution

a. We are given that $f(x)$ has a single positive root and note that, for example, $f(0) \times f(2) < 0$. A possible starting interval is then $[0, 2]$. The iterations shown in Table 13.2 lead to $x_r \simeq c_8 = 1.46094$.

b. We are given that $f(x)$ has a single negative root and note that, for example, $f(-2) \times f(-4) < 0$. A possible starting interval is then $[-4, -2]$. The iterations shown in Table 13.2 lead to $x_r \simeq c_8 = -3.46094$.

Table 13.2 Iterations of the bisection method in Example 13.2

	n	a_n	b_n	$x_r \simeq c_n$	ϵ_n	$g(a_n)$	$g(b_n)$	$g(c_n)$
a.	1	0.00000	2.00000	1.00000	1.00000	−5.04795	2.97405	−2.03695
	2	1.00000	2.00000	1.50000	0.50000	−2.03695	2.97405	0.21855
	3	1.00000	1.50000	1.25000	0.25000	−2.03695	0.21855	−0.97170
	4	1.25000	1.50000	1.37500	0.12500	−0.97170	0.21855	−0.39220
	5	1.37500	1.50000	1.43750	0.06250	−0.39220	0.21855	−0.09073
	6	1.43750	1.50000	1.46875	0.03125	−0.09073	0.21855	0.06293
	7	1.43750	1.46875	1.45313	0.01563	−0.09073	0.06293	−0.01414
	8	1.45313	1.46875	1.46094	0.00781			
b.	1	−4.00000	−2.00000	−3.00000	1.00000	2.90805	−5.06995	−2.08095
	2	−4.00000	−3.00000	−3.50000	0.50000	2.90805	−2.08095	0.16355
	3	−3.50000	−3.00000	−3.25000	0.25000	0.16355	−2.08095	−1.02120
	4	−3.50000	−3.25000	−3.37500	0.12500	0.16355	−1.02120	−0.44445
	5	−3.50000	−3.37500	−3.43750	0.06250	0.16355	−0.44445	−0.14436
	6	−3.50000	−3.43750	−3.46875	0.03125	0.16355	−0.14436	0.00862
	7	−3.46875	−3.43750	−3.45313	0.01563	0.00862	−0.14436	−0.06811
	8	−3.46875	−3.45313	−3.46094	0.00781			
c.	1	0.00000	2.00000	1.00000	1.00000	1.00000	−1.75680	0.84147
	2	1.00000	2.00000	1.50000	0.84147	0.50000	−1.75680	0.27807
	3	1.50000	2.00000	1.75000	0.27807	0.25000	−1.75680	−0.67099
	4	1.50000	1.75000	1.62500	0.27807	0.12500	−0.67099	−0.14473
	5	1.50000	1.62500	1.56250	0.27807	0.06250	−0.14473	0.08186
	6	1.56250	1.62500	1.59375	0.08186	0.03125	−0.14473	−0.02783
	7	1.56250	1.59375	1.57813	0.08186	0.01563	−0.02783	0.02795
	8	1.57813	1.59375	1.58594	0.00781			

c. We are given that $g(x)$ has a single positive root and note that, for example, $g(0) \times g(2) < 0$. A possible starting interval is then $[0, 2]$. The iterations shown in Table 13.2 lead to $x_r \simeq c_8 = 1.58594$.

Note that accurate roots for $f(x)$ can be obtained from the quadratic formula as $x_r = 1.456$ and -3.467; furthermore Wolfram Alpha determines the root of $g(x)$ to be $x_r = 1.58602$ to five decimal places.

An important aspect of any numerical method is the behavior of the error with repeated iteration. For the bisection method, it should be clear that the successive intervals are of width,

$$b_{n+1} - a_{n+1} = \frac{1}{2}(b_n - a_n) \quad \text{for } n \geq 1$$

Repeated back substitution of this expression leads to the relationship between the nth iteration and the initial bounds $a_1 = a$ and $b_1 = b$,

$$b_n - a_n = \frac{1}{2^{n-1}}(b - a) \quad \text{for } n \geq 2$$

This result is intuitively obvious as the essential feature of the bisection method is that it reduces the interval by a factor of two on each iteration. Since, by design, the root x_r must be in either half interval $[a_n, c_n]$ or $[c_n, b_n]$, it is clear that

$$\epsilon_n = |x_r - c_n| \leq \frac{1}{2^n}(b-a) \tag{13.1}$$

The estimate $x_r \simeq c_n$ is therefore seen converge to x_r as $n \to \infty$. That is, the bisection method is *certain* to converge to the actual root to within any error bound ϵ after a sufficiently large number of iterations. Furthermore, the accuracy of the prediction increases twofold with each iteration. This behavior can be observed in the ϵ_n column of Tables 13.1 and 13.2, for example. Equation 13.1 can be used to determine the number of iterations required under the bisection method to obtain the desired error tolerance, ϵ. In particular, setting $\epsilon_n = \epsilon$ and solving for n, we see that

$$n \geq \frac{\ln(b-a) - \ln \epsilon}{\ln 2} \tag{13.2}$$

The maximum number of iterations required therefore depends only the required error tolerance ϵ and the width of the initial interval $b-a$; it does not depend on the particular expression being solved. This is explored in the following examples.

EXAMPLE 13.3

Compare the theoretical number of iterations required under the bisection method against the actual number of iterations for each part of the solution to Example 13.2.

Solution

In each case, we began with an interval of width $b-a=2$ and worked toward an error tolerance of $\epsilon = 0.01$. Equation (13.2) therefore gives

$$n \geq \frac{\ln(2/0.01)}{\ln 2} = 7.64$$

This is interpreted as meaning that $n = 8$ will give an estimate of the root to within the required error tolerance. Table 13.2 shows that this was indeed true in each case.

EXAMPLE 13.4

Investigate the sensitivity of the required number of iterations required under the bisection method to

a. the initial interval width for fixed $\epsilon = 0.001$

b. the error tolerance for a fixed initial width of $b-a=1$.

Solution

a. This can be seen by plotting Eq. (13.2) with $\epsilon = 0.001$ as a function of $b-a$, as shown in Figure 13.3(a).

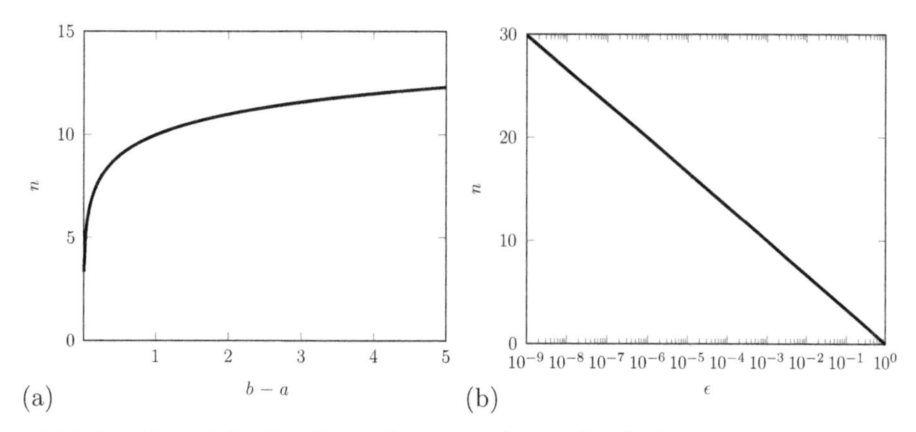

Figure 13.3 Sensitivity of the bisection method to n and $b - a$. Plots for Example 13.4. (a) n as a function of $b_1 - a_1$. (b) n as a function of ϵ.

b. This can be seen by plotting Eq. (13.2) with $b - a = 1$ as a function of ϵ, as shown in Figure 13.3(b).

We see that the convergence of the bisection method is particularly sensitive to ϵ.

The primary advantage of the bisection method is that it is certain to converge to within any desired error tolerance after a sufficiently large number of iterations. However, the results of Example 13.4 show the distinct disadvantage of the method: it can be very slow to converge if a high level of accuracy is required. We therefore look for an improvement to this method.

The regula falsi method

The *regula falsi method*, sometimes known as the *method of false position*, is a simple variation on the bisection method that benefits from some increase in the speed of convergence. The method is another example of an interval method and so does not need explicit information about the local gradient of the function close to the root. It is different to the bisection method in that, rather than using the midpoint of the interval $[a_n, b_n]$ during each iteration, the regula falsi method uses the point obtained from where the straight line joining $f(a_n)$ and $f(b_n)$ crosses the x-axis, c_n. This often provides a better estimate of the actual root, as shown in Figure 13.4.

The nth iteration of the root is approximated by $x_r \simeq c_n$. Once the point c_n has been determined within an iteration, the regula falsi method continues by testing for a sign change over the intervals $[a_n, c_n]$ and $[c_n, b_n]$ and keeps the interval within which the root is found to lie. The complication is of course in determining the value of c_n from the known values of a_n, b_n, $f(a_n)$, and $f(b_n)$. This is done by noting that the straight line, L, joining $f(a_n)$ and $f(b_n)$ has gradient given by both

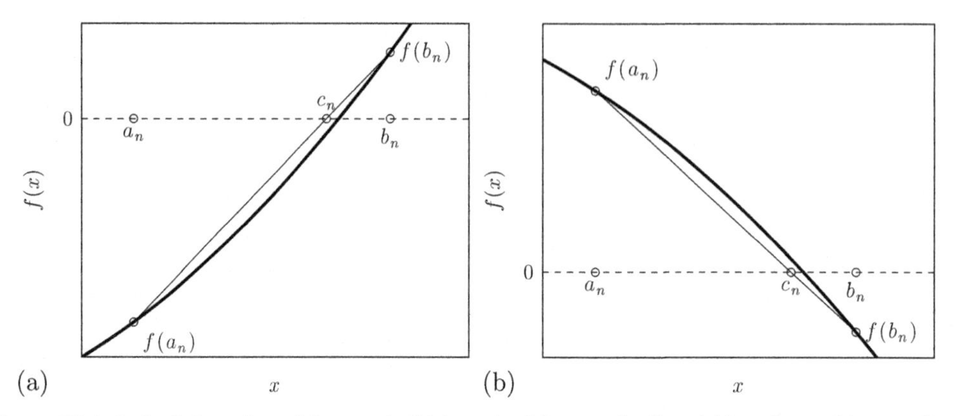

Figure 13.4 A single iteration of the regula falsi method for root finding. (a) Locally positive gradient with $f(a_n) < 0$ and $f(b_n) > 0$. (b) Locally negative gradient with $f(a_n) > 0$ and $f(b_n) < 0$.

$$m = \frac{f(b_n) - f(a_n)}{b_n - a_n} \quad \text{and} \quad m = \frac{f(b_n) - 0}{b_n - c_n}$$

These two expressions can be equated, leading to the formula

$$c_n = b_n - \frac{f(b_n) \times (b_n - a_n)}{f(b_n) - f(a_n)} \tag{13.3}$$

This expression can be used within each iteration of the regula falsi method. Note that since each b_n and a_n is either side of a single root, $b_n \neq a_n$ and we assume that $f(b_n) \neq f(a_n)$.

The definition of error used in the bisection method is not appropriate for the regula falsi method and we are forced to use an alternative. In particular, the termination criteria for the method is that

$$|f(c_n)| \leq \epsilon$$

You might actually consider this to be a more natural criteria than the criteria used in the bisection method.

EXAMPLE 13.5

Repeat Example 13.1 using the regula falsi method on the interval $[0, 1.5]$. You should use an error tolerance of 0.002.

Solution

We have $a_1 = 0$ and $b_1 = 1.5$, from which we use Eq. (13.3) to determine that

$$c_1 = 1.5 - \frac{g(1.5) \times (1.5 - 0)}{f(1.5) - f(0)} = 0.61538$$

Note that $|g(c_1)| = 1.1516 > \epsilon$ and we must proceed to the next iteration.

Table 13.3 Iterations of the regula falsi method in Example 13.5

n	a_n	b_n	$x_r \simeq c_n$	ϵ_n	$g(a_n)$	$g(b_n)$
1	0.00000	1.50000	0.61538	1.15157	−2.00000	2.87500
2	0.61538	1.50000	0.86838	0.47679	−1.15157	2.87500
3	0.86838	1.50000	0.95823	0.16193	−0.47679	2.87500
4	0.95823	1.50000	0.98711	0.05105	−0.16193	2.87500
5	0.98711	1.50000	0.99606	0.01571	−0.05105	2.87500
6	0.99606	1.50000	0.99880	0.00480	−0.01571	2.87500
7	0.99880	1.50000	0.99963	0.00146		

We have

$$g(b_1) \times g(c_1) = 2.87500 \times -1.15157 < 0$$

and so $x_r \in [c_1, b_1] = [0.61538, 1.5] = [a_2, b_2]$.

Using Eq. (13.3), we determine that

$$c_2 = 1.5 - \frac{g(1.5) \times (1.5 - 0.61538)}{f(1.5) - f(0.61538)} = 0.86838$$

Note that $|g(c_2)| = 0.4768 > \epsilon$ and we must proceed to the next iteration.

The seven iterations required to achieve the required accuracy are shown in Table 13.3.

Given a starting interval $[a, b]$ such that $f(a) \times f(b) < 0$ and a required error tolerance ϵ, the algorithm for the regula falsi method can be written as follows.

ALGORITHM FOR THE REGULA FALSI METHOD

1. $n = 1$, $a_n = a$, and $b_n = b$
2. $c_n = b_n - \frac{f(b_n) \times (b_n - a_n)}{f(b_n) - f(a_n)}$
3. if $|f(c_n)| \leq \epsilon$ accept $x_r \simeq c_n$ and end, otherwise
4. if $f(b_n) \times f(c_n) < 0$ then $a_{n+1} = c_n$ and $b_{n+1} = b_n$, otherwise $a_{n+1} = a_n$ and $b_{n+1} = c_n$
5. $n = n + 1$ and go to 2.

The algorithm is of course very similar to that for the bisection method.

EXAMPLE 13.6

Repeat Example 13.2 using the regula falsi method on the following intervals. You should use a fixed error tolerance of $\epsilon = 0.008$ in each case.

a. $[a, b] = [0, 2]$
b. $[a, b] = [-4, -2]$
c. $[a, b] = [0, 2]$

Solution

In each case, we apply the above algorithm with $a_1 = a$ and $b_1 = b$ until $|f(c_n)| < 0.008$. The iterations are summarized in Table 13.4.

Examples 13.5 and 13.6 were constructed so that direct comparisons can be made between the bisection and the regula falsi methods. Although the definition of error is different in each method, equivalent levels of accuracy in terms of the regula falsi's definition have been achieved between Examples 13.1 and 13.5, and 13.2 and 13.6. However, the reader should note that the regula falsi method has been quicker to converge in each case. Furthermore, note that Table 13.4 suggests that the number of iterations required is not simply a function of the width of the initial interval and the required error tolerance. This is in contrast to the bisection method, as demonstrated previously. In fact the convergence of the regula falsi method depends on the function $f(x)$ being solved, which is clear from Eq. (13.3). For this reason, it is not possible to perform the equivalent analysis that led us to Eq. (13.2) for the bisection method; a formal error analysis for the regula falsi method is beyond the scope of this book.

In addition to being a useful method to incorporate into computational procedures, the regula falsi method is often useful when finding the root of a function in "pen-and-paper" calculations. The key to its successful use by hand is in limiting the number of iterations that are required; ideally one should only have to compute a single iteration of Eq. (13.3). This can of course be achieved through the careful selection of a *narrow* starting interval.

Table 13.4 Iterations of the regula falsi method in Example 13.6

	n	a_n	b_n	$x_r \simeq c_n$	ϵ_n	$g(a_n)$	$g(b_n)$
a.	1	0.00000	2.00000	1.25853	−0.93316	−5.04795	2.97405
	2	1.25853	2.00000	1.43561	−0.09995	−0.93316	2.97405
	3	1.43561	2.00000	1.45396	−0.01002	−0.09995	2.97405
	4	1.45396	2.00000	1.45580	−0.00100		
b.	1	−4.00000	−2.00000	−3.27098	−0.92657	2.90805	−5.06995
	2	−4.00000	−3.27098	−3.44714	−0.09739	2.90805	−0.92657
	3	−4.00000	−3.44714	−3.46505	−0.00958	2.90805	−0.09739
	4	−4.00000	−3.46505	2.90805	−0.00958		
c.	1	0.00000	2.00000	0.72548	0.77688	1.00000	−1.75680
	2	0.72548	2.00000	1.11627	0.83146	0.77688	−1.75680
	3	1.11627	2.00000	1.40016	0.52487	0.83146	−1.75680
	4	1.40016	2.00000	1.53815	0.16206	0.52487	−1.75680
	5	1.53815	2.00000	1.57716	0.03135	0.16206	−1.75680
	6	1.57716	2.00000	1.58457	0.00517		

EXAMPLE 13.7

A business project requires an initial outlay of £10,000 and a further outlay of £20,000 after 2 years in order to generate an income of £40,000 after 5 years. Determine the approximate yield obtained from the project.

Solution

Expressed in units of £1000, the *equation of value* for the project is

$$10 + \frac{20}{(1+i)^2} = \frac{40}{(1+i)^5}$$

and we are required to solve this expression for i. This is equivalent to finding the relevant root of the function

$$f(i) = \frac{40}{(1+i)^5} - \frac{20}{(1+i)^2} - 10$$

We note that this expression is nonlinear and so could possibly have a number of real roots. However, note that the project returns £40,000 from a total outlay of £30,000 and so we expect a modest but positive yield, less than, say, 10% per annum. We therefore expect $f(i)$ to have a positive root, $i_r \in [0, 0.10]$, and this forms the focus of our search for a suitably narrow interval before applying a single iteration of the regula falsi method.

We begin by sampling $f(i)$ at values of i. In particular, $f(0.10) = -1.69207, f(0.09) = -0.83634$, and $f(0.08) = 0.07655$ which implies that $i_r \in [a, b] = [0.08, 0.09]$. The regula falsi method can then be used to obtain a more accurate estimate within this interval by using Eq. (13.3),

$$c = 0.09 - \frac{f(0.09) \times (0.09 - 0.08)}{f(0.09) - f(0.08)} = 0.08081$$

That is $i_r \approx 8.081\%$ which is confirmed from the observation that $f(0.08081) = 0.00038$ and is an acceptable level of accuracy. Note that Wolfram Alpha returns a value of $i_r \approx 8.08141\%$.

The above discussion of the bisection and regula falsi methods have assumed that we have prior knowledge of the approximation location of a single root. In practice, this is not always the case, particularly if the function is difficult to sketch. In such situations, the sampling technique used in Example 13.7 can be useful to find the number and approximate location of all roots over an extended interval. Either interval method can then be used to obtain more accurate estimates from within particular subintervals. Care needs to be taken to ensure that each subinterval found contains only one root, which may be a problem for highly oscillatory functions. This approach is demonstrated in the following example.

EXAMPLE 13.8

Determine all real roots of the following function over the interval $[-5, 5]$. You should use the regula falsi method with error tolerance $\epsilon = 0.00005$.

$$g(x) = x\sin(x) - 1.2$$

Solution

We note that $g(x)$ is an even function (symmetrical about $x = 0$) and so we can limit the effort required by working only in the positive interval $[0, 5]$. Any roots in this interval will be reflected about $x = 0$. We sample the function over the interval $[0, 5]$ in steps of 0.5, as shown in Table 13.5 and note that there appears to be two roots, contained in the subintervals $[1, 1.5]$ and $[2.5, 3]$.
The regula falsi method is then used in each subinterval, as shown in Table 13.6, to obtain both roots over $[0, 5]$. We therefore find the four roots of $g(x)$ over the interval $[-5, 5]$,

$$x_r = \pm 1.26029 \quad \text{and} \quad x_r = \pm 2.67671$$

Note that Wolfram Alpha returns roots at ± 1.26027 and ± 2.67672.

Table 13.5 Samples of $g(x)$ for Example 13.8

x	$g(x)$	x	$g(x)$
0	−1.20000	3	−0.77664
0.5	−0.96029	3.5	−2.42774
1	−0.35853	4	−4.22721
1.5	0.29624	4.5	−5.59889
2	0.61859	5	−5.99462
2.5	0.29618		

Table 13.6 Regula falsi method used to find the two roots in Example 13.8

n	a_n	b_n	$x_r \simeq c_n$	ϵ	$g(a_n)$	$g(b_n)$
1	1.00000	1.50000	1.27378	0.01801	−0.35853	0.29624
2	1.00000	1.27378	1.26069	0.00055	−0.35853	0.01801
3	1.00000	1.26069	1.26029	0.00002		
1	2.50000	3.00000	2.63804	0.07296	0.29618	−0.77664
2	2.63804	3.00000	2.66912	0.01468	0.07296	−0.77664
3	2.66912	3.00000	2.67526	0.00283	0.01468	−0.77664
4	2.67526	3.00000	2.67644	0.00054	0.00283	−0.77664
5	2.67644	3.00000	2.67667	0.00010	0.00054	−0.77664
6	2.67667	3.00000	2.67671	0.00002		

As both interval methods require the function to *cross* the x-axis at a root, neither method can be used to find a repeated root. Recall from Chapter 2 that, rather than crossing the x-axis, the function is tangential to it at a repeated root.

13.1.2 Gradient methods

Alternatives to the interval methods considered above can be developed that incorporate information about the continuous derivative of the function close to the root. Such approaches are often called *gradient methods*, for obvious reasons. As we shall see, gradient methods often converge much more quickly than interval methods and can in principle be used to find repeated roots. However, the compromise is that gradient methods are not always certain to converge.

The Newton-Raphson method

As in the previous discussions, we consider a single root, x_r, of the function $f(x)$. The Newton-Raphson method begins with an initial estimate of the root, denoted $x_0 \neq x_r$, and uses the tangent of $f(x)$ at x_0 to improve on the estimate of the root. In particular, the improvement, denoted x_1, is obtained from determining where the line tangent to $f(x)$ at x_0 crosses the x-axis. This represents a single iteration of the Newton-Raphson method and is illustrated in Figure 13.5(a). The next iteration uses the line tangent to $f(x)$ at x_1 to generate x_2 in exactly the same way. This is shown in Figure 13.5(b). The direction of the tangent line is determined by the sign of the local gradient at each x_n and, although the illustration has a positive local gradient, the method works analogously for functions with a negative local gradient.

Before we discuss the termination criteria for the Newton-Raphson method, it is important to consider the mathematical formulation of the first iteration of the process. Essentially, we require a expression that gives x_1 from x_0 and the properties of $f(x)$

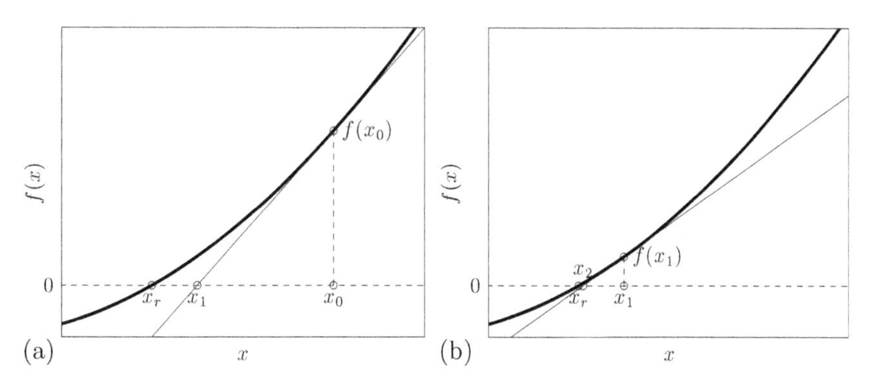

Figure 13.5 Illustration of the Newton-Raphson method. (a) First iteration leading to x_1 from the tangent line at $f(x_0)$. (b) Second iteration leading to x_2 from the tangent line at $f(x_1)$.

at x_0. The gradient of the line tangent to $f(x)$ at x_0 is, by definition, $f'(x_0)$ which can be determined from the expression of $f(x)$. We can also form an approximation to this gradient from the fact that the tangent line crosses the points $(x_0, f(x_0))$ and $(x_1, f(x_1)) \approx (x_1, 0)$; which of course arises from the definition of x_1. The gradient is then

$$\frac{f(x_0) - 0}{x_0 - x_1}$$

which is equated to $f'(x_0)$ and rearranged to give

$$x_1 = x_0 - \frac{f(x_0)}{f'(x_0)} \tag{13.4}$$

That is, we have an expression for estimate x_1 determined entirely from *known* properties of the function at x_0. Clearly the expression assumes that $f'(x_0) \neq 0$. If this was the case, the tangent line of the function at x_0 would be horizontal and not cross the x-axis; the Newton-Raphson method would then have failed to improve on x_0.

EXAMPLE 13.9

Use the Newton-Raphson method to determine an improvement on the initial estimate of the root in the following cases.

a. $f(x) = e^x - 4$ from initial estimate $x_0 = 1.5$.
b. $g(x) = x^3 - 2x - 4$ from initial estimate $x_0 = 2.5$.
c. $h(x) = \ln(x) - 2$ from initial estimate $x_0 = 8$.
d. $k(x) = \ln(x^3) - 5$ from initial estimate $x_0 = 5.5$.

Solution

In each case, we apply Eq. (13.4).

a. $f(x) = e^x - 4$ and $f'(x) = e^x$, therefore $x_1 = 1.5 - \frac{e^{1.5}-4}{e^{1.5}} = 1.3925$.
b. $g(x) = x^3 - 2x - 4$ and $g'(x) = 3x^2 - 2$, therefore $x_1 = 2.5 - \frac{2.5^3 - 2\times 2.5 - 4}{3\times 2.5^2 - 2} = 2.1045$.
c. $h(x) = \ln(x) - 2$ and $h'(x) = \frac{1}{x}$, therefore $x_1 = 8 - 8(\ln(8) - 2) = 7.3645$.
d. $k(x) = \ln(x^3) - 5$ and $k'(x) = \frac{3}{x}$, therefore $x_1 = 5.5 - \frac{5.5}{3}(\ln(5.5^3) - 5) = 5.2906$.

Equation (13.4) can be generalized to give an expression for the $(n+1)$th estimate under the Newton-Raphson method from properties of the nth iteration,

$$x_{n+1} = x_n - \frac{f(x_n)}{f'(x_n)} \tag{13.5}$$

for $n \in \mathbb{N}$ and $f'(x_n) \neq 0$.

The standard error estimate used in an implementation of the Newton-Raphson method is $\epsilon_n = |x_n - x_{n-1}|$. This means that an exit criteria is simply that $\epsilon_n < \epsilon$ for some predetermined tolerance, ϵ. That is, we terminate the iterative process when successive approximations become only marginally different.

EXAMPLE 13.10

Use the Newton-Raphson method with error tolerance $\epsilon = 0.0001$ to find the single real root of $h(x) = \ln(x) - 2$ from the initial estimate $x_0 = 8$.

Solution

As in Example 13.9(c), $f'(x) = \frac{1}{x}$ and so Eq. (13.5) is written as

$$x_{n+1} = x_n - x_n(\ln(x_n) - 2)$$

Table 13.7 shows the results of the first few iterations of this expression starting from $x_0 = 8$. We see that only three iterations are required to achieve the root to the required error tolerance and $x_r \simeq 7.38906$. The actual root is given by $x_r = e^2 = 7.38906$ which agrees with the estimate to the five decimal places reported.

Table 13.7 Iterations of the Newton-Raphson method for Example 13.10

n	x_n	$f(x_n)$	$f'(x_n)$	ϵ_n
0	8.00000	0.07944	0.12500	–
1	7.36447	−0.00333	0.13579	0.63553
2	7.38902	−0.00001	0.13534	0.02455
3	7.38906	0.00000	0.13534	0.00004

For a given initial estimate x_0 and a required error tolerance ϵ, the algorithm for the Newton-Raphson method can be written as follows.

ALGORITHM FOR THE NEWTON-RAPHSON METHOD

1. $n = 0$
2. $x_{n+1} = x_n - \frac{f(x_n)}{f'(x_n)}$
3. if $|x_{n+1} - x_n| \leq \epsilon$ accept $x_r \simeq x_{n+1}$ then end, otherwise
4. $n = n + 1$ and go to 2.

Although the description of the Newton-Raphson method has been given for functions with a single root, the method can be applied perfectly well to functions with multiple roots. The root on which the method converges is of course determined by the starting value, x_0. As with the interval methods, it is sensible to have a rough idea of the number of roots that a function may possess and the approximation location of each. There are no definitive rules on how an initial estimate should be determined such that the Newton-Raphson method approaches the particular root of interest and common sense should be applied in practice.

EXAMPLE 13.11

Use the Newton-Raphson method to determine all real roots of the function $f(x) = e^{(x-1)^2} - 2$. An error tolerance of $\epsilon = 0.0001$ should be used.

Solution

It should be clear that, in this case, $f(x)$ is an even function about $x = 1$ and has a root either side of this value. We might then guess that two initial values $x_0^1 = 0$ and $x_0^2 = 2.5$ would converge to the two distinct roots. In either case, the iterative formula (Eq. 13.5) is the same and given by

$$x_{n+1} = x_n - \frac{e^{(x_n-1)^2} - 2}{2(x_n - 1)\, e^{(x_n-1)^2}}$$

The iterations are shown in Table 13.8 where we see rapid converge to roots $x_r^1 \simeq x_4^1 = 1.83255$ and $x_r^2 \simeq x_4^2 = 0.16745$.

Table 13.8 Iterations of the Newton-Raphson method for the two roots in Example 13.11

$n^{1,2}$	$x_n^{1,2}$	$f(x_n^{1,2})$	$f'(x_n^{1,2})$	$\epsilon_n^{1,2}$
0	0.00000	0.71828	−5.43656	
1	0.13212	0.12382	−3.68643	0.13212
2	0.16571	0.00580	−3.34685	0.03359
3	0.16744	0.00001	−3.33026	0.00173
4	0.16745	0.00000	−3.33022	0.00000
0	2.00000	0.71828	5.43656	
1	1.86788	0.12382	3.68643	0.13212
2	1.83429	0.00580	3.34685	0.03359
3	1.83256	0.00001	3.33026	0.00173
4	1.83255	0.00000	3.33022	0.00000

EXAMPLE 13.12

Repeat Example 13.8 using the Newton-Raphson method with error tolerance $\epsilon = 0.0001$.

Solution

We are required to find the four roots of $g(x) = x\sin(x) - 1.2$ over $x \in [-5, 5]$. As in the solution to Example 13.8, we exploit the fact that the function is even and consider only $x \in [0, 5]$. We note that $g'(x) = \sin(x) + x\cos(x)$ and Eq. (13.5) becomes

$$x_{n+1} = x_n - \frac{x_n \sin(x_n) - 1.2}{\sin(x_n) + x_n \cos(x_n)}$$

Table 13.9 Iterations of the Newton-Raphson method for the two roots in [0, 5] in Example 13.12

$n^{1,2}$	$x_n^{1,2}$	$f(x_n^{1,2})$	$f'(x_n^{1,2})$	$\epsilon_n^{1,2}$
0	1.00000	−0.35853	1.38177	
1	1.25947	−0.00107	1.33773	0.25947
2	1.26027	0.00000	1.33726	0.00080
3	1.26027	0.00000	1.33726	0.00000
0	2.50000	0.29618	−1.40439	
1	2.71090	−0.06819	−2.04582	0.21090
2	2.67756	−0.00165	−1.94688	0.03333
3	2.67672	0.00000	−1.94435	0.00085
4	2.67672	0.00000	−1.94435	0.00000

Table 13.8 suggests that initial values $x_0^1 = 1$ and $x_0^2 = 2.5$ are sensible choices, although we could have equally chosen $x_0^1 = 1.5$ and $x_0^2 = 3$. The iterations are shown in Table 13.9 and we find four roots over the full interval $[-5, 5]$, $x_r \simeq \pm x_3^1 = \pm 1.26027$ and $x_r \simeq \pm x_4^2 = \pm 2.67672$.

Although the definition of the error tolerance is different between the regula falsi and Newton-Raphson methods, it should be clear from the second column of Table 13.9, for example, that the Newton-Raphson method can lead to more rapid convergence. However, the Newton-Raphson method is not without its drawbacks.

The most immediate problem with the Newton-Raphson method is that it requires an explicit expression for the derivative of the function. This may not be possible to determine in practice. A strategy for avoiding this requirement will lead us to the *secant method* in the next section. The Newton-Raphson method (and indeed the secant method) suffers from further disadvantages concerning their use with *ill-behaved* functions. In terms of these gradient methods, a function is said to be ill-behaved if it has repeated roots or a very small gradient at a particular x_n. In both cases, the convergence to the root may be very slow, or indeed impossible. This is of course due to the effect of the small gradient at x_n on the calculation of x_{n+1}. In particular, Eq. (13.5) shows that

$$x_{n+1} - x_n \to \infty \text{ as } f'(x_n) \to 0$$

We have previously noted that interval methods fail to cope with repeated roots, and so, even though gradient methods may be slow to converge to such roots, they are still superior in this respect.

EXAMPLE 13.13

Use the Newton-Raphson method to find the repeated root of $f(x) = (x-1)^3$ with an error tolerance $\epsilon = 0.0001$.

Solution

The Newton-Raphson method requires iteration of

$$x_{n+1} = x_n - \frac{(x_n - 1)^3}{3(x_n - 1)^2} = x_n - \frac{1}{3}(x_n - 1)$$

The process beginning at $x_0 = 1.5$ takes 20 iterations to converge. The reader is invited to confirm this using Excel, for example.

EXAMPLE 13.14

Explore the use of the Newton-Raphson method to determine all real roots of the function $g(x) = (x-2)^2 - 1$. Use an error tolerance of $\epsilon = 0.0002$.

Solution

The function clearly has two real roots at $x_r = 1$ and $x_r = 3$ but it also has a zero gradient at $x = 2$. The Newton-Raphson method requires iteration of

$$x_{n+1} = x_n - \frac{(x_n - 2)^2 - 1}{2(x_n - 2)}$$

While the Newton-Raphson method converges quickly to the relevant root for a reasonable estimate of $x_0 \in [3, \infty)$ or $x_0 \in (-\infty, 1]$, it can be thrown off for $x_0 \in (1, 3)$ and will not converge if $x_n = 2$ for some n. An example of where the convergence is relatively slow is given in Table 13.10. Note the large jumps in successive values when x_n is close to 2.

Table 13.10 Slow convergence of the Newton-Raphson method close to a turning point at $x = 2$

n	x_n	$g(x_n)$	$g'(x_n)$	ϵ_n
0	2.10000	−0.99000	0.20000	
1	7.05000	24.50250	10.10000	4.95000
2	4.62401	5.88543	5.24802	2.42599
3	3.50255	1.25767	3.00511	1.12146
4	3.08404	0.17515	2.16809	0.41851
5	3.00326	0.00653	2.00652	0.08079
6	3.00001	0.00001	2.00001	0.00325
7	3.00000	0.00000	2.00000	0.00001

The secant method

We begin by considering a single root x_r of the function $f(x)$. The *secant method* is similar to the Newton-Raphson method in that a straight line is used to determine the next approximation to the root. In contrast to the Newton-Raphson method, the secant method uses *two* initial guesses for the root, x_0 and $x_1 (\neq x_0)$, and a straight line is fitted between the evaluations of $f(x)$ at these positions. This line is called the *secant line* and an approximation of the root, x_2, is given by the intercept of the secant line with the x-axis. The secant method does not therefore need $f'(x)$, which is an advantage of this approach over the Newton-Raphson method. The secant method is illustrated in Figure 13.6, where we note the two situations that can occur regarding the location of x_0 and x_1 relative to x_r; namely that x_0 and x_1 are both on the same side of x_r, or they are on opposite sides of x_r.

The expression for implementing the secant method, that is, for computing x_2 from x_0 and x_1, can be derived by forming two different expressions for the gradient of the secant line, one involving x_1 and x_2 and another involving x_0 and x_1. This approach is similar to that used in the regula falsi method. In particular, we have gradients

$$\frac{f(x_1) - f(x_0)}{x_1 - x_0} \quad \text{and} \quad \frac{0 - f(x_1)}{x_2 - x_1}$$

Note that the second expression is an approximation to the gradient and replies on $f(x_2) \simeq 0$. These expressions can be equated and rearranged to give

$$x_2 = x_1 - f(x_1)\frac{x_1 - x_0}{f(x_1) - f(x_0)} \tag{13.6}$$

Note that the secant method requires $f(x_1) \neq f(x_0)$. Although we have assumed that $x_1 \neq x_0$, this constraint may be problematic for roots close to a turning point in $f(x)$.

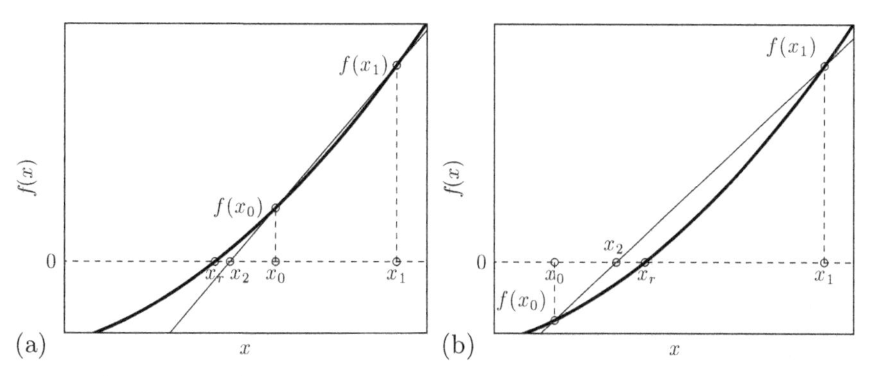

Figure 13.6 The first iteration of the secant method. (a) x_0 and x_1 are on the same side of x_r, either the right or the left. (b) x_0 and x_1 are on opposite sides of x_r, in either order.

EXAMPLE 13.15

Use the secant method to determine an improvement on the estimated root in the following cases.

a. $f(x) = e^x - 4$ from initial estimates $x_0 = 1.2$ and $x_1 = 1.5$.
b. $g(x) = x^3 - 2x - 4$ from initial estimates $x_0 = 2.3$ and $x_1 = 2.5$.
c. $h(x) = \ln(x) - 2$ from initial estimates $x_0 = 7$ and $x_1 = 8$.
d. $k(x) = \ln(x^3) - 5$ from initial estimates $x_0 = 5.5$ and $x_0 = 6.1$.

Solution

In each case, we apply Eq. (13.6).

a. $x_2 = 1.5 - f(1.5)\frac{1.5-1.2}{f(1.5)-f(1.2)} = 1.37559$
b. $x_2 = 2.5 - f(2.5)\frac{2.5-2.3}{f(2.5)-f(2.3)} = 2.06671$
c. $x_2 = 8 - f(8)\frac{8-7}{f(8)-f(7)} = 7.40507$
d. $x_2 = 6.1 - f(6.1)\frac{6.1-5.5}{f(6.1)-f(5.5)} = 5.27932$

We have seen that a single iteration of the secant method generates x_2 from x_0 and x_1. The next iteration of the method would then use x_1 and x_2 to generate x_3. In general, the nth iteration of the method requires x_n and x_{n-1} to generate x_{n+1} from the generalization of Eq. (13.6),

$$x_r \simeq x_{n+1} = x_n - f(x_n)\frac{x_n - x_{n-1}}{f(x_n) - f(x_{n-1})} \tag{13.7}$$

The error in the estimate of the root from the nth iteration $x_r \simeq x_{n+1}$ is given by $\epsilon_n = |x_{n+1} - x_n|$ for $n \in \mathbb{N} > 1$. The stopping criteria for the process is then $\epsilon_n \leq \epsilon$ for some predetermined error tolerance ϵ.

For a given pair of initial estimates, x_0 and x_1, and a required error tolerance ϵ the algorithm for the secant method can be written as follows.

ALGORITHM FOR THE SECANT METHOD

1. $n = 1$
2. $x_{n+1} = x_n - f(x_n)\frac{x_n - x_{n-1}}{f(x_n)-f(x_{n-1})}$
3. if $|x_{n+1} - x_n| \leq \epsilon$ accept $x_r \simeq x_{n+1}$ then end, otherwise
4. $n = n + 1$ and go to 2.

EXAMPLE 13.16

Use the secant method to determine the single root of $h(x) = \ln(x) - 2$ with an error tolerance $\epsilon = 0.00001$.

Solution

We use the initial values as stated in Example 13.15(c) and apply Eq. (13.7). The iterations are shown in Table 13.11 and we see converge to $x_r \simeq x_5 = 7.38906$ which agrees with the actual root $x_r = e^2$ to five decimal places.

Table 13.11 Iterations of the secant method in Example 13.16

n	x_{n-1}	x_n	$g(x_{n-1})$	$g(x_n)$	x_{n+1}	ϵ_n
1	7.00000	8.00000	−0.05409	0.07944	7.40507	0.59493
2	8.00000	7.40507	0.07944	0.00217	7.38840	0.01667
3	7.40507	7.38840	0.00217	−0.00009	7.38906	0.00065
4	7.38840	7.38906	−0.00009	0.00000	7.38906	0.00000

As a *two-point method*, care needs to taken when using the secant method for functions with multiple roots. For example, if initial estimates x_0 and x_1 are picked either side of the root of interest, care should be taken to ensure that they enclose only that root. Furthermore, if initial estimates are taken to one side of root, care must to taken to ensure that these do not enclose another root. Common sense should be used and, in cases when an unfamiliar function is being solved, there is some value in sampling that function at various points to get a feel for the number and spacing of the roots, as was done in Table 13.5, for example.

EXAMPLE 13.17

Repeat Example 13.12 using the secant method with error tolerance $\epsilon = 0.00001$.

Solution

We are required to find the four roots of $f(x) = x\sin(x) - 1.2$ over $x \in [-5, 5]$. As before we consider only the interval $[0, 5]$ where we know two roots to exist and reflect these about $x = 0$. We use two sets of initial values $x_0^1 = 1$, $x_1^1 = 1.1$ and $x_0^2 = 2.8$, $x_1^1 = 2.9$ which are known to be to one side of each root, respectively. Equation (13.7) can then be applied for both sets of initial data, as shown in Table 13.12. Four roots are found over the full interval $[-5, 5]$, $x_r \simeq \pm x_5^1 = \pm 1.26027$ and $x_r \simeq \pm x_6^2 = \pm 2.67672$.

Table 13.12 Iterations of the secant method for the two roots in [0, 5] in Example 13.17

$n^{1,2}$	$x_{n-1}^{1,2}$	$x_n^{1,2}$	$f(x_{n-1}^{1,2})$	$f(x_n^{1,2})$	$x_{n+1}^{1,2}$	$\epsilon_n^{1,2}$
1	1.00000	1.10000	−0.35853	−0.21967	1.25820	0.15820
2	1.10000	1.25820	−0.21967	−0.00277	1.26022	0.00202
3	1.25820	1.26022	−0.00277	−0.00007	1.26027	0.00005
4	1.26022	1.26027	−0.00007	0.00000	1.26027	0.00000
1	2.80000	2.90000	−0.26203	−0.50618	2.69267	0.20733
2	2.90000	2.69267	−0.50618	−0.03140	2.67896	0.01371
3	2.69267	2.67896	−0.03140	−0.00437	2.67675	0.00222
4	2.67896	2.67675	−0.00437	−0.00005	2.67672	0.00003
5	2.67675	2.67672	−0.00005	0.00000	2.67672	0.00000

It is generally true that the Newton-Raphson method requires fewer iterations than the secant method to converge to x_r with the same error tolerance. While it is possible to prove this with an analysis of the convergence speeds, this is beyond the scope of this book and we take this fact from observations of the above examples. However, as we have previously noted, the Newton-Raphson method requires an evaluation of both $f(x_n)$ and $f'(x_n)$ during each iteration. Even when an explicit expression for the derivative is known, the evaluation of these two distinct functions during each iteration is computationally expensive. Although the secant method requires the evaluation of $f(x)$ at both x_n and x_{n-1} during each iteration, it can be implemented in such a way that previous evaluations of $f(x_{n-1})$ are kept which greatly speeds up the program. Although typically taking more iterations than the Newton-Raphson method, the secant method is therefore often quicker in practice.

13.2 NUMERICAL DIFFERENTIATION

We now consider a number of methods for approximating numerically the derivative of a function. In practice, it is sometimes the case that, rather than having the explicit form of $f(x)$ from which we can directly determine $f'(x)$, we have a collection of data points representing $\{f(x_n)\}$. Numerical methods for approximating $\{f'(x_n)\}$ from $\{f(x_n)\}$ are therefore required. This is the motivation for this section. We begin with a discussion of the various methods in principle and return to their practical use for a given data set $\{f(x_n)\}$ at the end of this section.

13.2.1 Elementary numerical approximation

Recall from Eq. (3.4) and the surrounding discussion in Chapter 3 that the definition of $f'(x)$ is simply

$$f'(x) = \lim_{h \to 0} \frac{f(x+h) - f(x)}{h} \tag{13.8}$$

This expression implies an obvious strategy for obtaining $f'(x)$ numerically. Namely we consider a sequence $\{h_k\}$ such that $h_k \to 0$ as k increases and compute the limit of the resulting sequence $\{D_{h_k}f(x)\}$ as k increases, where

$$D_{h_k}f(x) = \frac{f(x+h_k) - f(x)}{h_k} \tag{13.9}$$

for $k = 1, 2, \ldots$. It should be clear that $D_h f(x)$ denotes the *numerical derivative of* $f(x)$ *with stepsize* h.

EXAMPLE 13.18

Compute $\{D_{h_k}f(x)\}$ at $x = 2$ for the following functions. You should use stepsize $h_k = 10^{-k}$ for $k = 1, 2, \ldots$ until an apparent convergence to five decimal places is observed.

a. $f(x) = 0.7x$
b. $g(x) = \ln(x)$
c. $k(x) = e^x$
d. $m(x) = \cos(x)$

Table 13.13 Numerical derivatives for Example 13.18

	k	h_k	$D_{h_k}f(x)$
a.	1	0.1	0.70000
	2	0.01	0.70000
b.	1	0.1	0.48790
	2	0.01	0.49875
	3	0.001	0.49988
	4	0.0001	0.49999
	5	0.00001	0.50000
	6	0.000001	0.50000
c.	1	0.1	7.77114
	2	0.01	7.42612
	3	0.001	7.39275
	4	0.0001	7.38943
	5	0.00001	7.38909
	6	0.000001	7.38906
	7	0.0000001	7.38906
d.	1	0.1	−0.88699
	2	0.01	−0.90720
	3	0.001	−0.90909
	4	0.0001	−0.90928
	5	0.00001	−0.90930
	6	0.000001	−0.90930

Solution

Equation (13.9) is used in each case, such that

$$D_{h_k}f(x)\,|_{x=2} = \frac{f(2+h_k) - f(2)}{h_k}$$

The calculations are shown in Table 13.13.

a. We find apparent convergence to $D_{h_k}f(x) = 0.7$ which is equal to the actual value of $f'(2) = 0.7$.

b. We find apparent convergence to $D_{h_k}g(x) = 0.5$ which is equal to the actual value of $g'(2) = 1/2$.

c. We find apparent convergence to $D_{h_k}k(x) = 0.738906$ which is equal to the actual value of $k'(2) = \mathrm{e}^2$ to five decimal places.

d. We find apparent convergence to $D_{h_k}m(x) = 0.5$ which is equal to the actual value of $m'(2) = -\sin(2)$ to five decimal places.

It should be clear from Example 13.18 that, while apparent convergence to the required accuracy has been reached in each case, the number of evaluations required depends on the form of $f(x)$. Furthermore, the number of evaluations required for apparent convergence may depend on the particular value of x for any given function, depending on properties of the local gradient of that function close to x. Unfortunately there is no clear way to determine the required value of h_k in advance to ensure an accurate calculation of the numerical derivative for a given function and location.

Some insight into these issues can gained from consideration of the Taylor expansion of $f(x)$. Recall from Chapter 5 that

$$f(x+h) = f(x) + hf'(x) + \frac{h^2}{2}f''(x) + O(h^3)$$

which can be substituted into Eq. (13.9) to yield

$$\begin{aligned} D_h f(x) &= \frac{1}{h}\left\{\left(f(x) + hf'(x) + \frac{h^2}{2}f''(x) + O(h^3)\right) - f(x)\right\} \\ &= f'(x) + \frac{h}{2}f''(x) + O(h^2) \end{aligned}$$

where $O(h^2)$ denotes the collection of all terms of order h^2 and above. We therefore see that the error in using the numerical derivative is given theoretically by

$$f'(x) - D_h f(x) = -\frac{h}{2}f''(x) + O(h^2) \tag{13.10}$$

The expression shows that this error is, at leading order, proportional to the product of h and the second derivative of the function evaluated at the particular x.

The situation is yet more complicated as illustrated by the following observations of an extension to the calculations for Example 13.18(d),

$$D_{10^{-6}} \cos(x)|_{x=2} = 0.90930 \quad \text{and} \quad D_{10^{-15}} \cos(x)|_{x=2} = -0.77716$$

This clearly shows that, although the value of the numerical derivative had apparently converged to the correct value for stepsize $h_k = 10^{-k} = 10^{-6}$, further increases in k cause the estimate to diverge from the accurate value. The reason for this behavior is that $h_k = 10^{-k}$ will eventually lose numerical significance within the computer the method is implemented on.

One therefore has to be careful in the picking the value of h when evaluating the numerical derivative by Eq. (13.8). A balance is required such that h is sufficiently small to allow an acceptable error for any given function $f(x)$, but not too small that numerical significance is lost within the implementation. In practice, the limiting procedure as demonstrated in Example 13.18 is required.

Note, however, that we are motivated by the need to evaluate $\{f'(x_n)\}$ from given $\{f(x_n)\}$ and $\{x_n\}$ and so we may not be free to vary $h = x_n - x_{x-1}$. Alternative methods for approximating $f'(x)$ which give significantly lower error for larger values of h are therefore required.

13.2.2 Central-difference formulas

Assuming that it is possible to evaluate $f(x)$ either side of the location of interest, the *central-difference formula of* $O(h^2)$ for obtaining the numerical derivative, denoted $D_h^{[2]}f(x)$, is given by

$$f'(x) \simeq D_h^{[2]}f(x) = \frac{f(x+h) - f(x-h)}{2h} \tag{13.11}$$

EXAMPLE 13.19

Use Taylor expansions of $f(x \pm h)$ to show that the theoretical error in using the central-difference formula (Eq. 13.11) to approximate $f'(x)$ is $O(h^2)$.

Solution

We note the following Taylor expansions

$$\begin{aligned} f(x+h) &= f(x) + hf'(x) + \frac{h^2}{2}f''(x) + O(h^3) \\ f(x-h) &= f(x) - hf'(x) + \frac{h^2}{2}f''(x) + O(h^3) \end{aligned}$$

and so

$$f(x+h) - f(x-h) = 2f'(x)h + O(h^3)$$

We therefore see that

$$f'(x) - \frac{f(x+h) - f(x-h)}{2h} = O(h^2)$$

as required. Furthermore, we note that, at leading order, this error depends on $f^{(3)}(x)$.

EXAMPLE 13.20

Repeat Example 13.18 using the central-difference formula of $O(h^2)$.

Solution

Equation (13.11) is used in each case and the results for increasing k are shown in Table 13.14. The converged numerical derivatives agree with the actual derivatives to five decimal places, as stated in Example 13.18.

We can go yet further and define the *central-difference formula of* $O(h^4)$ to be

$$f'(x) \simeq D_h^{[4]}f(x) = \frac{-f(x+2h) + 8f(x+h) - 8f(x-h) + f(x-2h)}{12h} \tag{13.12}$$

Taylor expansions of the functions in the numerator can again be taken to demonstrate that this expression has theoretical error $O(h^4)$ and, at leading order, depends on $f^{(5)}(x)$. This is left as an exercise for the interested reader.

Table 13.14 Numerical derivatives for Examples 13.20 and 13.21

	k	h_k	$D_{h_k}^{[2]}f(x)$	$D_{h_k}^{[4]}f(x)$
a.	1	0.1	0.7	0.7
	2	0.01	0.7	0.7
b.	1	0.1	0.50042	0.50000
	2	0.01	0.50000	0.50000
	3	0.001	0.50000	
c.	1	0.1	7.40138	7.38903
	2	0.01	7.38918	7.38906
	3	0.001	7.38906	7.38906
	4	0.0001	7.38906	

Equations (13.11) and (13.12) are therefore significant improvements on Eq. (13.8) in the sense that they are expected to given significantly lower errors when approximating $f(x)$ at particular h. Furthermore, the theoretical errors have been shown to depend on higher-order derivatives of $f(x)$ which might be expected to be more slowly varying.

EXAMPLE 13.21

Repeat Example 13.18 using the central-difference formula of $O(h^4)$.

Solution

Equation (13.12) is used in each case and the results for increasing k are shown in Table 13.14. The converged numerical derivatives agree with the actual derivatives to five decimal places, as stated in Example 13.18. Quicker convergence is seen for the $O(h^4)$ method.

EXAMPLE 13.22

Use Eqs. (13.8), (13.11), and (13.12) to obtain the numerical derivative of $f(x) = e^{x^2-x+2}$ at $x = 0.75$ with stepsize $h_k = 10^{-k}$ for $k = 1, \ldots, 10$.

Solution

The derivative can be determined explicitly as

$$f'(0.5) = (2x-1)\, e^{x^2-x+2}|_{x=0.75} = 3.06287$$

and this can be used to determine the error in the numerical approaches. Table 13.15 shows the computations for each method, together with the error calculated as the difference between the actual and the numerical derivatives.

The results of Example 13.22 show considerably quicker convergence from the central-difference methods, with particularly rapid convergence from the $O(h^4)$ method. The downside of using the higher-order central difference approaches is of course the number of values of $f(x)$ required to determine $f'(x)$ at some x.

Table 13.15 Numerical derivatives using the alternative methods for Example 13.22

k	h_k	$D_h f(x)$	|Error|	$D_h^{[2]} f(x)$	|Error$^{[2]}$|	$D_h^{[4]} f(x)$	|Error$^{[4]}$|
1	0.1	3.78795	−0.72508	3.09494	−0.03207	3.06220	0.00067
2	0.01	3.13211	−0.06924	3.06319	−0.00032	3.06287	0.00000
3	0.001	3.06977	−0.00690	3.06287	0.00000	3.06287	0.00000
4	0.0001	3.06356	−0.00069	3.06287	0.00000	3.06287	0.00000
5	0.00001	3.06294	−0.00007	3.06287	0.00000	3.06287	0.00000
6	0.000001	3.06288	−0.00001	3.06287	0.00000	3.06287	0.00000
7	0.0000001	3.06287	0.00000	3.06287	0.00000	3.06287	0.00000
8	0.00000001	3.06287	0.00000	3.06287	0.00000	3.06287	0.00000

13.2.3 Practical use

The previous discussion demonstrates that, given a data set $\{f(x_n)\}$ at evenly spaced *node points* $\{x_n\}$, the numerical derivatives $\{f'(x_n)\}$ can be estimated with reasonable accuracy using either the $O(h^2)$ or $O(h^4)$ central-difference methods given by Eqs. (13.11) and (13.12), respectively. Although it is possible to estimate the derivatives from the elementary approach given by Eq. (13.8), the resulting approximations would only be accurate from sufficiently tightly spaced $\{x_n\}$.

The node points $\{x_n\}$ will of course fix the stepsize h and we know that the $O(h^4)$ method will provide an accurate estimate of the derivatives using four data points, two either side of each location of interest, for most reasonable h. The downside of this method is that the derivative cannot be obtained close to the edges of the data set. For example, given $\{x_n\}$ for $n = 1, 2, \ldots, 50$ such that $x_{n+1} = x_n + h$, the left-most position that Eq. (13.12) can be used is x_3. In particular,

$$\begin{aligned} f'(x_3) &\simeq \frac{-f(x_3 + 2h) + 8f(x_3 + h) - 8f(x_3 - h) + f(x_3 - 2h)}{12h} \\ &= \frac{-f(x_5) + 8f(x_4) - 8f(x_2) + f(x_1)}{12h} \end{aligned}$$

The right-most position that this approach can be used is x_{48}.

The $O(h^2)$ approach also suffers from this restriction, but to a lesser extent. For example, the left-most position that Eq. (13.11) can be used is x_2. This downside of the $O(h^4)$ method may or may not be critical, but it is clear that the approach will give superior estimates of $\{f'(x_n)\}$ for $n = 3, 4, 5, \ldots, 47, 48$.

EXAMPLE 13.23

Table 13.16 gives $\{f(x_n)\}$ and $\{x_n\}$ for 21 regularly spaced node points over $[0, 3]$ for an unknown function. Use the $O(h^2)$ and $O(h^4)$ central-difference formula to approximate $\{f'(x_n)\}$. Plot the results and comment on your answer.

Table 13.16 Numerical derivatives using the $O(h^2)$ and $O(h^4)$ central-difference formula in Example 13.23

n	x_n	$f(x_n)$	$D_h^{[2]}f(x_n)$	$D_h^{[4]}f(x_n)$	n	x_n	$f(x_n)$	$D_h^{[2]}f(x_n)$	$D_h^{[4]}f(x_n)$
1	0.00	0.00000	–	–	12	1.65	0.40693	−3.01570	−2.92107
2	0.15	0.02250	0.29960	–	13	1.80	−0.09825	−3.57412	−3.40255
3	0.30	0.08988	0.59816	0.59540	14	1.95	−0.61383	−3.05820	−2.85460
4	0.45	0.20112	0.88242	0.87465	15	2.10	−0.95463	−1.22368	−1.08500
5	0.60	0.35227	1.12392	1.10728	16	2.25	−0.93933	1.56027	1.51662
6	0.75	0.53330	1.26938	1.24004	17	2.40	−0.49964	4.14189	3.85632
7	0.90	0.72429	1.24061	1.19679	18	2.55	0.21756	4.91928	4.48258
8	1.05	0.89234	0.94566	0.89057	19	2.70	0.84513	2.81171	2.48868
9	1.20	0.99146	0.30857	0.25384	20	2.85	0.96417	−1.44338	–
10	1.35	0.96849	−0.67871	−0.71128	21	3.00	0.41212	–	–
11	1.50	0.77807	−1.89042	−1.87186					

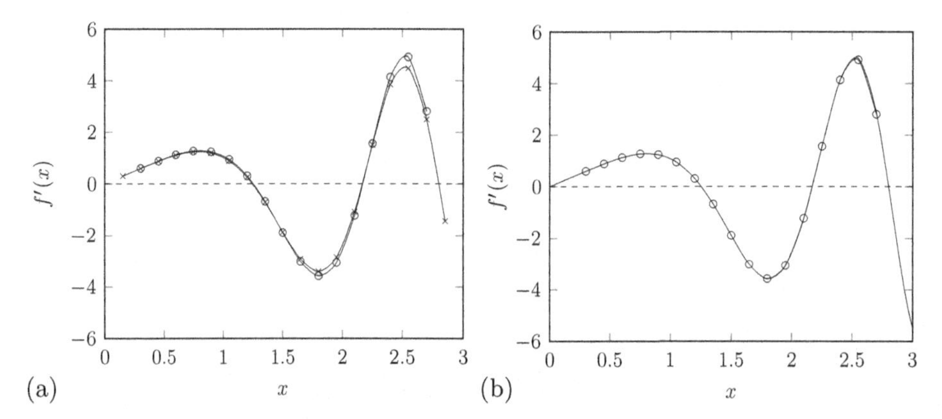

Figure 13.7 Numerical derivatives calculated from the central-difference approaches in Example 13.23. (a) Comparison between $D_h^{[2]}f(x)$ "-x" and $D_h^{[4]}f(x)$ "-o." (b) Comparison between $D_h^{[4]}f(x)$ "-o" and the actual $f'(x) = 3x\cos(x^2)$.

Solution

These data fix $h = 0.15$ and we apply Eqs. (13.11) and (13.12) with this value of h. The results are shown in Table 13.16 and Figure 13.7(a). We see some numerical differences between each $D_h^{[2]}f(x_n)$ and $D_h^{[4]}f(x_n)$, and expect $\{D_h^{[4}f(x_n)\}$ to give the better approximation to $\{f'(x_n)\}$.

These data for Example 13.23 were in fact generated from $f(x) = \sin(x^2)$ and so $f'(x) = 2x\cos(x^2)$. Figure 13.7(b) shows a comparison between the actual $\{f'(x_n)\}$ and the $O(h^4)$ central-difference approximations at each x_n. Excellent agreement is seen away from the edge points $x = 0$ and $x = 3$.

13.3 NUMERICAL INTEGRATION

It should be clear from Chapter 7 that, although many strategies exist for integration, many if not most integrals that occur in practice cannot be evaluated algebraically. For example, the definite *Gaussian* integral

$$\int_a^b \mathrm{e}^{-x^2}\,\mathrm{d}x$$

for some $a \neq b \in \mathbb{R}$, occurs regularly in probability theory but cannot be evaluated algebraically. Another example is, say,

$$\int_a^b \frac{\mathrm{e}^x}{x}\,\mathrm{d}x$$

The barrier to evaluating these and other integrals by hand is that the integrands cannot be expressed in terms of derivatives of elementary functions, such as those considered in Chapters 6 and 7. Numerical methods are therefore required.

In what follows, we will be concerned with definite integrals of the form

$$I = \int_a^b f(x)\,\mathrm{d}x \tag{13.13}$$

The two most fundamental approaches to numerical integration are the *trapezoidal approach* and *Simpson's approach* and we describe both in this final section on elementary numerical methods. Both methods require one to replace the integrand, $f(x)$, with a function that approximates it over the interval $[a, b]$. The approximating function is chosen such that it can be readily integrated. As we shall see, the two approaches are distinguished by the approximating function used.

13.3.1 The trapezoidal approach

The trapezoidal approach to definite integration approximates the integrand of Eq. (13.13) by a collection of *linear polynomials*, that is straight lines, over the domain of integration $[a, b]$.

This approximation of $f(x)$ is denoted $P^T_{[a,b],k}(f(x))$, where k represents the number of linear polynomials used over $[a, b]$. Each linear polynomial is fitted to $f(x)$ at consecutive node points, $\{x_i = a + hi\}$ where $i = 0, \ldots, k$ and $h = (b - a)/k$. Note that $x_0 = a$ and $x_k = b$.

The first level of approximation under the trapezoidal approach is shown in Figure 13.8(a). Here, we see the integrand being approximated by a single straight line, $P^T_{[a,b],1}(f(x))$. The straight line is simply that which passes through $f(x_0 = a)$ and $f(x_1 = b)$. We say that the approximation is *trapezoidal of order 1* and it is easy to show that

$$P^T_{[a,b],1}(f(x)) = \frac{1}{b-a}[(b - x)f(a) + (x - a)f(b)]$$

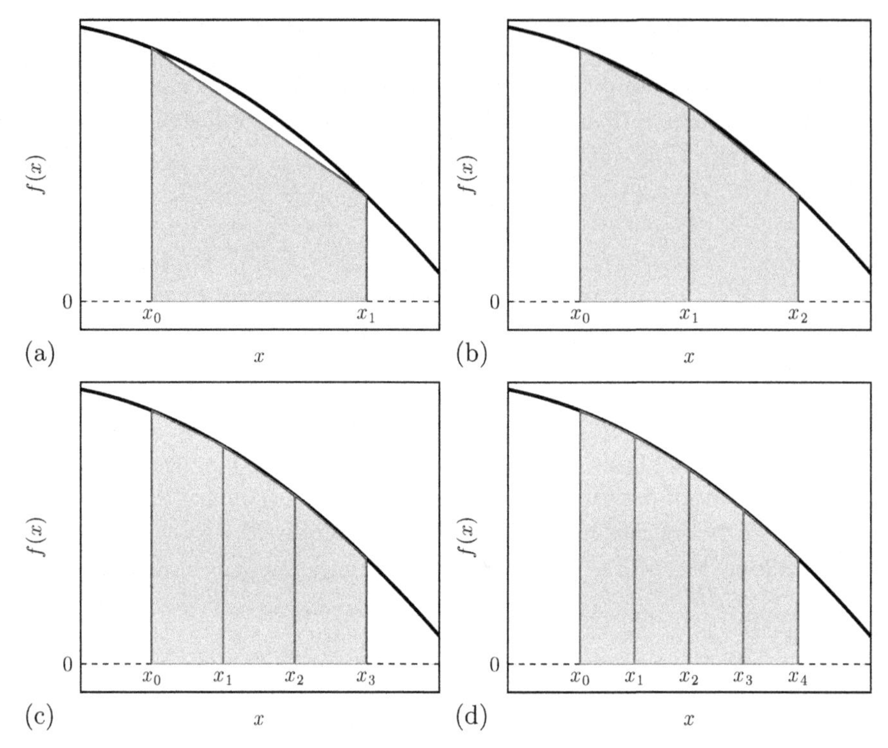

Figure 13.8 The trapezoidal approach for various k. (a) $k = 1$. (b) $k = 2$. (c) $k = 3$. (d) $k = 4$.

This approximation of the integrand can be readily integrated to give the *trapezoidal approximation of order 1 of the integral,*

$$T_{[a,b],1} = \int_a^b P^T_{[a,b],1}(f(x))(x)\,\mathrm{d}x$$

$$I \simeq T_{[a,b],1} = \left(\frac{b-a}{2}\right)\left(f(a) + f(b)\right) \tag{13.14}$$

The reader is invited to confirm the expression for I. The resulting $T_{[a,b],1}$ will be a reasonable approximation for I when the integrand is approximately linear over the interval $[a, b]$. However, when the integrand is not approximately linear, $T_{1,[a,b]}$ will be a poor approximation to I. This is demonstrated in the following example where we approximate the integral of a quadratic function.

EXAMPLE 13.24

Compute the trapezoidal approximation of order 1 for the integral

$$I = \int_2^5 (20 - 0.5x^2)\,\mathrm{d}x$$

Solution

We apply Eq. (13.14) with $f(a) = 20 - 0.5 \times 2^2 = 18$ and $f(b) = 20 - 0.5 \times 5^2 = 7.5$. In particular,

$$T_{[2,5],1} = \left(\frac{5-2}{2}\right)(18 + 7.5) = 38.25$$

Note that this integral is particularly straightforward to compute algebraically and its actual value is

$$I = \int_2^5 (20 - 0.5x^2)\,\mathrm{d}x = 40.5$$

A large discrepancy is therefore seen between I and $T_{[2,5],1}$.

An improved approximation for I can be obtained by using the trapezoidal approach with larger k. That is, we can fit numerous straight lines to the integrand over the interval $[a, b]$ and compute the sum of smaller integrals; this generates $T_{[a,b],k}$.

For example, $P^T_{[a,b],2}(f(x))$ is the trapezoidal approximation of order 2 for the integrand. It is obtained by fitting *two* straight lines; one between $f(x_0 = a)$ and $f(x_1)$ and another between $f(x_1)$ and $f(x_2 = b)$, where $x_1 = \frac{1}{2}(b - a)$. Similarly, $P^T_{[a,b],6}(f(x))$ is the trapezoidal approximation of order 6 for the integrand. It is formed from *six* straight lines fitted between $f(x)$ evaluated at regularly spaced node points $\{x_i = a + i\frac{(b-a)}{6}\}$. In either case, the approximating function can be integrated easily to generate an approximation for the integral. The approximations for various k are illustrated in Figure 13.8.

It should be clear that

$$T_{[a,b],k} = T_{[x_0,x_1],1} + T_{[x_1,x_2],1} + \cdots + T_{[x_{k-1},x_k],1}$$

That is, the kth order approximation for the integral over $[a, b]$ is simply the sum of the first-order approximations over each of the k sub-intervals. Therefore, by Eq. (13.14),

$$\begin{aligned} T_{[a,b],k} &= \left(\frac{x_1 - x_0}{2}\right)(f(x_1) + f(x_0)) + \cdots + \left(\frac{x_k - x_{k-1}}{2}\right)(f(x_k) + f(x_{k-1})) \\ &= \frac{h}{2}(f(x_1) + f(x_0) + f(x_2) + f(x_1) + \cdots + f(x_k) + f(x_{k-1})) \\ &= \frac{h}{2}(f(x_0) + 2f(x_1) + \cdots + 2f(x_{k-1}) + f(x_k)) \end{aligned}$$

From this expression we can write the general expression for the trapezoidal approximation of order k for a definite integral over $[a, b]$ as,

$$I \simeq T_{[a,b],k} = h\left(\frac{1}{2}f(a) + f(x_1) + f(x_2) + \cdots + f(x_{k-1}) + \frac{1}{2}f(b)\right) \tag{13.15}$$

where $h = \frac{b-a}{k}$ is the distance between consecutive node points. This expression is often referred to as the *trapezoidal rule of order k*.

EXAMPLE 13.25

Write down the trapezoidal rule of order k for the integral of $f(x)$ over $[-4, 6]$ for (a) $k = 1$, (b) $k = 2$, and (c) $k = 5$.

Solution

In each case, we use Eq. (13.15) with $a = -4$ and $b = 6$.

a. The $k + 1 = 2$ node points are $\{x_i = -4 + \frac{6+4}{1} i\}$ for $i = 0, 1$, leading to

$$\begin{aligned} T_{[-4,6],1} &= \left(\frac{6+4}{1}\right)\left(\frac{f(-4)}{2} + \frac{f(6)}{2}\right) \\ &= 5(f(-4) + f(6)) \end{aligned}$$

b. The $k + 1 = 3$ node points are $\{x_i = -4 + \frac{6+4}{2} i\}$ for $i = 0, 1, 2$, leading to

$$\begin{aligned} T_{[-4,6],2} &= \left(\frac{6+4}{2}\right)\left(\frac{f(-4)}{2} + f(1) + \frac{f(6)}{2}\right) \\ &= 5\left(\frac{f(-4)}{2} + f(1) + \frac{f(6)}{2}\right) \end{aligned}$$

c. The $k + 1 = 6$ node points are $\{x_i = -4 + \frac{6+4}{5} i\}$ for $i = 0, 1, \ldots, 5$, leading to

$$\begin{aligned} T_{[-4,6],5} &= \left(\frac{6+4}{5}\right)\left(\frac{f(-4)}{2} + f(-2) + f(0) + f(2) + f(4) + \frac{f(6)}{2}\right) \\ &= 2\left(\frac{f(-4)}{2} + f(-2) + f(0) + f(2) + f(4) + \frac{f(6)}{2}\right) \end{aligned}$$

EXAMPLE 13.26

Repeat Example 13.24 using trapezoidal approximations of order (a) 2, (b) 3, and (c) 4.

Solution

We are required to approximate

$$I = \int_2^5 f(x)\,dx = \int_2^5 (20 - 0.5x^2)\,dx$$

a. We have node points $\{x_i = 2 + 1.5i\}$ for $i = 0, 1, 2$. Equation (13.15) leads to

$$\begin{aligned} T_{[2,5],2} &= 1.5 \times \left(\frac{1}{2} f(2) + f(3.5) + \frac{1}{2} f(5)\right) \\ &= 1.5 \times (9 + 13.875 + 3.75) \\ I &\simeq 39.93750 \end{aligned}$$

b. We have node points $\{x_i = 2 + 1i\}$ for $i = 0, 1, 2, 3$. Equation (13.15) leads to

$$T_{[2,5],3} = 1 \times \left(\frac{1}{2}f(2) + f(3) + f(4) + \frac{1}{2}f(5)\right)$$
$$= 1 \times (9 + 15.5 + 12 + 3.75)$$
$$I \simeq 40.25000$$

c. We have node points $\{x_i = 2 + 0.75i\}$ for $i = 0, 1, 2, 3, 4$. Equation (13.15) leads to

$$T_{[2,5],4} = 0.75 \times \left(\frac{1}{2}f(2) + f(2.75) + f(3.5) + f(4.25) + f(4.25) + \frac{1}{2}f(5)\right)$$
$$= 0.75 \times (9 + 16.21875 + 13.875 + 10.96875 + 3.75)$$
$$I \simeq 40.35938$$

The error generated by the trapezoidal rule of order k is defined to be

$$E_k^T = |I - T_{[a,b],k}| \tag{13.16}$$

This is simply the difference between the actual value of the integral and the approximation obtained from the trapezoidal rule. A formal analysis of this error is beyond the scope of this book; however, we note that it is possible to show that there exists a theoretical upper bound on the error, $\hat{E}_k^T$, which is such that

$$E_k^T \leq \hat{E}_k^T = \left|\frac{h^2}{12}(f'(b) - f'(a))\right| \tag{13.17}$$

EXAMPLE 13.27

Determine the actual and theoretical errors of the approximations used in Examples 13.24 and 13.26 at various k.

Solution

Table 13.17 summarizes the value of $T_{[2,5],k}$ for various k and gives the value of the actual error E_k^T, defined by Eq. (13.16). We see that the error reduces with increased k but remains reasonably high at $k = 4$. Yet further increases in k demonstrate relatively slow convergence to the actual value $I = 40.5$. A comparison between the theoretical and the actual errors at each k is also shown. We see very close agreement between E_k^T and $\hat{E}_k^T$.

Equation (13.17) can be used to estimate the value of k required to ensure that the trapezoidal rule approximates an integral to within a desired accuracy. That is, given a, b, and a desired error tolerance, the expression can be used to determine k from $f'(a)$ and $f'(b)$. Equation (13.17) demonstrates that the error increases with the difference in $f'(b)$ and $f'(a)$. Note that the theoretical error bound obtained when integrating a linear function is zero. This is of course true because the approximating function will match precisely the linear integrand.

Table 13.17 Use of the trapezoidal rule in Examples 13.24 and 13.26

k	h	$T_{k,[2,5]}$	E_k^T	$\hat{E}_k^T$
1	3.00	38.25000	2.25000	2.25000
2	1.50	39.93750	0.56250	0.56250
3	1.00	40.25000	0.25000	0.25000
4	0.75	40.35938	0.14062	0.14063
5	0.60	40.41000	0.09000	0.09000
10	0.30	40.47750	0.02250	0.02250
15	0.20	40.49000	0.01000	0.01000
20	0.15	40.49438	0.00563	0.00563
50	0.06	40.49910	0.00090	0.00090

EXAMPLE 13.28

Use the trapezoidal rule to approximate the following integral to within an error of $E_k^T \leq 0.001$.

$$I = \int_0^1 \sin(x^2)\,\mathrm{d}x$$

Solution

We have $a = 0$ and $b = 1$ but not have a value for k. The theoretical error bound, Eq. (13.17), is given by

$$\hat{E}_k^T = \frac{h^2}{12}(f'(1) - f'(0)) = \frac{h^2}{12}(1.0806 - 0)$$

This is such that $\hat{E}_k^T = 0.001$ when $h = 0.10538 \Rightarrow k = \frac{1-0}{0.10538} = 9.48947$. We therefore estimate that $k = 10$ will give the required accuracy and $h = 0.1$.

The node points are therefore $\{x_i = 0 + 0.1i\}$ for $i = 1, \ldots, 10$ and Eq. (13.15) leads to

$$\begin{aligned} I \simeq T_{[0,1],10} &= 0.1 \times \left(\frac{1}{2}f(0) + f(0.1) + f(0.2) + \cdots + \frac{1}{2}f(1)\right) \\ &= 0.31117 \end{aligned}$$

Note that an accurate value of the integral is given by Wolfram Alpha as 0.31027 to five decimal places.

13.3.2 Simpson's approach

Improvements on the error generated by the trapezoidal method can be made using *Simpson's rule*. In contrast to the trapezoidal approach that fits linear functions, Simpson's approach fits *quadratic* functions to the integrand over each subinterval.

We begin by considering Simpson's approximation of order 2 over the interval $[a, b]$. Here, the integrand is approximated by a quadratic function, $P^S_{[a,b],2}(f(x))$, passing through $f(x_0), f(x_1)$, and $f(x_2)$, where $x_0 = a$, $x_1 = (b - a)/2$, and $x_2 = b$. Note that three points are required to define a quadratic function and so Simpson's approach requires three node points to define each subinterval; $k = 2$ is therefore the first-order use of Simpson's approach.

Some thought should convince you that the following is an appropriate form for the approximating quadratic function over $[a, b]$, and this is confirmed in Example 13.29.

$$P^S_{[a,b],2}(f(x)) = \frac{(x-x_2)(x-x_1)}{(x_0-x_2)(x_0-x_1)}f(x_0) + \frac{(x-x_2)(x-x_0)}{(x_1-x_2)(x_1-x_0)}f(x_1) + \frac{(x-x_1)(x-x_0)}{(x_2-x_1)(x_2-x_0)}f(x_2) \tag{13.18}$$

The integral of $f(x)$ over $[a, b]$ is then approximated by the integral of $P^T_{[a,b],2}(f(x))$ over the same interval; this defines Simpson's approximation of order 2 for the integral as

$$S_{[a,b],2} = \int_a^b \left(\frac{(x-x_2)(x-x_1)}{(x_0-x_2)(x_0-x_1)}f(x_0) + \frac{(x-x_2)(x-x_0)}{(x_1-x_2)(x_1-x_0)}f(x_1) + \frac{(x-x_1)(x-x_0)}{(x_2-x_1)(x_2-x_0)}f(x_2) \right) \mathrm{d}x$$

The reader is invited to perform the integral of each term using an appropriate change of variable in each case to show that

$$I \simeq S_{[a,b],2} = \frac{h}{3}(f(x_0) + 4f(x_1) + f(x_2)) \tag{13.19}$$

This expression will be accurate for integrands that are approximately quadratic over $[a, b]$.

EXAMPLE 13.29

Demonstrate that Eq. (13.18) is a quadratic that passes through $(x_o, f(x_0))$, $(x_1, f(x_1))$, and $(x_2, f(x_2))$.

Solution

Each of the three terms is of the form $c_1(x-c_2)(x-c_3)$ for some constant c_1, c_2, and c_3, $P^S_{[a,b],2}(f(x))$ is therefore quadratic. Furthermore,

$$\begin{aligned} P^S_{[a,b],2}(f(x_0)) &= \frac{(x_0-x_2)(x_0-x_1)}{(x_0-x_2)(x_0-x_1)}f(x_0) + \frac{(x_0-x_2)(x_0-x_0)}{(x_1-x_2)(x_1-x_0)}f(x_1) + \\ &\quad \frac{(x_0-x_1)(x_0-x_0)}{(x_2-x_1)(x_2-x_0)}f(x_2) \\ &= f(x_0) \end{aligned}$$

Similarly $P^S_{[a,b],2}(f(x_1)) = f(x_1)$ and $P^S_{[a,b],2}(f(x_2)) = f(x_2)$, as required.

EXAMPLE 13.30

Repeat Example 13.26 using Simpson's approximation of order 2 to approximate the integrand. Comment on the result.

Solution

We are required to approximate

$$I = \int_2^5 f(x)\, dx = \int_2^5 (20 - 0.5x^2)\, dx$$

Equation (13.19) with node points $\{x_i = 2 + hi\}$ where $h = \frac{5-2}{2} = \frac{3}{2}$ and $i = 0, 1, 2$ leads to

$$\begin{aligned} S_{[2,5],2} &= \frac{1}{2}(f(2) + 4f(3.5) + f(5)) \\ &= 1.5 \times (18 + 55.5 + 7.5) \\ &= 40.5 \end{aligned}$$

We note that in this case $S_{[2,5],2} \equiv I$, that is, there is zero error. This is because the integrand is quadratic and so Simpson's approximation of order 2 fits it exactly.

In cases where the integrand is not close to quadratic, a better approximation to I can be obtained by dividing the interval into subintervals $\{[x_i, x_{i+2}]\}$ for $i = 0, 2, \ldots, k-2$ and fitting $P^S_{[x_i,x_{i+2}],2}(f(x))$ to the integrand within each subinterval at nodes x_i, x_{i+1}, and x_{i+2}. Note that h denotes the width between node points, as with the trapezoidal method, that is $\{x_i = a + ih\}$ for $i = 1, 2, \ldots, k$. However, each subinterval within Simpson's method is of width $2h$; this is a consequence of needing three node points to fit a quadratic function, as shown in Figure 13.9.

A general definite integral over $[a, b]$ is then approximated by the sum of $k/2$ integrals as

$$S_{[a,b],k} = S_{[x_0,x_2],2} + S_{[x_2,x_4],2} + \cdots + S_{[x_{k-2},x_n],2}$$

Equation (13.19) can then be used for each term to give

$$\begin{aligned} I \simeq S_{[a,b],k} = \frac{h}{3}(f(a) &+ 4f(x_1) + 2f(x_2) + 4f(x_3) + 2f(x_4) + \cdots + 2f(x_{k-2}) \\ &+ 4f(x_{k-1}) + f(b)) \end{aligned} \tag{13.20}$$

where $h = \frac{b-a}{k}$. This expression is often referred to as *Simpson's rule of order k*. Note that k must be an even natural number.

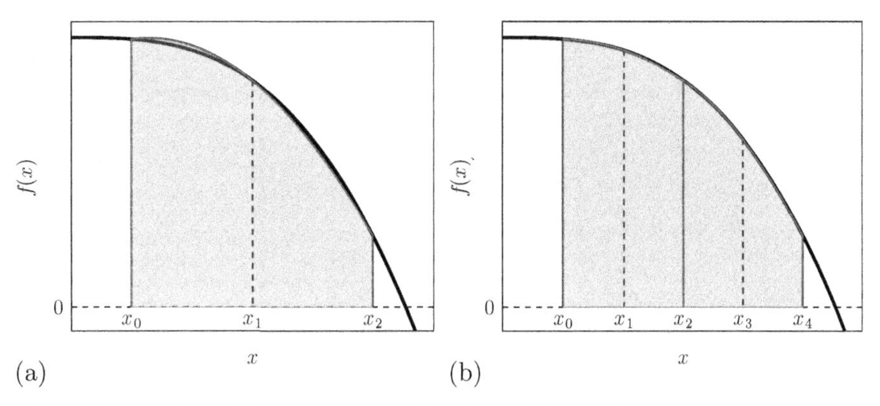

Figure 13.9 An illustration of Simpson's approach. (a) $k = 2$. (b) $k = 4$.

EXAMPLE 13.31

Write down Simpson's rule of order k for the integral of $f(x)$ over $[-4, 6]$ for (a) $k = 2$, (b) $k = 4$, and (c) $k = 6$.

Solution

In each case, we use Eq. (13.20) with $a = -4$ and $b = 6$.

a. The $k + 1 = 3$ node points are $x_i = -4 + hi$ for $h = \frac{6+4}{2} = 5$ and $i = 0, 1, 2$ which extend over the full interval $[-4, 6]$,

$$I \simeq S_{[-4,6],2} = \frac{h}{3}(f(x_0) + 4f(x_1) + f(x_2)) = \frac{5}{3}(f(-4) + 4f(1) + f(6))$$

b. The $k + 1 = 5$ node points are $x_i = -4 + hi$ for $h = \frac{6+4}{4} = \frac{5}{2}$ and $i = 0, 1, 2, 3, 4$ which forms two subintervals $[-4, 1]$ and $[1, 6]$,

$$\begin{aligned} I \simeq S_{[-4,6],4} &= \frac{h}{3}(f(x_0) + 4f(x_1) + 2f(x_2) + 4f(x_3) + f(x_4)) \\ &= \frac{5}{6}(f(-4) + 4f(-1.5) + 2f(1) + 4f(3.5) + f(6)) \end{aligned}$$

c. The $k + 1 = 7$ node points are $x_i = -4 + hi$ for $h = \frac{6+4}{6} = \frac{5}{3}$ and $i = 0, 1, \ldots, 6$ which forms three subintervals $[-4, -\frac{2}{3}]$, $[-\frac{2}{3}, \frac{8}{3}]$, and $[\frac{8}{3}, 6]$,

$$\begin{aligned} I \simeq S_{[-4,6],6} &= \frac{h}{3}(f(x_0) + 4f(x_1) + 2f(x_2) + 4f(x_3) + 2f(x_4) + 4f(x_5) + f(x_6)) \\ &= \frac{h}{3}\left(f(-4) + 4f\left(-\frac{7}{3}\right) + 2f\left(-\frac{2}{3}\right) + 4f(1)\right. \\ &\quad \left. + 2f\left(\frac{8}{3}\right) + 4f\left(\frac{13}{3}\right) + f(6)\right) \end{aligned}$$

EXAMPLE 13.32

Approximate the following definite integral using Simpson's rule for (a) $k = 2$, (b) $k = 4$, and (c) $k = 6$.

$$I = \int_{-4}^{6} \left(\frac{x}{10} + e^{-(x+2)} \right) dx$$

Solution

We use the results of Example 13.31 with $f(x) = 0.1x + e^{-(x+2)}$.

a. For $k = 2$,

$$\begin{aligned} S_{[-4,6],2} &= \frac{5}{3}(f(-4) + 4f(1) + f(6)) \\ &= \frac{5}{3}(6.98906 + 0.59915 + 0.60034) \\ &= 13.64757 \end{aligned}$$

b. For $k = 4$,

$$\begin{aligned} S_{[-4,6],4} &= \frac{5}{6}(f(-4) + 4f(-1.5) + 2f(1) + 4f(3.5) + f(6)) \\ &= \frac{5}{6}(6.98906 + 1.82612 + 0.29957 + 1.41635 + 0.60034) \\ &= 9.27620 \end{aligned}$$

c. For $k = 6$,

$$\begin{aligned} S_{[-4,6],6} &= \frac{h}{3}\left(f(-4) + 4f\left(-\frac{7}{3}\right) + 2f\left(-\frac{2}{3}\right) + 4f(1) + 2f\left(\frac{8}{3}\right) + 4f\left(\frac{13}{3}\right) + f(6) \right) \\ &= \frac{5}{9}(6.98906 + 4.64912 + 0.39386 + 0.59915 + 0.55214 + 1.74044 + 0.60034) \\ &= 8.62450 \end{aligned}$$

The error generated by Simpson's rule is defined to be

$$E_k^S = |I - S_{[a,b],k}| \tag{13.21}$$

As with trapezoidal rule, this is simply the difference between the actual value and the approximation. Again, a formal analysis of the error is beyond the scope of this book. However, note that in theory an upper bound on the error at order k can be shown to be given by

$$E_k^S \leq \hat{E}_k^S = \left| \frac{h^4}{180}(f^{(3)}(b) - f^{(3)}(a)) \right| \tag{13.22}$$

EXAMPLE 13.33

Determine the actual and theoretical errors of the approximations used in Example 13.32 at various k.

Solution

Table 13.18 shows the value of $S_{[2,5],k}$ for various k and the value of the actual error, defined by Eq. (13.21). We see that the error reduces with increased k but remains reasonably high at $k = 6$. Yet further increases in k, also shown in Table 13.18, demonstrate the relatively slow convergence to $I = 8.38872$. A comparison between the theoretical bounds and the actual errors in Example 13.32 is also given for various k; we confirm that $E_k^S \leq \hat{E}_k^S$ in each case.

Table 13.18 Use of Simpson's rule in Examples 13.32

k	h	$T_{k,[2,5]}$	E_k^S	$\hat{E}_k^S$
2	5.00	13.64756639	5.25885	25.65529
4	2.5	9.276196186	0.88748	1.60346
6	5/3	8.624497417	0.23578	0.31673
8	1.25	8.472932023	0.08421	0.10022
10	1	8.425347859	0.03663	0.04105
20	0.5	8.391211764	0.00249	0.00257
30	1/3	8.389220781	0.00050	0.00051

EXAMPLE 13.34

Repeat Example 13.28 using Simpson's rule.

Solution

We are required to approximate

$$I = \int_0^1 \sin(x^2)\,dx$$

using Simpson's rule. The theoretical error bound is given by Eq. (13.22) and is such that,

$$\hat{E}_k^S = \frac{h^4}{180}(f^{(3)}(1) - f^{(3)}(0)) = \frac{h^4}{180}(-14.4201 - 0)$$

Therefore, $\hat{E}_k^S = 0.001$ when $h = 0.33425$ and $k = \frac{1-0}{0.33425} = 2.99174$. We estimate that $k = 4$ (note that k must be even) will give the required accuracy and use $h = 0.25$.

The node points are therefore $x_i = 0 + 0.25i$ for $i = 0, 2, 3, 4$ and Eq. (13.20) leads to

$$\begin{aligned} I \simeq S_{[0,1],4} &= \frac{0.25}{3}\left(\frac{1}{2}f(0) + 4f(0.25) + 2f(0.5) + 4f(0.75) + f(1)\right) \\ &= 0.30994 \end{aligned}$$

It should be clear that the theoretical error bound obtained when using Simpson's rule to integrate a linear or quadratic function is zero for any k. This is because the approximating function $P^S_{[a,b],k}(f(x))$ will precisely match the integrand $f(x)$.

It is interesting to note that Eq. (13.17) indicates that the trapezoidal rule has an error bound of $O(h^2)$, whereas Eq. (13.22) indicates that Simpson's rule has an error bound of $O(h^4)$. Simpson's rule is therefore expected to give the same degree of accuracy as the trapezoidal rule for a relatively larger h, or equivalently, a relatively smaller k. The superior efficiency of Simpson's rule is demonstrated in the following example.

EXAMPLE 13.35

Approximate the following integral using both the trapezoidal rule and Simpson's rule at order $k = 2, 4, 6, 8$, and 10. Comment on the relative behavior of the errors with k.

$$I = \int_0^2 e^x \cos(x)\, dx$$

Solution

Note that this integral can be computed directly using integration by parts. This leads to

$$I = \int_0^2 e^x \cos(x)\, dx = \left[\frac{e^x}{2}(\sin(x) + \cos(x))\right]_0^2 = 1.32196$$

The required numerical approximations can be obtained with $f(x) = e^x \cos(x)$ and stepsize $h = \frac{2-0}{k}$ in Eqs. (13.15) and (13.20).

The results are shown in Table 13.19. We see rapid convergence of the results from Simpson's rule with increased k. While the results from the trapezoidal rule are converging with increased k, it is happening at a much slower rate.

Table 13.19 Comparison between the trapezoidal rule and the Simpson's rule for Example 13.35

k	h	$T_{[0,2],k}$	E^T_k	$S_{[0,2],k}$	E^S_k
2	1	0.43123	0.89073	1.26661	0.05535
4	0.5	1.09757	0.22439	1.31968	0.00228
6	$\frac{1}{3}$	1.22211	0.09985	1.32156	0.00040
8	0.25	1.26577	0.05619	1.32184	0.00012
10	0.2	1.28599	0.03597	1.32191	0.00005

13.4 ACTUARIAL APPLICATION: CONTINUOUS PROBABILITY DISTRIBUTIONS

A straightforward actuarial example of numerical root finding is, of course, obtaining an unknown interest rate (or yield) from the equation of value of a given cash flow.

The concepts "equation of value" and "yield" were introduced in Chapter 2, and Question 13.4 gives a brief illustration of this application of numerical root finding in that context. However, rather than elaborating on this very obvious application in this closing section, we instead choose to look at an application of numerical integration related to *continuous probability distributions*.

Probability theory in relation to *discrete distributions* was considered in Chapter 9. Recall that random quantities are said to be discrete if they can take one of a *finite number* of possible values. For example, if Y is a random quantity that can take either value -1, 0, or 1 with equal probability, then Y is said to have a discrete distribution. *Continuous distributions* are in contrast to discrete distributions. If X has a continuous distribution on $\mathbb{R}$, say, it can take *any* value within the interval $[-\infty, \infty]$.

As there is a continuum of possible values for X, the probability of the event that X takes the value 4.098, say, is zero. Although this may seem strange at first, it should be clear that in a continuum the value 4.098 is *not the same* as 4.098000001 or 4.0980000000000001, for example. Therefore, if X is to take the value 4.098, this actually means that it must take that value with infinite precision, and this has zero probability.

To avoid this fundamental complication for continuous distributions we instead work in terms of probabilities of the event that X takes some value *between* two points. For example, although $P(X = 4.098)$ does not make sense for a continuously distributed X, the probability $P(4.097 < X < 4.099)$ does make sense.

In order to determine such probabilities we must have information about the *probability density function* of X, $f_X(x)$, which is defined such that

$$P(a < X < b) = \int_a^b f_X(x)\,\mathrm{d}x$$

You should recall from Chapter 9 that the probability of an event occurring is a real number on the interval $[0, 1]$ and this imposes particular restrictions on the mathematical form of $f_X(x)$. In particular, it must be true $f_X(x) \geq 0$ for $x \in [-\infty, \infty]$ and also

$$P(-\infty < X < \infty) = \int_{-\infty}^{\infty} f_X(x)\,\mathrm{d}x = 1$$

Note that the distribution does not necessarily have to be defined across the entire domain of $\mathbb{R}$. For example, it is perfectly reasonable to define a probability density function $g_X(x)$ over $x \in [0, \infty]$ which indicates that the random variable must take nonnegative values. That is, the probability that X takes a negative value is zero. More formally we would interpret this to mean that

$$f_X(x) = \begin{cases} g_X(x) & \text{for } x \in [0, \infty) \\ 0 & \text{for } x \in (-\infty, 0) \end{cases}$$

and so the integral condition becomes

$$\int_{\infty}^{\infty} f_X(x)\,\mathrm{d}x = 1 \Rightarrow \int_{0}^{\infty} g_X(x)\,\mathrm{d}x + \int_{-\infty}^{0} 0\,\mathrm{d}x = \int_{0}^{\infty} g_X(x)\,\mathrm{d}x = 1$$

We can generalize the conditions for a valid probability density function $f_X(x)$ over $x \in [a_1, a_2]$ to be,

1. $f_X(x) \geq 0$ for all $x \in [a_1, a_2]$; and

2. $\int_{a_1}^{a_2} f_X(x)\,\mathrm{d}x = 1$

where $a_2 > a_1$ are any real constants.

EXAMPLE 13.36

State whether the following are valid probability density functions.

a. $f_X(x) = \sin(x)$ for $x \in [-\infty, \infty]$.

b. $g_Y(y) = \frac{1}{b-a}$ for $y \in [a, b]$ and $b > a$ are real constants.

c. $h_Z(z) = \lambda\,\mathrm{e}^{-\lambda z}$ for $z \in [0, \infty)$ and λ some positive constant.

Solution

a. We note that $f_X(x) = \sin(x)$ can take negative values for some $x \in [-\infty, \infty]$ and so is *not* a valid probability density function.

b. Since $b > a$, $g_Y(y) = \frac{1}{b-a} > 0$ for all $y \in [a, b]$. Furthermore, we have

$$P(a < Y < b) = \int_{a}^{b} \frac{1}{b-a}\,\mathrm{d}y = \frac{b-a}{b-a} = 1$$

We therefore see that $g_Y(y)$ for $y \in [a, b]$ is a valid probability density function. In fact, the associated distribution is often called the *Uniform distribution*.

c. We note that $h_Z(z) = \lambda\,\mathrm{e}^{-\lambda z} > 0$ for $z \in [0, \infty)$ and λ some positive constant. Furthermore, we have

$$P(0 < Z < \infty) = \int_{0}^{\infty} \lambda\,\mathrm{e}^{-\lambda z}\,\mathrm{d}z = 1$$

We therefore see that $h_Z(z)$ for $z \in [0, \infty)$ is a valid probability density function. In fact, the associated distribution is often called the *Exponential distribution*.

As is hinted in Example 13.36, there are a number of standard continuous probability distributions that are used in many areas of applied mathematics. The Uniform distribution and the Exponential distribution are two examples of such standard distributions. Other examples include the *Standard normal distribution*, the *Gaussian distribution*, and the *Beta distribution*, to name only a few. In particular, the Exponential distribution is often used to model the time between the occurrences of rare random events. In this case, the parameter λ is interpreted as the rate of the occurrence of the events within the basic time scale being used. For example, if an insurance company receives, on average, five claims per hour, the time between claims can be modeled by an Exponential distribution with

$\lambda = 5$. The Standard normal distribution is hugely important and will be considered later in this section.

Note, however, that you are not restricted to using standard distributions and are free to develop any weird and wonderful distribution to model a random process. You should be sure that the density function satisfies the conditions of validity stated above.

EXAMPLE 13.37

You are studying a newly discovered species of rodent. Observations suggest that a rodent selected at random with a tail of length described by a random variable X cm with probability density function

$$f_X(x) = \alpha \sin\left(\frac{\pi}{5}x\right) \quad \text{for } x \in [0, 5]$$

where α is some constant.

a. Determine the value of α such that $f_X(x)$ is a valid probability density function.

b. Calculate the probability that a randomly selected adult rodent will have a tail of length between 4 and 4.5 cm.

Solution

a. We require $f_X(x)$ to be nonnegative over $x \in [0, 5]$ which is true for $\alpha > 0$. Furthermore, we require

$$\int_0^5 f_X(x)\,\mathrm{d}x = \int_0^5 \alpha \sin\left(\frac{\pi}{5}x\right)\,\mathrm{d}x = 1 \Rightarrow \alpha = \frac{\pi}{10}$$

We therefore conclude that

$$f_X(x) = \frac{\pi}{10} \sin\left(\frac{\pi}{5}x\right) \quad \text{for } x \in [0.5]$$

b. We require $P(4 < X < 4.5)$ and, from the definition of a probability density function, this is obtained from the integral

$$\int_4^{4.5} f_X(x)\,\mathrm{d}x = \int_4^{4.5} \frac{\pi}{10} \sin\left(\frac{\pi}{5}x\right)\,\mathrm{d}x \approx 0.071$$

That is, under this assumed distribution, the probability that a randomly selected adult rodent will have a tail of length between 4 and 4.5 cm in length is approximately 7.1%.

Figure 13.10 gives a visual illustration of $P(4 < X < 4.5)$ as calculated in Example 13.37(b). Given that the integral of $f_X(x)$ between $x = 0$ and 5 has unit value, the calculation of $P(4 < X < 4.5)$ is essentially calculating the area bounded between $x = 4$ and 4.5 as a *proportion* of the full area under the curve. Since $f_X(x) \geq 0$, any calculation of $P(x_1 < X < x_2)$ will necessarily be such that to $P(x_1 < X < x_2) \in [0, 1]$, as is required for P to be a probability. This gives a useful interpretation of the calculation of probabilities for general continuous distributions.

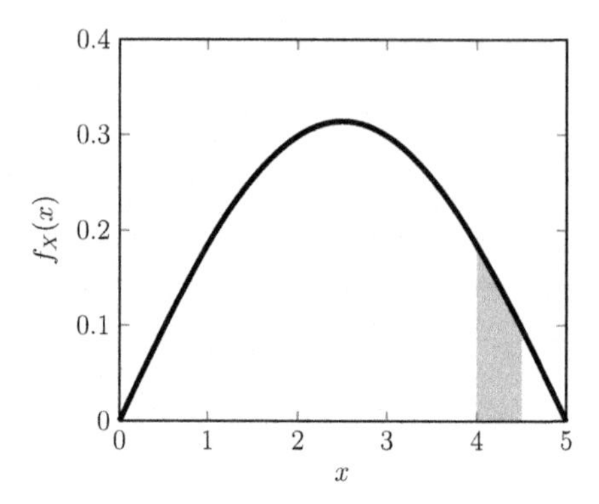

Figure 13.10 A visual illustration of $P(4 < X < 4.5)$ in Example 13.37.

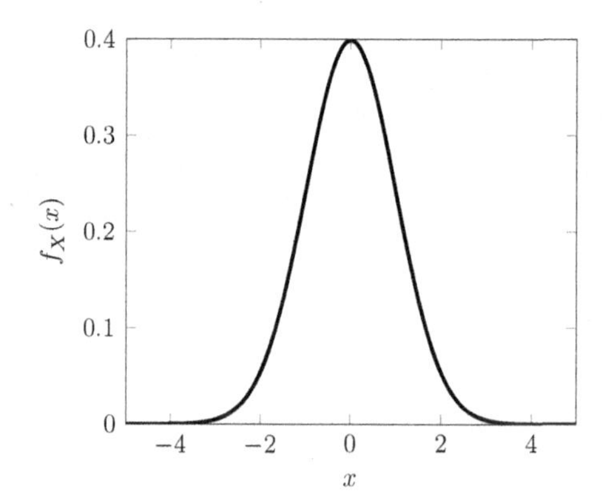

Figure 13.11 The probability density function for the Standard normal distribution on $x \in (-\infty, \infty)$.

An extremely important continuous probability distribution in many aspects of applied mathematics is the *Standard normal distribution*. This has a probability density function given by

$$f_X(x) = \frac{1}{\sqrt{2\pi}} \mathrm{e}^{-\frac{1}{2}x^2} \quad \text{for } x \in (-\infty, \infty)$$

and is illustrated in Figure 13.11. It should be clear from the properties of exponential functions that $f_X(x) \geq 0$, however, any attempt to confirm that

$$\int_{-\infty}^{\infty} \frac{1}{\sqrt{2\pi}} \mathrm{e}^{-\frac{1}{2}x^2} \, \mathrm{d}x = 1$$

leads to difficulty. In fact, it is impossible to determine the integral by hand and we must result to numerical integration techniques, such as those considered in this chapter. Furthermore, any attempt to calculate probabilities under this distribution also requires numerical integration. In practice, this difficultly is avoided by the use of widely published statistical tables that give values of $\Phi(x) = P(-\infty < X < x)$ for various x. From these it is possible to construct the value any general $P(x_1 < X < x_2)$ by exploiting the symmetry of the probability density function (note that it is an even function) and other additive properties of areas.

Prior to the advent of affordable computing power, statistical tables were the only way of working with Standard normal distributions. Of course we are now able to determine these probabilities using computational techniques, but the use of statistical tables is still an important skill, particularly for those taking written examinations. Note that $N(0, 1)$ is standard notation for the Standard normal distribution; that is, we would write $X \sim N(0, 1)$ to indicate that X has this particular distribution.

EXAMPLE 13.38

Given that,

$$\Phi(0.50) = 0.69146 \quad \text{and} \quad \Phi(1.00) = 0.84134$$

and $X \sim N(0, 1)$, determine the following probabilities.

a. $P(X > 0.50)$

b. $P(-1 < X < 0.50)$

Solution

Note that $\Phi(x) = P(-\infty < X < x) \equiv P(X < x)$.

a. $P(X > 0.50) = 1 - P(X < 0.5) = 1 - 0.69164 = 0.30854$

b. $P(-1 < X < 0.50) = P(X < 0.50) - P(X < -1) = P(X < 0.50) - P(X > 1) = P(X < 0.50) - (1 - P(X < 1)) = P(X < 0.5) + P(X < 1) - 1 = 0.69146 + 0.84134 - 1 = 0.53280$

A visual interpretation of these probabilities is given in Figure 13.12.

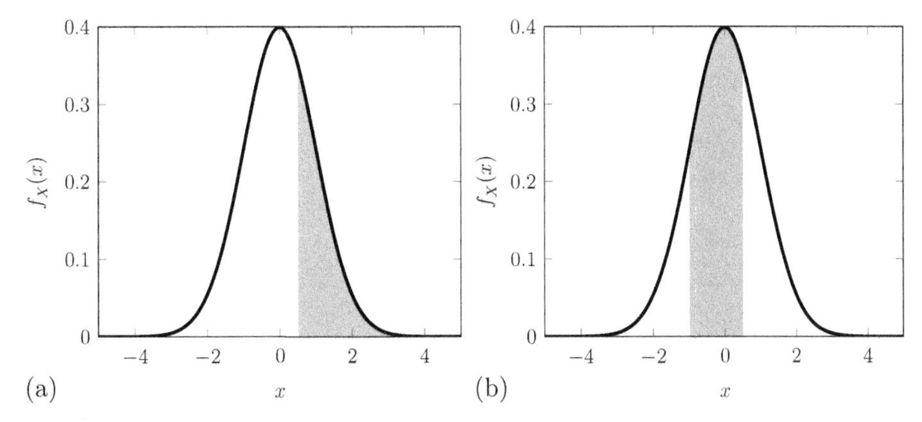

Figure 13.12 The probabilities under the Standard normal distribution in Example 13.38. (a) $P(X > 0.50)$. (b) $P(-1 < X < 0.50)$.

EXAMPLE 13.39

Repeat Example 13.38(b) using a numerical integration approach. You should use Simpson's rule of order 6.

Solution

We are required to evaluate

$$P(-1 < X < 0.50) = \int_{-1}^{0.5} f_X(x)\,\mathrm{d}x = \int_{-1}^{0.5} \frac{1}{\sqrt{2\pi}}\,\mathrm{e}^{-\frac{1}{2}x^2}\,\mathrm{d}x$$

Simpson's rule of order 6 uses 7 node points at $x_i = -1 + hi$ for $h = \frac{1.5}{6} = 0.25$ and $i = 0, 1, 2, \ldots, 6$. We therefore obtain the approximation,

$$\begin{aligned} P \simeq & \frac{h}{3}(f_X(x_0) + 4f_X(x_1) + 2f_X(x_2) + 4f_X(x_3) + 2f_X(x_4) + 4f_X(x_5) + f_X(x_6)) \\ \approx & 0.53283 \end{aligned}$$

Note that this value is very close to that calculated in Example 13.38(b).

Numerical integration techniques combined with affordable computing power now enable us to work with any valid probability density function. That is, we are not restricted to using $f_X(x)$ that are integrable by hand, nor are we restricted to using distributions that can related back to published statistical tables for the Standard normal distribution.

EXAMPLE 13.40

A random variable X has probability density function given by

$$g_X(x) = \frac{1}{2\sqrt{2\pi}} \exp\left(-\frac{1}{2}\left(\frac{x-1}{2}\right)^2\right)$$

for $-\infty < x < \infty$.

a. Demonstrate that $g_X(x)$ is a valid probability density function.

b. Calculate $P(1.5 < X < 4.5)$ using Simpson's rule of order 6.

c. Determine a transformation that takes X to a Standard normal distribution and use statistical tables to verify your answer to part b.

Hint. You are given that $\Phi(1.75) = 0.95994$ and $\Phi(0.25) = 0.59871$.

Solution

a. It is clear from the properties of exponential functions that $g_X(x) > 0$ for all $x \in (-\infty, \infty)$. This can be confirmed by the plot of $g_X(x)$ given in Figure 13.13(a). We

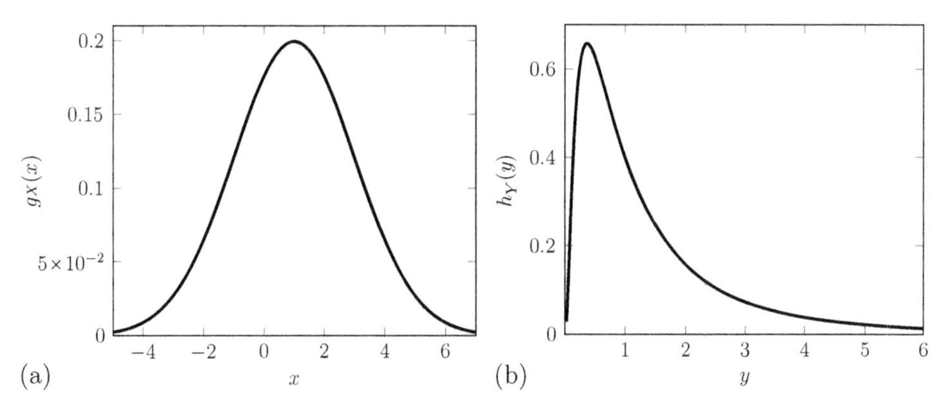

Figure 13.13 The probability density functions in Examples 13.40 and 13.41. (a) $g_X(x)$ in Example 13.40. (b) $h_Y(y)$ in Example 13.41.

are therefore required to show that

$$\int_{-\infty}^{\infty} \frac{1}{2\sqrt{2\pi}} \exp\left(-\frac{1}{2}\left(\frac{x-1}{2}\right)^2\right) \mathrm{d}x = 1$$

The plot shows that $g_X(x)$ rapidly decays outside of $x \in [-5, 7]$ and so it is appropriate to demonstrate that $P(-5 < X < 7) \approx 1$. Simpson's rule of order 6 uses 7 node points at $x_i = -5 + hi$ for $h = \frac{12}{6} = 2$ and $i = 0, 1, 2, \ldots, 6$. We therefore obtain the approximation,

$$\begin{aligned} P \simeq &\frac{h}{3}(g_X(x_0) + 4g_X(x_1) + 2g_X(x_2) + 4g_X(x_3) + 2g_X(x_4) + 4g_X(x_5) + g_X(x_6)) \\ \approx &1 \end{aligned}$$

We can therefore be reasonably sure that $P(-\infty < X < \infty) = 1$ and conclude that $g_X(x)$ is a valid probability function.

b. We are required to evaluate

$$P(1.5 < X < 4.5) = \int_{1.5}^{4.5} \frac{1}{2\sqrt{2\pi}} \exp\left(-\frac{1}{2}\left(\frac{x-1}{2}\right)^2\right) \mathrm{d}x$$

Simpson's rule of order 6 uses 7 node points at $x_i = 1.5 + hi$ for $h = \frac{3}{6} = 0.5$ and $i = 0, 1, 2, \ldots, 6$ and we obtain $P(1.5 < X < 4.5) \approx 0.361$.

c. Let us use Z to denote a random variable such that $Z \sim N(0, 1)$. We are therefore required to obtain a transformation between x and z such that

$$\int_{x_2}^{x_1} g_X(x)\,\mathrm{d}x \to \int_{z_2}^{z_1} \frac{1}{\sqrt{2\pi}} \exp\left(-\frac{1}{2}z^2\right) \mathrm{d}z$$

The clear choice is $z = \frac{x-1}{2}$ which shows that $\frac{X-1}{2} = Z \sim N(0,1)$. We can therefore proceed as follows,

$$\begin{aligned}
P(1.5 < X < 4.5) &= P(X < 4.5) - P(X < 1.5) \\
&= P\left(\frac{X-1}{2} < \frac{4.5-1}{2}\right) - P\left(\frac{X-1}{2} < \frac{1.5-1}{2}\right) \\
&= P(Z < 1.75) - P(Z < 0.25) \\
&= \Phi(1.75) - \Phi(0.25) \\
&= 0.36125
\end{aligned}$$

We see that the value obtained from numerical integration and the value obtained from statistical tables are in good agreement.

EXAMPLE 13.41

A random variable Y has probability density function given by

$$h_Y(y) = \frac{1}{\sqrt{2\pi}}\frac{1}{y}\exp\left(-\frac{1}{2}(\ln y)^2\right)$$

for $y > 0$.

a. Demonstrate that $h_Y(y)$ is a valid probability density function.

b. Calculate $P(1 < Y < 4)$ using Simpson's rule of order 6.

c. Given that this distribution is actually *Log-normal*, that is $\ln Y \sim N(0,1)$, use published statistical tables to verify your answer to part b.

Hint. You are given that $\Phi(1.39) = 0.91774$ and $\Phi(0) = 0.5$.

Solution

a. It is clear from the properties of exponential functions that $h_Y(y) > 0$ for all $y \in (0,\infty)$. This can be confirmed by the plot of $h_Y(y)$ given in Figure 13.13(b). We are therefore required to show that

$$\int_0^\infty \frac{1}{\sqrt{2\pi}}\frac{1}{y}\exp\left(-\frac{1}{2}(\ln y)^2\right)\,dy = 1$$

The plot shows that $h_Y(y)$ rapidly decays outside of $x \in (0,6]$ and so it is appropriate to demonstrate that $P(0 < X < 6) \approx 1$. Simpson's rule of order 6 uses 7 node points at $y_i = 0 + hi$ for $h = \frac{6}{6} = 1$ and $i = 0,1,2,\ldots,6$. We therefore obtain the approximation,

$$\begin{aligned}
P &\simeq \frac{h}{3}(h_Y(y_0) + 4h_Y(y_1) + 2h_Y(y_2) + 4h_Y(y_3) + 2h_Y(y_4) + 4h_Y(y_5) + h_Y(y_6)) \\
&\approx 1
\end{aligned}$$

We can therefore be reasonably sure that $P(0 < Y < \infty) = 1$ and conclude that $h_Y(y)$ is a valid probability density function.

b. We are required to evaluate

$$P(1 < Y < 4) = \int_1^4 \frac{1}{\sqrt{2\pi}} \frac{1}{y} \exp\left(-\frac{1}{2}(\ln y)^2\right) \mathrm{d}y$$

Simpson's rule of order 6 uses 7 node points at $y_i = 1 + hi$ for $h = \frac{3}{6} = 0.5$ and $i = 0, 1, 2, \ldots, 6$ and we obtain $P(1 < Y < 4) \approx 0.417$.

c. We are given that $\ln Y = Z \sim N(0, 1)$. We can therefore proceed as follows,

$$\begin{aligned} P(1 < Y < 4) &= P(Y < 4) - P(Y < 1) \\ &= P(\ln Y < \ln 4) - P(\ln Y < \ln 1) \\ &\approx P(Z < 1.39) - P(Z < 0) \\ &= \Phi(1.39) - \Phi(0) \\ &= 0.4177 \end{aligned}$$

We see that the value obtained from numerical integration and the value obtained from statistical tables are in agreement.

13.5 QUESTIONS

The following questions are intended to test your knowledge of the material discussed in this chapter. Full solutions are available in Chapter 13 Solutions of Part III.

Question 13.1. You are given that $f(x) = x^2 - 2\,\mathrm{e}^x - 2^x + 5$ has a single real root x_r on $[0, 1]$. Use the following methods to determine the root to within an error tolerance defined by $|f(x_r)| \leq 0.001$. Comment on your answers.

a. Bisection method

b. Regula falsi method

Question 13.2. Repeat Question 13.1 using the Newton-Raphson and Secant methods. Comment on your answers.

Question 13.3. Implement an appropriate numerical method to determine all real roots of

$$g(x) = -0.67 \times 2.5^{x^{1.3}} + 2.5^{x^{1.3}} \times x^{0.2} + 1.34x^{5.3} - 2x^{5.5}$$

Question 13.4. A saver has been depositing money in a fixed-interest bank account for the last 10 years. If the balance of his account is \$15,432 at time $t = 10$, calculate an approximate value of the annual interest rate earned. You are given that the investor deposited \$2000 at time $t = 0$, \$3000 at time $t = 3.5$, and \$5600 at $t = 7$. You should perform a simple "pen-and-paper" calculation.

Question 13.5. Implement the elementary definition of a derivative to determine a converged estimate of $f'(-1)$ for $f(x) = \sin(x^3)$.

Table 13.20 Samples of an unknown function $f(x)$ at various node points $\{x_n\}$

x_n	$f(x_n)$
−2	−0.6646
−1.9	−0.1671
−1.8	0.2639
−1.7	0.6276
−1.6	0.9252
−1.5	1.1592
−1.4	1.3331
−1.3	1.4521
−1.2	1.5218
−1.1	1.5489
−1	1.5403
−0.9	1.5035
−0.8	1.4459
−0.7	1.3748
−0.6	1.2971
−0.5	1.2194
−0.4	1.1474
−0.3	1.0860
−0.2	1.0392
−0.1	1.0100
0	1.0000
0.1	1.0100
0.2	1.0392
0.3	1.0860
0.4	1.1474
0.5	1.2194
0.6	1.2971
0.7	1.3748
0.8	1.4459
0.9	1.5035
1	1.5403
1.1	1.5489
1.2	1.5218
1.3	1.4521
1.4	1.3331
1.5	1.1592
1.6	0.9252
1.7	0.6276
1.8	0.2639
1.9	−0.1671
2	−0.6646

Question 13.6. Repeat Question 13.5 using the central-difference methods of orders 2 and 4. Comment on your answers.

Question 13.7. Use the data given in Table 13.20 to estimate the gradient of the unknown function $f(x)$ at each node point using the following approaches.
a. Elementary approximation.
b. Central-difference method of order 2.
c. Central-difference method of order 4.
You are given that $f(x) = 1 + x^2 \cos x$. Comment on your answers.

Question 13.8. For each of the following methods, determine the node points that should be used to evaluate the integral

$$I = \int_{-1.5}^{1.5} \sin(x^2)\, dx$$

to within a theoretical error of 0.001.
a. Trapezoidal approach
b. Simpson's approach
Implement each approximation.

Question 13.9. If Z is a random variable with Standard normal distribution, confirm that the probability that Z is between 0 and 4 is 49.997% to within 0.001%. You should use an appropriate implementation of Simpson's approach.

Question 13.10. An asset, currently priced at \$200, is expected to have a market price 1 year from now given by \$200S. Here, S is a random variable with density function

$$\frac{f_S(s)}{\$200} = \frac{a}{s} \exp\left\{ -\frac{1}{2} \left(\frac{\ln(s) - 1}{0.2} \right)^2 \right\} \quad \text{for } s \in [0.1, 6.1]$$

for some fixed constant a. Estimate the probability that the price 1 year from now will have increased by between 0% and 300%. You should use Simpson's approach of order 16 in any numerical integrals.

PART III

Worked Solutions to Questions

Chapter 1 Solutions

Solution 1.1. The numbers are identified as being members of the following real number systems.

a. $5 \in \mathbb{R}, \mathbb{Q}, \mathbb{Z}, \mathbb{N}$

b. $6.48763 \in \mathbb{R}, \mathbb{Q}$

c. $\pi^2 \in \mathbb{R}, \mathbb{J}$

d. $\frac{43}{7} \in \mathbb{R}, \mathbb{Q}$

e. $-6 \in \mathbb{R}, \mathbb{Q}, \mathbb{Z}$

Solution 1.2. The mathematical statements are "translated" as follows.

a. "$\forall x \in (-\infty, 0),\ x^2 \in \mathbb{R}^+$" reads as "for all x on the lower half of the real line, x^2 is a positive real number."

b. "$\forall p \in \mathbb{Z},\ q \in \mathbb{Z} \setminus \{0\},\ \frac{p}{q} \in \mathbb{Q}$" reads as "for all integer p and nonzero integer q, $\frac{p}{q}$ is a rational number."

c. "$\exists y \in \mathbb{Z} : y < 3$ and y is odd" reads as "there exists an integer y such that y is less than 3 and is odd."

d. "$\forall z \in \mathbb{R}^+,\ z > 0$" reads as "for all positive real z, z is greater than 0."

e. "$\forall a \in \mathbb{R},\ \exists b \in \mathbb{R} : a \times b = 2$" reads as "for all real a there exits a real b such that a times b equals 2."

Solution 1.3. We have $A = \{0, 1, 2, 3, 4, 5\}$, $B = \{-2, -1, 1, 2\}$, and $C = \{2, 3, 4, 5, 6\}$ and proceed by evaluating the left- and right-hand sides separately for comparison.

a. $A \cup (B \cap C) \equiv (A \cup B) \cap (A \cup C)$

LHS: $A \cup (B \cap C) = A \cup \{1, 2\} = \{0, 1, 2, 3, 4, 5\}$

RHS: $(A \cup B) \cap (A \cup C) = \{-2, -1, 0, 1, 2, 3, 4, 5\} \cap \{0, 1, 2, 3, 4, 5, 6\} = \{0, 1, 2, 3, 4, 5\}$

b. $A \cap (B \cup C) \equiv (A \cap B) \cup (A \cap C)$

LHS: $A \cap (B \cup C) = A \cap \{-2, -1, 1, 2, 3, 4, 5, 6\} = \{1, 2, 3, 4, 5\}$

RHS: $(A \cap B) \cup (A \cap C) = \{1, 2\} \cup \{2, 3, 4, 5\} = \{1, 2, 3, 4, 5\}$

c. $C \setminus (A \cap B) \equiv (C \setminus A) \cup (C \setminus B)$

LHS: $C \setminus (A \cap B) = C \setminus \{1, 2\} = \{3, 4, 5, 6\}$

RHS: $(C \setminus A) \cup (C \setminus B) = \{6\} \cup \{3, 4, 5, 6\} = \{3, 4, 5, 6\}$

d. $B \setminus \emptyset \equiv B$
LHS: $B \setminus \emptyset = \{-2, -1, 1, 2\}$
RHS: $\{-2, -1, 1, 2\}$

e. $(B \setminus A) \cap C \equiv (B \cap C) \setminus A$
LHS: $(B \setminus A) \cap C = \{-2, -1\} \cap C = \emptyset$
RHS: $(B \cap C) \setminus A = \{2\} \setminus A = \emptyset$

Note that, these expressions are examples of standard set identities.

Wolfram Alpha can be used to check these expressions using, for example, the commands

```
{0,1,2,3,4,5} union ( {-2,-1,1,2} intersection {2,3,4,5,6} )
```

for $A \cup (B \cap C)$ and

```
{2,3,4,5,6} setminus ( {0,1,2,3,4,5} intersection {-2,-1,1,2} )
```

for $C \setminus (A \cap B)$.

Solution 1.4. We attempt an algebraic approach in each case.

a. We note that $(y-4)(y+4) = y^2 - 4y + 4y - 16 = y^2 - 16$. The expression is therefore an identity.

b. The expression $y^4 + 9 = 10y^2$ is clearly not an identity. The equation is satisfied when $y^4 - 10y^2 + 9 = 0$ which is a quadratic in y^2. Therefore, we define $z = y^2$ and write $z^2 - 10z + 9 = 0$ which is solved for $z = 1$ and 9. These lead to $y = \pm 1$ and ± 3.

Note that, Wolfram Alpha can be used with the command

```
solve y^4 +9 = 10y^2
```

or

```
roots y^4 +9 - 10y^2
```

The use of Wolfram Alpha for solving equations is discussed in Chapter 2.

c. The expression $y^3 + 25ay = (25 + a)y^2$ is clearly not an identity. The equation is satisfied when $y^3 + 25ay - (25 + a)y^2 = 0$ which can be solved by factorizing the RHS as $y(y-25)(y-a)$ and so $y = 0, a,$ or 25. These can be confirmed with Wolfram Alpha if required.

d. The expression $-y^4 + 5y^3 - 9y^2 + 7y = 2$ is clearly not an identity. The equation is satisfied when $y^4 - 5y^3 + 9y^2 - 7y - 2 = 0$, which can be expressed as $(y-1)^3$ $(y-2) = 0$. It is therefore satisfied when $y = 1$ or 2. These can be confirmed with Wolfram Alpha if required.

Solution 1.5. As required, we attempt an algebraic approach in each case. Alternatively, the solutions can be visualized in Figure 1.1.

a. We work with the equation $z - 1 = 0$ and attempt to identify where $z - 1 > 0$. The equation clearly has a solution at $z = 1$ and therefore crosses the z-axis at this

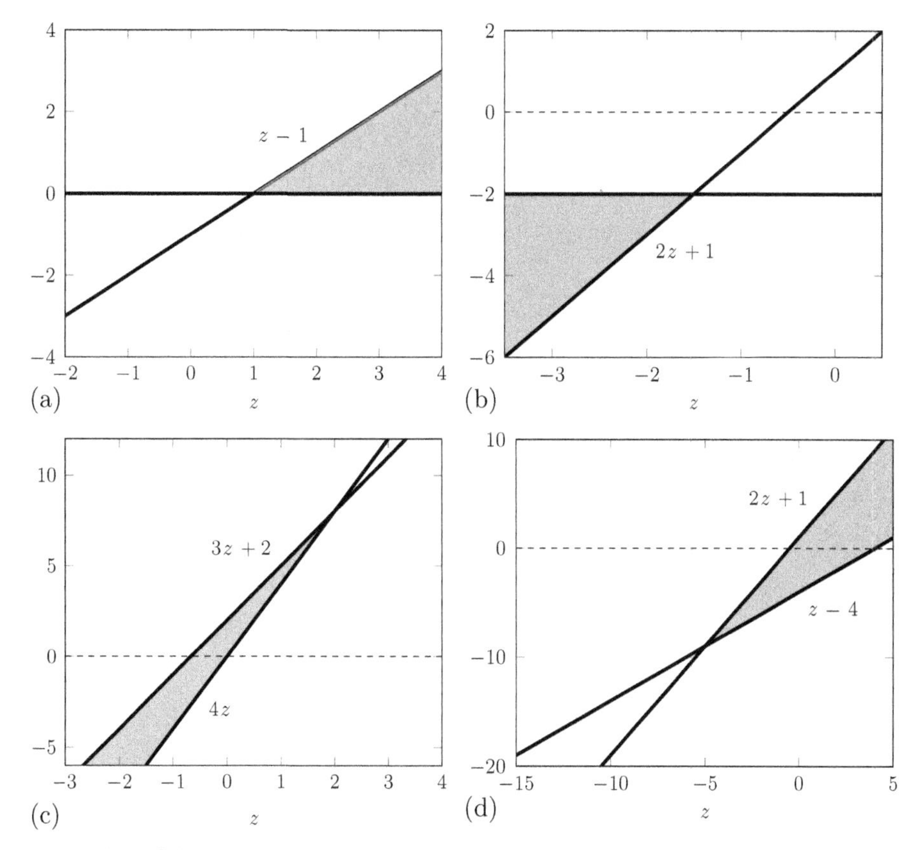

Figure 1.1 Plots of the inequalities in Solution 1.5. (a) $z - 1 > 0$, (b) $2z + 1 \leq -2$, (c) $4z < 3z + 2$, and (d) $z - 4 \leq 2z + 1$.

point. It also tends to $\pm\infty$ as $z \to \pm\infty$ and so we would expect that $z \in (1, \infty)$ solves $z - 1 > 0$.

b. We work with the equation $2z + 3 = 0$ and attempt to identify where $2z + 3 \leq 0$. The equation has a solution at $z = -1.5$ and tends to $\pm\infty$ as $z \to \pm\infty$. We would therefore expect that $z \in (-\infty, -1.5]$ solves $2z + 3 \leq 0$ and also $2z + 1 \leq -2$.

c. We work with the equation $z - 2 = 0$ and attempt to identify where $z - 2 < 0$. The equation has a solution at $z = 2$ and tends to $\pm\infty$ as $z \to \pm\infty$. We would therefore expect that $z \in (-\infty, 2)$ solves $z - 2 < 0$ and also $4z < 3z + 2$.

d. We work with the equation $z + 5 = 0$ and attempt to identify where $z + 5 \geq 0$. The equation has a solution at $z = -5$ and tends to $\pm\infty$ as $z \to \pm\infty$. We would therefore expect that $z \in [-5, \infty)$ solves $z + 5 \geq 0$ and also $z - 4 \leq 3z + 2$.

Note the appropriate use of open and closed intervals in each solution.

Chapter 2 Solutions

Solution 2.1. The following observations are made with knowledge of the individual terms of each mapping. For example, x^5 and x^3 are many-to-one mappings and x is a one-to-one mapping, therefore any combination of these terms must result in a many-to-one mapping.

a. $f(x) = x^5 - 29x^3 + 100x$ is a many-to-one mapping.

b. $g(z) = z^4 - 6z^3 + 4z^2 + 24z - 32$ is a many-to-one mapping.

c. $h(y) = y \times (0.5 - \cos(y))$ is a many-to-one-mapping.

d. $l(p) = (p^2 - 4) \times \cos(2p)$ is a many-to-one mapping.

e. $m(q) = \sqrt{q} - 2q\tan(q)$ is a one-to-many mapping owing to the square root that can take both positive and negative values unless explicitly stated.

Solution 2.2. One-to-one and many-to-one mappings are functions. One-to-many mappings are not. We can then use the results of Solution 2.1 to determine the following.

a. $f(x) = x^5 - 29x^3 + 100x$ is a function.

b. $g(z) = z^4 - 6z^3 + 4z^2 + 24z - 32$ is a function.

c. $h(y) = y \times (0.5 - \cos(y))$ is a function.

d. $l(p) = (p^2 - 4) \times \cos(2p)$ is a function.

e. $m(q) = \sqrt{q} - 2q\tan(q)$ is not a function.

Solution 2.3. We approach each part with general reasoning and confirm our conclusion in Figure 2.1.

a. We note that $f(x)$ consists of the sum of odd-powered polynomials and so $f(-x) = -x^5 + 29x^3 - 100x = -f(x)$. The function therefore has negative symmetry about $x = 0$. That is, it is an odd function. This is confirmed in Figure 2.1(a).

b. We note that $g(z)$ is a mixture of odd and even polynomial terms and so $g(\pm z) \neq \pm g(z)$ or $\mp g(z)$ for general z. It therefore has no symmetry about $z = 0$. This is confirmed in Figure 2.1(b).

c. We note that y and $\cos y$ are odd and even functions, respectively. Therefore, $h(-y) = -y(0.5 - \cos(y)) = -h(y)$ for general y and the function has negative symmetry about $y = 0$. That is, it is an odd function. This is confirmed in Figure 2.1(c).

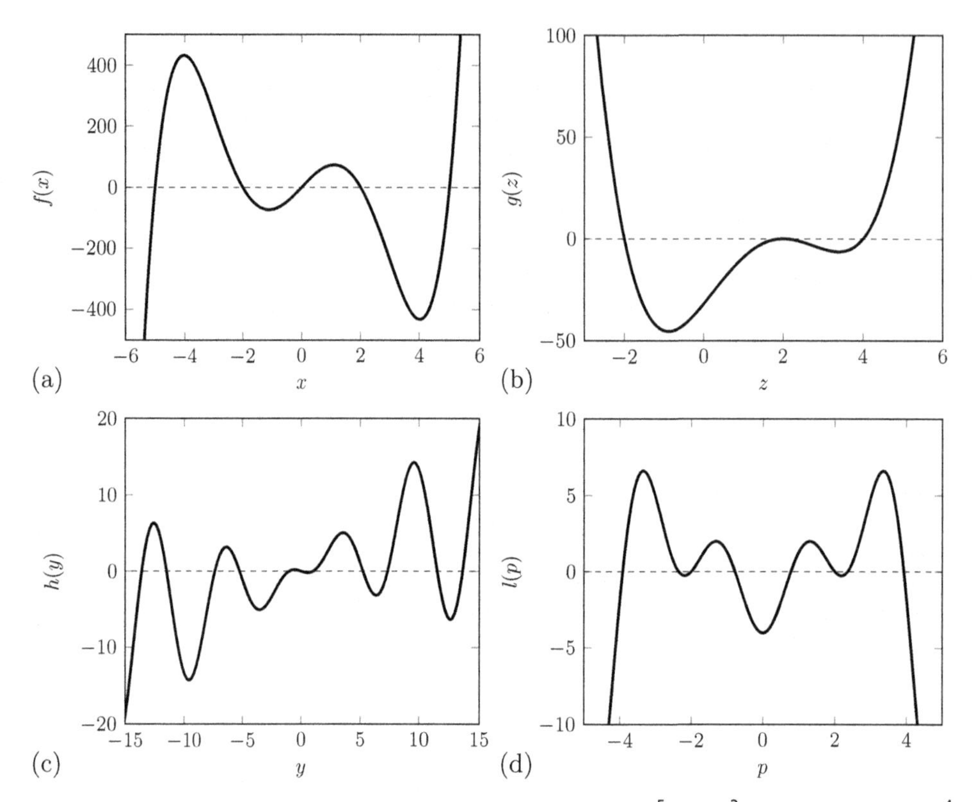

Figure 2.1 Plots of the functions in Questions 2.1–2.5. (a) $f(x) = x^5 - 29x^3 + 100x$, (b) $g(z) = z^4 - 6z^3 + 4z^2 + 24z - 32$, (c) $h(y) = y \times (0.5 - \cos(y))$, and (d) $l(p) = (p^2 - 4) \times \cos(2p)$.

d. We note that both p^2 and $\cos(2p)$ are even functions. Therefore, $l(-p) = (p^2 - 4) \times \cos(2p) = l(p)$ for general p and the function has positive symmetry about $p = 0$. That is, it is an even function. This is confirmed in Figure 2.1(d).

Solution 2.4. The plots in Figure 2.1 demonstrate that each function has more than one root.

a. We need x such that $x^5 - 29x^3 + 100x = 0$. It is immediately clear that $x = 0$ is a root and so we proceed to consider $x^4 - 29x^2 + 100 = 0$. Trial and error leads to $x = 2$ and, since this is an odd function, $x = -2$ is also a root. The factors $(x - 2)(x + 2)$ can then be divided out (see Appendix B) resulting in $x^2 - 25 = 0$. Therefore, $x = \pm 5$ are also roots. That is, we have found five roots $x = 0, \pm 2$, and ± 5.

b. We need z such that $z^4 - 6z^3 + 4z^2 + 24z - 32 = 0$. We note that $x = 2$ solves the expression and Figure 2.1 suggests that it is a repeated root; that is, $(x - 2)^2$ is a factor. Removing this factor results in the quadratic $z^2 - 2z - 8 = 0$ and this is easily solved, leading to $z = -2$ and 4. That is, we have found three distinct roots $z = 2$ (repeated), 4, and -2.

c. We need y such that $y \times (0.5 - \cos(y)) = 0$. It is immediately clear that $y = 0$ is a root, and $\cos y = 0.5$ determines the others. We know that cos is a cyclic function such that $\cos(y + 2n\pi) = \cos y$ for $n \in \mathbb{Z}$, therefore $y + 2n\pi = \cos^{-1}(0.5)$. Now $\cos y$ is a many-to-one mapping, therefore the inverse, $\cos^{-1} y$, must be a one-to-many mapping. In fact, it is sufficient to take $\cos^{-1}(0.5) = \pm\frac{\pi}{3}$ which leads to additional roots at $y = \pi\left(\pm\frac{1}{3} - 2n\right)$. That is, we have found roots at $y = 0$ and $y = \pi\left(\pm\frac{1}{3} - 2n\right)$ for $n \in \mathbb{Z}$.

d. We need p such that $(p^2 - 4) \times \cos(2p) = 0$. The bracketed factor leads to roots at $p = \pm 2$ and it remains to solve $\cos(2p) = 0$. We therefore have $2p = \frac{(2n\pm1)\pi}{2}$ and so have additional roots at $\frac{(2n\pm1)\pi}{4}$. That is, we have found roots at $p = \pm 2$ and at $p = \frac{(2n\pm1)\pi}{4}$ for $n \in \mathbb{Z}$.

Wolfram Alpha can be used to confirm the roots in each case using, for example, the command

```
roots z^4-6z^3+4z^2+24z-32
```

Solution 2.5. The following observations can be made from Figure 2.1.

a. We see that $f(x)$ has no singularities or discontinuities and is defined across the domain $\mathbb{R}$. It is clear that, $f(x) \to \pm\infty$ as $x \to \pm\infty$ and so the range is also $\mathbb{R}$. The function displays no horizontal or vertical asymptotes. Any straight horizontal line passes through the curve at up to five points; furthermore, any vertical line passes through the curve only once. The mapping is therefore many-to-one.

b. We see that $g(z)$ has no singularities or discontinuities and is defined across the domain $\mathbb{R}$. It is clear that $g(x) \to \infty$ as $z \to \pm\infty$ and the function has a minimum close to $g(z) = g_{min} \approx -45$. The range is therefore $\{g \in \mathbb{R} : g \in [g_{min}, \infty)\}$. Any straight horizontal line passes through the curve at least once; furthermore, any vertical line passes through the curve only once. The mapping is therefore many-to-one.

c. We see that $h(y)$ has no singularities or discontinuities and is defined across the domain $\mathbb{R}$. It is clear that $h(y) \to \pm\infty$ as $y \to \pm\infty$ and so the range is also $\mathbb{R}$. The function displays no horizontal or vertical asymptotes. Any straight horizontal line passes through the curve at least once; furthermore, any vertical line passes through the curve only once. The mapping is therefore many-to-one.

d. We see that $l(p)$ has no singularities or discontinuities and is defined on the domain $\mathbb{R}$. It is clear that $l(p)$ oscillates as a cos function of increasing amplitude as p increases in either direction; it therefore has range $\mathbb{R}$. Any straight horizontal line passes through the curve at least once; furthermore, any vertical line passes through the curve only once. The mapping is therefore many-to-one.

It is useful to note that Wolfram Alpha can be used to determine the domain and range of a function using the commands `domain` and `range`. For example,

```
domain x^5 - 29x^3+100x
```

and

```
range x^5 - 29x^3+100x
```

can be used to confirm the solution to part a. The reader is invited to confirm the solution to parts b-d.

Solution 2.6. In this case, the square root operation is restricted to nonnegative values and so is a one-to-one mapping and a function. The square root of a negative number is undefined on $\mathbb{R}$ (but see Chapter 8) and there are no singularities for positive numbers, the domain is therefore the interval $\{x \in \mathbb{R} : x \in [0, \infty)\}$. By definition of this function, the range is limited to nonnegative real numbers and we would write $\{f \in \mathbb{R} : f \in [0, \infty)\}$.

Solution 2.7. We attempt an algebraic approach in each case.

a. We work with equation $z^3 - 2z^2 - z + 2 = 0$ and attempt to identify where $z^3 - 2z^2 - z + 2 > 0$. The equation can be factorized as $(z-2)(z-1)(z+1) = 0$ leading to solutions being identified at $z = -1, 1$, and 2; the curve of $z^3 - 2z^2 - z + 2$ therefore crosses the z-axis at these points. The highest power of the equation is 3 and so we would expect the curve to tend to $\pm\infty$ as $z \to \pm\infty$. Given these properties, we would expect that $z \in (2, \infty)$ and $z \in (-1, 1)$ solves $z^3 - 2z^2 - z + 2 > 0$ and also $z^3 - 2z^2 - z > -2$.

b. We work with the equation $z^2 + 2z + 5 = 0$ and note that this does not cross the z-axis. The inequality therefore holds for all $z \in \mathbb{R}$.

c. We work with the equivalent inequality $z^3 + 2z^2 + 2z < 0$ and note that the equation $z^2 + 2z^2 + 2z = 0$ is solved only for $z = 0$. The solution to the inequality is therefore $z \in (-\infty, 0)$.

d. We work with the equation $z^2 - 4 = 0$ which has solutions at $z = 2$ and -2. The curve of $z^2 - 4$ tends to ∞ as $z \to \pm\infty$ and so the inequality is solved within $z \in [-2, 2]$.

Note the appropriate use of open and closed intervals in each solution. The solutions can be visualized in Figure 2.2 and also confirmed with Wolfram Alpha using, for example, the command

```
solve z^3 - 2z^2 -z > -2
```

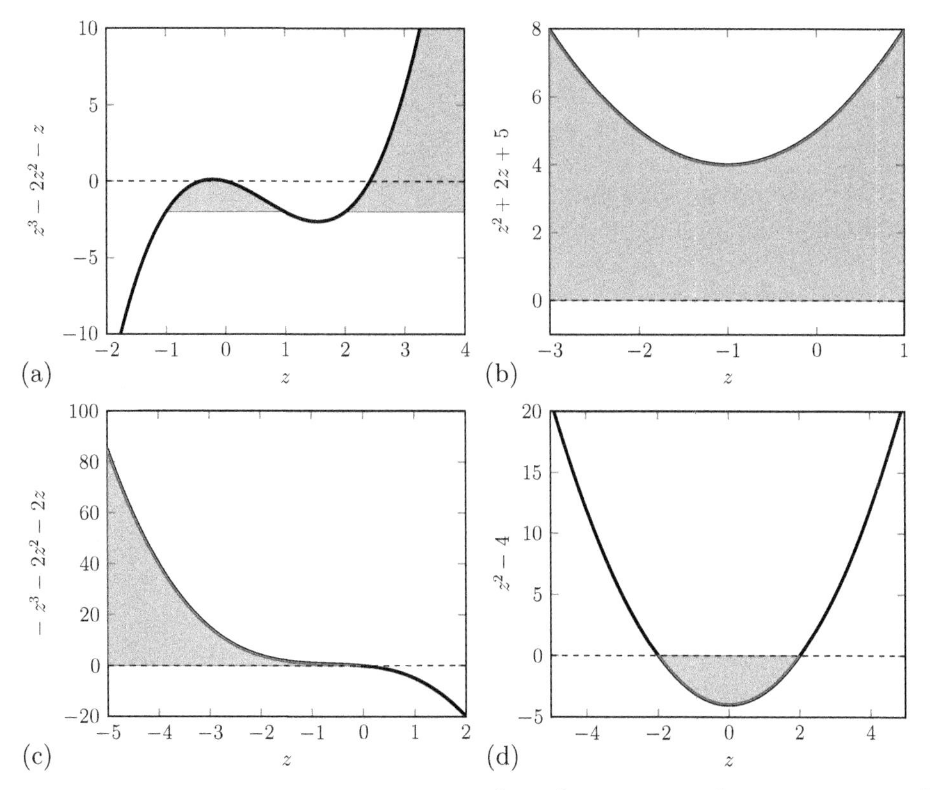

Figure 2.2 Plots of the inequalities in Solution 2.7. (a) $z^3 - 2z^2 - z > -2$, (b) $z^2 + 2z + 5 \geq 0$, (c) $-z^3 - 2z^2 - 2z > 0$, and (d) $z^2 - 4 \leq 0$.

Solution 2.8. Note that the question requires a *discussion using words* rather than a mathematical justification.

i. A second-order polynomial is dominated by the leading term a_2x^2 with $a_2 \neq 0$. For $a_2 > 0$, the polynomial tends to ∞ as $x \to \pm\infty$ and has a single minimum value at some finite x at which the curve "turns." The curve will cross the x-axis twice if this minimum value is negative, will not cross if the minimum is positive, or will touch the axis if this minimum is zero. These cases correspond to two real roots, zero real roots, and one repeated real root, respectively. If $a_2 < 0$, the polynomial tends to $-\infty$ as $x \to \pm\infty$ and has a single turning point at some finite x that is a maximum. The curve will cross the x-axis twice if this maximum value is positive, will not cross if the maximum is negative, or will touch the axis if this maximum is zero.

ii. An odd-ordered polynomial is dominated by the leading term a_nx^n for $a_n \neq 0$ and $n \in \{1, 3, 5, \ldots\}$. For $a_n > 0$, the polynomial tends to $\pm\infty$ as $x \to \pm\infty$ and so must cross the x-axis at least once. For $a_n < 0$, the polynomial tends to $\mp\infty$ as $x \to \pm\infty$ and so must cross the x-axis at least once. In both cases, the polynomial has at least one root.

Solution 2.9. Let us introduce four distinct functions $f(x)$, $g(x)$, $h(x)$, and $k(x)$ which are such that

$$f(-x) = f(x) \quad g(-x) = g(x) \quad h(-x) = -h(x) \quad k(-x) = -k(x)$$

That is, $f(x)$ and $g(x)$ are even functions, and $g(x)$ and $h(x)$ are odd functions. Consider now, the following functions obtained from various products

$$l(x) = f(x)g(x) \quad m(x) = f(x)h(x) \quad n(x) = h(x)k(x)$$

Note that this does not include all possible combinations but is sufficient for our needs. We then see that

$$\begin{aligned} l(-x) =& f(-x)g(-x) = f(x)g(x) = l(x) \\ m(-x) =& f(-x)h(-x) = -f(x)h(x) = -m(x) \\ n(-x) =& h(-x)k(-x) = (-1)^2 h(x)m(x) = n(x) \end{aligned}$$

This leads us to the following conclusions,

- a function formed from the product of any two even functions is even.
- a function formed from the product of an odd and an even function is odd.
- a function formed from the product of any two even functions is even.

Solution 2.10. In each case, we begin by noting a sensible substitution that simplifies the expression.

a. Using the substitution $u = e^x$, we obtain $f(u) = u^2 + u - 30$, which has roots at $u = -6$ and 5. We then require x such that $e^x = -6$ (not possible) and $e^x = 5$. The only real root of $f(x) = e^{2x} + e^x - 30$ is then $x = \ln 5 \approx 1.6094$.

b. Using the substitution $u = \cos(5z)$, we obtain $g(u) = u^2 - 4$, which has roots at $u = \pm 2$. We then require z such that $\cos(5z) = \pm 2$ which is not possible. The function $g(z) = \cos^2(5z) - 4$ therefore has no real roots.

c. Using the substitution $u = e^{2y}$, we obtain $h(u) = u^3 + u^2 - u - 1 = (u-1)(u+1)^2$, which has roots at $u = \pm 1$. We then require y such that $e^{2y} = -1$ (not possible) and $e^{2y} = 1$. The only real root of $h(y) = e^{6y} + e^{4y} - e^{2y} - 1$ is then $y = 0$.

d. Using the substitution $u = 2q^2$, we obtain $n(u) = \sin u$, which has roots at $u = n\pi$ for $n \in \mathbb{Z}$. We then require $2q^2 = n\pi$ with $n \in \mathbb{N}$. The function $n(q) = \sin(2q^2)$ therefore has roots at $q = \pm\sqrt{\frac{n\pi}{2}}$ for $n \in \mathbb{N}$.

e. Using the substitution $u = e^{2p} + 2e^p$, we obtain $m(u) = \ln u$, which has a real root at $u = 1$. We then require $e^{2p} + 2e^p = 1$ and use a further substitution $v = e^p$ which leads to the equation $v^2 + 2v - 1 = 0$. This has solutions at $v = -1 - \sqrt{2}$ and $-1 + \sqrt{2}$. Note that $e^p = -1 - \sqrt{2}$ is not possible and the function $m(p) = \ln\left(e^{2p} + 2e^p\right)$ therefore has a single root at $p = \ln(-1 + \sqrt{2}) \approx -0.8814$.

These can be checked with Wolfram Alpha using the command `roots` or `real roots`. For example, the result in part e. can be confirmed by the command

```
real roots ln(e^(2p)+2e^p)
```

Solution 2.11. We conduct a visual exploration of the functions' properties, based on Figure 2.3.

i. Each function demonstrates singularities arising from roots of the function in the denominator; in particular, $\operatorname{cosec} x$ and $\cot x$ have singularities at $x = n\pi$ and $\cos x$ has singularities at $x = \frac{(2n\pm1)\pi}{2}$ for $n \in \mathbb{Z}$. In each case, the domain is $\mathbb{R}$ with the singular points removed.

ii. The functions $\operatorname{cosec} x$ and $\sec x$ take a minimum/maximum value of ±1 between each pair of vertical asymptotes. The range of each of these functions is therefore

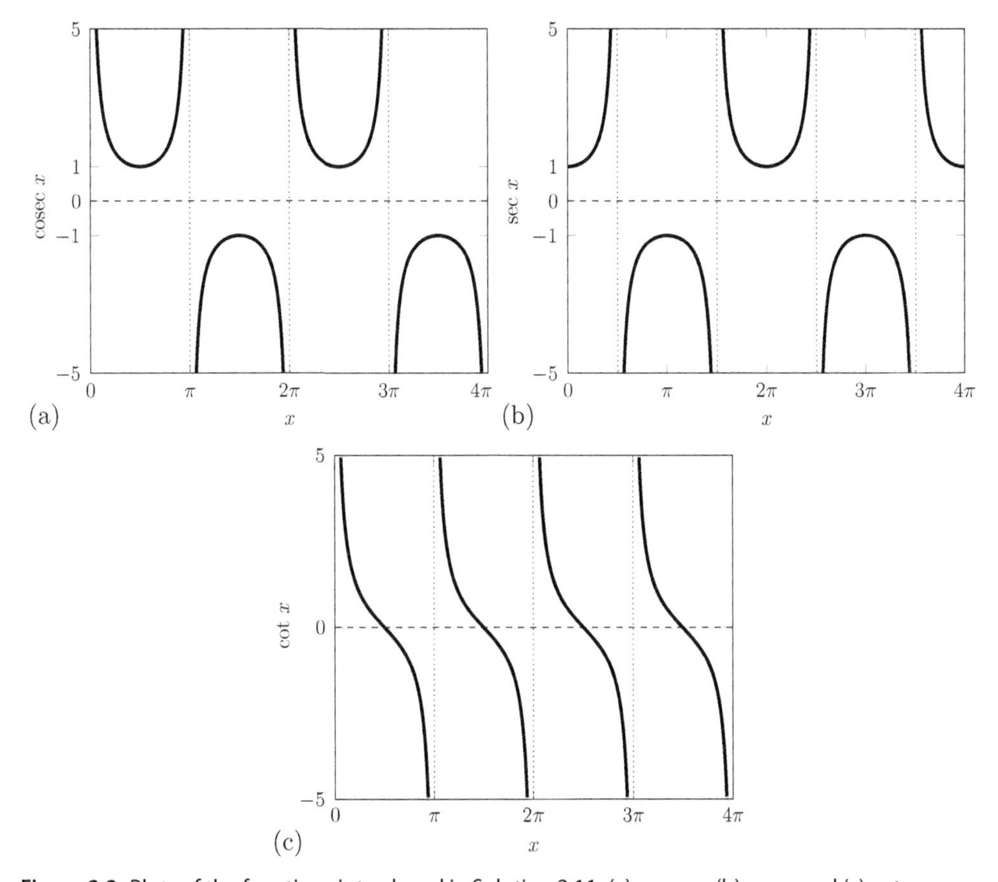

Figure 2.3 Plots of the functions introduced in Solution 2.11. (a) $\operatorname{cosec} x$, (b) $\sec x$, and (c) $\cot x$.

$\{y \in \mathbb{R} : y \in (-\infty, 1] \cup [1, \infty)\}$. The $\cot x$ function has not such restrictions and has a range equal to $\mathbb{R}$.

iii. The functions $\operatorname{cosec} x$ and $\sec x$ have no roots. The $\cot x$ function has roots corresponding to roots of $\cos x$, that is at $x = \frac{(2n+1)\pi}{2}$ for $n \in \mathbb{Z}$.

iv. Each function has a parity dictated by the parity of particular standard circular functions. In particular, $\operatorname{cosec} x$ and $\cot x$ are odd and $\sec x$ is even. Note also that the functions $\operatorname{cosec} x$ and $\sec x$ repeat every 2π and $\cot x$ every π.

v. Each function has vertical asymptotes at the locations of the singular points described in part i.

Solution 2.12. Using

$$f(x) = \sin x \quad g(x) = e^x \quad h(x) = \frac{2x+1}{2x^2+3}$$

the following composites can be easily written down.

a. $(h \circ f)(x) = h(f(x)) = \frac{2\sin x+1}{2\sin^2 x+3}$

b. $(g \circ h)(x) = g(h(x)) = \exp\left(\frac{2x+1}{2x^2+3}\right)$

c. $(f \circ g \circ h)(x) = f(g(h(x))) = \sin\left(e^{\frac{2x+1}{2x^2+3}}\right)$

d. $(g \circ f \circ g)(x) = g(f(g(x))) = \exp(\sin e^x)$

Solution 2.13. We determine the domain and range of the following functions over $\mathbb{R}$.

a. This is a rational function and so has domain equal to $\mathbb{R}$ with any singular points removed. The denominator has roots at $x = -1$ and 0.5 with the latter coinciding with a root of the numerator. We therefore see that the function is not written in its basic form and we rewrite as $f(x) = \frac{1}{x+1}$. The domain is then clearly $\{x \in \mathbb{R} : x \neq -1\}$. Note that $f(x) \to 0$ as $x \to \pm\infty$ and so we expect a horizontal asymptote at $x = 0$. The range is therefore $\{f \in \mathbb{R} : f \neq 0\}$.

b. Note that $g(z) = \ln(f(z))$ with $f(z)$ from part a. We therefore rewrite the function in its basic form $g(z) = \ln\left(\frac{1}{z+1}\right)$. The basic $\ln(z)$ function has a domain $\mathbb{R}^+$ and so we must restrict the domain of $f(z)$ such that $f(z) > 0$. The domain of $g(z)$ is then $\{z \in \mathbb{R} : z > -1\}$. The range of $g(z)$ is the range of $\ln(z)$, that is $\mathbb{R}$.

c. Note that $h(y) = \sin(f(y))$ and is rewritten in its basic form $h(y) = \sin\left(\frac{1}{y+1}\right)$. The domain of $\sin y$ is simply $\mathbb{R}$ but we must remove the singular point at $y = -1$, this means that the domain of $h(y)$ is $\{y \in \mathbb{R} : y \neq -1\}$. The range of $h(y)$ is the range of $\sin y$, that is $\{h \in \mathbb{R} : y \in [-1, 1]\}$.

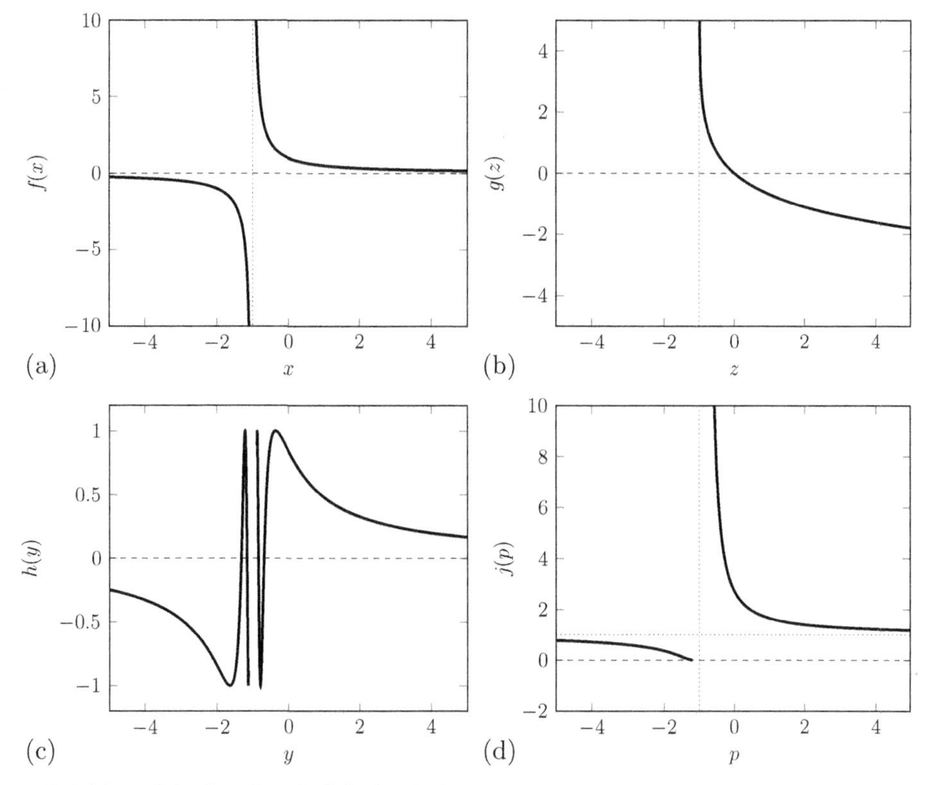

Figure 2.4 Plots of the functions in Solution 2.13.

d. Note that $j(p) = \exp(f(p))$ and is rewritten in its basic form $j(p) = \exp\left(\frac{1}{1+p}\right)$. The domain of e^p is $\mathbb{R}$ but we must remove the singular point at $p = -1$, this means that the domain of $j(p)$ is $\{p \in \mathbb{R} : p \neq -1\}$. We note that $j(p)$ tends to 1 as $p \to \pm\infty$ and so has a horizontal asymptote. The basic exponential function e^p has range $\mathbb{R}^+$ from which we must remove the location of the asymptote, $j(p)$ therefore has range $\{j \in \mathbb{R}^+ : j \neq 1\}$.

A plot of each function is shown in Figure 2.4. In each case, the range and domain can be confirmed in Wolfram Alpha with the commands `range` and `domain`.

Solution 2.14. Let us define $u = \log a$ and $v = \log b$.

a. The definition of u is such that $a = 10^u$ and this is our stating point.

$$\begin{aligned}
& a = 10^u \\
\Rightarrow\ & a^b = 10^{ub} && \text{(after raising both sides by } b\text{)} \\
\Rightarrow\ & \log\left(a^b\right) = \log\left(10^{ub}\right) = ub && \text{(after taking logs of both sides)} \\
\Rightarrow\ & \log\left(a^b\right) = b \log a && \text{(after using } u = \log a\text{)}
\end{aligned}$$

b. The definitions of u and v are such that $a = 10^u$ and $b = 10^v$, therefore

$$\begin{aligned} & ab = 10^{u+v} \\ \Rightarrow\ & \log(ab) = \log\left(10^{u+v}\right) = u + v && \text{(after taking logs of both sides)} \\ \Rightarrow\ & \log(ab) = \log a + \log b && \text{(after using } u = \log a \text{ and } v = \log b\text{)} \end{aligned}$$

c. The definitions of u and v are such that $a = 10^u$ and $b = 10^v$, therefore

$$\begin{aligned} & \tfrac{a}{b} = 10^{u-v} \\ \Rightarrow\ & \log\left(\tfrac{a}{b}\right) = \log\left(10^{u-v}\right) = u - v && \text{(after taking logs of both sides)} \\ \Rightarrow\ & \log\left(\tfrac{a}{b}\right) = \log a - \log b && \text{(after using } u = \log a \text{ and } v = \log b\text{)} \end{aligned}$$

Note that the proof of the equivalent expressions for the natural log follow in exactly the same way.

Solution 2.15. We attempt to find $f^{-1}(y)$ in each case and can then replace the dummy variable y with x.

a. We define $y = (x-2)^{\frac{1}{4}}$ and so $x = y^4 + 2$. Therefore, $f^{-1}(y) = y^4 + 2$ and $f^{-1}(x) = x^4 + 2$.

b. We define $y = \frac{x-2}{3x+9}$ and so $x = \frac{9y+2}{1-3y}$. Therefore, $f^{-1}(y) = \frac{9y+2}{1-3y}$ and $f^{-1}(x) = \frac{9x+2}{1-3x}$.

c. We define $y = \mathrm{e}^{4x^2}$ and so $x = \pm\frac{\sqrt{\ln(y)}}{2}$. Therefore, $f^{-1}(y) = \pm\frac{\sqrt{\ln(y)}}{2}$ and $f^{-1}(x) = \pm\frac{\sqrt{\ln(x)}}{2}$.

d. We define $y = \ln(x+6)$ and so $x = \mathrm{e}^y - 6$. Therefore, $f^{-1}(y) = \mathrm{e}^y - 6$ and $f^{-1}(x) = \mathrm{e}^x - 6$.

Solution 2.16. Working in time units of years, the accumulation is the exponential function $A(t) = 1.1^t$ for $t \in \mathbb{R}$.

a. The accumulated value of \$100 after 17 years is given by $100A(17) \approx \$505.45$.

b. The accumulated value at $t = 10$ for a deposit of \$40 made at $t = 2$ and another of \$60 at $t = 7$ is given by $40A(8) + 60A(3) \approx \$165.60$.

c. The present value of a \$150,000 debt due 3 months in the future is given by $150,000A(-\frac{3}{12}) \approx \$146,668.11$.

d. The net present value of cash flows \$40, \$50, and \$60 received at times $t = 10$, 11, and 12 is given by $40A(-10) + 50A(-11) + 60A(-12) \approx \$52.06$.

e. The accumulated value at $t = 15$ of the cash flows in part d. is given by $40A(5) + 50A(4) + 60A(3) \approx \$217.49$.

Solution 2.17. We work in terms of money units of £m and time units of years.

a. The equation of value (at $t = 0$) for this opportunity is,

$$-12 + \frac{10}{(1+i)^{3.5}} + \frac{5}{(1+i)^7} = 0$$

that is, $f(i) = -12 + \frac{10}{(1+i)^{3.5}} + \frac{5}{(1+i)^7}$. We use the substitution $x = \frac{1}{(1+i)^{3.5}}$ which leads to the quadratic function,

$$-12 + 10x + 5x^2 = 0 \Rightarrow x = \frac{1}{5}\left(-5 \pm \sqrt{85}\right)$$

Since we are after real values of i, we take the positive value of x and so

$$\frac{1}{(1+i)^{3.5}} = \frac{1}{5}\left(-5 \pm \sqrt{85}\right) \Rightarrow i \approx 0.0497$$

b. Part a. demonstrates that the opportunity has an internal rate of return of 4.97% per annum. A loan rate of 7% is far in excess of this internally generated return and so the opportunity will not be profitable if the loan is used to fund the initial investment. This can be confirmed by noting that $f(0.07) = -0.9948$. That is, the net present value at $i = 7\%$ is negative.

Chapter 3 Solutions

Solution 3.1. The equation of a straight line is $y = mx + c$, where m, the gradient, and c, the y-intercept, are to be determined. We proceed by establishing two simultaneous equations, one from each data point, and solve for the unknown m and c.

a. We have $0 = m \times 0 + c$ and $3 = 3m + c$ and see that $c = 0$ and $m = 1$. The equation is therefore $y = x$.

b. We have $-5 = -m + c$ and $0 = 4m + c$ and see that $m = 1$ and $c = -4$. The equation is therefore $y = x - 4$.

c. We have $0 = m\pi + c$ and $-\pi = m\pi^2 + c$ and see that $m = \frac{-1}{\pi - 1}$ and $c = \frac{\pi}{\pi - 1}$. The equation is therefore $y = -\frac{x - \pi}{\pi - 1}$.

d. We have $-56 = -10m + c$ and $-129 = -100m + c$ and see that $m = -\frac{37}{18}$ and $c = -\frac{689}{9}$. The equation is therefore $y = -\frac{37x}{18} - \frac{689}{9}$.

Note that Wolfram Alpha can be used to confirm these results with, for example, the command `straight line through(-1,-5) and (4,0)`.

Solution 3.2. A function $f(x)$ is continuous at $x = c$ if the left and right limits exist and are equal to $f(c)$.

a. Note that $f(x) = x^3 + x^2 + 2$ in the neighborhood of $x = 5$ and so the function is continuous at that point.

b. We have

$$\lim_{y \to 2^-} g(y) = \lim_{y \to 2^-} (y^5 - y^2 - 20) = 8$$

and

$$\lim_{y \to 2^+} g(y) = \lim_{y \to 2^+} y^3 = 8$$

Therefore, the left and right limits are equal (that is, the limit exists). Furthermore, $g(2) = 2^5 - 2^3 - 20 = 8$. The function $g(y)$ is therefore continuous at $y = 2$.

c. We have

$$\lim_{z \to 0^-} h(z) = \lim_{z \to 0^-} e^{-2z} = 1$$

and

$$\lim_{z \to 0^+} h(z) = \lim_{z \to 0^+} e^{-2z} = 1$$

Therefore, the left and right limits are equal (that is, the limit exists). However, $h(0) = 1.1 \neq 1$ and the function $h(z)$ is therefore not continuous at $z = 0$.

Solution 3.3. We approach each n in turn.

- $n = 1$: This is simply $\lim_{x \to a} f(x) = L_1$.
- $n = 2$: Note that $f(x)$ and $g(x)$ are general functions. In the particular case that $g(x) \equiv f(x)$ we can write

$$\lim_{x \to a} f(x)g(x) = \lim_{x \to a} f(x)f(x) = L_1 \times L_1 = \left(\lim_{x \to a} f(x)\right)^2$$

- $n = 3$: A similar approach to $n = 2$ is taken. In particular, if $g(x) \equiv f^2(x)$, we can use the result for $n = 2$ and write

$$\lim_{x \to a} f(x)g(x) = \lim_{x \to a} f(x)f^2(x) = L_1 \times L_1^2 = \left(\lim_{x \to a} f(x)\right)^3$$

It should then be clear that this process can be repeated to justify the general result as required.

Solution 3.4. The results of the tabular approach can be seen in Table 3.1. We see that,

$$\lim_{\delta \to 0} \left(\frac{\sin \delta}{\delta}\right) = 1 \quad \lim_{\delta \to 0} \left(\frac{1 - \cos \delta}{\delta}\right) = 0$$

We return to an algebraic method for determining these and similar limits in Chapter 5.

Table 3.1 Tabular method for $\lim_{\delta \to 0} \left(\frac{\sin(\delta)}{\delta}\right)$ and $\lim_{\delta \to 0} \left(\frac{1-\cos(\delta)}{\delta}\right)$ in Solution 3.4

δ	1	0.01	0.001	0.0001	0.00001	0.000001
$\frac{\sin \delta}{\delta}$	0.84147	0.99833	1.00000	1.00000	1.00000	1.00000
δ	1	0.01	0.001	0.0001	0.00001	0.000001
$\frac{\sin \delta}{\delta}$	0.84147	0.99833	1.00000	1.00000	1.00000	1.00000
δ	1	0.01	0.001	0.0001	0.00001	0.000001
$\frac{1-\cos \delta}{\delta}$	−0.45970	−0.04996	−0.00050	−0.00005	−0.00001	0.00000
δ	1	0.01	0.001	0.0001	0.00001	0.000001
$\frac{1-\cos \delta}{\delta}$	0.45970	0.04996	−0.00050	0.00005	0.00001	0.00000

Solution 3.5. In each case, we first obtain the instantaneous gradient of the function at the stated point and then fit a straight line with this gradient at that point.

a. We have $m(2\pi) = 4\pi(2\pi^2 - 1)$ and so $m'(p) = 3p^2 - 2 + 2\cos(2p)$ which is such that $m'(2\pi) = 12\pi^2$. The equation of the tangent is $y = 12\pi^2 p + c$ with c such that $4\pi(2\pi^2 - 1) = 24\pi^3 + c$. Therefore, $y = 4\pi\left(3\pi p - 4\pi^2 - 1\right)$.

b. We have $n(0) = 1$ and so $n'(q) = -e^{-q}(\sin^2 q + \pi \sin(2\pi q) + \cos^2(\pi q) - 2\sin q \cos q)$ which is such that $n'(0) = -1$. The equation of the tangent is $y = -q + c$ with c such that $1 = -1 \times 0 + c$. Therefore, $y = -q + 1$.

The tangent lines are visually confirmed in Figure 3.1.

Solution 3.6. We use the definition of a derivative, Eq. (3.4), and attempt to perform the limit in each case.

a. We require $f'(x)$ where $f(x) = x^2 - 2x + 1$.

$$\begin{aligned} f'(x) &= \lim_{\delta \to 0} \left(\frac{(x+\delta)^2 - 2(x+\delta) + 1 - x^2 + 2x - 1}{\delta} \right) \\ &= \lim_{\delta \to 0} (\delta + 2x - 2) \\ &= 2x - 2 \end{aligned}$$

That is, $f'(x) = 2x - 2$.

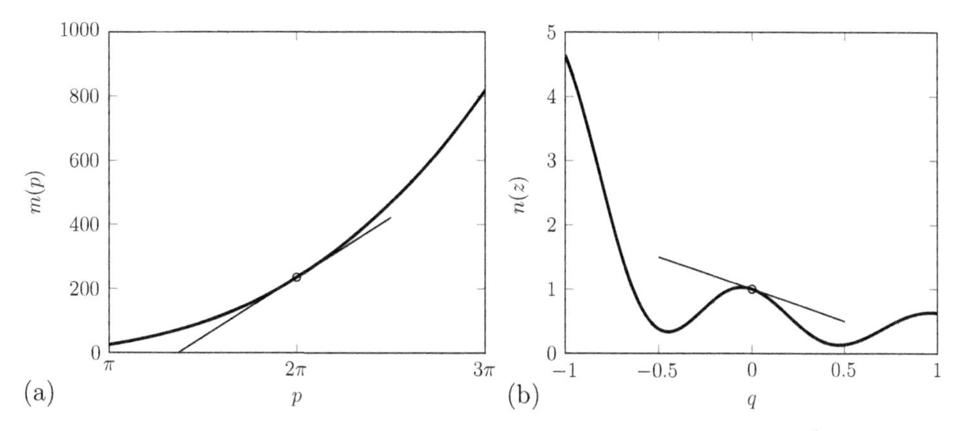

Figure 3.1 Visual confirmation of tangent lines derived in Solution 3.5. (a) $m(p) = p^3 - 2p + \sin(2p)$ and tangent line at $p = 2\pi$. (b) $n(q) = e^{-q}\left(\sin^2(q) + \cos^2(\pi q)\right)$ and tangent line at $q = 0$.

b. We require $g'(y)$ where $g(y) = e^{2y}$.

$$\begin{aligned} g'(y) &= \lim_{\delta \to 0} \left(\frac{e^{2(y+\delta)} - e^{2y}}{\delta} \right) \\ &= e^{2y} \lim_{\delta \to 0} \left(\frac{e^{2\delta} - 1}{\delta} \right) \\ &= e^{2y} \times 2 \end{aligned}$$

That is, $g'(y) = 2e^{2y}$. Note that the intermediate limit can be confirmed using the tabular method.

c. We require $h'(z)$ where $h(z) = \frac{1}{z^2}$.

$$\begin{aligned} h'(z) &= \lim_{\delta \to 0} \left(\frac{\frac{1}{(z+\delta)^2} - \frac{1}{z^2}}{\delta} \right) \\ &= \lim_{\delta \to 0} \left(-\frac{\delta + 2z}{z^2(z+\delta)^2} \right) \\ &= -\frac{2}{z^3} \end{aligned}$$

That is, $h'(z) = -\frac{2}{z^3}$.

d. We require $m'(p)$ where $m(p) = 4e^{p^2}$.

$$\begin{aligned} m'(p) &= \lim_{\delta \to 0} \left(\frac{4e^{(p+\delta)^2} - 4e^{p^2}}{\delta} \right) \\ &= 4e^{p^2} \lim_{\delta \to 0} \left(\frac{e^{2\delta p + \delta^2} - 1}{\delta} \right) \\ &= 4e^{p^2} \times 2p \end{aligned}$$

That is, $m'(p) = 8pe^{p^2}$. Note that the intermediate limit can be confirmed using the tabular method for various fixed p.

Note that an algebraic approach to the intermediate limits in parts b. and d. is discussed in Chapter 5. The limits can be confirmed using Wolfram Alpha with, for example, the command

```
limit (e^(2(y+d))-e^(2y))/d as d->0
```

Solution 3.7. Possible approaches include the product, quotient, and chain rules.

a. The derivative of a rational function can be performed by either the product or quotient rules (recall that these are equivalent). We proceed to use the product rule as follows:

$$
\begin{aligned}
f'(x) &= \frac{1}{x^4+25}\frac{\mathrm{d}}{\mathrm{d}x}(x^2-2x+1) + (x^2-2x+1)\frac{\mathrm{d}}{\mathrm{d}x}(x^4+25)^{-1} \\
&= \frac{2x-2}{x^4+25} + (x^2-2x+1) \times -(x^4+2)^{-2} \times 4x^3 \\
&= -\frac{2(x^5-3x^4+2x^3-25x+25)}{(x^4+25)^2}
\end{aligned}
$$

where the last line follows after some manipulation.

b. Note that $g(z)$ is a composite function and the chain rule is required. In particular,

$$
\begin{aligned}
g'(z) &= \left(\frac{\mathrm{d}}{\mathrm{d}z}\sin z\right)\left.\frac{\mathrm{d}e^u}{\mathrm{d}u}\right|_{u=\sin z} \\
&= \cos z e^{\sin z}
\end{aligned}
$$

c. Note that $h(y)$ is a quadratic in $\cos y$, that is, $h(y)$ a composite function and the chain rule is required. In particular,

$$
\begin{aligned}
h'(y) &= \left(\frac{\mathrm{d}}{\mathrm{d}y}\cos y\right)\left.\frac{\mathrm{d}}{\mathrm{d}u}(u^2-4u+2)\right|_{u=\cos y} \\
&= -\sin y\,(2\cos y - 4)
\end{aligned}
$$

d. A combination of the product rule and chain rule is required. In particular,

$$
\begin{aligned}
m'(p) &= \sin(\mathrm{e}^p)\frac{\mathrm{d}}{\mathrm{d}p}\cos p + \cos(p)\frac{\mathrm{d}}{\mathrm{d}p}\sin(\mathrm{e}^p) \\
&= -\sin(p)\sin(\mathrm{e}^p) + \cos(p)\left(\frac{\mathrm{d}}{\mathrm{d}p}\mathrm{e}^p\right)\left(\left.\frac{\mathrm{d}}{\mathrm{d}u}\sin u\right|_{u=\mathrm{e}^p}\right) \\
&= -\sin(p)\sin(\mathrm{e}^p) + \mathrm{e}^p\cos(p)\cos(\mathrm{e}^p)
\end{aligned}
$$

Note that these can be confirmed using Wolfram Alpha with, for example, the command

```
d/dp  cos(p)sin(e^p)
```

Solution 3.8. The function is a composite such that

$$
\sin\left(\mathrm{e}^{x^2-4}\right) = (f \circ g \circ h)(x)
$$

with $f(x) = \sin x$, $g(x) = \mathrm{e}^x$, and $h(x) = x^2 - 4$. It is then appropriate to make repeated use of the chain rule. In particular, we would write

$$
\frac{\mathrm{d}}{\mathrm{d}x}\left(\sin\left(\mathrm{e}^{x^2-4}\right)\right) = \frac{\mathrm{d}}{\mathrm{d}x}(x^2-4)\left(\left.\frac{\mathrm{d}}{\mathrm{d}u}\sin(\mathrm{e}^u)\right|_{u=x^2-4}\right)
$$

$$= 2x \left[\frac{\mathrm{d}e^u}{\mathrm{d}u} \left(\left. \frac{\mathrm{d}\sin v}{\mathrm{d}v} \right|_{v=e^u} \right) \right] \Bigg|_{u=x^2-4}$$

$$= 2x \left(e^u \cos\left(e^u\right) \right) \big|_{u=x^2-4}$$

$$= 2xe^{x^2-4} \cos\left(e^{x^2-4} \right)$$

This result can be confirmed using Wolfram Alpha with the command

```
d/dx sin(e^(x^2-4))
```

Solution 3.9. In each case, we obtain the gradient function, that is the derivative, and find its roots. Each root of the first derivative corresponds to a point of zero local gradient.

a. We have $f'(z) = z^2 - 1$ which clearly has roots at $z = \pm 1$. A plot of $f(z)$ is given in Figure 3.2(a) and shows that the roots of the derivative correspond to points at which the curve changes direction. We have a local maximum at $z = -1$ and a local minimum at $z = 1$.

b. We have $g'(y) = -\sin y$ which has roots at $y = n\pi$ for $n \in \mathbb{Z}$. A plot of $g(y)$ is given in Figure 3.2(b) and again shows that the roots of the derivative correspond to points at which the curve changes direction. We have a local maxima for n even and minima for n odd.

Note that Wolfram Alpha can be used to confirm these results using, for example, the command

```
turning points cos(y)+2
```

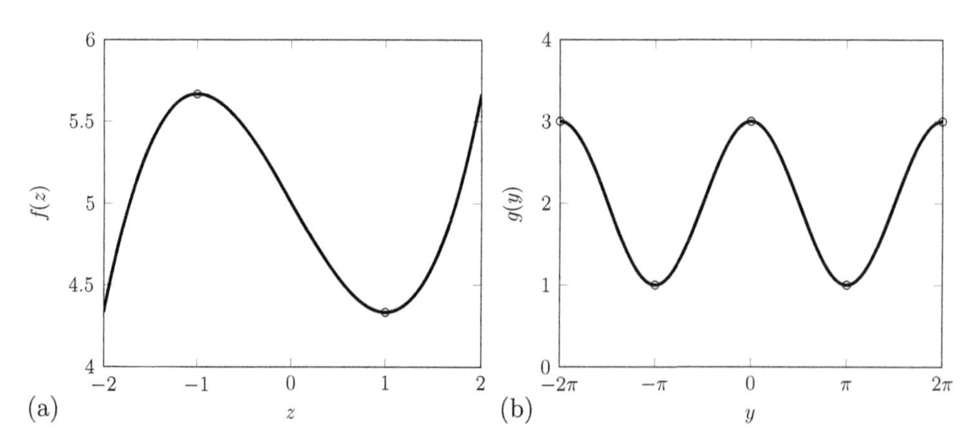

Figure 3.2 "Turning points" identified in Solution 3.9. (a) $f(z) = \frac{z^3}{3} - z + 5$, (b) $g(y) = \cos(y) + 2$.

Solution 3.10. The chain rule can be used in each case.

a. We proceed as follows:

$$\begin{aligned}\frac{\mathrm{d}}{\mathrm{d}x}\arcsin(4x) &= \frac{\mathrm{d}}{\mathrm{d}x}(4x)\left(\frac{\mathrm{d}}{\mathrm{d}u}\arcsin u\bigg|_{u=4x}\right)\\ &= \frac{4}{\sqrt{1-16x^2}}\end{aligned}$$

b. We proceed as follows:

$$\begin{aligned}\frac{\mathrm{d}}{\mathrm{d}x}\arccos(x^2-1) &= \frac{\mathrm{d}}{\mathrm{d}x}(x^2-1)\left(\frac{\mathrm{d}}{\mathrm{d}u}\arccos u\bigg|_{u=x^2-1}\right)\\ &= -\frac{2x}{\sqrt{1-(x^2-1)^2}}\\ &= -\frac{2}{\sqrt{2-x^2}}\end{aligned}$$

c. We proceed as follows:

$$\begin{aligned}\frac{\mathrm{d}}{\mathrm{d}x}\arctan(\mathrm{e}^x) &= \frac{\mathrm{d}}{\mathrm{d}x}(\mathrm{e}^x)\left(\frac{\mathrm{d}}{\mathrm{d}u}\arctan u\bigg|_{u=\mathrm{e}^x}\right)\\ &= \frac{\mathrm{e}^x}{\mathrm{e}^{2x}+1}\end{aligned}$$

These results can all be confirmed using Wolfram Alpha.

Solution 3.11. We are required to determine the present value of \$5600 due at $t = 7$ under the stated force of interest. This is given by

$$\frac{5600}{A(0,7)}$$

The problem is that δ is piecewise constant with

$$\delta = \begin{cases} 5\% & \text{for } t \in [0,4] \\ 7\% & \text{for } t \in (4,10] \end{cases}$$

and so the form of $A(0,7)$ is unclear. However, we can invoke the principle of consistency and state that

$$A(0,7) = A(0,4)A(4,7)$$

It is clear that $A(0,4) = \mathrm{e}^{4\times 0.05} = \mathrm{e}^{0.2}$ and $A(4,7) = \mathrm{e}^{3\times 0.07} = \mathrm{e}^{0.21}$. The required present value is therefore given by,

$$\frac{5600}{A(0,7)} = \frac{5600}{A(0,4)A(4,7)} = 5600\mathrm{e}^{-0.41} \approx \$3716.44$$

Solution 3.12. We are given that $A(0, t) = te^{0.05t^3}$ and are required to determine the underlying $\delta(t)$. Using Eq. (3.22),

$$\delta(t) = \frac{\mathrm{d}}{\mathrm{d}t} \ln A(0, t)$$

we determine that

$$\delta(t) = \frac{\mathrm{d}}{\mathrm{d}t} \ln \left(te^{0.05t^3} \right)$$

$$\Rightarrow \delta(t) = \frac{1 + 0.15t^3}{t}$$

Chapter 4 Solutions

Solution 4.1. The *strict* smoothness condition is that the derivatives at *all* orders are continuous.

a. We have $f(x) = \cos(2x)$ which has no singular points. The first few derivatives are obtained as

$$\begin{aligned} f'(x) &= -2\sin(2x) \\ f''(x) &= -4\cos(2x) \\ f^{(3)}(x) &= 8\sin(2x) \end{aligned}$$

We therefore see that $f^{(n)}(x)$ is some multiple of $\sin(2x)$ (for n odd) or $\cos(2x)$ (for n even). Both circular functions are known to be continuous across $\mathbb{R}$ and so we can conclude that $f(x)$ is smooth in the strict sense.

b. We have $h(y) = e^{y^2}$ which has no singular points. The first few derivatives are obtained as

$$\begin{aligned} h'(y) &= 2ye^{2y^2} \\ h''(y) &= 2(2y^2 + 1)e^{y^2} \\ h^{(3)}(y) &= 4y(2y^2 + 3)e^{y^2} \end{aligned}$$

We see that $h^{(n)}(y)$ is a product of an nth order polynomial and e^{y^2} for all $n \in \mathbb{N}^+$. Both polynomial and exponential functions are known to be continuous over $\mathbb{R}$ and so we conclude that $h(y)$ is smooth in the strict sense.

c. We have $g(z) = z^2 - 2z + 2$ which has no singular points. The first few derivatives are obtained as

$$\begin{aligned} g'(z) &= 2z - 2 \\ g''(z) &= 2 \\ g^{(3)} &= 0 \end{aligned}$$

We therefore see that $g^{(n)}(z)$ is a straight line for all $n \in \mathbb{N}^+$. Straight lines are of course continuous over $\mathbb{R}$ and so we conclude that $g(z)$ is smooth in the strict sense.

d. We have $k(p) = \frac{p^2}{p^2+2}$ which has no singular points. The first few derivatives are obtained as

$$\begin{aligned} k'(p) &= \frac{4p}{(p^2+2)^2} \\ k''(p) &= \frac{-12p^2+8}{(p^2+2)^3} \\ k^{(3)}(p) &= \frac{48p(p^2-2)}{(p^2+2)^4} \end{aligned}$$

We see that $k^{(n)}(p)$ is a rational function with denominator $(p^2+4)^{n+1}$ which has no real roots for any $n \in \mathbb{N}^+$. We therefore expect each $k^{(n)}(p)$ to be continuous over $\mathbb{R}$ and so conclude that $k(p)$ is smooth in the strict sense.

Solution 4.2. The location of each turning point of a function is given by a root of the first derivative of that function.

a. We have $f'(x) = -2\sin(2x)$ and so require x such that $\sin(2x) = 0$. The sin function is cyclic and is such that

$$2x = n\pi \Rightarrow x = \frac{n}{2}\pi$$

for $n \in \mathbb{Z}$.

b. We have $h'(y) = 2ye^{y^2}$ and find a single turning point at $y = 0$.

c. We have $g'(z) = 2z - 2$ and find a single turning point at $z = 1$.

d. We have $k'(p) = \frac{4p}{(p^2+2)^2}$ and find a single turning point at $p = 0$.

Solution 4.3. We attempt to apply the second-derivative test at each turning point found in Solution 4.2.

a. We have turning points at $x = \frac{n}{2}\pi$ and see that $f''(x) = -4\cos(2x)$. $f''\left(\frac{n}{2}\pi\right)$ takes the value 4 for n odd and -4 for n even which correspond to minima and maxima, respectively.

b. We have a turning point at $y = 0$ and see that $h''(y) = 2(2y^2+1)e^{y^2}$. $h''(0) = 2$ and so corresponds to a minimum.

c. We have a turning point at $z = 1$ and see that $g''(z) = 2$ for all z. The turning point therefore corresponds to a minimum.

d. We have a turning point at $p = 0$ and see that $k''(p) = \frac{-12p^2+8}{(p^2+2)^3}$. $k''(0) = 1$ and so corresponds to a minimum.

These results, and those of Solution 4.1, are confirmed in Figure 4.1. They can also be confirmed with Wolfram Alpha using, for example, the command

```
turning points cos(2x)
```

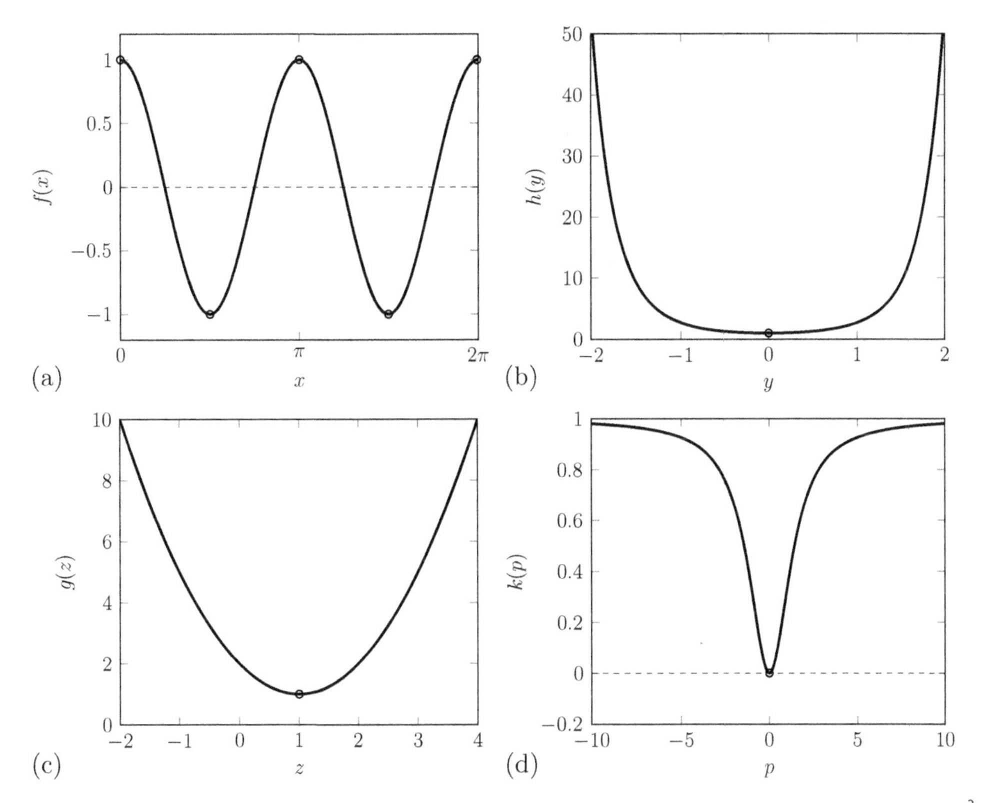

Figure 4.1 Illustrations of the functions in Solutions 4.1–4.5. (a) $f(x) = \cos(2x)$, (b) $h(y) = e^{y^2}$, (c) $g(z) = z^2 - 2z + 2$, and (d) $k(p) = \frac{p^2}{p^2+2}$.

Solution 4.4. In each case, we begin with the second derivative previously found in Solution 4.3 and repeatedly apply the standard rules for differentiation.

a. We have $f''(x) = -4\cos(2x)$ and so

$$\begin{aligned} f^{(3)} &= 8\sin(2x) \\ f^{(4)} &= 16\cos(2x) \end{aligned}$$

b. We have $h''(y) = 2(2y^2 + 1)e^{y^2}$ and the repeated use of the product rule and some manipulation leads to

$$\begin{aligned} h^{(3)} &= 4y(2y^2 + 3)e^{y^2} \\ h^{(4)} &= 4(4y^4 + 12y^2 + 3)e^{y^2} \end{aligned}$$

c. We have $g''(z) = 2$ and so all higher derivatives are 0.

d. We have $k''(p) = \frac{-12p^2+8}{(p^2+2)^3}$ and the repeated use of the product or quotient rules and some manipulation leads to

$$k^{(3)} = \frac{48p(p^2-2)}{(p^2+2)^4}$$

$$k^{(4)} = \frac{-48(5p^4-20p^2+4)}{(p^2+2)^5}$$

Note that these derivatives can be confirmed with Wolfram Alpha using, for example, the command

```
d^4/dx^4 cos(2x)
```

Solution 4.5. We look at the asymptotic behavior and piece together the information from the previous solutions.

a. $f(x)$ remains between -1 and 1 as $x \to \pm\infty$. The range is therefore $\{f \in \mathbb{R} : -1 \leq f \leq 1\}$.

b. We have $h(y) \to \infty$ as $y \to \pm\infty$. There is a single local minimum at $y = 0$ such that $h(0) = 1$ which, given the asymptotic behavior, is also a global minimum. The range is therefore $\{h \in \mathbb{R} : h \in [1, \infty)\}$.

c. We have $g(z) \to \infty$ as $z \to \pm\infty$. There is a single local minimum at $z = 1$ which is such that $z(1) = 1$, which, given the asymptotic behavior, is also a global minimum. The range is therefore $\{g \in \mathrm{Z} : g \in [1, \infty)\}$.

d. The rational function is such that $k(p) \to 1$ as $p \to \pm\infty$ and we have a horizontal asymptote. The single local minimum at $p = 0$ is such that $k(0) = 0$ is a global minimum. The range is therefore $\{k \in \mathbb{R} : k \in [0, 1)\}$. Note that the interval is open at $p = 1$ reflecting that the value is never actually reached for finite p.

The ranges are visually confirmed in Figure 4.1. Note that these ranges can also be confirmed with Wolfram Alpha using, for example, the command

```
range (p^2)/(p^2+2)
```

Solution 4.6. We note that the two functions are very similar and differ only by a sign in the denominator. This slight difference is important as the denominator of $f(x)$ has no real roots but the denominator of $h(x)$ has two real roots not shared with the numerator; $f(x)$ therefore has no singularities but $g(x)$ has two. Both functions have horizontal asymptotes at 0.2 as $x \to \pm\infty$.

a. $f(x)$ is such that

$$f'(x) = \frac{30^4 - 72x^3 - 4}{(5x^4+2)^2}$$

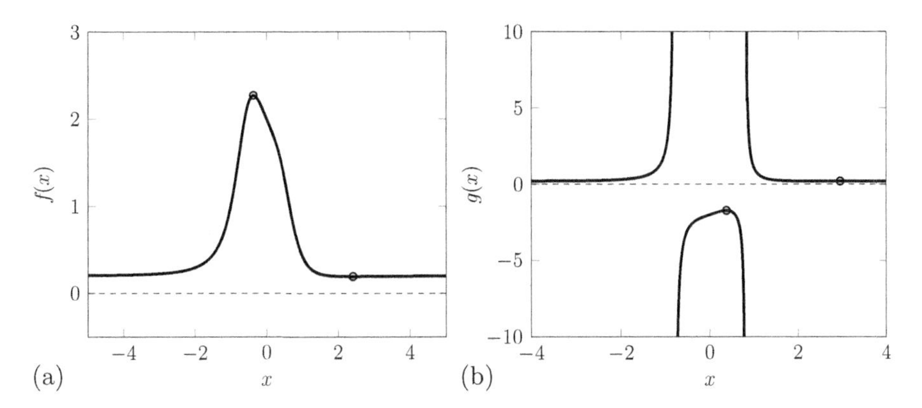

Figure 4.2 Visual confirmation of the ranges in Solution 4.6. (a) $f(x) = \dfrac{x^4 - 2x + 4}{5x^4 + 2}$, (b) $g(x) = \dfrac{x^4 - 2x + 4}{5x^4 - 2}$.

which has roots $f'(-0.364) = 0$ and $f'(2.410) = 0$. Furthermore, $f''(-0.364) < 0$ and $f''(2.410) > 0$ and we find local maxima and minima at $x \approx -0.364$ and $x \approx 2.410$, respectively. These are such that $f(-0.364) = 2.273$ and $f(2.410) = 0.193$ which are above and below the horizontal asymptote and so represent global extremes. The range is therefore $\{f \in \mathbb{R} : f \in [0.193, 2.273]\}$.

b. A similar analysis leads to local a maximum and minimum at $x \approx 0.373$ and $x \approx 2.928$ which are such that $g(0.373) \approx -1.719$ and $g(2.928) \approx 0.196$. The presence of the vertical asymptotes means that these are not globally extreme values but define a "gap" in the range. In particular, the range is $\{g \in \mathbb{R} : g \in (-\infty, -1.719] \cup [0.196, \infty)\}$.

The ranges are visually confirmed in Figure 4.2. Wolfram Alpha can also be used to verify these. For example, for $g(x)$, the command

```
range (x^4 - 2*x+4)/(5*x^4-2)
```

leads to the output "$\{y \in \mathbb{R} : 20(y + 2) \leq 5.60185$ or $20(y + 2) \geq 43.9203\}$." While this is an unusual representation, some manipulation should convince you that this is the same as our result in part b.

Solution 4.7. We denote the market price (in \$) of the asset with $P(t)$ such that

$$P(t) = 150.4891 - (t - 7)^2 - \frac{e^{5-t^2}}{100}$$

for $t \in [0, \infty)$. Assuming that the investor is rational, the optimal time to sell the investment is when the model predicts the *maximum* price. We therefore find t such that $P'(t) = 0$ and can confirm that this is a maximum with the second-derivative test. In particular,

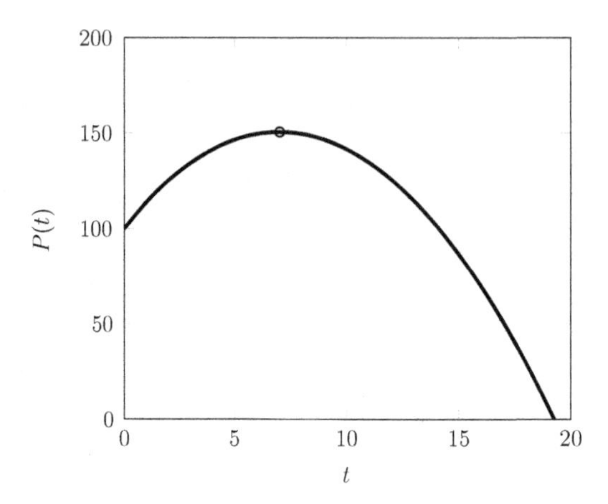

Figure 4.3 The asset price $P(t)$ in Solution 4.7.

$$P'(t) = 2(t-7) + 2t\frac{e^{5-t^2}}{100} = 14 + \left(\frac{e^{5-t^2}}{50} - 2\right)t$$

which has a root at $t \approx 7$. The second derivative test results in $P''(7) \approx -2$ and so we have a maximum with $P(7) \approx 150.49$. We have therefore predicted a maximum profit of \$50.49 if the asset is sold after 7 years. The price evolution is shown in Figure 4.3.

Solution 4.8. We have two functions, $O(t)$ for the output (in units of 1000) and $P(t)$ for the unit price (in units of \$),

$$O(t) = 2 + \sin\left(\frac{2\pi}{365}t\right) \quad \text{and} \quad P(t) = 4 - \left(1 - \frac{2}{365}t\right)^2$$

Both are defined on $t \in [0, 365]$ and we ignore that days are actually discrete.

a. The maximum output is therefore given by the maximum value of $O(t)$ on $t \in [0, 365]$. This could be obtained from investigating the roots of $O'(t)$. However, note that the variation in $O(t)$ comes from a sin function and we know that $\sin(x)$ has a maximum of 1 at $x = \frac{\pi}{2}$. The output $O(t)$ will therefore have a maximum of 3 at t such that

$$\frac{2\pi}{365}t = \frac{\pi}{2} \Rightarrow t = \frac{365}{4}$$

That is, the maximum output is $O(91.25) = 3$ on the 92nd day. This can be confirmed in Figure 4.4(a).

b. The maximum price that can be charged is given by the maximum of $P(t)$ on $t \in [0, 365]$. We note that

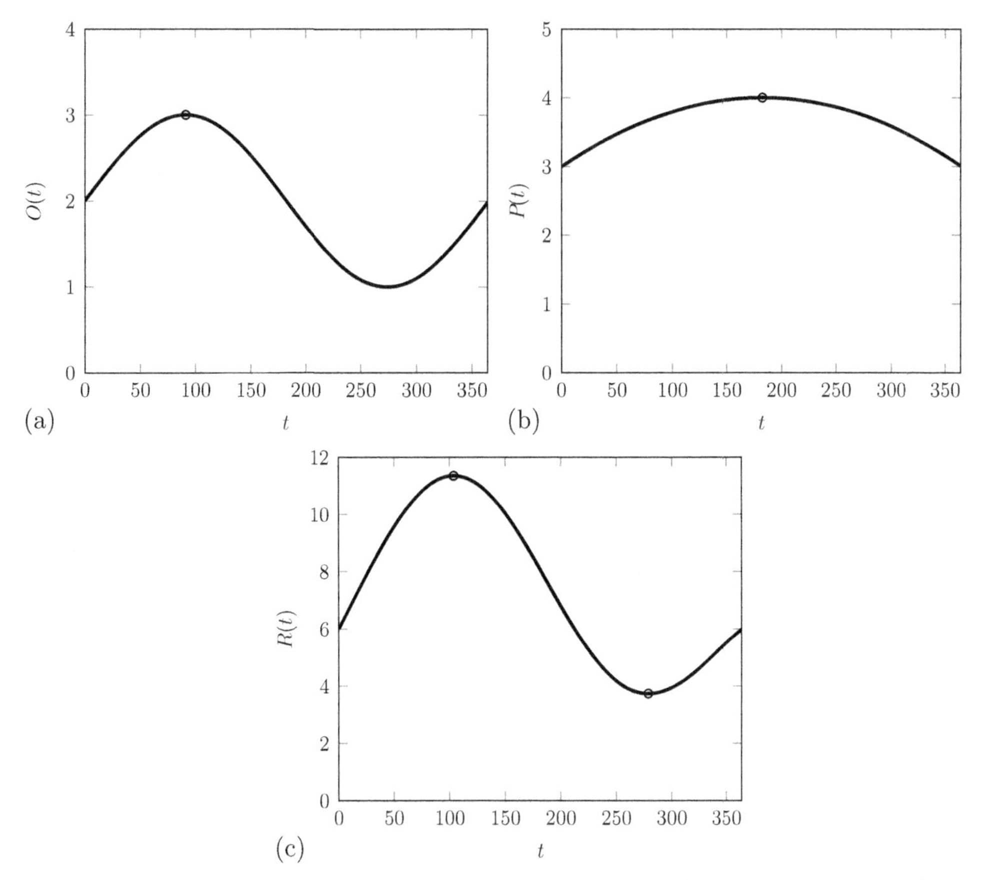

Figure 4.4 The functions relevant to Solution 4.8. (a) The output function $O(t)$ in units of 1000 sandwiches, (b) the price function $P(t)$ in units of \$ per sandwich, and (c) the revenue function $R(t) = O(t) \times P(t)$ in units of \$1000.

$$P'(t) = \frac{4}{365}\left(1 - \frac{2t}{365}\right) = 0 \Rightarrow t = \frac{365}{2}$$

and $P''(182.5) < 0$. The function $P(t)$ therefore has a maximum value of $P(182.5) = 4$ on the 183rd day. This can be confirmed in Figure 4.4(b).

c. The daily revenue $R(t)$ is given by the product of the output and unit price. That is,

$$R(t) = \left(2 + \sin\left(\frac{2\pi}{365}t\right)\right)\left(4 - \left(1 - \frac{2}{365}t\right)^2\right)$$

which is plotted in Figure 4.4(c). The revenue is then maximized when $R'(t) = 0$ and $R''(0) < 0$. Some manipulation leads to a maximum value of $R(t) \approx 11.3535$

at $t \approx 103.80$. That is, the maximum revenue is \$11,354 and this is obtained on the 104th day. We note that this value does *not* correspond with the value of $O_{\max} \times P_{\max} = 12$ and this reflects that the two functions are not maximized at the same time. "On balance" the revenue is maximized at some other date.

d. A rational manufacturer would opt to shut down production on the day that gives the minimum revenue. That is, we are required to find t such that $R(t) = O(t)P(t)$ is minimized. Some work shows a further stationary point at $t \approx 279.04$ and this can be shown to be a minimum with $R(279.04) \approx 3.74$. This is confirmed in Figure 4.4(c). The manufacturer should therefore shut down production at $t = 279.04$, that is, on the 280th day of the year.

Note that $t = 0$ corresponds to the 1st day of the year, and so, for example, $t = 279.04$ corresponds to the 280th day.

Solution 4.9. In each case, we use $f(i)$ to denote the price (that is, the present value of proceeds), $D(i)$ to be the effective duration, and $\tau(\delta)$ to be the duration. We have $i = 0.06$ per annum in all calculations.

a. Working in money units of \$ and time units of years,

$$f(i) = 50\left(\frac{1}{1+i} + \frac{1.05}{(1+i)^2} + \frac{1.05^2}{(1+i)^3} + \frac{1.05^3}{(1+i)^4} + \frac{1.05^4}{(1+i)^5}\right) \Rightarrow f(0.06) \approx 231.441$$

$$f'(i) = -50\left(\frac{1}{(1+i)^2} + \frac{2\times 1.05}{(1+i)^3} + \frac{3\times 1.05^2}{(1+i)^4} + \frac{4\times 1.05^3}{(1+i)^5} + \frac{5\times 1.05^4}{(1+i)^6}\right)$$
$$\Rightarrow f'(0.06) \approx -650.882$$

The price is therefore \$231.44. The effective duration and duration are then given by

$$D(0.06) = -\frac{f'(0.06)}{f(0.06)} \approx 2.81 \quad \text{and} \quad \tau(i = 0.06) = 1.06 \times D(0.06) \approx 2.98 \text{ years}$$

b. Working in money units of £ and time units of years,

$$f(i) = \frac{100}{(1+i)^3} + \frac{100}{(1+i)^6} + \frac{100}{(1+i)^9} + \frac{1000}{(1+i)^{10}} \Rightarrow f(0.06) \approx 772.043$$

$$f'(i) = -\frac{300}{(1+i)^4} - \frac{600}{(1+i)^7} - \frac{900}{(1+i)^{10}} - \frac{10,000}{(1+i)^{11}} \Rightarrow f'(0.06) \approx -6407.09$$

The price is therefore £772.04. The effective duration and duration are then given by

$$D(0.06) = -\frac{f'(0.06)}{f(0.06)} \approx 8.30 \quad \text{and} \quad \tau(i = 0.06) = 1.06 \times D(0.06) \approx 8.77 \text{ years}$$

Solution 4.10. We are given that $f(i_0) = 1010$ and $\tau(\delta_o) = 10$ where $i_0 = 0.03$ and $\delta_0 = \ln(1.03)$ per annum. Therefore, $D(i_0) = \frac{\tau(\delta_0)}{1+i_0} \approx 9.71$.

a. We now have $i_1 = 0.04$ and so $\Delta i = h = 0.01$ and so estimate the new price to be

$$f(i_1) \simeq f(i_0)\big(1 - hD(i_0)\big) \approx \$911.94$$

b. We now have $\delta_1 = 0.025$ and so $\Delta\delta = h = 0.025 - \ln(1.03) \approx -0.00456$ and so estimate the new price to be

$$f(\delta_1) \simeq f(\delta_0)\big(1 - h\tau(\delta_0)\big) \approx \$1056.60$$

Chapter 5 Solutions

Solution 5.1. Some thought should convince you that the following mappings are correct. In each case, $n \in \mathbb{N}^+$, that is, $n = 1, 2, \ldots$.

a. $x_n = \frac{1}{n^2}$ and the sequence continues $\left\{1, \frac{1}{4}, \frac{1}{9}, \frac{1}{16}, \frac{1}{25}, \frac{1}{36}, \frac{1}{49} \ldots\right\}$

b. $x_n = n^2 + 2$ and the sequence continues $\{3, 6, 11, 18, 27, 38, 51 \ldots\}$

c. $x_n = (-1)^{n-1}2^n$ or $x_n = (-1)^{n+1}2^n$ and the sequence continues $\{-2, 4, -8, 16, -32, 64, -128 \ldots\}$

d. $x_n = 3n + (-1)^n$ and the sequence continues $\{2, 7, 8, 13, 14, 19, 20 \ldots\}$

Solution 5.2. In each case, we attempt to manipulate the summation into some combination of the given series and standard results.

a. We proceed as,

$$\begin{aligned}
\sum_{i=1}^{100}(1 + (0.2i)^2) &= \sum_{i=1}^{100} 1 + \sum_{i=1}^{100}(0.2i)^2 \\
&= \sum_{i=1}^{100} 1 + 2^2 \sum_{i=1}^{100}(0.1i)^2 \\
&= 100 + 4 \times \frac{6767}{2} \\
&= 13,634
\end{aligned}$$

b. We proceed as,

$$\begin{aligned}
\sum_{k=1}^{100}(-0.2)^{i+10} &= 0.2^{10} \sum_{k=1}^{100}(-0.2)^i \\
&= 0.2^{10} \times \left(-\frac{1,627,604}{9,765,625}\right) \\
&\approx -1.707 \times 10^{-8}
\end{aligned}$$

c. We proceed as,

$$\begin{aligned}
\sum_{i=1}^{20}(-0.2)^i &= \sum_{i=1}^{10}(-0.2)^i + \sum_{i=11}^{20}(-0.2)^i \\
&= -\frac{1,627,604}{9,765,625} + \sum_{k=1}^{10}(-0.2)^{k+10} \\
&= -\frac{1,627,604}{9,765,625} + (-0.2)^{10}\sum_{k=1}^{10}(-0.2)^k \\
&= -\frac{1,627,604}{9,765,625}\left(1+(-0.2)^{10}\right) \\
&\approx -0.1667
\end{aligned}$$

d. We proceed as,

$$\begin{aligned}
\sum_{j=-100}^{-1} j^2 + \sum_{k=6}^{15}(-0.2)^{k-6} &= \sum_{i=100}^{1}(-i)^2 + \sum_{i=1}^{10}(-0.2)^{i+5-6} \\
&= \sum_{i=1}^{100} i^2 + \sum_{i=1}^{10}(-0.2)^{i-1} \\
&= 100\sum_{i=1}^{100}(0.1i)^2 - 5\sum_{i=1}^{10}(-0.2)^i \\
&= 100 \times \frac{6767}{2} + 5 \times \frac{1,627,604}{9,765,625} \\
&\approx 338,350.833
\end{aligned}$$

The value of each summation can be checked with Wolfram Alpha using, for example, the command

```
sum (-0.2)^(i+10) between i=1 and 100
```

Solution 5.3. We attempt to relate each series to either a geometric or arithmetic progression.

a. The series is a finite geometric progression and therefore has a finite evaluation. It is already expressed in standard form and has initial value $a = 10$ and common factor

$r = 1.2$. The standard expression for the sum then gives,

$$S_{G,10} = \frac{a(1 - r^{10})}{1 - r} = \frac{10(1 - 1.2^{10})}{1 - 1.2} \approx 259.587$$

b. The series is a finite arithmetic progression and therefore has a finite evaluation. It is not expressed in standard form and we rearrange it to be $\sum_{i=1}^{10}(7 + (i - 1)2)$. We therefore see that the series has an initial value of $a = 7$ and common difference $d = 2$. The standard expression for the sum then gives,

$$S_{A,15} = \frac{15}{2}(2a + (k - 1)d) = \frac{15}{2}(2 \times 7 + 14 \times 2) = 315$$

c. The series is an infinite geometric progression. We begin by expressing it in standard form

$$\sum_{i=1}^{\infty} \frac{\pi}{2} \times \left(\frac{1}{2}\right)^{i} = \sum_{i=1}^{\infty} \frac{\pi}{4} \times \left(\frac{1}{2}\right)^{i-1}$$

and see that $a = \frac{\pi}{4}$ and $r = \frac{1}{2}$. The progression is convergent since $|r| < 1$ and has value,

$$S_{G,\infty} = \frac{a}{1 - r} = \frac{\pi}{2}$$

d. This is an infinite arithmetic progression with standard form $x_i = 2 + (i - 1)2$. Since we are adding 2 at each step, we can tell immediately that, as with all infinite arithmetic progressions, it will not converge.

The values can be confirmed with Wolfram Alpha using, for example, the command

```
sum 10*1.2^(i-1) for i=1 to 10
```

Solution 5.4. We use the general form of the ratio test in each case. That is, we require $|L| < 1$ for convergence where

$$L = \lim_{n \to \infty} \left(\frac{x_{n+1}}{x_n}\right)$$

a. We have

$$L = \lim_{n \to \infty} \left(\frac{n + 2}{n + 1}\right) = \lim_{n \to \infty} \left(\frac{1 + \frac{2}{n}}{1 + \frac{1}{n}}\right) = 1 \not< 1$$

The series is therefore not convergent.

b. We have

$$L = \lim_{n\to\infty}\left(\frac{2^{n+2}}{2^{n+1}}\right) = 2 \not< 1$$

The series is therefore not convergent.

c. We have

$$L = \lim_{n\to\infty}\left(\frac{(-1)^{n+1}0.99^{n-1}}{(-1)^{n}0.99^{n-2}}\right) = -0.99$$

The series is therefore convergent.

d. We have

$$L = \lim_{n\to\infty}\left(\frac{2(n+1)+0.2^{n+1}}{2n+0.2^{n}}\right) = \lim_{n\to\infty}\left(\frac{2(1+\frac{1}{n})+0.2^{n}}{2+0.2^{n-1}}\right) = 1 \not< 1$$

The series is therefore not convergent.

Solution 5.5. In Example 5.19, we derived the Taylor expansion of $\ln(x)$ about $x = 1.5$ to be

$$\ln(1+x) = \ln(2.5) + \sum_{n=1}^{\infty}\frac{(-1)^{n+1}}{n\times 2.5^{n}}(x-1.5)^{n}$$

The ratio test therefore requires us to consider the limit

$$L = \lim_{n\to\infty}\left(\frac{\frac{(-1)^{n+2}}{(n+1)\times 2.5^{n+1}}(x-1.5)^{n+1}}{\frac{(-1)^{n+1}}{n\times 2.5^{n}}(x-1.5)^{n}}\right) = \lim_{n\to\infty}\left(\frac{-(x-1.5)}{2.5(1+\frac{1}{n})}\right) = -0.4(x-1.5)$$

We require $|L| < 1$ for the series to be convergent and so

$$-1 < -0.4(x-1.5) < 1 \Rightarrow -1 < x < 4$$

Solution 5.6. We begin by determining the relevant expansion of $\cos\delta$. Since the limit to $\delta = 0$ is required, it is appropriate to work with the Maclaurin expansion, that is, the Taylor expansion about $\delta = 0$. We therefore write that,

$$\cos\delta = 1 - \frac{\delta^2}{2} + \frac{\delta^4}{24} - \frac{\delta^6}{720} + \cdots$$

The required limit is therefore written as

$$\lim_{\delta\to 0}\left(\frac{\cos\delta - 1}{\delta}\right) = \lim_{\delta\to 0}\left(-\frac{\delta}{2} + \frac{\delta^3}{24} - \frac{\delta^5}{720} + \cdots\right) = 0$$

This is consistent with the value obtained in Solution 3.4.

Solution 5.7. In each case, we determine the nth term of the Maclaurin expansion and apply the ratio test.

a. As a short cut, we begin with the Maclaurin expansion of e^y

$$e^y = 1 + \frac{y}{1!} + \frac{y^2}{2!} + \frac{y^3}{3!} + \cdots$$

and so, for $y = 2x^3$, we can immediate write that

$$e^{2x^3} = 1 + \frac{2x^3}{1!} + \frac{2^2x^6}{2!} + \frac{2^3x^9}{3!} + \cdots = \sum_{n=0}^{\infty} 2^n \frac{x^{3n}}{n!}$$

The ratio test for convergence requires $|L| < 1$ where

$$L = \lim_{n\to\infty} \left(\frac{2^{n+1}\dfrac{x^{3(n+1)}}{(n+1)!}}{2^n \dfrac{x^{3n}}{n!}} \right) = \lim_{n\to\infty} \left(\frac{2x^3}{(n+1)} \right) = 0$$

The expansion is therefore convergent for all finite x.

b. We begin with the Maclaurin expansion of $\sin x$ which is

$$\sin x = x - \frac{x^3}{6} + \frac{x^5}{120} + \cdots = \sum_{n=0}^{\infty} \frac{(-1)^n x^{2n+1}}{(1+2n)!}$$

and can immediately write down

$$\frac{\sin x}{x} = \frac{1}{x} \sum_{n=0}^{\infty} \frac{(-1)^n x^{2n+1}}{(1+2n)!} = \sum_{n=0}^{\infty} \frac{(-1)^n x^{2n}}{(1+2n)!}$$

The ratio test for convergence requires $|L| < 1$ where

$$L = \lim_{n\to\infty} \left(\frac{\dfrac{(-1)^{n+1}x^{2n+2}}{(3+2n)!}}{\dfrac{(-1)^n x^{2n}}{(1+2n)!}} \right) = \lim_{n\to\infty} \left(-\frac{x^2}{(3+2n)(2+2n)} \right) = 0$$

The expansion is therefore convergent for all finite x

Solution 5.8. Since the annuity is in advance, the unit cash flows begin at $t = 0$ and end at $t = n - 1$. The present value is therefore given by,

$$\ddot{a}_{\overline{n|}} = 1 + \frac{1}{(1+i)} + \frac{1}{(1+i)^2} + \cdots + \frac{1}{(1+i)^{n-1}}$$

This is the sum of geometric progression with starting value $a = 1$ and common factor $r = \frac{1}{1+i}$. Using the standard expression for the sum of a geometric progression, we find that

$$\ddot{a}_{\overline{n}|} = \frac{1 - \frac{1}{(1+i)^n}}{1 - \frac{1}{1+i}} = (1+i)\left(\frac{1-(1+i)^{-n}}{i}\right)$$

Alternatively, we might express this as $\ddot{a}_{\overline{n}|} = (1+i)a_{\overline{n}|}$.

Solution 5.9. The present value of the dividends can be constructed as a level 5-year annuity of £10,000 per annum paid in arrears and a deferred 5-year annuity with payments that start at £20,000 and increase by 5% per annum. Working in units of £10,000, the price, P, is given by

$$\begin{aligned} P &= a_{\overline{5}|} + \frac{2}{(1+i)^5}\left[\frac{1}{(1+i)} + \frac{1.05}{(1+i)^2} + \frac{1.05^2}{(1+i)^3} + \frac{1.05^3}{(1+i)^4} + \frac{1.05^4}{(1+i)^5}\right] \\ &= a_{\overline{5}|} + \frac{2}{1.05 \times (1+i)^5}\left[\left(\frac{1.05}{1+i}\right) + \left(\frac{1.05}{1+i}\right)^2 + \left(\frac{1.05}{1+i}\right)^3 + \left(\frac{1.05}{1+i}\right)^4 + \left(\frac{1.05}{1+i}\right)^5\right] \end{aligned}$$

with $i = 0.06$. The terms in the square brackets represent the sum of a geometric progression with $a = r \approx 0.9906$ and so we have

$$\begin{aligned} P &\approx 4.2123 + 1.4233 \times \left[0.9906 \times \frac{1 - 0.9906^5}{1 - 0.9906}\right] \\ &\approx 11.1306 \end{aligned}$$

The maximum price that the investor should pay is therefore approximately £111,306.

Solution 5.10. We begin by equating the present value of the investment to the present value of the proceeds. Working in units of \$10,000 and assuming that the proceeds continue in perpetuity, we have

$$\begin{aligned} 50 &= \frac{2}{(1+i)} + \frac{2 \times 1.04}{(1+i)^2} + \frac{2 \times 1.04^2}{(1+i)^3} + \cdots \\ &= \frac{2}{1.04}\left[\left(\frac{1.04}{1+i}\right) + \left(\frac{1.04}{1+i}\right)^2 + \left(\frac{1.04}{1+i}\right)^3 + \cdots\right] \end{aligned}$$

To simplify the analysis, we define $1 + j = \frac{1+i}{1.04}$ and express the equation as

$$50 = \frac{2}{1.04}\left[\left(\frac{1}{1+j}\right) + \left(\frac{1}{1+j}\right)^2 + \left(\frac{1}{1+j}\right)^3 + \cdots\right]$$

The terms in the square brackets clearly represent a perpetuity at interest rate j and we can rewrite the expression as

$$50 = \frac{2}{1.04} \times \frac{1}{j} \Rightarrow j = \frac{1}{26}$$

Therefore,

$$1 + \frac{1}{26} = \frac{1+i}{1.04} \Rightarrow i = \frac{2}{25} = 0.08$$

The yield obtained is therefore found to be 8% per annum.

Chapter 6 Solutions

Solution 6.1. We note that these are integrals of standard functions but with some linear argument. A simple change of variables is therefore required. Alternatively, you might be able to write down the answer directly from observation.

a. We use $u = 5x$ and so have $du \equiv 5dx$, therefore

$$\begin{aligned}\int \cos(5x)dx &= \int \frac{\cos u}{5}du \\ &= \frac{1}{5}\sin(5x) + c\end{aligned}$$

b. We use $u = 0.7 + \frac{y}{2}$ and so have $du = \frac{dy}{2}$, therefore

$$\begin{aligned}\int 10\sin\left(0.7 + \frac{y}{2}\right)dy &= \int 20\sin(u)dy \\ &= -20\cos\left(0.7 + \frac{y}{2}\right) + c\end{aligned}$$

c. We split the integral as

$$\begin{aligned}\int \left(e^{5z} + \ln(3z)\right)dz &= \int e^{5z}dz + \int \ln(3z)dz \\ &= \frac{1}{5}e^{5z} + z\left(\ln(3z) - 1\right) + c\end{aligned}$$

d. We split the integral as

$$\begin{aligned}\int \left(\mathrm{cosec}^2(3p) - \sec^2(8p)\right)dp &= \int \mathrm{cosec}^2(3p)dp - \int \sec^2(8p)dp \\ &= -\frac{1}{3}\cot(3p) - \frac{1}{8}\tan(8p) + c\end{aligned}$$

e. In this case, we note that the integrand can be simplified as follows:

$$\sin(\pi q)\cot(\pi q) = \sin(\pi q)\frac{\cos(\pi q)}{\sin(\pi q)} = \cos(\pi q)$$

The integral is then simply

$$\int \sin(\pi q)\cot(\pi q)\mathrm{d}q = \int \cos(\pi q)\mathrm{d}q = \frac{1}{\pi}\sin(\pi q) + c$$

These integrals can be confirmed using Wolfram Alpha's `int` command.

Solution 6.2. We note that these integrals are amenable to a change of variables approach.

a. We define $u = x^4 - 2$, which is such that $\mathrm{d}u \equiv 4x^3\mathrm{d}x$. Therefore,

$$\begin{aligned}\int x^3(x^4-2)^3\mathrm{d}x &= \int \frac{u^3}{4}\mathrm{d}u \\ &= \frac{1}{16}(x^4-2)^4 + c\end{aligned}$$

b. We define $u = \sin y$, which is such that $\mathrm{d}u \equiv \cos(y)\mathrm{d}y$. Therefore,

$$\begin{aligned}\int \cos(y)\mathrm{e}^{\sin y}\mathrm{d}y &= \int \mathrm{e}^u\mathrm{d}u \\ &= \mathrm{e}^{\sin y} + c\end{aligned}$$

c. We define $u = z^2 + 5$, which is such that $\mathrm{d}u \equiv 2z\mathrm{d}z$. Therefore,

$$\begin{aligned}\int \frac{2z}{z^2+5}\mathrm{d}z &= \int \frac{1}{u}\mathrm{d}u \\ &= \ln\left(z^2+5\right) + c\end{aligned}$$

d. We define $u = \tan p$, which is such that $\mathrm{d}u = \sec^2 p\mathrm{d}p$. Therefore,

$$\begin{aligned}\int \tan^4 p\sec^2 p\mathrm{d}p &= \int u^4\mathrm{d}u \\ &= \frac{1}{5}\tan^5 p + c\end{aligned}$$

e. We define $u = 1 + q^{n+1}$, which is such that $\mathrm{d}u = (n+1)q^n\mathrm{d}q$. Therefore,

$$\begin{aligned}\int q^n\sqrt{1+q^{n+1}}\mathrm{d}q &= \int \frac{u^{\frac{1}{2}}}{n+1}\mathrm{d}u \\ &= \frac{2}{3(n+1)}(1+q^{n+1})^{\frac{3}{2}} + c\end{aligned}$$

These integrals can be confirmed using Wolfram Alpha's `int` command.

Solution 6.3. We note that these integrals are amenable to the integration by parts approach.

a. We define $u = x$ and $v' = \sin x$. Therefore, $u' = 1$, $v = -\cos x$, and

$$\int x \sin x dx = -x \cos x + \int \cos x dx$$
$$= -x \cos x + \sin x + c$$

b. We define $u = y^2$ and $v' = e^{6y}$. Therefore, $u' = 2y$, $v = \frac{1}{6}e^{6y}$, and

$$\int y^2 e^{6y} dy = \frac{y^2 e^y}{6} - \frac{1}{3} \int y e^{6y} dy$$

The intermediate integral also requires integration by parts, and we obtain

$$\int y^2 e^{6y} dy = \frac{y^2 e^y}{6} - \frac{1}{3}\left(\frac{1}{36} e^{6y}(6y - 1)\right)$$
$$= \frac{e^{6y}}{108}\left(18y^2 - 6y + 1\right)$$

c. We define $u = e^{2z}$ and $v' = \sin z$. Therefore, $u' = 2e^{2z}$, $v = -\cos z$, and

$$\int e^{2z} \sin z dz = -e^{2z} \cos z + \int 2e^{2z} \cos(z)\mathrm{z}$$

The intermediate integral is also amenable to integration by parts and is

$$\int 2e^{2z} \cos(z)\mathrm{z} = 2e^{2z} \sin z - 4 \int e^{2z} \sin(z) dz$$

Substituting this back into the evaluation of the main integral, we find

$$\int e^{2z} \sin z dz = -e^{2z} \cos z + 2e^{2z} \sin z - 4 \int e^{2z} \sin(z) dz$$
$$\Rightarrow \int e^{2z} \sin z dz = \frac{e^{2z}}{5}(2 \sin z - \cos z)$$

Note that this is an example of a cyclical integration by parts.

d. We define $u = p$ and $v' = \sqrt{1 + p}$. Therefore, $u' = 1$, $v = \frac{2}{3}(1 + p)^{3/2}$, and

$$\int p\sqrt{1 + p} dp = \frac{2}{3}p(1 + p)^{3/2} - \int \frac{2}{3}(1 + p)^{3/2} dp$$
$$= \frac{2}{3}p(1 + p)^{3/2} - \frac{4}{15}(1 + p)^{5/2} + c$$
$$= \frac{2}{15}(3p - 2)(1 + p)^{3/2} + c$$

e. We define $u = \sin q$ and $v' = \cos q$. Therefore, $u' = \cos q$, $v = \sin q$, and

$$\int \sin(q)\cos(q)\mathrm{d}q = \sin^2 q - \int \sin(q)\cos(q)\mathrm{d}q$$
$$\Rightarrow \int \sin(q)\cos(q)\mathrm{d}q = \frac{1}{2}\sin^2 q + c$$

Alternatively, since $\sin^2 q + \cos^2 q = 1$, we could represent this as $\int \sin(q)\cos(q)\mathrm{d}q = -\frac{1}{2}\cos^2 q + c$.

These integrals can be confirmed using Wolfram Alpha's `int` command.

Solution 6.4. In each case, we attempt to "complete the square" and relate these integrals back to those resulting in inverse trigonometric functions. For example,

$$\int \frac{1}{1+x^2}\mathrm{d}x = \arctan(x) + c \quad \text{and} \quad \int \frac{1}{\sqrt{1-x^2}}\mathrm{d}x = \arcsin(x) + c$$

a. We note that this is already close to standard form and we can immediately write that

$$\int \frac{4}{1+x^2}\mathrm{d}x = 4\arctan(x) + c$$

b. The completed-square form of the denominator is $y^2 + 4y + 5 = (y+2)^2 + 1$ and so we can rewrite the integral and proceed as

$$\int \frac{-2}{y^2+4y+5}\mathrm{d}y = -2\int \frac{1}{(y+2)^2+1}\mathrm{d}y$$
$$= -2\arctan(y+2) + c$$

c. The appropriate completed-square form is $4 - 8z - 4z^2 = 8 - 4(z+1)^2$ and so we rewrite the integral and proceed as

$$\int \frac{1}{\sqrt{4-8z-4z^2}}\mathrm{d}z = \int \frac{1}{\sqrt{8-4(z+1)^2}}\mathrm{d}z$$
$$= \int \frac{1}{2\sqrt{2-(z+1)^2}}\mathrm{d}z$$
$$= \frac{1}{2}\arcsin\left(\frac{z+1}{\sqrt{2}}\right) + c$$

d. The appropriate completed-square form is $25p^2 - 5p + 1 = 25\left(p - \frac{1}{10}\right)^2 + \frac{3}{4}$ and so we rewrite the integral and proceed as

$$\int \frac{6}{25p^2-5p+1}\mathrm{d}p = \int \frac{6}{25\left(p-\frac{1}{10}\right)^2+\frac{3}{4}}\mathrm{d}p$$

$$= \frac{6}{25}\int \frac{1}{\left(p-\frac{1}{10}\right)^2+\frac{3}{100}}\mathrm{d}p$$

$$= \frac{4}{5}\sqrt{3}\arctan\left(\frac{10p-1}{\sqrt{3}}\right)$$

e. The appropriate completed-square form is $q^2+4q-5=(q+2)^2-9$ which is not of the form $(q+a)^2+b^2$. Therefore, despite the similarity with part a., this integral is of a different class and cannot be related to an "arctan" solution. Instead, we attempt to express the integrand as partial fractions. In particular,

$$\frac{-2}{q^2+4q-5} = \frac{1}{(q+5)(q-1)} = \frac{1}{3(q+5)} - \frac{1}{3(q-1)}$$

The integral is then

$$\int \frac{-2}{q^2+4q-5}\mathrm{d}q = \int \frac{1}{3(q+5)} - \frac{1}{3(q-1)}\mathrm{d}q$$
$$= \frac{1}{3}\left(\ln(q+5)-\ln(1-q)\right)+c$$

As before, these integrals can be confirmed using Wolfram Alpha's `int` command.

Solution 6.5. In each case, we begin by manipulating the integrand to a more useful form.

a. We can express the integral as

$$I = \int \sec^3 x\mathrm{d}x = \int \sec x\sec^2 x\mathrm{d}x$$

Integration by parts is then used with $u=\sec x$ and $v'=\sec^2 x$, which are such that $u'=\sec x\tan x$ and $v=\tan x$. This gives,

$$I = \sec x\tan x - \int \sec x\tan^2 x\mathrm{d}x$$
$$= \sec x\tan x + \int \sec x(1-\sec^2 x)\mathrm{d}x$$
$$\Rightarrow I = \sec x\tan x + \int \sec x\mathrm{d}x - I$$

Using the result for $\int \sec x\mathrm{d}x$ as stated in the question, the last line can be rearranged to give the required result,

$$I = \frac{1}{2}\sec x\tan x + \frac{1}{2}\ln|\sec x+\tan x| + c$$

b. We express the integral as

$$I = \int \sec^3 x \mathrm{d}x = \int \frac{\cos x}{\cos^4 x}\mathrm{d}x = \int \frac{\cos x}{(1-\sin^2 x)^2}\mathrm{d}x$$

As instructed, we proceed to use the substitution $u = \sin x$ which is such that $\mathrm{d}u \equiv \cos x \mathrm{d}x$ and obtain,

$$\begin{aligned}
I &= \int \frac{1}{(1-u^2)^2}\mathrm{d}u \\
&= \frac{1}{4}\int \left(\frac{1}{1-u} + \frac{1}{(1-u)^2} + \frac{1}{1+u} + \frac{1}{(1+u)^2}\right)\mathrm{d}u \\
&= \frac{1}{4}\left(-\ln(1-u) + \frac{1}{1-u} + \ln(1+u) - \frac{1}{1+u}\right) + c \\
&= \frac{1}{4}\left(\ln\left|\frac{1+u}{1-u}\right| + \frac{2u}{1-u^2}\right) + c
\end{aligned}$$

Back substituting $u = \sin x$ leads to the required result after a little further manipulation.

Solution 6.6. Note that the general quadratic polynomial of the form $a_1x^2 + a_2x + a_3$ can be expressed in completed-square form,

$$\begin{aligned}
a_1x^2 + a_2x + a_3 &= a_1\left(x^2 + \frac{a_2}{a_1}x + \frac{a_3}{a_1}\right) \\
&= a_1\left(x + \frac{a_2}{2a_1}\right)^2 - \left(\frac{a_2^2}{4a_1} - a_3\right)
\end{aligned}$$

The roots of the quadratic are therefore given by x such that

$$\begin{aligned}
a_1\left(x + \frac{a_2}{2a_1}\right)^2 - \left(\frac{a_2^2}{4a_1} - a_3\right) &= 0 \\
\Rightarrow \left(x + \frac{a_2}{2a_1}\right)^2 &= \frac{1}{a_1}\left(\frac{a_2^2}{4a_1} - a_3\right) \\
\Rightarrow x &= -\frac{a_2}{2a_1} \pm \sqrt{\frac{1}{a_1}\left(\frac{a_2^2}{4a_1} - a_3\right)} \\
\Rightarrow x &= \frac{-a_2 \pm \sqrt{a_2^2 - 4a_1a_3}}{2a_1}
\end{aligned}$$

This is of course the standard expression for calculating the roots of a quadratic polynomial, as required.

Solution 6.7. We begin by using the completed-square version of the quadratic, as obtained in Solution 6.6, to write the integral as

$$\frac{1}{a_1}\int \frac{1}{\left(x+\frac{a_2}{2a_1}\right)^2+\left(\frac{a_3}{a_1}-\frac{a_2^2}{4a_1^2}\right)}\mathrm{d}x$$

We define $u = x + \dfrac{a_2}{2a_1}$, which is such that $\mathrm{d}u \equiv \mathrm{d}x$, and also define $A^2 = \dfrac{a_3}{a_1} - \dfrac{a_2^2}{4a_1^2}$ to rewrite the integral as

$$\frac{1}{a_1}\int \frac{1}{u^2+A^2}\mathrm{d}u = \frac{1}{Aa_1}\arctan\left(\frac{u}{A}\right)+c$$

Back substituting u and A leads to

$$\int \frac{1}{a_1x^2+a_2x+a_3}\mathrm{d}x = \frac{1}{a_1\sqrt{\frac{a_3}{a_1}-\frac{a_2^2}{4a_1^2}}}\arctan\left(\frac{x+\frac{a_2}{2a_1}}{\sqrt{\frac{a_3}{a_1}-\frac{a_2^2}{4a_1^2}}}\right)+c$$

which, after some manipulation, leads to

$$\int \frac{1}{a_1x^2+a_2x+a_3}\mathrm{d}x = \frac{2}{\sqrt{4a_1a_3-a_2^2}}\arctan\left(\frac{2a_1x+a_2}{\sqrt{4a_1a_3-a_2^2}}\right)+c$$

as required.

Solution 6.8. While we have not looked at the integration of Taylor or Maclaurin expansions explicitly, the integration of an infinite series involves no new mathematics. Writing out the first few terms of the integrand leads to

$$\begin{aligned}\int \sum_{k=0}^{\infty}\frac{(-1)^k x^{2k}}{(2k)!}\mathrm{d}x &= \int \frac{1}{1}-\frac{x^2}{2!}+\frac{x^4}{4!}-\cdots\mathrm{d}x\\ &= \int \frac{1}{1}\mathrm{d}x - \int \frac{x^2}{2!}\mathrm{d}x + \int \frac{x^4}{4!}\mathrm{d}x - \cdots\end{aligned}$$

That is, we integrate each term separately. This observation should convince you that it is possible to *bring the integration inside the summation* without expanding it,

$$\begin{aligned}\int \sum_{k=0}^{\infty}\frac{(-1)^k x^{2k}}{(2k)!}\mathrm{d}x &= \sum_{k=0}^{\infty}\int \frac{(-1)^k x^{2k}}{(2k)!}\mathrm{d}x\\ &= c+\sum_{k=0}^{\infty}\frac{(-1)^k x^{1+2k}}{(2k)!(1+2k)}\end{aligned}$$

$$= c + \sum_{k=0}^{\infty} \frac{(-1)^k x^{1+2k}}{(1+2k)!}$$

where c is a single arbitrary constant. Note that

$$\sum_{k=0}^{\infty} \frac{(-1)^k x^{2k}}{(2k)!} \equiv \cos x$$

and

$$\sum_{k=0}^{\infty} \frac{(-1)^k x^{1+2k}}{(1+2k)!} \equiv \sin x$$

We have therefore simply demonstrated that

$$\int \cos(x)\mathrm{d}x = \sin(x) + c$$

by integrating the Maclaurin expansion of $\cos x$.

Chapter 7 Solutions

Solution 7.1. We consider each case in turn.

a. This is a straightforward integration of a polynomial expression.

$$\int_3^5 (x^3 - 2x + 1)\mathrm{d}x = \left[\frac{x^4}{4} - x^2 + x\right]_3^5 = 122$$

b. We use the substitution $u = y^3 + 2$, which is such that $\mathrm{d}u \equiv 3y^3\mathrm{d}y$ and so

$$\begin{aligned}\int_0^1 y^2(y^3+2)^5\mathrm{d}y &= \int_{u=2}^{u=3} \frac{u^5}{3}\mathrm{d}u \\ &= \left[\frac{u^6}{18}\right]_2^3 \\ &= \frac{665}{18}\end{aligned}$$

c. We use integration by parts with $u = z$ and $v' = \mathrm{e}^{-2z}$ and so $u' = 1$ and $v = \frac{-\mathrm{e}^{-2z}}{2}$. Therefore,

$$\begin{aligned}\int_4^5 z\mathrm{e}^{-2z}\mathrm{d}z &= \left[-\frac{z}{2}\mathrm{e}^{-2z}\right]_4^5 + \frac{1}{2}\int_4^5 \mathrm{e}^{-2z}\mathrm{d}z \\ &= \left[-\frac{z}{2}\mathrm{e}^{-2z}\right]_4^5 - \frac{1}{4}\left[\mathrm{e}^{-2z}\right]_4^5 \\ &\approx 0.00063\end{aligned}$$

d. We use the substitution $u = \cos p$, which is such that $\mathrm{d}u \equiv -\sin(p)\mathrm{d}p$. Therefore,

$$\begin{aligned}\int_{-\frac{\pi}{2}}^0 \sin p\sqrt{\cos p}\,\mathrm{d}p &= \int_{u=0}^{u=1} -u^{1/2}\mathrm{d}u \\ &= \left[-\frac{2}{3}u^{3/2}\right]_0^1 \\ &= -\frac{2}{3}\end{aligned}$$

e. We use the substitution $u = 4q^3 - 1$, which is such that $du \equiv 12q^2 dq$. Therefore,

$$\begin{aligned}\int_9^{10} \frac{q^2}{4q^3 - 1} &= \int_{u=2915}^{u=3999} \frac{1}{12u} du \\ &= \left[\frac{1}{12} \ln u\right]_{2915}^{3999} \\ &= \frac{1}{12} \ln\left(\frac{3999}{2915}\right) \\ &\approx 0.02635\end{aligned}$$

These definite integrals can be confirmed using Wolfram Alpha with, for example, the command

```
int q^2/(4q^3-1) dq between 9 and 10
```

Solution 7.2. We note that the limits involve ∞ which we did not consider explicitly in the chapter. However, as we shall see, there is no new mathematics involved.

a. This is a straightforward integral of an exponential function, and we obtain

$$\int_0^{\infty} e^{-\pi x} dx = \left[\frac{-1}{\pi} e^{-\pi x}\right]_0^{\infty}$$

The upper limit is such that this has value

$$\int_0^{\infty} e^{-\pi x} dx = \frac{1}{\pi}\left(e^0 - e^{-\pi \times \infty}\right) = \frac{1}{\pi}$$

We note that this integral is over an infinite interval but has a *finite* value. This is because the integrand is such that $e^{-\pi x} \to 0$ as $x \to \infty$ and so the area under the curve of $e^{-\pi x}$, in some sense, converges to a finite value.

b. The integrand is the same as in part a., and we obtain

$$\begin{aligned}\int_{-\infty}^{2} e^{-\pi x} dx &= \left[\frac{-1}{\pi} e^{-\pi x}\right]_{-\infty}^{2} \\ &= \frac{1}{\pi}\left(e^{\pi \times \infty} - e^{2\pi}\right)\end{aligned}$$

which has an *infinite* value. This definite integral is therefore *not* defined. The interpretation of this is that the integrand $e^{-\pi x}$ is not bounded as $x \to -\infty$ and so the area under the curve is infinite.

These integrals are illustrated in Figure 7.1. Both definite integrals can be confirmed using Wolfram Alpha with, for example, the command

```
int e^(-pi*x) dx between x=0 and infty
```

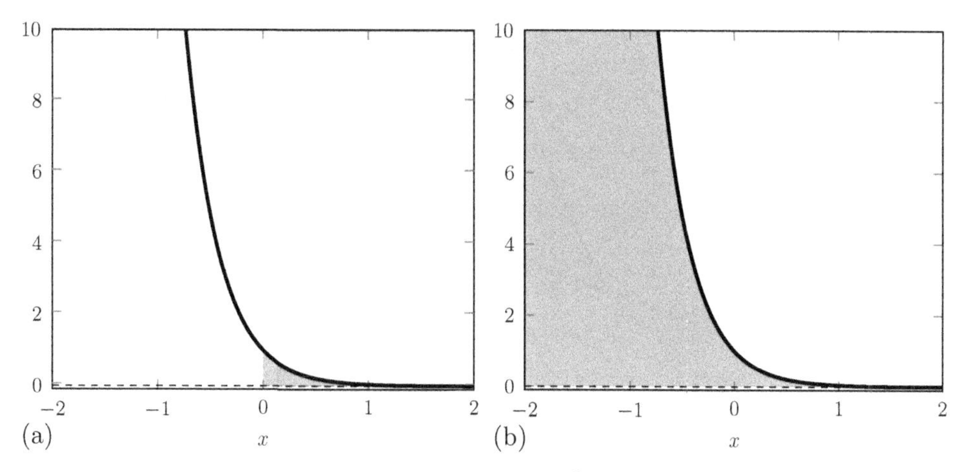

Figure 7.1 Illustration of Solution 7.2. (a) $\int_0^{\infty} e^{-\pi x} dx$ and (b) $\int_{-\infty}^{2} e^{-\pi x} dx$.

Note that in the case of part b., Wolfram Alpha does not return a numerical value. This reflects that the definite integral does not exist.

Solution 7.3. The integrand is a rational function. The denominator has no real roots and so we do not have to consider singular points. Note that the integrand is related to the standard result,

$$\int_a^b \frac{1}{y^2+1} dy = \Big[\arctan y \Big]_a^b$$

and so we attempt to manipulate the integral to this form. We begin with

$$\int_0^1 \frac{1}{5+(x-2)^2} dx = \frac{1}{5} \int_0^1 \frac{1}{1+\left(\frac{x-2}{\sqrt{5}}\right)^2} dx$$

and use a change of variable $u = \frac{x-2}{\sqrt{5}}$ which is such that $du \equiv \frac{1}{\sqrt{5}} dx$. Then,

$$\begin{aligned}
\int_0^1 \frac{1}{5+(x-2)^2} dx &= \frac{1}{\sqrt{5}} \int_{u=\frac{-2}{\sqrt{5}}}^{u=\frac{-1}{\sqrt{5}}} \frac{1}{1+u^2} du \\
&= \left[\frac{1}{\sqrt{5}} \arctan u \right]_{\frac{-2}{\sqrt{5}}}^{\frac{-1}{\sqrt{5}}} \\
&\approx 0.13828
\end{aligned}$$

The can be confirmed using Wolfram Alpha with the command

```
int 1/(5+(x-2)^2) dx between x=0 and 1
```

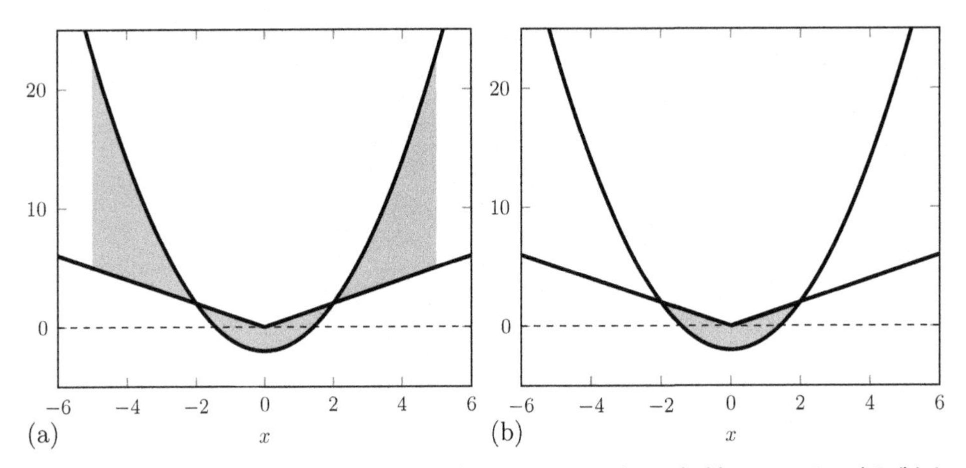

Figure 7.2 Illustration of Solution 7.4. (a) Area between curves bounded by $x = -5$ and 5. (b) Area bounded between the two curves.

Solution 7.4. The solution to both parts relies on the intersection points of $f(x)$ and $g(x)$. For $x \geq 0$, this is when $x = x^2 - 2$ and so $x = 2$; similarly, for $x < 0$, this is when $-x = x^2 - 2$ and so $x = -2$. These are illustrated in Figure 7.2.

a. It is necessary to partition the integral over $[-5, -2]$, $[-2, 0]$, $[0, 2]$, and $[2, 5]$, as shown in Figure 7.2(a). That is,

$$\begin{aligned} A_a &= \int_{-5}^{-2} (x^2 - 2 + x)\mathrm{d}x + \int_{-2}^{0} (-x - x^2 + 2)\mathrm{d}x + \int_{0}^{2} (x - x^2 + 2)\mathrm{d}x \\ &\quad + \int_{2}^{5} (x^2 - 2 - x)\mathrm{d}x \\ &= \frac{45}{2} + \frac{10}{3} + \frac{10}{3} + \frac{45}{2} \\ &= \frac{155}{3} \end{aligned}$$

b. Figure 7.2(b) indicates that the area bounded between the two curves is obtained from the partition

$$\begin{aligned} A_b &= \int_{-2}^{0} (-x - x^2 + 2)\mathrm{d}x + \int_{0}^{2} (x - x^2 + 2)\mathrm{d}x \\ &= \frac{10}{3} + \frac{10}{3} \\ &= \frac{20}{3} \end{aligned}$$

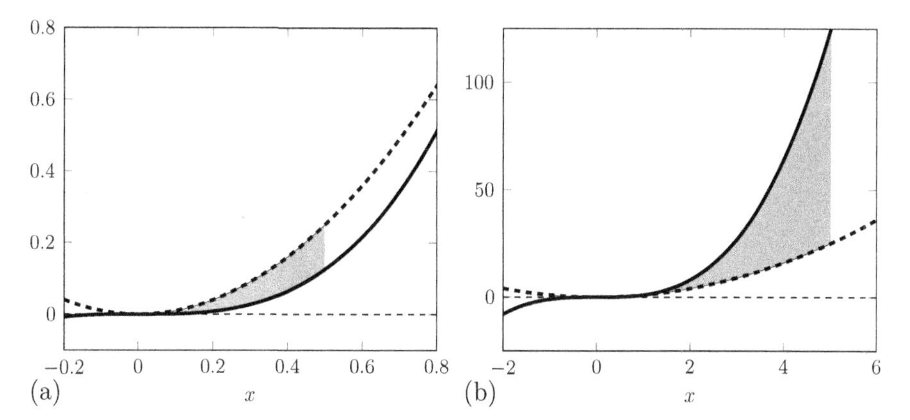

Figure 7.3 Illustration of Solution 7.5. (a) $\int_0^\infty e^{-\pi x}dx$. (b) Area between x^3 and x^2 (dashed) bounded between $x = -1$ and 5.

These can be confirmed with Wolfram Alpha using the respective commands

```
area between |x| and x^2-2 between x=-5 and 5
```

and

```
area bounded between |x| and x^2-2
```

Solution 7.5. We begin by computing the points of intersection.

a. The intersection points are x such that $x^3 = x^2$ and so are $x = 0$ and 1. The interval $[0, 1]$ is such $x^2 > x^3$ throughout and the required area is simply

$$\int_0^{0.5} (x^2 - x^3)dx \approx 0.02604$$

b. The intersection points are as in part a., however, we note that a different partition is required to reflect that $x^3 > x^2$ for $x > 1$. In particular, the area is given by

$$\int_{-1}^{1} (x^2 - x^3)dx + \int_0^1 (x^2 - x^3)dx + \int_1^5 (x^3 - x^2)dx = \frac{346}{3}$$

These are illustrated in Figure 7.3 and can be confirmed numerically using Wolfram Alpha.

Solution 7.6. We begin by determining the points of intersection.

a. The points of intersection are such that

$$2x^2 = 4 - x \Rightarrow x = x_\pm = \frac{1}{4}(-1 \pm \sqrt{33})$$

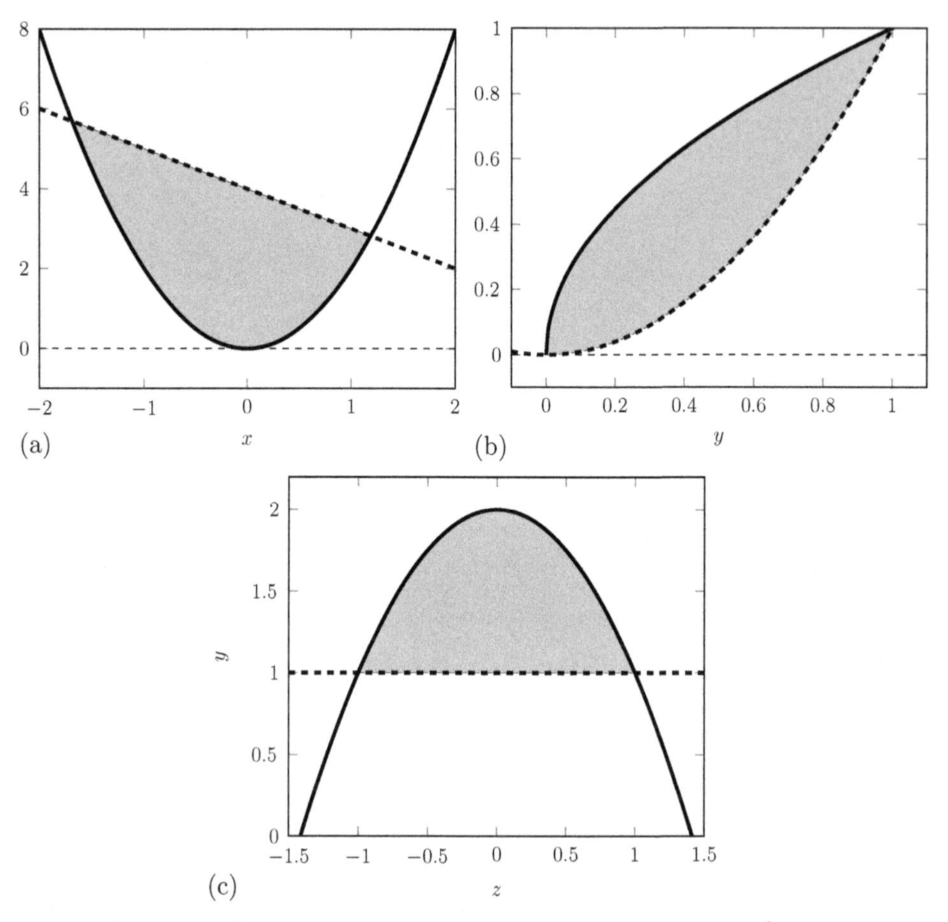

Figure 7.4 Illustration of Solution 7.6. (a) Area bounded between $f(x) = 2x^2$ and $g(x) = 4 - x$ (dashed). (b) Area bounded between $h(y) = y^2$ (dashed) and $k(y) = \sqrt[+]{y}$. (c) Area bounded between $l(x) = 1$ (dashed) and $m(z) = 2 - z^2$.

The functions are plotted in Figure 7.4(a) and we see a single-bounded region with area given by

$$\int_{x_-}^{x_+} (4 - x - 2x^2)\mathrm{d}x = \frac{11\sqrt{33}}{8}$$

b. The points of intersection are such that

$$y^2 = \sqrt{y} \Rightarrow y = 0, 1$$

The functions are plotted in Figure 7.4(b) and we see a single-bounded region with area given by

$$\int_0^1 (y^{\frac{1}{2}} - y^2)\mathrm{d}y = \frac{1}{3}$$

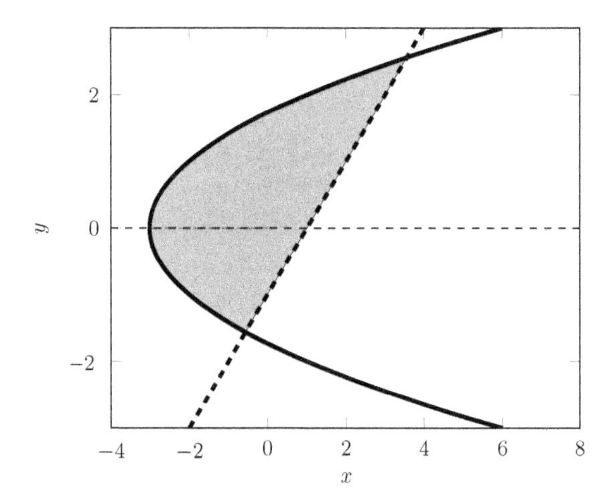

Figure 7.5 Illustration of Solution 7.7. (a) $\int_0^\infty e^{-\pi x}dx$ and (b) $\int_{-\infty}^2 e^{-\pi x}dx$.

c. The points of intersection are such that

$$2 - z^2 = 1 \Rightarrow z = \pm 1$$

The functions are plotted in Figure 7.4(c) and we see a single-bounded region with area given by

$$\int_{-1+}^{1} (1 - z^2)\mathrm{d}z = \frac{4}{3}$$

Each area can be confirmed numerically using Wolfram Alpha.

Solution 7.7. Both functions can be considered to be of the form $y(x)$ and the bounded areas can therefore be obtained by integrating with respect to y. This is illustrated in Figure 7.5. The required points of intersection are then y such that

$$y^2 - 3 = y + 1 \Rightarrow y = y_\pm = \frac{1}{2}(1 \pm \sqrt{17})$$

The required area is then given by

$$\int_{y_-}^{y_+} (y + 1 - y^2 + 3)\mathrm{d}y = \int_{y_-}^{y_+} (y - y^2 + 4)\mathrm{d}y = \frac{17\sqrt{17}}{6}$$

This can be confirmed using Wolfram Alpha with the command

```
area between y^2-3 and y+1
```

Solution 7.8. The accumulated value is given by $500A(0, 12)$ for an appropriate accumulation factor $A(0, 12)$. Given that the force of interest is a piecewise function, it is sensible to invoke the principle of consistency on the subintervals $[0, 5)$, $[5, 9)$, and $[9, 12]$, that is

$$A(0,12) \equiv A(0,5) \times A(5,9) \times A(9,12)$$

where each A is obtained from

$$A(t_1, t_2) = \exp\left(\int_{t_1}^{t_2} \delta(s)\mathrm{d}s\right)$$

for the appropriate $\delta(t)$. We therefore have,

$$A(0,5) = \exp\left(\int_0^5 0.04\mathrm{d}s\right) = \mathrm{e}^{0.2}$$

$$A(5,9) = \exp\left(\int_5^9 0.05\mathrm{d}s\right) = \mathrm{e}^{0.2}$$

$$A(9,12) = \exp\left(\int_9^{12} 0.006s\mathrm{d}s\right) = \mathrm{e}^{0.189}$$

The required accumulation is

$$500A(0,12) = 500\mathrm{e}^{0.2}\mathrm{e}^{0.2}\mathrm{e}^{0.189} \approx \$901.09$$

Solution 7.9. Note that we are required to determine an expression for the present value at some general time $t \in [0, 20]$. The complication here is that the force of interest is a piecewise function and we will necessarily need a piecewise function for the present value. In particular, if $V(t)$ denotes the present value (at $t = 0$) of the \$100 due at time t, we can use the principle of consistency to determine that

$$V(t) = \begin{cases} \dfrac{100}{A_1(0,t)} & \text{for } t \in [0,5) \\ \dfrac{100}{A_1(0,5)A_2(5,t)} & \text{for } t \in [5,9) \\ \dfrac{100}{A_1(0,5)A_2(5,9)A_3(9,t)} & \text{for } t \in [9,20) \end{cases}$$

The individual accumulation factors are given by

$$A_1(0,t) = \exp\int_0^t 0.04\mathrm{d}s = \exp(0.04t) \;\text{ for } t \in [0,5)$$

$$A_2(5,t) = \exp\int_5^t 0.05\mathrm{d}s = \exp(0.05(t-5)) \;\text{ for } t \in [5,9)$$

$$A_5(9,t) = \exp\int_9^t 0.06s\mathrm{d}s = \exp\left(0.03(t^2 - 81)\right) \;\text{ for } t \in [9,20)$$

We therefore have

$$V(t) = \begin{cases} 100\mathrm{e}^{-0.04t} & \text{for } t \in [0,5) \\ 100\mathrm{e}^{-0.04\times 5}\mathrm{e}^{-0.05(t-5)} = 100\mathrm{e}^{-0.05(t-1)} & \text{for } t \in [5,9) \\ 100\mathrm{e}^{-0.05(9-1)}\mathrm{e}^{-0.03(t^2-81)} = 100\mathrm{e}^{-(0.03t^2-2.03)} & \text{for } t \in [9,20) \end{cases}$$

Solution 7.10. We consider a continuous payment stream of rate 1 over $[0, n]$. That is, in any year, at total of 1 is paid from a continuous stream, and this stream lasts for n years. At some time, $t \in [0, n]$, we place a small time element $\mathrm{d}t$ which is such that the total amount paid over this interval is $1 \times \mathrm{d}t = \mathrm{d}t$. The present value of this element is then given by

$$\frac{\mathrm{d}t}{A(0,t)} = \mathrm{e}^{-\delta t}\mathrm{d}t$$

The present value of the *total* payment stream over $[0, n]$ is therefore given by

$$\int_0^n \mathrm{e}^{-\delta t}\mathrm{d}t = \left[\frac{\mathrm{e}^{-\delta t}}{-\delta}\right]_0^n = \frac{1-\mathrm{e}^{-\delta n}}{\delta}$$

Similarly to the discrete unit annuities discussed in Chapter 5, the present value of this continuous annuity is typically denoted by $\bar{a}_{\overline{n}|}$. We have therefore shown that

$$\bar{a}_{\overline{n}|} = \frac{1-\mathrm{e}^{-\delta n}}{\delta} = \frac{1-(1+i)^{-n}}{\delta}$$

Note the similarity between this expression and that given in Eq. (5.11).

Chapter 8 Solutions

Solution 8.1. The Argand diagram is given in Figure 8.1.

a. $z_1 = 3$ is a real number. Note that real numbers are contained in the set of complex numbers and so, technically, it is also a complex number.

b. $z_2 = 2 + 4\mathrm{i}$ is a complex number.

c. $z_3 = 2\mathrm{i}$ is an imaginary number. Note that imaginary numbers are contained in the set of complex numbers and so, technically, it is also a complex number.

d. $z_4 = -1 + 3\mathrm{i}$ is a complex number.

e. $z_5 = -2 - 4\mathrm{i}$ is a complex number.

Solution 8.2. In each case, we obtain the modulus r and principle argument θ and express the complex number in the form $r\mathrm{e}^{\mathrm{i}\theta}$.

a. $r = 3$ and $\theta = 0$, therefore $z_1 = 3\mathrm{e}^{\mathrm{i}0}$.

b. $r = \sqrt{2^2 + 4^2} = 2\sqrt{5}$ and $\theta = \arctan\left(\frac{4}{2}\right) = \arctan(2)$, therefore $z_2 = 2\sqrt{5}\mathrm{e}^{\mathrm{i}\arctan 2}$

c. $r = 2$ and $\theta = \frac{\pi}{2}$, therefore $z_3 = 2\mathrm{e}^{\mathrm{i}\frac{\pi}{2}}$.

d. $r = \sqrt{1^2 + 3^2} = \sqrt{10}$ and $\theta = \arctan(-3) = \pi - \arctan(3)$, therefore $z_4 = \sqrt{10}\mathrm{e}^{\mathrm{i}(\pi - \arctan(3))}$.

e. $r = \sqrt{2^2 + 4^2} = 2\sqrt{5}$ and $\theta = \arctan(2) - \pi$, therefore $z_5 = 2\sqrt{5}\mathrm{e}^{\mathrm{i}(\arctan(2) - \pi)}$.

Solution 8.3. We use the Cartesian form.

a. $z_1 + z_2 + z_3 + z_4 - z_5 = 3 + 2 + 4\mathrm{i} + 2\mathrm{i} - 1 + 3\mathrm{i} - (-2 - 4\mathrm{i}) = 6 + 13\mathrm{i}$

b. $z_2 z_3 = (2 + 4\mathrm{i}) \times 2\mathrm{i} = -8 + 4\mathrm{i}$

c. $\frac{z_1}{z_2} = \frac{3}{2+4\mathrm{i}} = \frac{3}{2+4\mathrm{i}} \times \frac{2-4\mathrm{i}}{2-4\mathrm{i}} = \frac{3}{10}(1 - 2\mathrm{i})$

d. $z_2^2 z_3 = (2 + 4\mathrm{i}) \times (2 + 4\mathrm{i}) \times (2\mathrm{i}) = (-12 + 16\mathrm{i}) \times (2\mathrm{i}) = -32 - 24\mathrm{i}$

e. $\frac{z_1 z_4 + z_4}{z_2 z_3} = \frac{3(-1+3\mathrm{i})-1+3\mathrm{i}}{(2+4\mathrm{i})(2\mathrm{i})} = \frac{-4+12\mathrm{i}}{-8+4\mathrm{i}} = \frac{-4+12\mathrm{i}}{-8+4\mathrm{i}} \times \frac{-8-4\mathrm{i}}{-8-4\mathrm{i}} = 1 - \mathrm{i}$

Solution 8.4. Recall that an nth order polynomial will have n roots in the complex plane.

a. We solve $x^2 - 4 = 0$ and obtain two real roots $x_{1,2} = \pm 2$. In polar form these are $x_1 = 2\mathrm{e}^{0\mathrm{i}}$ and $x_2 = 2\mathrm{e}^{\pi\mathrm{i}}$.

b. We factorize the function as $g(y) = (y - 2)(y + 1)(y + 4)$ and obtain three real roots $y_1 = 2 = 2\mathrm{e}^{0\mathrm{i}}$, $y_2 = -1 = 1\mathrm{e}^{\pi\mathrm{i}}$, and $y_3 = -4 = 4\mathrm{e}^{\pi\mathrm{i}}$.

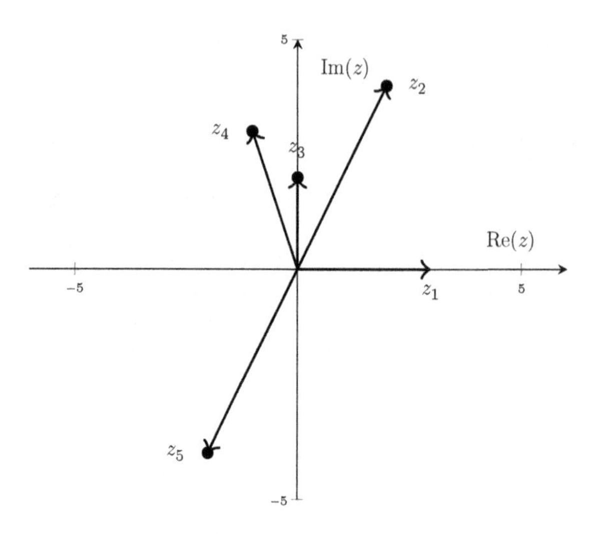

Figure 8.1 Argand diagram for Solution 8.1.

c. We consider each factor in turn and expect four roots in total.
- The factor $z^2 + 1 = 0$ leads to $z_{1,2} = \pm\mathrm{i}$, that is, $z_1 = 1\mathrm{e}^{\frac{\pi}{2}\mathrm{i}}$ and $z_2 = 1\mathrm{e}^{-\frac{\pi}{2}\mathrm{i}}$.
- The factor $z^2 + 2z + 2 = 0$ leads to $z_{3,4} = -1 \pm \mathrm{i}$, that is, $z_3 = \sqrt{2}\mathrm{e}^{\frac{3\pi}{4}\mathrm{i}}$ and $z_4 = \sqrt{2}\mathrm{e}^{-\frac{3\pi}{4}\mathrm{i}}$.

d. We solve $k^5 = -32$, that is, $k = (32\mathrm{e}^{\mathrm{i}\pi + 2\pi n})^{1/5} = 2\mathrm{e}^{\mathrm{i}(\frac{\pi}{5} + \frac{2\pi}{5}n)}$ where $n \in \mathbb{Z}$. This leads to five distinct values $k_1 = 2\mathrm{e}^{-\frac{3}{5}\mathrm{i}}$, $k_2 = 2\mathrm{e}^{-\frac{\pi}{5}\mathrm{i}}$, $k_3 = 2\mathrm{e}^{\frac{\pi}{5}\mathrm{i}}$, $k_4 = 2\mathrm{e}^{\frac{3}{5}\mathrm{i}}$, and $k_5 = 2\mathrm{e}^{\pi\mathrm{i}}$, these correspond to $n = -2, -1, 0, 1$, and 2, respectively.

Solution 8.5. We work with either the Cartesian or polar form of $z = 1 + \mathrm{i} = \sqrt{2}\mathrm{e}^{\frac{\pi}{4}\mathrm{i}}$, whichever is convenient.

a. $zz^* = (1 + \mathrm{i})(1 - \mathrm{i}) = 2$

b. $z^{10} = (1 + \mathrm{i})^{10} = \sqrt{2}^{10}\mathrm{e}^{10\frac{\pi}{4}\mathrm{i}}$, which has principal form $32\mathrm{e}^{\frac{\pi}{2}\mathrm{i}} = 32\mathrm{i}$.

c. $\frac{1}{z} = \frac{1}{\sqrt{2}}\mathrm{e}^{-\mathrm{i}\frac{\pi}{4}} = \frac{1}{2}(1 - \mathrm{i})$

d. $\frac{1}{z^5} = \frac{1}{\sqrt{2}^5}\mathrm{e}^{-5\frac{\pi}{4}\mathrm{i}}$, which has principal form $\frac{1}{4\sqrt{2}}\mathrm{e}^{\frac{3\pi}{4}\mathrm{i}} = \frac{1}{8}(-1 + \mathrm{i})$.

Solution 8.6. We are attempting to prove that

$$\mathrm{e}^{\mathrm{i}x} = \cos(x) + \mathrm{i}\sin(x)$$

The Maclaurin series for a general exponential function is

$$\mathrm{e}^y = \sum_{k=0}^{\infty} \frac{y^k}{k!}$$

In the particular case that $y = \mathrm{i}x$, we have

$$
\begin{aligned}
e^{\mathrm{i}x} &= \sum_{k=0}^{\infty} \frac{(\mathrm{i}x)^k}{k!} \\
&= 1 + \mathrm{i}\frac{x}{1!} - \frac{x^2}{2!} - \mathrm{i}\frac{x^3}{3!} + \frac{x^4}{4!} + \mathrm{i}\frac{x^5}{5!} + \cdots \\
&= \left(1 - \frac{x^2}{2!} + \frac{x^4}{4!} + \cdots\right) + \mathrm{i}\left(\frac{x}{1!} - \frac{x^3}{3!} + \frac{x^5}{5!} + \cdots\right) \\
&= \sum_{n=0}^{\infty} \frac{(-1)^n x^{2n}}{(2n)!} + \mathrm{i}\sum_{n=0}^{\infty} \frac{(-1)^n x^{2n+1}}{(2n+1)!} \\
&= \sin x + \mathrm{i}\cos x
\end{aligned}
$$

This is as required.

Solution 8.7. We approach this from two directions. First, note that

$$
\begin{aligned}
(\cos x + \mathrm{i}\sin x)^4 &= \left(e^{x\mathrm{i}}\right)^4 \\
&= e^{4x\mathrm{i}} = \cos(4x) + \mathrm{i}\sin(4x)
\end{aligned}
$$

Second, we can expand $(\cos x + \mathrm{i}\sin x)^4$ as

$$
\begin{aligned}
(\cos x + \mathrm{i}\sin x)^4 = \;& \sin^4(x) + \cos^4(x) - 6\sin^2(x)\cos^2(x) + \mathrm{i}\left(4\sin(x)\cos^3(x)\right. \\
& \left. -4\sin^3(x)\cos(x)\right)
\end{aligned}
$$

Note that $\sin(x), \cos(x) \in \mathbb{R}$ and so we can equate the real and imaginary parts across the two approaches. These leads directly to the following identities.

a. $\sin(4x) \equiv 4\sin(x)\cos^3(x) - 4\sin^3(x)\cos(x)$

b. $\cos(4x) \equiv \sin^4(x) + \cos^4(x) - 6\sin^2(x)\cos^2(x)$

Solution 8.8. We use the complex exponential form of the cos function

$$
\cos u = \frac{1}{2}\left(e^{u\mathrm{i}} + e^{-u\mathrm{i}}\right)
$$

a. Working with the left-hand side of the identity,

$$
\begin{aligned}
\cos^2(u) &= \frac{1}{4}\left(e^{u\mathrm{i}} + e^{-u\mathrm{i}}\right)\left(e^{u\mathrm{i}} + e^{-u\mathrm{i}}\right) \\
&= \frac{1}{4}\left(e^{2u\mathrm{i}} + e^{-2u\mathrm{i}} + 2\right)
\end{aligned}
$$

$$= \frac{1}{4}\left(e^{2ui} + e^{-2ui}\right) + \frac{2}{4}$$
$$= \frac{1 + \cos(2u)}{2}$$

as required.

b. Working with the left-hand side of the identity,

$$\begin{aligned}
\cos^3(u) &= \frac{1}{8}\left(e^{ui} + e^{-ui}\right)\left(e^{ui} + e^{-ui}\right)\left(e^{ui} + e^{-ui}\right) \\
&= \frac{1}{8}\left(e^{2ui} + e^{-2ui} + 2\right)\left(e^{ui} + e^{-ui}\right) \\
&= \frac{1}{8}\left(e^{3ui} + e^{-3ui} + 3e^{ui} + 3e^{-ui}\right) \\
&= \frac{1}{4}\left(3\cos(u) + \cos(3u)\right)
\end{aligned}$$

as required.

Chapter 9 Solutions

Solution 9.1. We attempt to formulate the problems in terms of formal set notation. Note that this might not be necessary to answer the questions, but it is an opportunity to explore the formal use of the mathematics.

a. The sample space is the collection of all balls. That is,

$$\Omega = \left\{ \textcircled{1}, \textcircled{2}, \ldots, \textcircled{20} \right\}$$

Let A and B denote the events that a multiple of 3 and 4 is drawn, respectively. That is

$$A = \left\{ \textcircled{3}, \textcircled{6}, \textcircled{9}, \textcircled{12}, \textcircled{15}, \textcircled{18} \right\} \text{ and } B = \left\{ \textcircled{4}, \textcircled{8}, \textcircled{12}, \textcircled{16}, \textcircled{20} \right\}$$

It is clear that $P(A) = \frac{6}{20} = \frac{3}{10}$ and $P(B) = \frac{5}{20} = \frac{1}{4}$. We are required to compute $P(A \cap B)$ and note that A and B are not mutually exclusive. Two approaches can be taken.

i. Defining the event, $C = A \cup B = \left\{ \textcircled{3}, \textcircled{4}, \textcircled{6}, \textcircled{8}, \textcircled{9}, \textcircled{12}, \textcircled{15}, \textcircled{16}, \textcircled{18}, \textcircled{20} \right\}$ which is such that $P(C) = \frac{10}{20} = \frac{1}{2}$. The addition rule then gives

$$\begin{aligned} P(A \cap B) = P(A) + P(B) - P(A \cup B) &= P(A) + P(B) - P(C) \\ &= \frac{3}{10} + \frac{1}{4} - \frac{1}{2} \\ &= \frac{1}{20} \end{aligned}$$

ii. Alternatively, we note that we simply require $P(D)$ where D is the event $B \cap C = \left\{ \textcircled{12} \right\}$. That is, $P(D) = \frac{1}{20}$.

b. We use the same events A, B, C, and D as in part a. Again, two approaches can be taken.

i. We now require $P(A \cup B) = P(C) = \frac{1}{2}$.

ii. $P(A \cup B) = P(A) + P(B) - P(A \cap B) = P(A) + P(B) - P(D) = \frac{1}{2}$

c. This question can be answered using combinations. That is, since order does not matter, we require one particular combination of arranging 5 distinct balls from

20 distinct balls. The total number of combinations is ${}_{20}C_5 = \frac{20!}{15!5!} = 15,504$. The probability is then simply $\frac{1}{15,504} \approx 0.006\%$.

d. The question can be answered using permutations. That is, since order does matter, we require one particular permutation of arranging 3 distinct balls from 20 distinct balls. The total number of permutations is ${}_{20}P_3 = \frac{20!}{17!} = 6840$. The probability is then simply $\frac{1}{6840} \approx 0.015\%$.

Solution 9.2. We begin by defining two events

$$A = \{\text{unit 1 is defective}\} \quad \text{and} \quad B = \{\text{unit 2 is defective}\}$$

We are given that $P(A) = 0.1$ and the question requires us to calculate $P(A \cap B)$ under the assumptions of independence and dependence.

a. If unit 1 is replaced before unit 2 is selected, events A and B are independent with $P(A) = P(B) = 0.1$. We then have $P(A \cap B) = P(A) \cdot P(B) = 0.1^2 = 1\%$.

b. If unit 1 is not replaced before unit 2 is selected, events A and B are not independent. We then have $P(A \cap B) = P(B|A)P(A)$. Note that 10% defective corresponds to 5 units from the 50. If unit 1 was defective, there are 4 units defective from 49 just prior to unit 2 begin selected, therefore $P(B|A) = \frac{4}{49}$. We than have $P(A \cap B) = 0.1 \times \frac{4}{49} = \frac{2}{245} \approx 0.82\%$.

Solution 9.3. We define the events

$$A = \{\text{has a blackboard}\} \text{ and } B = \left\{\text{has a computer terminal}\right\}$$

We are given that $P(A) = 0.6$, $P(B) = 0.7$, and $P(A^c \cap B^c) = 0.1$ and are interested in the probability of event $A^c \cup B^c$. Using the addition rule, we have

$$P(A^c \cup B^c) = P(A^c) + P(B^c) - P(A^c \cap B^c)$$

where $P(A^c) = 1 - P(A) = 0.4$ and $P(B^c) = 1 - P(B) = 1 - 0.7 = 0.3$. Therefore,

$$P(A^c \cup B^c) = 0.4 + 0.3 - 0.1 = 0.6$$

That is, there is a 60% probability that at least either the blackboard or the computer terminal will be missing from an office.

Solution 9.4. We have a sample space $\Omega = \left\{Ⓡ, Ⓖ, Ⓑ, Ⓨ, Ⓚ\right\}$. Recall that a valid partition $\{B_i\}$ of Ω has three requirements,

1. $B_i \cap B_j = \emptyset$ for any $i \neq j$
2. $B_1 \cup B_2 \cup \ldots \cup B_n = \Omega$
3. $P(B_k) > 0$ for any k

a. The event of drawing (Y) is not in any A_i. Requirement 2 therefore fails and $\{A_i\}$ is not a valid partition.
b. All requirements are satisfied and $\{B_i\}$ is a valid partition.
c. The event of drawing (R) appears in both C_1 and C_2, that is, $C_1 \cap C_3 \neq \emptyset$. Requirement 1 fails and $\{C_i\}$ is not a valid partition.
d. (P) $\notin \Omega$. Therefore, both requirements 2 and 3 fail, and $\{D_i\}$ is not a valid partition.

Solution 9.5. Let us define the events

$$A = \{\text{you have the condition}\} \quad \text{and} \quad B = \{\text{you test positive}\}$$

Currently, all you know is that you have a positive test result, that is, event B has happened. We then require the probability that, given that you test positive, you do not have the condition, that is, $P(A^c|B)$. We are that given $P(B|A) = P(B^c|A^c) = 0.99$ and so $P(B|A^c) = 0.01$. We use Bayes' law with the partition A and A^c and form,

$$P(A^c|B) = \frac{P(B|A^c)P(A^c)}{P(B|A)P(A) + P(B|A^c)P(A^c)}$$

a. If $P(A) = 0.01 \Rightarrow P(A^c) = 0.99$, and so

$$P(A^c|B) = \frac{0.01 \times 0.99}{0.99 \times 0.01 + 0.01 \times 0.99} = \frac{1}{2}$$

That is, there is a 50% change that you do not have the relatively common condition, despite testing positive.
b. If $P(A) = 0.001 \Rightarrow P(A^c) = 0.999$, and so

$$P(A^c|B) = \frac{0.01 \times 0.999}{0.99 \times 0.001 + 0.01 \times 0.999} = \frac{111}{122}$$

That is, there is an approximately 91% change that you do not the rare condition, despite testing positive.

Solution 9.6. We use the expression derived in Solution 9.5

$$P(A^c|B) = \frac{P(B|A^c)P(A^c)}{P(B|A)P(A) + P(B|A^c)P(A^c)}$$

Our unknown is now $P(A)$ and we have that $P(A^c) = 1 - P(A)$, $P(A^c|B) = 0.10$, $P(B|A) = 0.99$, and $P(B|A^c) = 0.01$. The expression then becomes

$$0.10 = \frac{0.01(1 - P(A))}{0.99P(A) + 0.01(1 - P(A))}$$

After some manipulation we find that $P(A) = \frac{1}{12}$. That is, despite the test being 99% accurate, the condition has to be very common and suffered by 1 in 12 for you to be 90% sure that the positive result is correct.

Solution 9.7. We begin by formulating the problem in general. The sample space Ω consists of the results of infinitely many dice rolls, and we define the events

$$S_1 = \{\text{⚅ on first roll}\} \quad \text{and} \quad E = \{\text{first ⚅ on roll } n\}$$

where n is an even number. We denote the probability of rolling ⚅ from a single roll by p, therefore $P(S_1) = p$. Note that the value of p distinguishes parts a.-c. of the question and in each case we require $P(E)$.

The events S_1 and S_1^c form a partition of Ω and we can express the required probability in terms of this partition as

$$P(E) = P(E|S_1)P(S_1) + P(E|S_1^c)P(S_1^c)$$

It should be clear that $P(E|S_1) = 0$, that is, we cannot score our first ⚅ on both the first roll (event S_1) and after an even number of rolls (event E). Furthermore,

$$P(E|S_1^c) = P(E^c) = 1 - P(E)$$

That is, the probability that the first ⚅ appears after an even number of rolls (event E) given that it does not appear at the first roll (event S_1^c), is equal to the probability that it occurs after an *additional odd number* of rolls. Since each roll is independent, this must have probability equal to $P(E^c)$.

Therefore, back to $P(E)$,

$$\begin{aligned} P(E) &= P(E|S_1)P(S_1) + P(E|S_1^c)P(S_1^c) \\ &= 0 \times P(S_1) + (1 - P(E))P(S_1^c) \\ &= (1 - P(E))(1 - p) \\ \Rightarrow P(E) &= \frac{1-p}{2-p} \end{aligned}$$

a. Since the dice is fair, $p = \frac{1}{6}$ and $P(E) = \frac{5}{11} \approx 45\%$.
b. We have $p = 0.3$ and so $P(E) = \frac{7}{17} \approx 41\%$.
c. We have $p = 1$ and so $P(E) = 0$. This reflects that we are certain to roll ⚅ on the first roll.

Solution 9.8. We use standard ${}_tp_t$ and ${}_tq_t$ notation in each case.
a. The probability that a life aged 50 exact will survive to age 56 exact is given by ${}_6p_{50}$.
b. The probability that a life aged 40 exact will die aged either 90, 91, or 92 is given by ${}_{50}p_{40} \cdot {}_3q_{90}$.

c. The probability that a newborn will die either aged 60, 61, or 78 is given by ${}_{60}p_0 \cdot {}_2q_{60} +_{78} p_0 \cdot q_{78}$

Solution 9.9. We need to express each probability in terms of the 1-year survival probabilities.

a. The probability that an animal aged 5 exact will survive another year is given by

$$p_5 = \frac{0.9}{(1+5)^2} = 0.025$$

b. The probability that a newborn animal will die within the first 3 years of life is obtained from the probability that it dies within a year of being born, or survives to age 1 exact and dies within a year, or survives to age 2 exact and dies within a year. This is given by,

$$\begin{aligned} {}_3q_0 &= q_0 + p_0 \cdot q_1 + {}_2p_0 \cdot q_2 \\ &= (1 - p_0) + p_0 \cdot (1 - p_1) + p_0 \cdot p_1 \cdot (1 - p_2) \\ &\approx 0.9798 \end{aligned}$$

Solution 9.10. The financial value of the gift (car + contract) can be estimated from their *expected present value*.

- The car is certain to be received immediately and so has expected present value equal to \$11,000.
- The \$10,000 will be paid only if he survives a further 4 years, and this has a probability of ${}_4p_{16}$. This probability is evaluated from the product of each 1-year survival probability, that is

$$ {}_4p_{16} = p_{16} \cdot p_{17} \cdot p_{18} \cdot p_{19}$$

which can each be evaluated using $p_x = 0.8 + 0.006 \times (x - 20)^2$. We therefore find that

$$ {}_4p_{16} \approx 0.5082$$

The expected present value of the payout due 4 years in the future is then obtained from

$$\$10,000 \times \frac{{}_4p_{16}}{1.05^4} \approx \$4180.91$$

An estimate of the total value of the gift is therefore \$15,180.91. Although the financial incentive to drive carefully may be considered much more valuable!

Chapter 10 Solutions

Solution 10.1. Since A, B, and C are matrices of the same dimension and no matrix multiplication is involved, we can proceed as if they were scaler quantities. That is,

$$\begin{aligned}
&4(A+4B+C)+3(2B-A)-4(5(2B+A-C)-2(A+2B-C))\\
&\quad=4(A+4B+C)+3(2B-A)-4(10B+5A-5C-2A-4B+2C)\\
&\quad=(4A+16B+4C)+(-3A+6B)+(-12A-24B+12C)\\
&\quad=A(4-3-12)+B(16+6-24)+C(4+12)\\
&\quad=11A-2B+16C
\end{aligned}$$

Solution 10.2. This question is related to the properties of transposed matrices.

a. Let us define $B=A-A^{\mathrm{T}}$ and we need to prove that B is skew symmetric, that is, $B^{\mathrm{T}}=-B$. We begin with B^{T} as proceed as follows:

$$\begin{aligned}
B^{\mathrm{T}}&=(A-A^{\mathrm{T}})^{\mathrm{T}}\\
&=A^{\mathrm{T}}-(A^{\mathrm{T}})^{\mathrm{T}}\\
&=A^{\mathrm{T}}-A\\
&=-(A-A^{\mathrm{T}})\\
&=-B
\end{aligned}$$

This is as required.

b. We are required to obtain C and D such that $A=C+D$ where $C^{\mathrm{T}}=C$ and $D^{\mathrm{T}}=-D$. Note that

$$A=C+D\Rightarrow A^{\mathrm{T}}=(C+D)^{\mathrm{T}}=C^{\mathrm{T}}+D^{\mathrm{T}}=C-D$$

That is, we have two simultaneous equations for C and D in terms of A and A^{T}. We solve these as follows:

$$\left.\begin{aligned}A&=C+D\\A^{\mathrm{T}}&=C-D\end{aligned}\right\}\Rightarrow\left.\begin{aligned}A+A^{\mathrm{T}}&=2C\\A-A^{\mathrm{T}}&=2D\end{aligned}\right\}\Rightarrow\left.\begin{aligned}C&=\tfrac{1}{2}\left(A+A^{\mathrm{T}}\right)\\D&=\tfrac{1}{2}\left(A-A^{\mathrm{T}}\right)\end{aligned}\right\}$$

It should be clear that the resulting C and D are symmetric and skew symmetric, respectively, as required.

c. Consider the skew symmetric square matrix $S = [s_{ij}]$ with dimension $n \times n$. By definition, $S^{\mathrm{T}} = -S$ and so $[s'_{ij}] = [s_{ji}] = [-s_{ij}]$. In the particular cases that $i = j$, that is on the diagonal, we require $s_{ii} = -s_{ii} \Rightarrow s_{ii} = 0$ for all $i = 1, \ldots, n$.

Solution 10.3. We work at the level of matrix entries in each case.

a. If A is a general matrix then $A^2 = 0_2$ can be written as

$$\begin{bmatrix} a & b \\ c & d \end{bmatrix}\begin{bmatrix} a & b \\ c & d \end{bmatrix} = \begin{bmatrix} 0 & 0 \\ 0 & 0 \end{bmatrix}$$

$$\Rightarrow \begin{bmatrix} a^2 + bc & ab + bd \\ ac + cd & cb + d^2 \end{bmatrix} = \begin{bmatrix} 0 & 0 \\ 0 & 0 \end{bmatrix}$$

We therefore have four simultaneous equations,

$$a^2 + bc = 0 \quad b(a + d) = 0$$
$$c(a + d) = 0 \quad cb + d^2 = 0$$

Some manipulation leads to two solutions for $b \neq 0$,

$$A = \begin{bmatrix} 0 & 0 \\ 0 & 0 \end{bmatrix} = 0_2 \quad \text{or} \quad \begin{bmatrix} a & b \\ -\frac{a^2}{b} & -a \end{bmatrix}$$

b. We begin with the solution of part a. Clearly, $A = 0_2$ is a symmetric matrix. The nontrivial solution requires $A^{\mathrm{T}} = A$, that is

$$\begin{bmatrix} a & b \\ -\frac{a^2}{b} & -a \end{bmatrix} = \begin{bmatrix} a & -\frac{a^2}{b} \\ b & -a \end{bmatrix}$$

This leads to a further four simultaneous equations

$$a = a \qquad b = -\frac{a^2}{b}$$
$$-\frac{a^2}{b} = b \qquad -a = -a$$

which, for any given a, require $b^2 = -a^2$. If we insist that $b \in \mathbb{R}$, this can only be true for $a = 0 = b$ and we again return the trivial solution $A = 0_2$.

Note that if we broaden our discussion to matrices with complex entries, that is $a, b \in \mathbb{C}$, we can have the nontrivial solution

$$A = \begin{bmatrix} a & a\mathrm{i} \\ a\mathrm{i} & -a \end{bmatrix} = a\begin{bmatrix} 1 & \mathrm{i} \\ \mathrm{i} & -1 \end{bmatrix}$$

The reader is invited to confirm that $A^{\mathrm{T}} = A$ and $A^2 = 0_2$. No new mathematics is required to manipulate matrices with complex entries.

Solution 10.4. In each case, we manipulate the expressions in an attempt to make B the subject.

a. We proceed as,

$$2(B^{\mathrm{T}} - 2I_2)^{-1} = \begin{bmatrix} 0 & 2 \\ 1 & -1 \end{bmatrix}$$
$$\Rightarrow B^{\mathrm{T}} - 2I_2 = \begin{bmatrix} 0 & 1 \\ \frac{1}{2} & -\frac{1}{2} \end{bmatrix}^{-1}$$

We are therefore required to invert a matrix before proceeding. Using the adjoint method we have

$$\begin{bmatrix} 0 & 1 \\ \frac{1}{2} & -\frac{1}{2} \end{bmatrix}^{-1} = \begin{bmatrix} 1 & 2 \\ 1 & 0 \end{bmatrix}$$

and proceed as

$$B^{\mathrm{T}} - 2I_2 = \begin{bmatrix} 1 & 2 \\ 1 & 0 \end{bmatrix}$$
$$\Rightarrow B^{\mathrm{T}} = \begin{bmatrix} 3 & 2 \\ 1 & 2 \end{bmatrix}$$
$$\Rightarrow B = \begin{bmatrix} 3 & 1 \\ 2 & 2 \end{bmatrix}$$

b. This expression is similar that in part a. and, proceeding along the same lines, we find the requirement to obtain

$$\begin{bmatrix} 0 & 6 \\ 0 & -6 \end{bmatrix}^{-1}$$

Note, however, that this has a determinant equal to 0. That is, we cannot invert this matrix and the matrix equation *cannot* be solved.

Solution 10.5. We first confirm that A can be inverted by computing its determinant. Taking the Laplace expansion along the 4th row,

$$\begin{vmatrix} 1 & 2 & 1 & 2 \\ 1 & 0 & 1 & 2 \\ 2 & 1 & 2 & 1 \\ 0 & 0 & 1 & 2 \end{vmatrix} = -1 \times \begin{vmatrix} 1 & 2 & 2 \\ 1 & 0 & 2 \\ 2 & 1 & 1 \end{vmatrix} + 2 \times \begin{vmatrix} 1 & 2 & 1 \\ 1 & 0 & 1 \\ 2 & 1 & 2 \end{vmatrix} = -6 + 0 = -6$$

that is, $\det(A) \neq 0$ and so it is invertible.

The next stage is to calculate $\operatorname{adj}(A)$. This is obtained from the transpose of the cofactor matrix $C(A)$, which is in turn obtained from the minor $M(A)$, such that $[c_{ij}(A)] = [(-1)^{i+j} m_{ij}(A)]$. After some effort, we obtain the minor and cofactor as

$$M(A) = \begin{bmatrix} 0 & 3 & 2 & 1 \\ 6 & 3 & -6 & -3 \\ 0 & 0 & -4 & -2 \\ -6 & 0 & 6 & 0 \end{bmatrix} \Rightarrow C(A) = \begin{bmatrix} 0 & -3 & 2 & -1 \\ -6 & 3 & 6 & -3 \\ 0 & 0 & -4 & 2 \\ 6 & 0 & -6 & 0 \end{bmatrix}$$

and so

$$\text{adj}(A) = \begin{bmatrix} 0 & -6 & 0 & 6 \\ -3 & 3 & 0 & 0 \\ 2 & 6 & -4 & -6 \\ -1 & -3 & 2 & 0 \end{bmatrix} \Rightarrow A^{-1} = \frac{1}{-6}\begin{bmatrix} 0 & -6 & 0 & 6 \\ -3 & 3 & 0 & 0 \\ 2 & 6 & -4 & -6 \\ -1 & -3 & 2 & 0 \end{bmatrix}$$

We therefore obtain that

$$A^{-1} = \frac{1}{6}\begin{bmatrix} 0 & 6 & 0 & -6 \\ 3 & -3 & 0 & 0 \\ -2 & -6 & 4 & 6 \\ 1 & 3 & -2 & 0 \end{bmatrix}$$

By the definition of A^{-1}, we need to check that $AA^{-1} = I_4 = A^{-1}A$ and the reader is invited to confirm these. Alternatively, the inverse can be confirmed directly with Wolfram Alpha. The relevant command is

```
inverse {{1,2,1,2},{1,0,1,2},{2,1,2,1},{0,0,1,2}}
```

Intermediate steps in the manual calculation can also be checked with Wolfram Alpha using the commands `cofactor`, `minor` and `adjugate`.

Solution 10.6. We begin by expressing the system of linear simultaneous in terms of matrices. In particular, we have

$$\begin{bmatrix} 1 & 2 & 1 & 1 \\ 1 & -2 & 3 & 0 \\ 2 & 1 & 0 & -1 \\ 0 & 0 & 1 & 1 \end{bmatrix}\begin{bmatrix} x \\ y \\ z \\ a \end{bmatrix} = \begin{bmatrix} 4 \\ 2 \\ -3 \\ 2 \end{bmatrix}$$

*Pre*multiplying both sides by the inverse of the coefficient matrix leads to

$$\begin{bmatrix} x \\ y \\ z \\ a \end{bmatrix} = \begin{bmatrix} 1 & 2 & 1 & 1 \\ 1 & -2 & 3 & 0 \\ 2 & 1 & 0 & -1 \\ 0 & 0 & 1 & 1 \end{bmatrix}^{-1}\begin{bmatrix} 4 \\ 2 \\ -3 \\ 2 \end{bmatrix}$$

If we denote the coefficient matrix as A, the problem is now to obtain A^{-1}. We proceed using the adjoint method and first confirm that $\det(A) \neq 0$. We can take the Laplace expansion along the 4th row and, after some work, obtain that $\det(A) = 5$. Furthermore, the cofactor matrix is obtained from the minor as

$$C(A) = \begin{bmatrix} -5 & 5 & 5 & 5 \\ -2 & 1 & 3 & -3 \\ 6 & -3 & -4 & 4 \\ 11 & -8 & -9 & 14 \end{bmatrix}$$

and so we obtain that

$$A^{-1} = \frac{1}{5}C(A)^{\mathrm{T}} = \frac{1}{5}\begin{bmatrix} -5 & -2 & 6 & 11 \\ 5 & 1 & -3 & -8 \\ 5 & 3 & -4 & -9 \\ -5 & -3 & 4 & 14 \end{bmatrix}$$

Finally, the solution to the system of simultaneous equations is obtained from

$$\begin{bmatrix} x \\ y \\ z \\ a \end{bmatrix} = \frac{1}{5}\begin{bmatrix} -5 & -2 & 6 & 11 \\ 5 & 1 & -3 & -8 \\ 5 & 3 & -4 & -9 \\ -5 & -3 & 4 & 14 \end{bmatrix}\begin{bmatrix} 4 \\ 2 \\ -3 \\ 2 \end{bmatrix} = \begin{bmatrix} -4 \\ 3 \\ 4 \\ -2 \end{bmatrix}$$

Note that this solution can be confirmed by Wolfram Alpha with the command

```
solve x+2y+z+a=4, x-2y+3z=2,2x+y-a=-3, a+z=2
```

Solution 10.7. Cramer's rule gives

$$x = x_1 = \frac{\det(A_1)}{\det(A)}, \quad y = x_2 = \frac{\det(A_2)}{\det(A)}, \quad z = x_3 = \frac{\det(A_3)}{\det(A)}, \quad a = x_4 = \frac{\det(A_4)}{\det(A)}$$

with A as defined in Solution 10.6 and each A_i given by replacing the ith column of A with the column matrix

$$\begin{bmatrix} 4 \\ 2 \\ -3 \\ 2 \end{bmatrix}$$

That is, for example,

$$A_1 = \begin{bmatrix} 4 & 2 & 1 & 1 \\ 2 & -2 & 3 & 0 \\ -3 & 1 & 0 & -1 \\ 2 & 0 & 1 & 1 \end{bmatrix}$$

After some work we obtain that

$$\det(A) = 5, \quad \det(A_1) = -20, \quad \det(A_2) = 15, \quad \det(A_3) = 20, \quad \det(A_4) = -10$$

and so $x = -4$, $y = 3$, $z = 4$, and $a = -2$. These agree with the results of Solution 10.7.

Solution 10.8. The 1-step transition matrix is given by

$$P = \begin{bmatrix} 0.5 & 0.5 & 0 & 0 \\ 0.4 & 0 & 0.6 & 0 \\ 0 & 0.3 & 0 & 0.7 \\ 0 & 0 & 0.2 & 0.8 \end{bmatrix}$$

The 5-step matrix is obtained, after some effort, as

$$P^{(5)} = P^5 = \begin{bmatrix} 0.22725 & 0.21355 & 0.1791 & 0.3801 \\ 0.17084 & 0.12116 & 0.2292 & 0.4788 \\ 0.07164 & 0.1146 & 0.16136 & 0.6524 \\ 0.04344 & 0.0684 & 0.1864 & 0.70176 \end{bmatrix}$$

The probability that a policyholder initially in State 1 will be in State 3 after 5 years is therefore given by 0.1791. That is, there is a probability of approximately 18%.

Solution 10.9. The revenue from a single policyholder in each state is given by

- State 1: 0% discount $\Rightarrow P = \$600$
- State 2: 10% discount $\Rightarrow P = \$540$
- State 3: 40% discount $\Rightarrow P = \$360$
- State 4: 70% discount $\Rightarrow P = \$180$

To obtain the total revenue from the 27,000 policyholders, we first assume that the movement of the policyholders between states has settled to the *stationary* distribution given by $\{n_i\}$ for $i = 1, \ldots, 4$. That is, $\sum n_i = 27,000$ and

$$\begin{bmatrix} n_1 & n_2 & n_3 & n_4 \end{bmatrix} \begin{bmatrix} 0.5 & 0.5 & 0 & 0 \\ 0.4 & 0 & 0.6 & 0 \\ 0 & 0.3 & 0 & 0.7 \\ 0 & 0 & 0.2 & 0.8 \end{bmatrix} = \begin{bmatrix} n_1 \\ n_2 \\ n_3 \\ n_4 \end{bmatrix}$$

We therefore have five simultaneous equations

$$\begin{cases} 0.5n_1 + 0.4n_2 = n_1 \\ 0.5n_1 + 0.3n_3 = n_2 \\ 0.6n_2 + 0.2n_4 = n_3 \\ 0.7n_3 + 0.8n_4 = n_4 \\ n_1 + n_2 + n_3 + n_4 = 27,000 \end{cases}$$

and the system is solved to give

$$n_1 = 2000 \quad n_2 = 2500 \quad n_3 = 5000 \quad n_4 = 17,500$$

The total annual revenue is therefore given by

$$2000 \times \$600 + 2500 \times \$540 + 5000 \times \$360 + 17,500 \times 180 = \$7,500,000$$

Solution 10.10. The frog's movement along the chain of lily pads can be considered as the Markov process illustrated in Figure 10.1. Note we have labeled the central pad as State 3. The 1-step transition matrix is therefore

$$P = \begin{bmatrix} 1 & 0 & 0 & 0 & 0 \\ 0.5 & 0 & 0.5 & 0 & 0 \\ 0 & 0.5 & 0 & 0.5 & 0 \\ 0 & 0 & 0.5 & 0 & 0.5 \\ 0 & 0 & 0 & 0 & 1 \end{bmatrix}$$

a. The 3-step transition matrix is given by

$$P^{(3)} = P^3 = \begin{bmatrix} 1 & 0 & 0 & 0 & 0 \\ 0.625 & 0 & 0.25 & 0 & 0.125 \\ 0.25 & 0.25 & 0 & 0.25 & 0.25 \\ 0.125 & 0 & 0.25 & 0 & 0.625 \\ 0 & 0 & 0 & 0 & 1 \end{bmatrix}$$

If the frog begins in State 3, the probability that it will be in either end state (State 1 or 5) after three jumps is $0.25 + 0.25 = 0.50$. Since the end states are *absorbing*, there is therefore a 50% probability that the frog has not visited an end state (and so has *not* eaten) within three jumps.

b. We look at the stationary distribution of the chain. This is obtained from solving

$$\begin{bmatrix} p_1 & p_2 & p_3 & p_4 & p_5 \end{bmatrix} \begin{bmatrix} 1 & 0 & 0 & 0 & 0 \\ 0.5 & 0 & 0.5 & 0 & 0 \\ 0 & 0.5 & 0 & 0.5 & 0 \\ 0 & 0 & 0.5 & 0 & 0.5 \\ 0 & 0 & 0 & 0 & 1 \end{bmatrix} = \begin{bmatrix} p_1 \\ p_2 \\ p_3 \\ p_4 \\ p_5 \end{bmatrix}$$

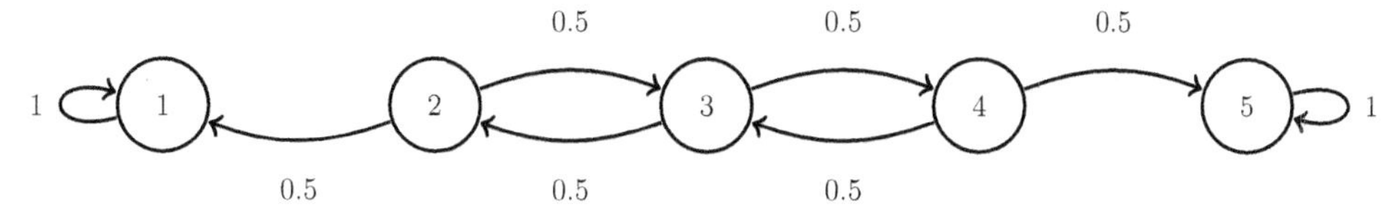

Figure 10.1 The chain of five lily pads in Solution 10.10.

which forms a system of six simultaneous equations

$$\begin{cases} p_1 + 0.5p_2 = p_1 \\ 0.5p_3 = p_2 \\ 0.5p_2 + 0.5p_4 = p_3 \\ 0.4p_3 = p_4 \\ 0.5p_4 + p_5 = p_5 \\ p_1 + p_2 + p_3 + p_4 + p_5 = 1 \end{cases}$$

The first equation is such that $p_2 = 0$ and it then follows that $p_3 = 0 = p_4$. Therefore, from the last equation, $p_1 + p_5 = 1$. That is, irrespective of where the frog starts, it is *certain* to reach either end of the chain (and so eat) in the long term. The probability that it will *never* eat is therefore 0, assuming of course that he does not starve before the long-term state is reached.

Note that this can be confirmed by computing P^n for increasing large n. For example, P^{100} is obtained from Wolfram Alpha with the command

```
{{1,0,0,0,0},{0.5,0,0.5,0,0},{0,0.5,0,0.5,0},{0,0,0.5,0,0.5},{0,0,0,0,1}}
^(100)
```

and we find

$$P^{(100)} = P^{100} = \begin{bmatrix} 1 & 0 & 0 & 0 & 0 \\ 0.75 & 0 & 0 & 0 & 0.25 \\ 0.5 & 0 & 0 & 0 & 0.5 \\ 0.25 & 0 & 0. & 0 & 0.75 \\ 0 & 0 & 0 & 0 & 1 \end{bmatrix}$$

This shows that the particular end state it is likely to end in depends on the starting state, but it is certain to end in either one or the other.

Chapter 11 Solutions

Solution 11.1. We are required to perform implicit differentiation with respect to x and obtain $y'(x)$ at appropriate (x, y). We then fit the line $y = mx + c$ with $m = y'$ at these points.

If $x = 1$, the associated values of y are obtained from solving $1 + y^2 = 2 - y$. That is, $y = y_\pm = \frac{1}{2}(-1 \pm \sqrt{5})$ and so $y_+ \approx 0.6180$ and $y_- \approx -1.6180$.

The gradients at these points can be found by first taking the implicit derivative with respect to x,

$$\begin{aligned} x^2 + y^2 &= 2 - xy \\ \Rightarrow 2x + 2yy' &= -xy' - y \\ \Rightarrow y' &= -\frac{y + 2x}{x + 2y} \end{aligned}$$

- At $(x, y_+) = (1, 0.6180)$ we have $y' \approx -1.1708$. The tangent line is then found to be $y \approx -1.1708x + 1.788$.
- At $(x, y_-) = (1, -1.6180)$ we have $y' \approx 0.1708$. The tangent line is then found to be $y \approx 0.1708x - 1.788$.

Solution 11.2. We classify the ODEs in terms of order, linearity, and coefficients.

a. The ODE $\frac{df}{dx} + f = 0$ is first order, linear with constant coefficients.

b. The ODE $g'' + xg = 2x$ is second order, linear with variable coefficients.

c. The ODE $x^2h^2 + h^{(4)} = \sin x$ is fourth order, nonlinear with variable coefficients.

d. The ODE $\frac{d^2y}{dz^2} + y\cos z = e^y$ is second order with variable coefficients. Note that the term e^y means that it is also nonlinear.

Solution 11.3. The ODEs in this question are all separable.

a. We rewrite the ODE and proceed as

$$\begin{aligned} y' - \frac{y^2 - 1}{x} &= 0 \\ \Rightarrow \frac{1}{y^2 - 1}\frac{dy}{dx} &= \frac{1}{x} \end{aligned}$$

$$\Rightarrow \int \frac{1}{y^2-1}\mathrm{d}y = \int \frac{1}{x}\mathrm{d}x$$

$$\Rightarrow \int \left(\frac{1}{2(y-1)} - \frac{1}{2(y+1)}\right)\mathrm{d}y = \ln(x)$$

$$\Rightarrow \frac{1}{2}\ln\left(\frac{1-y}{1+y}\right) = \ln(x) + c$$

Since $y(5) = 0$, we determine that $c = \ln(5)$ and write the particular solution as

$$\ln\left(\frac{1-y}{1+y}\right) = 2\ln(5x)$$

$$\Rightarrow \frac{1-y}{1+y} = 25x^2$$

$$\Rightarrow y(x) = \frac{25-x^2}{25+x^2}$$

b. We rewrite the ODE and proceed as

$$\cos(y)\frac{\mathrm{d}y}{\mathrm{d}z} = \frac{3}{z^2}$$

$$\Rightarrow \int \cos(y)\mathrm{d}y = \int \frac{3}{z^2}\mathrm{d}z$$

After some manipulation we find that

$$\sin y = -\frac{3}{z} + c$$

and the condition $y(10) = 0$ leads to $c = \frac{3}{10}$. The particular solution is then

$$y(z) = \arcsin\left(\frac{3}{10} - \frac{3}{z}\right)$$

c. We rewrite the ODE and proceed as

$$\frac{1}{g}\frac{\mathrm{d}g}{\mathrm{d}t} = 5t + 1$$

$$\Rightarrow \ln(g) = \frac{5}{2}t^2 + t + c$$

The condition $g(0) = 1$ leads to $c = 0$ and we write the particular solution is

$$g(t) = \exp\left(\frac{5}{2}t^2 + t\right)$$

d. Note that the right-hand side can be factorized. We therefore proceed as

$$\begin{aligned}\frac{dk}{dy} &= 5(ky + 2y + k + 2)\\ \Rightarrow \frac{dk}{dy} &= 5(k + 2)(y + 1)\\ \Rightarrow \frac{1}{k+2}\frac{dk}{dy} &= 5(y + 1)\\ \Rightarrow \int \frac{1}{k+2}dk &= \int 5(y + 1)dy\end{aligned}$$

Some manipulation then leads to

$$\ln(k + 2) = 5\left(\frac{y^2}{2} + y\right) + c$$

Since $k(12) = 3$, we find that $c = \ln(5) - 420$ and write the particular solution as

$$k(y) = 5\exp\left(\frac{5}{2}y(y + 2) - 420\right) - 2$$

Note that Wolfram Alpha can be used to confirm your solutions using, for example, the command

```
solve k'=5(ky+2y+k+2), k(12)=3
```

Solution 11.4. We note that the ODEs in this question can be solved using an integrating factor approach.

a. We express the ODE in standard from as $f' - \frac{f}{x} = x$ and note that an appropriate integrating factor is $e^{-\ln(x)} = -\frac{1}{x}$. We therefore proceed as

$$\begin{aligned}\frac{f'}{x} - \frac{f}{x^2} &= 1\\ \Rightarrow \frac{d}{dx}\left(\frac{f}{x}\right) &= 1\\ \Rightarrow f &= x^2 + cx\end{aligned}$$

The value of c is obtained from $f(1) = \pi$, that is $c = \pi - 1$, and we determine the particular solution to be

$$f(x) = x(x + \pi - 1)$$

b. We express the ODE in standard form as $h' - 2hy = 5y$ and note that an appropriate integrating factor is e^{-y^2}. We therefore proceed as

$$\begin{aligned} e^{-y^2}h' - 2yhe^{-y^2} &= 5ye^{-y^2} \\ \Rightarrow \frac{d}{dy}\left(he^{-y^2}\right) &= 5ye^{-y^2} \\ \Rightarrow he^{-y^2} &= -\frac{5}{2}e^{-y^2} + c \end{aligned}$$

Note that integration by parts was required on the right-hand side. The value of c is determined from $h(-4) = 2$, that is $c = \frac{9}{2}e^{-16}$, and we determine the particular solution to be

$$h(y) = \frac{1}{2}\left(9e^{y^2-16} - 5\right)$$

c. We express the ODE in standard form as $g' + 2g = 10$ and choose to denote the independent variable by x. Note that an appropriate integrating factor is e^{2x}. We therefore proceed as

$$\begin{aligned} g'e^{2x} + 2e^{2x}g &= 10e^{2x} \\ \Rightarrow \frac{d}{dx}\left(ge^{2x}\right) & 10e^{2x} \\ \Rightarrow ge^{2x} &= 5e^{2x} + c \end{aligned}$$

The value of c is determined by the condition $g(0) = 0$, that is $c = -5$, and we determine the particular solution to be

$$g(x) = 5\left(1 - e^{-2x}\right)$$

d. We express the ODE in standard form $y' + y = \sin(x)$ and note that an appropriate integrating factor is e^x. We therefore proceed as

$$\begin{aligned} y'e^x + ye^x &= e^x\sin(x) \\ \Rightarrow \frac{d}{dx}(ye^x) &= e^x\sin(x) \\ \Rightarrow ye^x &= \int e^x\sin(x)dx \end{aligned}$$

The integral on the right-hand side is an example of a cyclic integration by parts and, after some effort, we can determine that

$$ye^x = \frac{e^x}{2}(\sin(x) - \cos(x)) + c$$

The value of c is determined by the condition $y(\pi) = 1$, that is $c = \frac{1}{2}e^{\pi}$, and we determine the particular solution to be

$$y = \frac{1}{2}\left(e^{\pi - x} + \sin(x) - \cos(x)\right)$$

Note that Wolfram Alpha can be used to confirm your solutions using, for example, the command

```
solve y'+y=sin(x), y(pi)=1
```

Solution 11.5. In each case, we begin by testing to see if it is an exact ODE.

a. We express the ODE in standard form $P(x, y) + Q(x, y)y' = 0$,

$$y^2 - 2x + (2xy + 1)y' = 0$$

Therefore,

$$P(x, y) = y^2 - 2x \Rightarrow \frac{\partial P}{\partial y} = 2y \quad \text{and} \quad Q(x, y) = 2xy + 1 \Rightarrow \frac{\partial Q}{\partial y} = 2y$$

and the ODE is confirmed as exact. We define $\Lambda(x, y)$ such that the original ODE can be expressed as $\frac{d\Lambda}{dx} = 0$. That is, $\partial_x \Lambda = P$ and $\partial_y \Lambda = Q$. Working with P,

$$\Lambda = \int (y^2 - 2x)\mathrm{d}x = y^2 x - x^2 + \tilde{c}(y)$$

where $\tilde{c}(y)$ is an arbitrary function of y. Therefore,

$$\frac{\partial \Lambda}{\partial y} = 2xy + \tilde{c}'(y)$$

and, comparing to Q, we find that $\tilde{c}'(y) = 1$ and $\tilde{c} = y$. Λ is then confirmed to be $xy^2 - x^2 + y$ and the original ODE is written as

$$\frac{\mathrm{d}}{\mathrm{d}x}\left(xy^2 - x^2 + y\right) = 0$$
$$\Rightarrow xy^2 - x^2 + y = c$$

where c is found from the condition $y(0) = 1$. That is $c = 1$, and we determine an implicit solution to the ODE as

$$xy^2 + y - (1 + x^2) = 0$$

The solution has an explicit form

$$y(x) = \frac{-1 + \sqrt[+]{4x^3 + 4x + 1}}{2x}$$

b. We express the ODE in standard form as

$$(f + \cos f - \cos y) + (y - y\sin f)f' = 0$$

and note that

$$P(f,y) = f + \cos f - \cos y \Rightarrow \partial_f P = 1 - \sin f \quad \text{and}$$
$$Q(f,y) = \dot{y} - y\sin f \Rightarrow \partial_y Q = 1 - \sin f$$

we therefore confirm that the ODE is exact. Following the process used in part a., it is possible to express the ODE as

$$\frac{\mathrm{d}}{\mathrm{d}y}\left(yf + y\cos f - \sin y\right) = 0$$
$$\Rightarrow yf + y\cos f - \sin y = c$$

where c is found from the condition $f(\pi) = 1$. That is $c = \pi(1 + \cos(1))$, and we determine an implicit solution to the ODE as

$$\sin y + \pi(1 + \cos(1)) = yf(y) + y\cos f(y)$$

c. We express the ODE in standard form as

$$(g+1)\mathrm{e}^z + (\mathrm{e}^z - 2g)g' = 0$$

and can confirm that it is exact. Following the standard procedure, as in part a. for example, we find that the ODE can be expressed as

$$\frac{\mathrm{d}}{\mathrm{d}z}\left((g+1)\mathrm{e}^z - g^2\right) = 0$$
$$\Rightarrow (g+1)\mathrm{e}^z - g^2 = c$$

where c can be found from the condition $g \to 4$ as $z \to -\infty$. Therefore, $c = -16$, and we determine an implicit solution to the ODE as

$$g^2 - g\mathrm{e}^z - (\mathrm{e}^z + 16) = 0$$

The solution has explicit form

$$g = \frac{\mathrm{e}^z + \sqrt[+]{\mathrm{e}^{2z} + 4(\mathrm{e}^z + 16)}}{2}$$

d. We express the ODE in standard form as

$$t\arctan(z) + \frac{t^2}{2(1+z^2)}\frac{\mathrm{d}z}{\mathrm{d}t} = 0$$

and can confirm it is exact. Following the standard procedure, we find that the ODE can be expressed as

$$\frac{\mathrm{d}}{\mathrm{d}t}\left(\frac{1}{2}t^2\arctan(z)\right) = 0$$
$$\Rightarrow \frac{1}{2}t^2\arctan(z) = c$$

where c can be found from the condition $z(1) = 1$. Therefore, $c = \frac{1}{2}\arctan(1) = \frac{\pi}{8}$, and we determine an implicit solution to the ODE

$$t^2 \arctan(z) = \frac{\pi}{4}$$

The solution has explicit form

$$z = \tan\left(\frac{\pi}{4t^2}\right)$$

Note that Wolfram Alpha can be used to confirm your solutions using, for example, the command

```
z'/t=-2(1+z^2)arctan(z)/t^2, z(1)=1
```

Solution 11.6. We require a function that gives the beetle population in the research institute as a function of time, say $P(t)$. It is natural that we work in time units of weeks. The information given in the question enables us to form an ODE for $P(t)$ that we will solve to determine $P(t)$. Let us take each piece of information in turn,

- The population increases at a rate proportional to the population. The birth rate is therefore kP where $k > 0$.
- The birth rate is such that the population will quadruple every 5 weeks. This will help to determine k.
- We start with 50 beetles at $t = 0$.

Putting these together, we can form an ODE that models the population growth from birth effects *only*, $P_B(t)$. That is, we initially neglect all other affects. Note that this is not the required $P(t)$, but it gives us a route to finding k. In particular,

$$\frac{dP_B(t)}{dt} = kP_B(t)$$

which can be solved to give $P_B(t) = ce^{kt}$. The values of c and k can be found from imposing the conditions $P_B(0) = 50$ and $P_B(5) = 4 \times 50 = 200$. That is, $c = 50$ and $e^{5k} = 4$ and so

$$k = \frac{\ln(4)}{5}$$

Note that this value of k will hold throughout the full model for $P(t)$. In order to get there, we note one further piece of information

- Each week 20 beetles are posted away and 5 die. That is, each week the population reduces by 25.

The initial value problem for the *actual* system of interest is therefore

$$\frac{dP(t)}{dt} = kP - 25 \quad \text{such that } P(0) = 50$$

where k is as given above. This can be solved with use of the integrating factor e^{-kt},

$$\frac{\mathrm{d}}{\mathrm{d}t}\left(\mathrm{e}^{-kt}P\right) = -\,25\mathrm{e}^{-kt}$$

$$\Rightarrow \mathrm{e}^{-kt}P = \frac{25}{k}\mathrm{e}^{-kt} + c_1$$

We note c_1 can be determined from $P(0) = 50$. That is, $c_1 = 50 - \frac{25}{k}$ and we write the solution

$$P(t) = \frac{25}{k} + \left(50 - \frac{25}{k}\right)\mathrm{e}^{kt}$$

Substituting back the value of k, we see that

$$P(t) = \frac{75}{\ln(4)} - \left(\frac{75}{\ln(4)} - 50\right)\mathrm{e}^{\frac{\ln(4)}{5}t}$$

Alternatively, we can write $P(t) \approx 54.101 - 4.101\mathrm{e}^{0.277t}$.

A plot of this population is shown in Figure 11.1 and we see that it *decreases* with time. In fact, it will be zero at some finite time and a little work shows that $P(t) = 0$ at $t \approx 9.3$ weeks, as can be confirmed in the figure. That is, we expect the population to last 9.3 weeks in San Diego.

Solution 11.7. Although we have not considered second-order ODEs in this chapter, this question is included as a demonstration that it is sometimes possible to extend familiar concepts to unfamiliar topics.

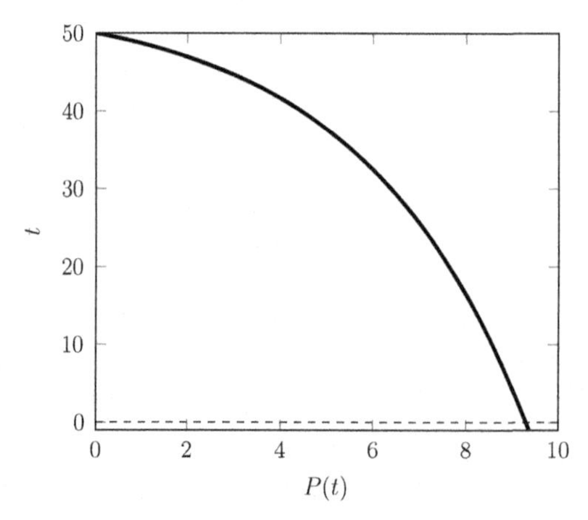

Figure 11.1 Plot of $P(t)$ obtained in Solution 11.6.

As suggested in the hint, we note that the second-order ODE can be factorized as

$$\left(\frac{\mathrm{d}}{\mathrm{d}x}+5\right)\left(\frac{\mathrm{d}}{\mathrm{d}x}-2\right)y=0$$

Although this representation may be unfamiliar, the reader is invited to expand the brackets to confirm that this is indeed true as long as the y remains premultiplied by the bracketed terms. Since, however, the order of the factorization does not matter, it is reasonable to understand that the factorization means that either

$$\left(\frac{\mathrm{d}}{\mathrm{d}x}+5\right)y=0 \quad \text{or} \quad \left(\frac{\mathrm{d}}{\mathrm{d}x}-2\right)y=0$$

These represent simple first-order ODEs with respective solutions $y = c_1 \mathrm{e}^{-5x}$ and $y = c_2 \mathrm{e}^{2x}$. Either solution therefore solves the second-order ODE and so it is reasonable to suppose that the general solution is

$$y = c_1 \mathrm{e}^{-5x} + c_2 \mathrm{e}^{2x}$$

The reader is invited to confirm that this is indeed true.

Now that we have the general solution, the two unknowns can be calculated from the two initial boundary conditions, $y'(0) = 1$ and $y(0) = 0$. A little work then shows that the second-order ODE has particular solution,

$$y(x) = \frac{1}{7}\left(\mathrm{e}^{2x} - \mathrm{e}^{-5x}\right)$$

The solution can be confirmed with Wolfram Alpha using the command

```
solve y"+3y'-10y=0, y'(0)=1, y(0)=0
```

Solution 11.8. Note that Solution 11.7 implies that the ODE will have general solution given by

$$f(x) = c_1 \mathrm{e}^{\lambda_1 x} + c_2 \mathrm{e}^{\lambda_2 x} + c_3 \mathrm{e}^{\lambda_3 x} + c_4 \mathrm{e}^{\lambda_4 x}$$

where $\{c_i\}$ are arbitrary constants obtained from the boundary conditions, and $\{\lambda_i\}$ are constants that result from an appropriate factorization of the ODE. To obtain $\{\lambda_i\}$, let us substitute the basic solution $f(x) = \mathrm{e}^{\lambda x}$. This leads to

$$\lambda^4 - 10\lambda^3 + 35\lambda^2 - 50\lambda + 24 = 0$$

which can be solved to give $\lambda = 1, 2, 3,$ and 4. We therefore see that the ODE can actually be written as

$$\left(\frac{\mathrm{d}}{\mathrm{d}x}-1\right)\left(\frac{\mathrm{d}}{\mathrm{d}x}-2\right)\left(\frac{\mathrm{d}}{\mathrm{d}x}-3\right)\left(\frac{\mathrm{d}}{\mathrm{d}x}-4\right)f=0$$

Using the same reasoning as in Solution 11.7, we therefore expect the general solution to be

$$f(x) = c_1 e^x + c_2 e^{2x} + c_3 e^{3x} + c_4 e^{4x}$$

After a some effort, we find that the boundary values fix the values of $\{c_i\}$ and the particular solution is

$$f(x) = \frac{e^x}{6}\left(23e^{3x} - 84e^{2x} + 105e^x - 44\right)$$

This solution can be confirmed with Wolfram Alpha using the command

```
f""-10f"'+35f"-50f'+24f=0 where f"'(0)=0, f"(0)=-2, f'(0)=1 and f(0)=0
```

Solution 11.9. The quantity ${}_tp_{50}$ can simply be interpreted as a function of $t \in [0, \infty)$ and we have a separable ODE to solve. We note that ${}_0p_{50} = 1$ (that is, he is certain to survive a further 0 units of time) and so we proceed to solve the ODE as follows:

$$\begin{aligned} \frac{d}{dt}\ln\left({}_tp_{50}\right) &= -\alpha\beta t^{\beta-1} \\ \Rightarrow \int_0^t \frac{d}{ds}\ln\left({}_sp_{50}\right) ds &= \int_0^t -\alpha\beta s^{\beta-1} ds \\ \Rightarrow \ln({}_tp_{50}) - \ln 1 &= -\alpha t^\beta \end{aligned}$$

We therefore find that

$${}_tp_{50} = e^{-\alpha t^\beta}$$

The values of α and β are still unknown, but can be determined from the probabilities stated. In particular, we have

$${}_{10}p_{50} = 0.6 \quad \text{and} \quad {}_{20}p_{50} = 0.5$$

and after some manipulation we find

$$\alpha \approx 0.185328 \quad \text{and} \quad \beta \approx 0.440331$$

That is,

$${}_tp_{50} \approx \exp\left(-0.185328 \times t^{0.440331}\right)$$

a. The probability that the individual will *not* survive to his 90th birthday is equal to $1 - {}_{40}p_{50}$. Therefore, we have

$$1 - \exp\left(-0.185328 \times 40^{0.440331}\right) \approx 0.6096$$

That is, there is a 61% probability that the individual will die before reaching his 90th birthday.

b. We are required to calculate t such that ${}_tp_{50} = 0.95$. After some manipulation we find that

$$\exp\left(-0.185328 \times t^{0.440331}\right) = 0.95 \Rightarrow t \approx 0.054 \text{ years}$$

The individual can therefore be 95% sure that he will survive only a further 19 days! In this case, the maximum age reached would be his current age, 50.

Solution 11.10. We proceed along similar lines to Solution 11.9 but with a different ODE. The ODE is still separable and we proceed as follows:

$$\frac{\mathrm{d}}{\mathrm{d}t}\ln\left({}_tp_{50}\right) = -\,Bc^{50}c^t$$

$$\Rightarrow \int_0^t \frac{\mathrm{d}}{\mathrm{d}s}\ln\left({}_sp_{50}\right)\mathrm{d}s = \int_0^t -Bc^{50}c^s\mathrm{d}s$$

$$\Rightarrow \ln({}_tp_{50}) - \ln 1 = -\,Bc^{50}\left(\frac{c^t-1}{\ln c}\right)$$

We therefore find that

$${}_tp_{50} = \exp\left(Bc^{50}\left(\frac{c^t-1}{\ln c}\right)\right) \equiv \mathrm{e}^{A(c^t-1)}$$

Note that we have simplified the expression with the use of a new constant A. The constants A and c can be determined from ${}_{10}p_{50} = 0.6$ and ${}_{20}p_{50} = 0.5$ and, after some work, we arrive at

$$A \approx 0.794337 \quad \text{and} \quad c \approx 0.902104$$

That is,

$${}_tp_{50} \approx \exp\left(0.794337 \times (0.902104^t - 1)\right)$$

a. As in Solution 11.9(a), the probability that the individual will *not* survive to his 90th birthday is equal to $1 - {}_{40}p_{50}$. Therefore, we have

$$1 - \exp\left(0.794337 \times (0.902104^{40} - 1)\right) \approx 0.5423$$

That is, there is a 54% probability that the individual will die before reaching his 90th birthday.

b. As in Solution 11.9(b), we are required to calculate t such that ${}_tp_{50} = 0.95$. After some manipulation we find that

$$\exp\left(0.794337 \times (0.902104^t - 1)\right) = 0.95 \Rightarrow t \approx 0.648 \text{ years}$$

The individual can therefore be 95% sure that he will survive only a further 236 days. In this case, the maximum age reached would be his current age, 50.

Note that whilst the different models have been calibrated to the same ${}_{10}p_{50} = 0.6$ and ${}_{20}p_{50} = 0.5$, the behavior away from $t = 10$ and 20 is different. This leads to different numerical values across Solutions 11.9 and 11.10.

Chapter 12 Solutions

Solution 12.1. We note that, in each case, the order in which we perform the partial derivatives does not matter. This simplifies our approach.

a. We proceed as follows:

$$\begin{aligned}\frac{\partial^3}{\partial x \partial y \partial z}\left(x^2yz+\sin(xy+x)\right) &= \frac{\partial^2}{\partial x \partial y}\left(\frac{\partial}{\partial z}\left(x^2yz+\sin(xy+x)\right)\right)\\ &= \frac{\partial}{\partial x}\left(\frac{\partial}{\partial y}\left(x^2y\right)\right)\\ &= \frac{\partial}{\partial x}x^2\\ &= 2x\end{aligned}$$

b. We proceed as follows:

$$\begin{aligned}\frac{\partial^3}{\partial x^2 \partial y}\left(e^{x^2+\cos(y)}\right) &= \frac{\partial^2}{\partial x \partial y}\left(\frac{\partial}{\partial x}e^{x^2+\cos(y)}\right)\\ &= \frac{\partial}{\partial y}\left(\frac{\partial}{\partial x}\left(2xe^{x^2+\cos(y)}\right)\right)\\ &= \frac{\partial}{\partial y}\left(2(2x^2+1)e^{x^2+\cos(y)}\right)\\ &= -2(2x^2+1)\sin(y)e^{x^2+\cos(y)}\end{aligned}$$

c. We proceed as follows:

$$\begin{aligned}\partial_{zz}(fg) &= \frac{\partial^2}{\partial z \partial z}\left(f(x,y,z)g(x,y)\right)\\ &= g(x,y)\frac{\partial}{\partial z}\left(\frac{\partial}{\partial z}f(x,y,z)\right)\\ &= e^{x^2+\cos(y)}\frac{\partial}{\partial z}\left(\frac{\partial}{\partial z}(x^2yz+\sin(xy+x))\right)\\ &= e^{x^2+\cos(y)}\frac{\partial}{\partial z}\left(x^2y\right)\\ &= 0\end{aligned}$$

Note that these answers can be confirmed with Wolfram Alpha using, for example, the command

```
d^3/dxdydz (x^2yz+sin(xy+x))
```

Solution 12.2. We attempt to locate the stationary points and apply the second-derivative test in each case.

a. We note that $f_x = 5 - 4x$ and $f_y = 10 - 2y$ and these simultaneously equal zero at $(x, y) = \left(\frac{5}{4}, 5\right)$. Furthermore, we obtain that

$$f_{xx} = -4 < 0 \quad f_{yy} = -2 \quad f_{xy} = 0$$

That is, $f_{xx}f_{yy} > f_{xy}^2$ and we have a maximum at $\left(\frac{5}{4}, 5\right)$ where $f(x, y) = \frac{225}{8}$.

b. We note that $g_x = -(1 + z^2)e^{2-x}$ and $g_z = 2ze^{2-x}$ which cannot simultaneously equal zero for real z. The function $g(x, z)$ therefore has no critical points.

c. We note that

$$h_x = \frac{-x^2 - 2xz + 2z^2 + 6}{(x^2 + 2z^2 + 6)^2} \quad \text{and} \quad h_z = \frac{x^2 - 4xz - 2z^2 + 6}{(x^2 + 2z^2 + 6)^2}$$

Some manipulation confirms that these are simultaneously equal to zero when $(x, z) = \pm(2, 1)$.

- At $(x, z) = (-2, -1)$,

$$h_{xx} = \frac{1}{24} \quad h_{zz} = \frac{1}{12} \quad h_{xz} = 0$$

that is, $h_{xx} > 0$ and $h_{xx}h_{zz} > h_{xz}^2$ and so we have a minimum where $h(x, z) = -\frac{1}{4}$.

- At $(x, z) = (2, 1)$,

$$h_{xx} = -\frac{1}{24} \quad h_{zz} = -\frac{1}{12} \quad h_{xz} = 0$$

that is, $h_{xx} < 0$ and $h_{xx}h_{zz} > h_{xz}^2$ and so we have a maximum where $h(x, z) = \frac{1}{4}$.

d. We note that

$$k_a = 2\sin(a)\cos(a) \quad \text{and} \quad k_b = 2b$$

These simultaneously equal zero when $b = 0$, $\sin(a) = 0 \Rightarrow a = n\pi$, and $\cos(a) = 0 \Rightarrow a = \left(n + \frac{1}{2}\right)\pi$ for $n \in \mathbb{Z}$. We note that

$$k_{aa} = 2\cos(2a) \quad h_{bb} = 2 \quad h_{xz} = 0$$

and so the sign of k_{aa} alternates with n.

- At $(a, b) = (n\pi, 0)$, $k_{aa} = 2 > 0$, and $k_{aa}h_{bb} > h_{xz}^2$ and so we have minima where $k(a, b) = 0$.
- At $(a, b) = \left(\left(n + \frac{1}{2}\right)\pi, 0\right)$, $k_{aa} = -2 < 0$, and $k_{aa}k_{bb} < k_{ab}^2$ and so we have saddle points where $k(a, b) = 1$.

Solution 12.3. We use Lagrange multipliers in each case.

a. We have $f(x, y) = x^2 + 2y^2$ and $g(x, y) - c \equiv x^2 + y^2 - 4$. The Lagrange function is then

$$L(x, y, \lambda) = x^2 + 2y^2 + \lambda \left(x^2 + y^2 - 4\right)$$

and we need to simultaneously solve

$$L_x = 2x + 2x\lambda = 0 \quad L_y = 4y + 2y\lambda = 0 \quad L_\lambda = x^2 + y^2 - 4 = 0$$

We obtain $(x, y, \lambda) = (\pm 2, 0, -1)$ and $(0, \pm 2, -2)$ and note that $f(\pm 2, 0) = 4$ and $f(0, \pm 2) = 8$. Therefore, the required constrained maxima are located at $(x, y) = (0, \pm 2)$.

b. We have $f(x, y) = x + 2y^2$ and $g(x, y) - c \equiv x + y^2 - 1$. The Lagrange function is then

$$L(x, y, \lambda) = x + 2y^2 + \lambda \left(x + y^2 - 1\right)$$

and we need to simultaneously solve

$$L_x = 1 + \lambda = 0 \quad L_y = 4y + 2y\lambda = 0 \quad L_\lambda = x + y^2 - 1 = 0$$

We obtain $(x, y, \lambda) = (1, 0, -1)$ and note that $f(1, 0) = 1$ and, for example, $f(0, 1) = 2$. Therefore, $(x, y) = (1, 0)$ is the location of the constrained minimum.

c. We have $f(x, y, z) = 2x^2 + y^2 - z^2$ and $g(x, y, z) - c \equiv 3xy - 4z^2$. The Lagrange function is then

$$L(x, y, z, \lambda) = 2x^2 + y^2 - z^2 + \lambda \left(3xy - 4z^2\right)$$

and we need to simultaneously solve

$$L_x = 4x + 3y\lambda = 0 \quad L_y = 2y + 3x\lambda \quad L_z = -2z - 8z\lambda = 0 \quad L_\lambda = 3xy - 4z^2 = 0$$

We obtain $(x, y, z, \lambda) = (0, 0, 0, \lambda)$ and note that $f(0, 0, 0) = 0$ and, for example, $f(1, 4, \sqrt{3}) = 15$. Therefore, $(x, y, z) = (0, 0, 0)$ is the location of the constrained minimum.

d. We have $f(x, y, z) = -y + z - x + 6$, $g_1(x, y, z) - c_1 \equiv x - y - z - 2$, and $g_2(x, y, z) \equiv x^2 + y^2 - 1$. The Lagrange function is then

$$L(x, y, z, \lambda, \mu) = -y + z - x + 6 + \lambda (x - y - z - 2) + \mu \left(x^2 + y^2 - 1\right)$$

and we need to simultaneously solve

$$L_x = -1 + \lambda + 2x\mu = 0 \quad L_y = -1 - \lambda + 2y\mu = 0 \quad L_z = 1 - \lambda = 0$$
$$L_\lambda = x - y - z - 2 = 0, \quad L_\mu = x^2 + y^2 - 1$$

We obtain $(x, y, x, \lambda, \mu) = (0, -1, -1, 1, -1)$ and $(0, 1, -3, 1, 1)$ and note that $f(0, -1, -1) = 6$ and $f(0, 1, -3) = 2$. Therefore, $(x, y, z) = (0, -1, -1)$ is the location of the constrained maximum.

Note that these results can be confirmed with Wolfram Alpha using, for example, the command

```
maximise x^2+2y^2 such that x^2+y^2=4
```

Solution 12.4. This is a constrained optimization problem. We form the Lagrange function

$$L(x, y, z, \lambda) = x^2 + \lambda(y^2 + z^2 - x)$$

and need to simultaneously solve

$$L_x = 2x - \lambda = 0 \quad L_y = 2y\lambda = 0 \quad L_z = 2z\lambda = 0 \quad L_\lambda = y^2 + z^2 - x = 0$$

If $\lambda = 0$, then $x = 0$ and so $y = z = 0$. If $\lambda \neq 0$, then $z = 0$, $y = 0$ and so $x = 0$. In either case, the critical point is at $(x, y, z) = (0, 0, 0)$ with $f(x, y, z) = x^2 = 0$. Note that $f \geq 0$ and so the critical point at $(0, 0, 0)$ must be a minimum.

Solution 12.5. Consider a rectangular box of length x, width y, and height 0.1. The volume is therefore given by $V(x, y) = 0.1xy$ and the surface area by $A(x, y) = 2xy + 0.2x + 0.2y$. The question can then be cast as a constrained optimization problem under which we maximize V subject to the constraint that $A = 2 \Rightarrow xy + 0.1x + 0.1y = 1$. The Lagrange function is therefore

$$L(x, y) = 0.1xy + \lambda(xy + 0.1x + 0.1y - 1)$$

and we are required to simultaneously solve

$$L_x = 0.1y + \lambda(y + 0.1) = 0 \quad L_y = 0.1x + \lambda(x + 0.1) = 0$$
$$L_\lambda = xy + 0.1x + 0.1y - 1 = 0$$

Some manipulation leads to two solutions

- $\lambda \approx -0.11$ and $x = y \approx -1.105$
- $\lambda \approx -0.09$ and $x = y \approx 0.905$

Clearly, we must have x and y positive and so the constrained problem is solved by $x = y \approx 0.905$ m. This corresponds to a volume $V = 0.1xy \approx 0.082\,\text{m}^3$.

We note that $x = 0.5$ and $y \approx 1.58$, for example, also solves $A = 2$ and corresponds to $V \approx 0.079\,\text{m}^3$. The solution $x = y \approx 0.905$ must therefore be the constrained maximum, as required.

Solution 12.6. Each definite integral is over a rectangular domain and so we can evaluate the iterative integral in any order.

a. We have $D_a = \{x, y|x \in [-1, 2], y \in [2, 4]\}$. Note that the integrand is separable and so we can proceed as

$$\iint_{D_a} 2xy\mathrm{d}x\mathrm{d}y = \int_2^4 \int_{-1}^2 2xy\mathrm{d}x\mathrm{d}y$$
$$= 2\int_2^4 y\mathrm{d}y \int_{-1}^2 x\mathrm{d}x$$
$$= 2\left[\frac{y^2}{2}\right]_2^4 \left[\frac{x^2}{2}\right]_{-1}^2$$
$$= 2 \times 6 \times \frac{3}{2}$$
$$= 18$$

b. We have $D_b = \{x, z|x \in [0, 10], z \in [-1, 1]\}$. The integrand is not separable and we proceed as follows:

$$\iint_{D_b} \left(x^2 - z^2 + 10\right) \mathrm{d}x\mathrm{d}z = \int_{-1}^1 \int_0^{10} \left(x^2 - z^2 + 10\right) \mathrm{d}x\mathrm{d}z$$
$$= \int_{-1}^1 \left[\frac{x^3}{3} - xz^2 + 10x\right]_{x=0}^{10} \mathrm{d}z$$
$$= \int_{-1}^1 \left(\frac{1300}{3} - 10z^2\right) \mathrm{d}z$$
$$= \left[\frac{1300}{3}z - \frac{10}{3}z^3\right]_{-1}^1$$
$$= 860$$

c. We have $D_c = \{p, q|p \in [-2, -1], q \in [0, \pi]\}$. Note that the integrand is separable and we can continue as follows:

$$\iint_{D_c} 0.1p\cos(q)\mathrm{d}p\mathrm{d}q = 0.1\int_{-2}^{-1} p\mathrm{d}p \int_0^{\pi} \cos(q)\mathrm{d}q$$
$$= 0.1\left[\frac{p^2}{2}\right]_{-2}^{-1} \left[\sin q\right]_0^{\pi}$$
$$= 0.1 \times \frac{-5}{2} \times 0$$
$$= 0$$

d. We have $D_d = \{x, y | x \in [1,2], y \in [1,2]\}$. Note that $e^{x+y} = e^x e^y$ and $\sin(x+y) = \sin(x)\cos(y) + \cos(x)\sin(y)$, the integrand is then written as

$$\begin{aligned}
&\iint_{D_d} e^{x+y}\sin(x+y)dxdy \\
&= \iint_{D_d} e^x e^y(\sin(x)\cos(y) + \cos(x)\sin(y))dxdy \\
&= \iint_{D_d} e^x e^y(\sin(x)\cos(y))dxdy + \iint_{D_d} e^x e^y(\cos(x)\sin(y))dxdy \\
&= \int_1^2\int_1^2 e^x e^y(\sin(x)\cos(y))dxdy + \int_1^2\int_1^2 e^x e^y(\cos(x)\sin(y))dxdy \\
&= \int_1^2 e^x\sin(x)dx\int_1^2 e^y\cos(y)dy + \int_1^2 e^x\cos(x)dx\int_1^2 e^y\sin(y)dy \\
&= 2\int_1^2 e^u\sin(u)du\int_1^2 e^v\cos(v)dv
\end{aligned}$$

Note that the final step is possible because a variable's "name" does not matter in the definite integral of a univariate function. Each of the remaining integrals can be performed using integration by parts as explained in Chapters 6 and 7. That is,

$$\int_1^2 e^u\sin(u)du = \left[\frac{1}{2}e^u(\sin(u) - \cos(u))\right]_1^2 \approx 4.4876$$

$$\int_1^2 e^v\cos(v)dv = \left[\frac{1}{2}e^v(\sin(v) + \cos(v))\right]_1^2 \approx -0.0561$$

We therefore find that

$$\iint_{D_d} e^{x+y}\sin(x+y)dxdy \approx 2\times 4.4876\times -0.0561 \approx -0.05032$$

Note that Wolfram Alpha can be used to confirm each integral using, for example, the command

```
int int e^(x+y) sin(x+y) dx dy between x=1 and 2 and between y=1 and 2
```

Solution 12.7. We note the domains are *not* rectangular and so we need to be careful about the order in which we evaluate the integrals. The domains are illustrated in Figure 12.1.

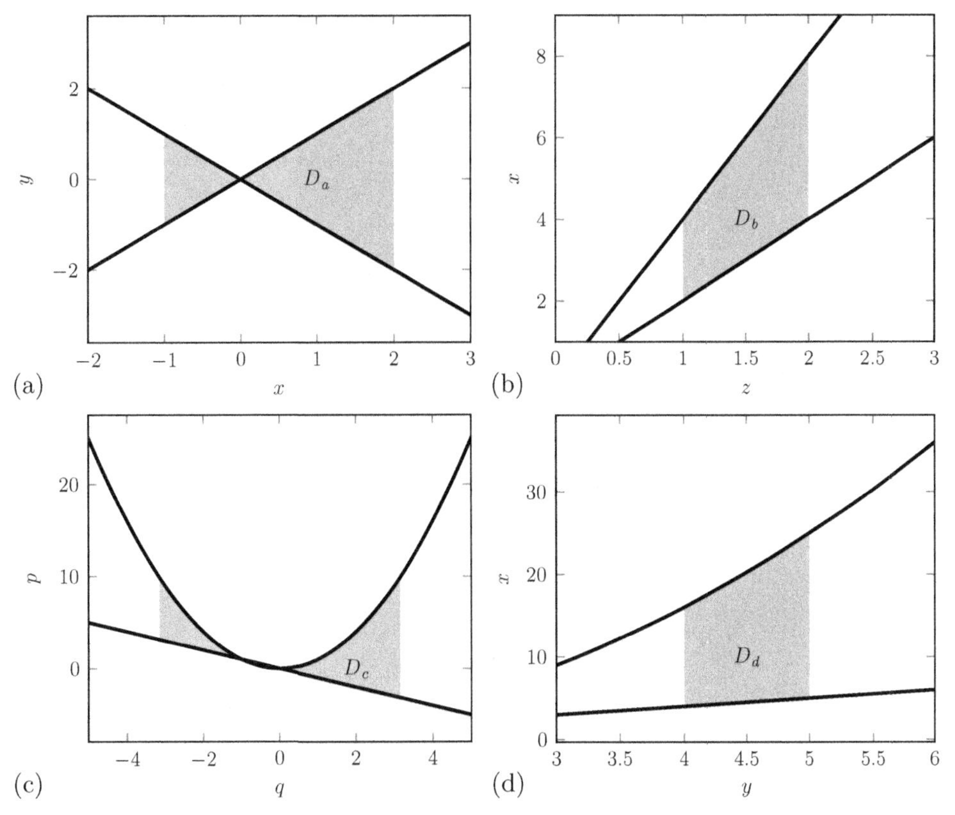

Figure 12.1 Domains of integration for Solution 12.7. Note the axes labels in each case. (a) $D_a = \{x, y | x \in [-1, 2], y \in [-x, x]\}$, (b) $D_b = \{x, z | x \in [2z, 4z], z \in [1, 2]\}$, (c) $D_c = \{p, q | p \in [-q, q^2], q \in [-\pi, \pi]\}$, and (d) $D_d = \{x, y | x \in [y, y^2], y \in [4, 5]\}$.

a. We have $D_a = \{x, y | x \in [-1, 2], y \in [-x, x]\}$ and so must proceed as follows:

$$
\begin{aligned}
\iint\limits_{D_a} 2xy\mathrm{d}x\mathrm{d}y &= \int_{-1}^{2} \int_{-x}^{x} 2xy\mathrm{d}y\mathrm{d}x \\
&= \int_{-1}^{2} \left[xy^2\right]_{y=-x}^{x} \mathrm{d}x \\
&= \int_{-1}^{2} x(x^2 - (-x)^2)\mathrm{d}x \\
&= \int_{-1}^{2} 0\mathrm{d}x \\
&= 0
\end{aligned}
$$

b. We have $D_b = \{x, z | x \in [2z, 4z], z \in [1, 2]\}$ and so must proceed as follows:

$$\iint_{D_b} \left(x^2 - z^2 + 10\right) \mathrm{d}x\mathrm{d}z = \int_1^2 \int_{2z}^{4z} \left(x^2 - z^2 + 10\right) \mathrm{d}x\mathrm{d}z$$

$$= \int_1^2 \left[\frac{x^3}{3} - z^2 x + 10x\right]_{2z}^{4z} \mathrm{d}z$$

$$= \int_1^2 \left(\frac{50z^3}{3} + 20z\right) \mathrm{d}z$$

$$= \left[\frac{50}{12}z^4 + 10z^2\right]_1^2$$

$$= \frac{185}{2}$$

c. We have $D_c = \{p, q | p \in [-q, q^2 + 1], q \in [-\pi, \pi]\}$ and so must proceed as follows:

$$\iint_{D_c} 0.1p\sin(q)\mathrm{d}p\mathrm{d}q = \int_{-\pi}^{\pi} \int_{-q}^{q^2} 0.1p\sin(q)\mathrm{d}p\mathrm{d}q$$

$$= \int_{-\pi}^{\pi} 0.05(q^4 - q^2)\sin(q)\mathrm{d}q$$

$$= 0$$

For the final step, we note that it is a symmetric integral of an odd integrand and so must have value 0.

d. We have $D_d = \{x, y | x \in [y, y^2], y \in [4, 5]\}$ and must proceed as follows:

$$\iint_{D_d} \mathrm{e}^{\frac{x}{y}} \mathrm{d}x\mathrm{d}y = \int_4^5 \int_y^{y^2} \mathrm{e}^{\frac{x}{y}} \mathrm{d}x\mathrm{d}y$$

$$= \int_4^5 \left[y\mathrm{e}^{\frac{x}{y}}\right]_y^{y^2} \mathrm{d}y$$

$$= \int_4^5 y(\mathrm{e}^y - e)\mathrm{d}y$$

$$= \mathrm{e}^4(4e - 3) - \frac{9}{2}e$$

$$\approx 417.6259$$

Integration by parts was used to perform the final integral.

Note that Wolfram Alpha can be used to confirm each integral using, for example, the command

```
int int e^(x/y) dx dy between x=y and y^2 and between y=4 and 5
```

Solution 12.8. This is an application of double integrals.

a. The value of a can be determined by noting that the total probability must be 1. That is,

$$\iint_D \frac{1}{a}(x - y + 5)\mathrm{d}x\mathrm{d}y = 1$$

for $D = \{x, y | x \in [0, 5], y \in [0, 7]\}$. The integral is over a rectangular domain and we proceed as

$$\int_0^7 \int_0^5 \frac{1}{a}(x - y + 5)\mathrm{d}x\mathrm{d}y = 1$$
$$\Rightarrow \int_0^7 \left[\frac{x^2}{2} - yx + 5x\right]_{x=0}^5 \mathrm{d}x\mathrm{d}y = a$$
$$\Rightarrow \int_0^7 \left(\frac{75}{2} - 5y\right)\mathrm{d}y = a$$
$$\Rightarrow a = 140$$

The joint distribution is therefore,

$$p(x, y) = \begin{cases} \frac{1}{140}(x - y + 5) & \text{for } 0 \leq x \leq 5,\ 0 \leq y \leq 7 \\ 0 & \text{otherwise} \end{cases}$$

b. We require the probability that $x + y > 6$ subject to the constrains imposed on x and y by the domain of the joint distribution function. This can be obtained from the integral of the joint distribution function over the domain D_p illustrated in Figure 12.2. In particular, we have $D_p = \{x, y | x \in [0, 5], y \in [6 - x, 7]\}$ and the probability P is obtained from

$$P = \iint_{D_p} \frac{1}{140}(x - y + 5)\mathrm{d}x\mathrm{d}y$$
$$= \int_0^5 \int_{6-x}^7 \frac{1}{140}(x - y + 5)\mathrm{d}y\mathrm{d}x$$
$$= \int_0^5 \frac{1}{140}\left[xy - \frac{y^2}{2} + 5y\right]_{y=6-x}^7 \mathrm{d}x$$

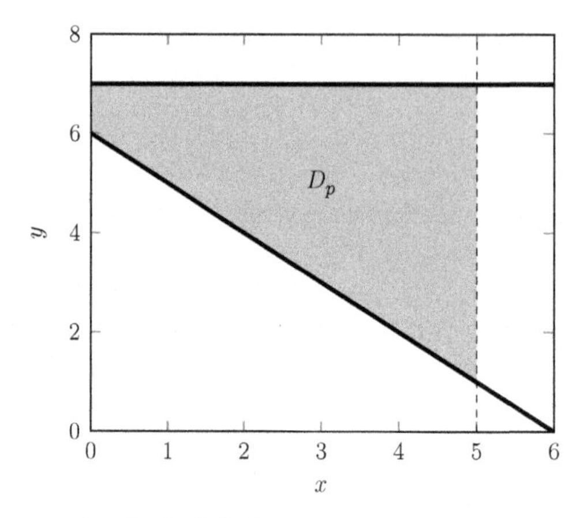

Figure 12.2 Domain of integration D_p for Solution 12.8.

$$= \int_0^5 \frac{3}{280}\left(x^2 - 1\right) \mathrm{d}x$$
$$= \frac{11}{28} \approx 0.3929$$

The probability of the total claims exceeds \$60M is therefore around 39.3%. Each double integral can be confirmed with Wolfram Alpha.

Solution 12.9. This is a constrained optimization problem. In particular, we are required to maximize $H(s, c) = s^{\frac{1}{2}} c^{\frac{2}{5}}$ subject to the constraint that $3.50s + 1.25c = 10$. The Lagrange function is formed as,

$$L(s, c, \lambda) = s^{\frac{1}{2}} c^{\frac{2}{5}} + \lambda\,(3.50s + 1.25c - 10)$$

and we are required to simultaneously solve

$$L_s = \frac{1}{2} s^{-\frac{1}{2}} c^{\frac{2}{5}} + 3.50\lambda = 0 \quad L_c = \frac{2}{5} s^{\frac{1}{2}} c^{-\frac{3}{5}} + 1.25\lambda = 0 \quad L_\lambda = 3.50s + 1.25c - 10 = 0$$

Some manipulation leads to $s = \frac{100}{63} \approx 1.59$ and $c = \frac{32}{9} \approx 3.56$ which is such that $H(s, c) \approx 2.09$. We note that, for example, $s = 2$ and $c = 2.4$ satisfies $3.50s + 1.25c = 10$ and is such that $H(s, c) \approx 2.01 < 2.09$. Therefore, $(s, c) = (1.59, 3.56)$ is the required constrained maximum. That is, we should purchase 1.59 sandwiches and 3.56 chocolate bars to maximize our afternoon happiness.

Solution 12.10. This is a constrained optimization problem.

a. The factory should aim to maximize the profit generated from its activities, subject to the constraint that $x + y = 12$. The Lagrange function is formed as,

$$L(x, y, z, \lambda) = 6x + 8y + 4z - 0.4x^2 - y^2 - z^2 + \lambda(x + y - 12)$$

and we are required to simultaneously solve

$$L_x = 6 - 0.8x + \lambda = 0 \quad L_y = 8 - 2y + \lambda = 0 \quad L_z = 4 - 2z = 0 \quad x + y - 12 = 0$$

Some manipulation leads to $(x, y, z) \approx \left(\frac{55}{7}, \frac{29}{7}, 2\right)$ which is such that $P(x, y, z) = \frac{297}{7} \approx 42.429$. We note that, for example, $x = 10$, $y = 2$, and $z = 2$ satisfies the constraints and are such that $P(x, y, z) \approx 36 < 42.429$. We can therefore conclude that the factory should manufacture 78,572 of X, 41,428 of Y, and 20,000 of Z to maximize its profits.

b. Now we have the constraint that $x + y = 10$. Following exactly the same process as in part a., we find that $(x, y, z) = \left(\frac{45}{7}, \frac{25}{7}, 2\right)$ give the optimal values with $P(x, y, z) = \frac{293}{7} \approx 41.857 < 42.429$. The factory would therefore make a *smaller* profit with this particular restructuring of its contracts.

Chapter 13 Solutions

Solution 13.1. We are required to find the single root x_r of $f(x) = x^2 - 2e^x - 2^x + 5$ and are given that $x_r \in [0, 1]$. It is sensible to use starting values $a_1 = 0$ and $b_1 = 1$ in both parts of the solution.

a. Note that we are required to use a definition of the error that is different from the standard definition for the bisection method. Otherwise, the algorithm proceeds in the standard way. The results of the bisection method are given in Table 13.1 and we see that c_7 gives $x_r \simeq 0.6563$ which is such that $|f(c_7)| = 0.0004$.

b. We use the standard form of the regula falsi method and each c_n is obtained from the expression

$$c_n = b_n - \frac{f(b_n) \times (b_n - a_n)}{f(b_n) - f(a_n)}$$

The results are given in Table 13.1 where see that c_4 gives $x_r \simeq 0.6560$ which is such that $|f(c_4)| = 0.0006$.

We note that the regula falsi method required fewer iterations to obtain convergence to the root within the required error tolerance. Note also that Wolfram Alpha's internal algorithms return $x_r \simeq 0.656135$ which is consistent with the values obtained from both methods. This can be confirmed by using the command

```
roots x^2-2e^x-2^x+5
```

Solution 13.2. We repeat Question 13.1 using the Newton-Raphson and secant methods. Note that, in both cases, we are required to use a measure of the error that is different to standard forms for these methods. The methods otherwise proceed in the standard way.

a. We have $f(x) = x^2 - 2e^x - 2^x + 5 = x^2 - 2e^x - e^{x\ln 2} + 5$ and so $f'(x) = 2x - 2e^x - \ln(2)2^x$. The iterative expression for the method is therefore given by

$$x_{n+1} = x_n - \frac{f(x_n)}{f'(x_n)} = x_n - \frac{x^2 - 2e^x - 2^x + 5}{2x - 2e^x - \ln(2)2^x}$$

We require only a single starting value and opt to take the midpoint of the interval, that is $x_0 = 0.5$. The results of the process are given in Table 13.2 and we see that $x_r \simeq x_2 = 0.6562$.

Table 13.1 Iterations of a. the bisection and b. the regula falsi methods for Solution 13.1

	n	a_n	b_n	$x_r \simeq c_n$	$f(a_n)$	$f(c_n)$	$f(b_n)$
a.	1	0.0000	1.0000	0.5000	2.0000	0.5383	−1.4366
	2	0.5000	1.0000	0.7500	0.5383	−0.3533	−1.4366
	3	0.5000	0.7500	0.6250	0.5383	0.1119	−0.3533
	4	0.6500	0.7500	0.7000	0.0223	−0.1620	−0.3533
	5	0.6500	0.7000	0.6750	0.0223	−0.0690	−0.1620
	6	0.6500	0.6750	0.6625	0.0223	−0.0232	−0.0690
	7	0.6500	0.6625	0.6563	0.0223	−0.0004	−0.0232
	n	a_n	b_n	$x_r \simeq c_n$	$f(a_n)$	$f(c_n)$	$f(b_n)$
b.	1	0.0000	1.0000	0.5820	2.0000	0.2627	−1.4366
	2	0.5820	1.0000	0.6466	0.2627	0.0346	−1.4366
	3	0.6466	1.0000	0.6549	0.0346	0.0045	−1.4366
	4	0.6549	1.0000	0.6560	0.0045	0.0006	−1.4366

These figures were obtained by implementing the methods in Excel.

Table 13.2 Iterations of a. the Newton-Raphson method and b. the secant method for Solution 13.2

	n	x_n	x_{n+1}	$f(x_r \simeq x_{n+1})$			
a.	0	0.5000	0.6642	−0.0296			
	1	0.6642	0.6562	−0.0001			
	n	x_{n-1}	x_n	$f(x_{n-1})$	$f(x_n)$	x_{n+1}	$f(x_{n+1})$
b.	1	0.0000	1.0000	2.0000	−1.4366	0.5820	0.2627
	2	1.0000	0.5820	−1.4366	0.2627	0.6466	0.0346
	3	0.5820	0.6466	0.2627	0.0346	0.6564	−0.0009

These figures were obtained by implementing the methods in Excel.

b. The iterative process for the secant method is given by

$$x_{n+1} = x_n - f(x_n)\frac{x_n - x_{n-1}}{f(x_n) - f(x_{n-1})}$$

Two starting values are required and we take these to be the boundary values of the interval, that is $x_0 = 0$ and $x_1 = 1$. The results of the process are given in Table 13.2 and we see that $x_r \simeq x_4 = 0.6564$.

Both gradient methods quickly converged to very similar estimates of x_r. These values were consistent with those obtained in Solution 13.1 and Wolfram Alpha. Out of all the methods used here and in Solution 13.1, the Newton-Raphson method converged most quickly. The downside of that method is, of course, that it requires the evaluation of $f'(x)$ at each x_n.

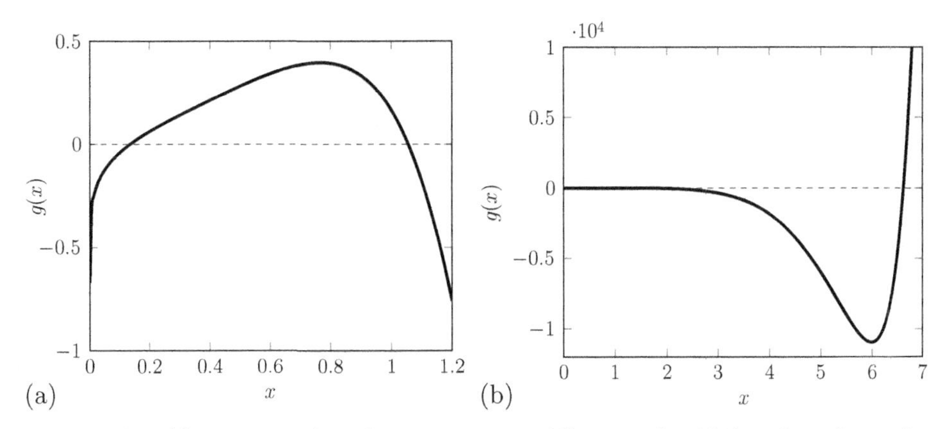

Figure 13.1 Plot of function $g(x)$ for Solution 13.3 at two different scales. (a) Plot of $g(x)$ for $x \in [0, 1.2]$ and (b) plot of $g(x)$ for $x \in [0, 7]$.

Solution 13.3. The presence of fractional powers means that, if we wish $g(x)$ to remain in $\mathbb{R}$, we must restrict the domain to $x \in \mathbb{R}^+$. Computational plots of $g(x)$ at two different scales can be seen in Figure 13.1 and we identify three roots. It is clear from the figures that $x_a \in [0, 0.2]$, $x_b \in [1, 1.2]$, and $x_c \in [6, 7]$. Given the difficulty in obtaining $f'(x)$, we opt not to use the Newton-Raphson, and instead implement the secant method. The secant method is expected to provide more rapid convergence than an interval method (that is, the bisection or regula falsi methods).

Table 13.3 shows the convergence of the secant method obtained for each of the three roots. Note that in each case we have taken the initial two estimates x_0 and x_1 to be the end points of the intervals highlighted from the computational plot. Using the error tolerance that $f(x_r) = 0$ to 4 decimal places, we obtain $x_a \simeq 0.1350$, $x_b \simeq 1.0564$, and $x_c \simeq 6.6326$.

These values can be confirmed by Wolfram Alpha's internal root solving algorithms from the command

```
roots -0.67*2.5^(x^1.3)+2.5^(x^1.3)*x^0.2+1.34*x^5.3-2*x^5.5
```

Solution 13.4. This question is related to the time value of money, as discussed in Chapter 2. It is sensible to look at the accumulation of each investment to $t = 10$ using a fixed annual compounding interest rate i. We have that,

- \$2000 is accumulated for 10 years
- \$3000 is accumulated for 6.5 years
- \$5600 is accumulated for 3 years

collectively leading to \$15,432. Therefore, working in money units of \$1000 and time units of years, we form the expression

$$2(1+i)^{10} + 3(1+i)^{6.5} + 5.6(1+i)^3 = 15.432$$

Table 13.3 The secant method for obtaining the three roots identified in Solution 13.3

	n	x_{n-1}	x_n	$g(x_{n-1})$	$g(x_n)$	x_{n+1}	$g(x_{n+1})$
x_a	1	0.0000	0.2000	−0.6700	0.0613	0.1832	0.0467
	2	0.2000	0.1832	0.0613	0.0467	0.1298	−0.0056
	3	0.1832	0.1298	0.0467	−0.0056	0.1355	0.0005
	4	0.1298	0.1355	−0.0056	0.0005	0.1350	0.0000
	n	x_{n-1}	x_n	$g(x_{n-1})$	$g(x_n)$	x_{n+1}	$g(x_{n+1})$
x_b	1	1.0000	1.2000	0.1650	−0.7571	1.0358	0.0673
	2	1.2000	1.0358	−0.7571	0.0673	1.0492	0.0245
	3	1.0358	1.0492	0.0673	0.0245	1.0569	−0.0016
	4	1.0492	1.0569	0.0245	−0.0016	1.0564	0.0000
	n	x_{n-1}	x_n	$g(x_{n-1})$	$g(x_n)$	x_{n+1}	$g(x_{n+1})$
x_c	1	6.0000	7.0000	−10, 958.6459	30,906.2491	6.2618	−9643.0324
	2	7.0000	6.2618	30,906.2491	−9643.0324	6.4373	−6683.4845
	3	6.2618	6.4373	−9643.0324	−6683.4845	6.8338	12,926.1294
	4	6.4373	6.8338	−6683.4845	12,926.1294	6.5724	−2548.7704
	5	6.8338	6.5724	12,926.1294	−2548.7704	6.6155	−777.4108
	6	6.5724	6.6155	−2548.7704	−777.4108	6.6344	81.3765
	7	6.6155	6.6344	−777.4108	81.3765	6.6326	−2.2368
	8	6.6344	6.6326	81.3765	−2.2368	6.6326	−0.0062
	9	6.6326	6.6326	−2.2368	−0.0062	6.6326	0.0000

These figures were obtained by implementing the methods in Excel.

and attempt to solve this for i. We therefore define the polynomial function

$$f(i) = 2(1+i)^{10} + 3(1+i)^{6.5} + 5.6(1+i)^3 - 15.432$$

and seek its roots. It is clear that $2 + 3 + 5.6 < 15.432$ and so the interest earned on the account has been positive; we therefore require $i \in \mathbb{R}^+$.

We note that $f(0.10) = 2.7832$ therefore $i = 10\%$ is too high. Also, $f(0.05) = -1.5719$ and so $i \in (0.05, 0.10)$. We use these two values and implement a single iteration of the regula falsi method to gain a rough approximation of the root. In particular,

$$\begin{aligned} i &\approx 0.10 - \frac{f(0.10) \times (0.10 - 0.05)}{f(0.10) - f(0.05)} \\ &= 0.10 - \frac{0.05 \times 2.7832}{2.7832 + 1.5719} \\ &= 0.0680 \end{aligned}$$

We have therefore found that the interest rate is approximately equal to 6.80%.

Implementing the method on a computer and iterating leads to $i \approx 6.98\%$, which is reasonably close to our approximation. If a more accurate approximation was required from the pen and paper approach, we could have sampled $f(i)$ at $i = 0.06$ and $i = 0.07$ and applied the method between these two values.

Note that the accurate value can be confirmed using Excel's Goalseek or Wolfram Alpha.

Solution 13.5. We are seeking a converged value of

$$D_{h_k}f(-1) = \frac{f(-1 + h_k) - f(-1)}{h_k}$$

for $f(x) = \sin(x^3)$ and some appropriately small h_k. We proceed by defining $h_k = 10^{-k}$ for $k = 1, 2, \ldots$ and compute $D_{h_k}f(-1)$ as shown in Table 13.4. Note that the value has converged to $f'(-1) \simeq D_{h_7}f(-1) = 1.6209$ at $k = 7$.

Algebraically, we note that $f'(x) = 3x^2 \cos(x^3)$ and so $f'(-1) = 3\cos(1) \approx 1.6209$. This is equal to that obtained from the numerical approximation.

Table 13.4 Implementations of the elementary and higher-order central difference methods for computing the derivative of $f(x) = \sin(x^3)$ at $x = -1$ in Solutions 13.5 and 13.6

	k	$h_k = 10^{-k}$	$D_{h_k}f(-1)$
D_{h_k}	1	0.1	1.7535
	2	0.01	1.6416
	3	0.001	1.6231
	4	0.0001	1.6211
	5	0.00001	1.6209
	6	0.000001	1.6209
	k	$h_k = 10^{-k}$	$D^{[2]}_{h_k}f(-1)$
$D^{[2]}_{h_k}$	1	0.1	1.5263
	2	0.01	1.6200
	3	0.001	1.6209
	4	0.0001	1.6209
	k	$h_k = 10^{-k}$	$D^{[2]}_{h_k}f(-1)$
$D^{[4]}_{h_k}$	1	0.1	1.6203
	2	0.01	1.6209
	3	0.001	1.6209

These figures were obtained by implementing the methods in Excel.

Solution 13.6. The relevant central-difference formulae of orders h^2 and h^4, respectively, are

$$D_{h_k}^{[2]} f(-1) = \frac{f(-1+h_k) - f(-1-h_k)}{2h_k}$$

$$D_{h_k}^{[4]} f(-1) = \frac{-f(-1+2h_k) + 8f(-1+h_k) - 8f(-1-h_k) + f(-1-2h_k)}{12h_k}$$

where $f(x) = \sin(x^2)$ and we choose $h_k = 10^{-k}$ for $k = 1, 2, \ldots$.

The values of both estimates are shown in Table 13.4 with increased k. We find converged values of $f'(-1) \simeq D_{h_4}^{[2]} f(-1) = 1.6209$ and $f'(-1) \simeq D_{h_3}^{[4]} f(-1) = 1.6209$. Note that these values agree with the actual value and, as might be expected, convergence was found much sooner than for the elementary approximation in Solution 13.5. Of the two estimates used here, the 4th order approach achieved quicker convergence, that is, required fewer reductions in step size.

Solution 13.7. Note that the data fixes the step size at $h = x_n - x_{n-1} = 0.1$.

a. The elementary approach approximates the derivative at each x_n using

$$D_h f(x_n) = \frac{f(x_n + h) - f(x_n)}{h}$$

This expression can be implemented in Excel, for example, at all x_n except the end point $x_n = 2$. The results are shown in Table 13.5.

b. The central-difference method of order 2 approximates the derivative at each x_n using

$$D_h^{[2]} f(x_n) = \frac{f(x_n + h) - f(x_n - h)}{2h}$$

This expression can be implemented in Excel, for example, at all x_n except the two end points $x_n = \pm 2$. The results are shown in Table 13.5.

c. The central-difference method of order 4 approximates the derivative at each x_n using

$$D_h^{[4]} f(x_n) = \frac{-f(x_n + 2h) + 8f(x_n + h) - 8f(x_n - h) + f(x_n - 2h)}{12h}$$

This expression can be implemented in Excel, for example, at all x_n except the first two and last two points, that is $x_n = \pm 2$ and ± 1.9. The results are shown in Table 13.5.

We are given that $f(x) = 1 + x^2 \cos x$ and so can determine that $f'(x) = x(2\cos(x) - x\sin(x))$. Accurate values of $f'(x_n)$ can therefore be computed and are included in Table 13.5 for comparison. Despite the relatively large step size used, remarkable agreement is seen between the values of $D_h^{[4]} f(x_n)$ and $f'(x_n)$. The downside is of course

Table 13.5 Derivatives of $f(x)$ for Solution 13.7

x_n	$f(x_n)$	$D_h f(x_n)$	$D_h^{[2]} f(x_n)$	$D_h^{[4]} f(x_n)$	$f'(x_n)$
−2	−0.6646	4.9751	–	–	5.3018
−1.9	−0.1671	4.3094	4.6423	–	4.6446
−1.8	0.2639	3.6377	3.9736	3.9732	3.9732
−1.7	0.6276	2.9761	3.3069	3.3040	3.3040
−1.6	0.9252	2.3391	2.6576	2.6524	2.6523
−1.5	1.1592	1.7398	2.0394	2.0322	2.0322
−1.4	1.3331	1.1894	1.4646	1.4556	1.4556
−1.3	1.4521	0.6972	0.9433	0.9330	0.9329
−1.2	1.5218	0.2706	0.4839	0.4725	0.4725
−1.1	1.5489	−0.0855	0.0925	0.0805	0.0804
−1	1.5403	−0.3680	−0.2267	−0.2391	−0.2391
−0.9	1.5035	−0.5761	−0.4721	−0.4843	−0.4844
−0.8	1.4459	−0.7112	−0.6437	−0.6556	−0.6556
−0.7	1.3748	−0.7765	−0.7439	−0.7551	−0.7551
−0.6	1.2971	−0.7773	−0.7769	−0.7871	−0.7871
−0.5	1.2194	−0.7203	−0.7488	−0.7577	−0.7577
−0.4	1.1474	−0.6139	−0.6671	−0.6745	−0.6745
−0.3	1.0860	−0.4678	−0.5408	−0.5466	−0.5466
−0.2	1.0392	−0.2925	−0.3802	−0.3841	−0.3841
−0.1	1.0100	−0.0995	−0.1960	−0.1980	−0.1980
0	1.0000	0.0995	0.0000	0.0000	0.0000
0.1	1.0100	0.2925	0.1960	0.1980	0.1980
0.2	1.0392	0.4678	0.3802	0.3841	0.3841
0.3	1.0860	0.6139	0.5408	0.5466	0.5466
0.4	1.1474	0.7203	0.6671	0.6745	0.6745
0.5	1.2194	0.7773	0.7488	0.7577	0.7577
0.6	1.2971	0.7765	0.7769	0.7871	0.7871
0.7	1.3748	0.7112	0.7439	0.7551	0.7551
0.8	1.4459	0.5761	0.6437	0.6556	0.6556
0.9	1.5035	0.3680	0.4721	0.4843	0.4844
1	1.5403	0.0855	0.2267	0.2391	0.2391
1.1	1.5489	−0.2706	−0.0925	−0.0805	−0.0804
1.2	1.5218	−0.6972	−0.4839	−0.4725	−0.4725
1.3	1.4521	−1.1894	−0.9433	−0.9330	−0.9329
1.4	1.3331	−1.7398	−1.4646	−1.4556	−1.4556
1.5	1.1592	−2.3391	−2.0394	−2.0322	−2.0322
1.6	0.9252	−2.9761	−2.6576	−2.6524	−2.6523
1.7	0.6276	−3.6377	−3.3069	−3.3040	−3.3040
1.8	0.2639	−4.3094	−3.9736	−3.9732	−3.9732
1.9	−0.1671	−4.9751	−4.6423	–	−4.6446
2	−0.6646	–	–	–	−5.3018

that the method is unable to produce estimates close to the end points. The other two estimates, whilst resulting in approximations at or close to the end points, are seen to be less accurate for general x_n; indeed the elementary approximation is seen to be particularly poor.

These comments are only qualitative. If required, a quantitative analysis is possible from the data in Table 13.5.

Solution 13.8. The kth order trapezoidal and Simpson's approaches for approximating the definite integral I are, respectively, given by

$$I \simeq T_{[a,b],k} = h\left(\frac{1}{2}f(a) + f(x_1) + f(x_2) + \cdots + f(x_{k-1}) + \frac{1}{2}f(b)\right)$$

$$I \simeq S_{[a,b],k} = \frac{h}{3}\Big(f(a) + 4f(x_1) + 2f(x_2) + 4f(x_3) + 2f(x_4) + \cdots + 2f(x_{k-2}) + 4f(x_{k-1}) + f(b)\Big)$$

where $f(x)$ is the integrand, a, b are the lower and upper values of the domain of integration, and $\{x_k\}$ are the equally spaced node points with step size $h = \frac{b-a}{k}$.

In this question we have $f(x) = \sin(x^2)$, $a = -1.5$, and $b = 1.5$, and are required to determine h, and so $\{x_n\}$, such that the theoretical error is less than 0.001.

a. The theoretical error under the trapezoidal approach is given by

$$\hat{E}_k^T = \left|\frac{h^2}{12}\left(f'(b) - f'(a)\right)\right|$$

where, in this case, $f'(x) = 2x\cos(x^2)$. We therefore have

$$\left|f'(b) - f'(a)\right| = \left|-1.88452 - 1.88452\right| \approx 3.76904$$

and can determine h from

$$\hat{E}_k^T = 0.001 = \frac{h^2}{12} \times 3.76904 \Rightarrow h = 0.0564$$

In practice, we are then required to use $h = 0.05$ over the interval $[-1.5, 1.5]$. This leads to 61 node points $\{x_n\} = \{-1.5 + 0.05n\}$ for $n = 0, 1, \ldots, 60$. The expression for the trapezoidal approximation to I stated above can be implemented in Excel, for example, and leads to

$$I \simeq T_{[-1.5,1.5],60} = 0.05\left(\frac{1}{2}f(-1.5) + f(-1.45) + f - 1.40) + \cdots + f(1.45) + \frac{1}{2}f(1.50)\right) = 1.5557$$

b. The theoretical error under Simpson's approach is given by

$$\hat{E}_k^S = \left| \frac{h^4}{180} \left(f^{(3)}(b) - f^{(3)}(a) \right) \right|$$

where, in this case, $f^{(3)}(x) = -4x(3\sin(x^2) + 2x^2\cos(x^2))$. We therefore have

$$\left| f^{(3)}(b) - f^{(3)}(a) \right| = \left| 2.95537 + 2.95537 \right| \approx 5.91074$$

and can determine h from

$$\hat{E}_k^S = 0.001 = \frac{h^4}{180} \times 5.91074 \Rightarrow h = 0.41774$$

We note that the domain (of width 3) is not divisible by 0.4 and so opt to use the step size $h = 0.375$. This leads to $k = 8$ and we have 9 node points $\{x_n\} = \{-1.5 + 0.375n\}$ for $n = 0, 1, \ldots, 8$. The expression for Simpson's approximation to I stated above can be implemented in Excel, for example, and leads to

$$\begin{aligned} I \simeq S_{[-1.5,1.5],8} &= \frac{0.375}{3} \Big(f(-1.5) + 4f(-1.125) + 2f(-0.75) + 4f(-0.375) \\ &+ 2f(0) + 4f(0.375) + 2f(0.75) + 4f(1.125) + f(1.5) \Big) \\ &= 1.5551 \end{aligned}$$

Using, in Wolfram Alpha, the command

```
int sin(x^2) dx between x=-1.5 and 1.5
```

its numerical integration algorithms are seen to give $I \simeq 1.55648$. Good agreement is seen between this value and the results of both the trapezoidal and Simpson's approach.

Solution 13.9. Since $Z \sim N(0, 1)$ we know that it has distribution function

$$f(z) = \frac{1}{\sqrt{2\pi}} e^{-\frac{z^2}{2}}$$

for $z \in (-\infty, \infty)$. This is such that

$$P(0 \leq z \leq 4) = \int_0^4 \frac{1}{\sqrt{2\pi}} e^{-\frac{z^2}{2}} dz$$

and we attempt to use Simpson's rule to evaluate this integral. An error tolerance of 0.001% is required and so we must choose node points such that the theoretical error is $\hat{E}_k^S < 0.00001$. That is,

$$\left| \frac{h^4}{180} \left(f^{(3)}(4) - f^{(3)}(0) \right) \right| = 0.00001$$

We have $f^{(3)}(z) = \frac{1}{\sqrt{2\pi}} z(z^2 - 3)e^{-\frac{z^2}{2}}$ and so $f^{(3)}(0) = 0$ and $f^{(3)}(4) \approx -0.00696$ which leads to $h = 0.71$. We therefore find it appropriate to use $k = 6$ with step size $h = \frac{2}{3}$. The node points are therefore $\{z_n\} = \{\frac{2n}{3}\}$ for $n = 0, 1, \ldots, 6$ and Simpson's rule can be implemented with the expression

$$P(0 \leq z \leq 4) \simeq \frac{h}{3}\Big(f(0) + 4f(z_1) + 2f(z_2) + 4f(z_3) + 2f(z_4) + 4f(z_5) + f(z_6)\Big)$$

$$= 49.996\%$$

This agrees with the value stated in the question to within the desired error tolerance.

Solution 13.10. We are required to determine the probability that s is between 1 and 3. However, it is first necessary to determine the value of a in the distribution function. This can be done by noting that the total probability should be 1, and so

$$\int_{0.1}^{6.1} \frac{1}{s} \exp\left\{-\frac{1}{2}\left(\frac{\ln(s) - 1}{0.2}\right)^2\right\} ds = \frac{1}{a}$$

We use Simpson's approach of order $k = 16$ to evaluate this. Therefore, $h = \frac{6}{16} = 0.375$ and we have node points $\{s_n\} = \{0.1 + 0.375n\}$ for $n = 0, 1, \ldots, 16$. A formal error analysis can be performed, but experience shows that this order is usually sufficient for very good accuracy. We therefore have

$$\frac{1}{a} \simeq \frac{h}{3}\Big(f(s_0) + 4f(s_1) + 2f(s_2) + 4f(s_3) + 2f(s_4) + \ldots + f(s_{16})\Big)$$

$$\approx 0.50099$$

The required probability is then given by

$$P(1 < s < 3) \approx \int_1^3 \frac{1}{0.50099s} \exp\left\{-\frac{1}{2}\left(\frac{\ln(s) - 1}{0.2}\right)^2\right\} ds$$

and again we use Simpson's approach of order 16, that is $h = 0.125$ and at node points $\{s_n\} = \{1 + 0.125n\}$ for $n = 0, 1, \ldots, 16$. After some work with obtain that

$$P(1 < s < 3) \simeq 68.95\%$$

Under this model, an asset currently priced at \$200 has a 69% probability that it will have a market price between \$200 and \$600 in 1 year's time.

PART IV

Appendices

Appendix A

Mathematical Identities

Contents

A.1 TRIGONOMETRIC IDENTITIES

Pythagorean

$$\sin^2(\alpha) + \cos^2(\alpha) = 1$$

$$\tan^2(\alpha) + 1 = \sec^2(\alpha)$$

$$1 + \cot^2(\alpha) = \operatorname{cosec}^2(\alpha)$$

Addition and subtraction

$$\sin(\alpha + \beta) = \sin(\alpha)\cos(\beta) + \cos(\alpha)\sin(\beta)$$

$$\cos(\alpha + \beta) = \cos(\alpha)\cos(\beta) - \sin(\alpha)\sin(\beta)$$

$$\tan(\alpha + \beta) = \frac{\tan(\alpha) + \tan(\beta)}{1 - \tan(\alpha)\tan(\beta)}$$

$$\sin(\alpha - \beta) = \sin(\alpha)\cos(\beta) - \cos\alpha\sin(\beta)$$

$$\cos(\alpha - \beta) = \cos(\alpha)\cos(\beta) + \sin(\alpha)\sin(\beta)$$

$$\tan(\alpha - \beta) = \frac{\tan(\alpha) - \tan(\beta)}{1 + \tan(\alpha)\tan(\beta)}$$

Product-sum

$$\sin(\alpha) + \sin(\beta) = 2\sin\left(\frac{\alpha + \beta}{2}\right)\cos\left(\frac{\alpha - \beta}{2}\right)$$

$$\sin(\alpha) - \sin(\beta) = 2\cos\left(\frac{\alpha + \beta}{2}\right)\sin\left(\frac{\alpha - \beta}{2}\right)$$

$$\cos(\alpha) + \cos(\beta) = 2\cos\left(\frac{\alpha+\beta}{2}\right)\cos\left(\frac{\alpha-\beta}{2}\right)$$

$$\cos(\alpha) - \cos(\beta) = -2\sin\left(\frac{\alpha+\beta}{2}\right)\sin\left(\frac{\alpha-\beta}{2}\right)$$

Double/half-angle

$$\sin(2\alpha) = 2\sin(\alpha)\cos(\alpha)$$

$$\cos(2\alpha) = \cos^2(\alpha) - \sin^2(\alpha)$$

$$\tan(2\alpha) = \frac{2\tan(\alpha)}{1-\tan^2(\alpha)}$$

$$\sin\left(\frac{\alpha}{2}\right) = \pm\sqrt{\frac{1-\cos^2(2\alpha)}{2}}$$

$$\cos\left(\frac{\alpha}{2}\right) = \pm\sqrt{\frac{1+\cos^2(2\alpha)}{2}}$$

$$\tan\left(\frac{\alpha}{2}\right) = \pm\frac{\sin(\alpha)}{1+\cos(\alpha)}$$

Sum-product

$$\sin(\alpha)\sin(\beta) = \frac{\cos(\alpha-\beta)-\cos(\alpha+\beta)}{2}$$

$$\cos(\alpha)\cos(\beta) = \frac{\cos(\alpha-\beta)+\cos(\alpha+\beta)}{2}$$

$$\sin(\alpha)\cos(\beta) = \frac{\sin(\beta+\alpha)-\sin(\beta-\alpha)}{2}$$

A.2 DERIVATIVES OF STANDARD FUNCTIONS

Let $c, a \in \mathbb{R}$ be constants with $a > 0$, and $x \in \mathbb{R}$ be an independent variable.

Polynomial

$$\frac{\mathrm{d}}{\mathrm{d}x}c = 0$$

$$\frac{\mathrm{d}}{\mathrm{d}x}x = 1$$

$$\frac{\mathrm{d}}{\mathrm{d}x}cx = c$$

$$\frac{\mathrm{d}}{\mathrm{d}x}x^c = cx^{c-1}$$

$$\frac{\mathrm{d}}{\mathrm{d}x}x^{-c} = -cx^{-(c+1)}$$

Exponential/logarithmic

$$\frac{\mathrm{d}}{\mathrm{d}x}\mathrm{e}^x = \mathrm{e}^x$$

$$\frac{\mathrm{d}}{\mathrm{d}x}a^x = a^x \ln(a)$$

$$\frac{\mathrm{d}}{\mathrm{d}x}x^x = x^x(1 + \ln(x))$$

$$\frac{\mathrm{d}}{\mathrm{d}x}\ln|x| = \frac{1}{|x|}$$

$$\frac{\mathrm{d}}{\mathrm{d}x}\log_a(x) = \frac{1}{x\ln(a)}$$

Trigonometric

$$\frac{\mathrm{d}}{\mathrm{d}x}\sin(x) = \cos(x)$$

$$\frac{\mathrm{d}}{\mathrm{d}x}\cos(x) = -\sin(x)$$

$$\frac{\mathrm{d}}{\mathrm{d}x}\tan(x) = \sec^2(x)$$

$$\frac{\mathrm{d}}{\mathrm{d}x}\sec(x) = \sec(x)\tan(x)$$

$$\frac{\mathrm{d}}{\mathrm{d}x}\mathrm{cosec}(x) = -\mathrm{cosec}(x)\cot(x)$$

$$\frac{\mathrm{d}}{\mathrm{d}x}\cot(x) = -\mathrm{cosec}^2(x)$$

Inverse trigonometric

$$\frac{\mathrm{d}}{\mathrm{d}x}\sin^{-1}(x) = \frac{1}{\sqrt{1-x^2}}$$

$$\frac{\mathrm{d}}{\mathrm{d}x}\cos^{-1}(x) = -\frac{1}{\sqrt{1-x^2}}$$

$$\frac{\mathrm{d}}{\mathrm{d}x}\tan^{-1}(x) = \frac{1}{1+x^2}$$

$$\frac{\mathrm{d}}{\mathrm{d}x}\cot^{-1}(x) = -\frac{1}{1+x^2}$$

$$\frac{\mathrm{d}}{\mathrm{d}x}\sec^{-1}(x) = \frac{1}{x\sqrt{x^2-1}}$$

$$\frac{\mathrm{d}}{\mathrm{d}x}\mathrm{cosec}^{-1}(x) = -\frac{1}{x\sqrt{x^2-1}}$$

Appendix B

Long Division of Polynomials

Contents

B.1 MOTIVATION

You are likely to be familiar from the mathematics learned at school with the concept (if not the details) of *long division*. For example, the following calculations should look familiar.

$$\begin{array}{r} 56 \\ 23\overline{)1288} \\ \underline{1150} \\ 138 \\ \underline{138} \\ 0 \end{array} \qquad \begin{array}{r} 96 \\ 5316\overline{)510336} \\ \underline{478440} \\ 31896 \\ \underline{31896} \\ 0 \end{array} \qquad \begin{array}{r} 2365 \\ 42\overline{)99331} \\ \underline{84000} \\ 15331 \\ \underline{12600} \\ 2731 \\ \underline{2520} \\ 211 \\ \underline{210} \\ 1 \end{array}$$

However, you may not be aware that the same idea can be generalized for dividing polynomial expressions. This is the topic of this appendix and, as we shall see, the technique is extremely useful for factorizing polynomial expressions and simplifying rational functions.

Consider the quadratic polynomial $x^2 + x - 2$. It is clear from direct observation that $x = 1$ is a root and so $(x - 1)$ must be a factor. Using long division it is possible to "divide out" this factor and obtain,

$$\frac{x^2 + x - 2}{x - 1} = x + 2 \quad \Rightarrow \quad x^2 + x - 2 = (x - 1)(x + 2) \tag{B.1}$$

We have therefore factorized the quadratic polynomial into two linear factors.

Consider now the cubic polynomial $x^3 + 5x^2 - x - 5$. It is clear from direct observation that $x = 1$ is a root and so $(x - 1)$ must be a factor. Using long division we "divide out" this factor and obtain,

$$\frac{x^3 + 5x^2 - x - 5}{x - 1} = x^2 + 6x + 5 \quad \Rightarrow \quad x^3 + 5x^2 - x - 5 = (x - 1)(x^2 + 6x + 5)$$

We have therefore factorized the cubic polynomial into the product of a linear polynomial and a quadratic polynomial. Going further, it is clear that the quadratic factor $x^2 + 6x + 5$ has a root at $x = -1$ and so $(x + 1)$ is a factor. Therefore, we can use long division to obtain,

$$\frac{x^2 + 6x + 5}{x + 1} = x + 5$$

and determine that

$$x^3 + 5x^2 - x - 5 = (x - 1)(x + 1)(x + 5)$$

We have therefore factorized the cubic polynomial into three linear factors.

It should be clear that the ability to factorize an nth-order polynomial is very useful in root finding. For example, knowing that $x^3 + 5x^2 - x - 5 = (x - 1)(x + 1)(x + 5)$, we immediately see that $x = \pm 1$ and $x = -5$ are the three roots of $x^3 + 5x^2 - x - 5$.

But how do we perform these long divisions?

B.2 PERFORMING THE LONG DIVISION

The technique is best illustrated by example and we begin with the division given in Eq. (B.1), $\frac{x^2+x-2}{x-1}$. The first stage is to express the problem in terms of the usual long-division notation,

$$x - 1 \overline{\big) \; x^2 \;\; + x - 2}$$

which is entirely analogous to that used when dividing integers. In order to explain the process of performing the division, it is useful to use the terms *divisor* and *divided* which, in this case, are $(x - 1)$ and $x^2 + x - 2$, respectively.

The leading term of the divided is x^2 and divisor is x. The ratio of these two terms is $\frac{x^2}{x} = x$ and this is the leading term of our answer, recorded as

$$\begin{array}{r} x \qquad\quad \\ x - 1 \overline{\big) \; x^2 \;\; + x - 2} \end{array}$$

We now take this quantity, x, and multiply it by the divisor, $x - 1$, which leads to $x(x - 1) = x^2 - x$. This is then subtracted from the associated terms (i.e., those of the same polynomial order) of the divided, namely $x^2 + x$; that is, $x^2 + x - (x^2 - x) = 2x$. We write this interim calculation as follows and also bring down to next term in the divided, which is -2

$$\begin{array}{r|l} & x \\ \hline x-1) & x^2 + x - 2 \\ & -x^2 + x \\ \cline{2-2} & \phantom{-x^2 +{}} 2x - 2 \end{array}$$

We now repeat the process with $(x - 1)$ remaining as the divisor and $2x - 2$ giving a new divided. The ratio of the leading terms is $\frac{2x}{x} = 2$ and we add this to answer as

$$\begin{array}{r|l} & x + 2 \\ \hline x-1) & x^2 + x - 2 \\ & -x^2 + x \\ \cline{2-2} & \phantom{-x^2 +{}} 2x - 2 \end{array}$$

Multiplying this result, 2, by the divisor, $x - 1$, and subtracting that product from the associated terms in the new divided leads to,

$$\begin{array}{r|l} & x + 2 \\ \hline x-1) & x^2 + x - 2 \\ & -x^2 + x \\ \cline{2-2} & \phantom{-x^2 +{}} 2x - 2 \\ & - 2x + 2 \\ \cline{2-2} & \phantom{-x^2 + 2x - {}} 0 \end{array}$$

The process is now complete and we have no remainder. We have therefore demonstrated that,

$$\frac{x^2 + x - 2}{x - 1} = x + 2$$

EXAMPLE B.1

Use long division to demonstrate that

$$\frac{x^3 + 5x^2 - x - 5}{x - 1} = x^2 + 6x + 5$$

Solution

We repeat the process detailed above as follows,

$$
\begin{array}{r|l}
 & x^2 + 6x + 5 \\ \hline
x - 1 & x^3 + 5x^2 - x - 5 \\
 & -x^3 + x^2 \\ \hline
 & 6x^2 - x \\
 & -6x^2 + 6x \\ \hline
 & 5x - 5 \\
 & -5x + 5 \\ \hline
 & 0
\end{array}
$$

That is, we have obtained

$$\frac{x^3 + 5x^2 - x - 5}{x - 1} = x^2 + 6x + 5$$

as required.

EXAMPLE B.2

Simplify the rational function

$$\frac{x^4 - 27x^2 + 14x + 120}{x + 5}$$

Solution

Note that $x = -5$ is a root of both the numerator and the denominator. Therefore, $(x + 5)$ is a factor of the numerator and can be "divided out." The division is performed as follows,

$$
\begin{array}{r|l}
 & x^3 - 5x^2 - 2x + 24 \\ \hline
x + 5 & x^4 - 27x^2 + 14x + 120 \\
 & -x^4 - 5x^3 \\ \hline
 & -5x^3 - 27x^2 \\
 & 5x^3 + 25x^2 \\ \hline
 & -2x^2 + 14x \\
 & 2x^2 + 10x \\ \hline
 & 24x + 120 \\
 & -24x - 120 \\ \hline
 & 0
\end{array}
$$

That is,

$$\frac{x^4 - 27x^2 + 14x + 120}{x + 5} = x^3 - 5x^2 - 2x + 24$$

Note that some care is required since the numerator has no x^3 term. As illustrated in the division process, we must treat it as $x^4 + 0x^3 - 27x^2 + 14x + 120$ and proceed in the standard way.

EXAMPLE B.3

Factorize the cubic polynomial

$$x^3 + 5x^2 + 7x + 2$$

Solution

We note that $x = -2$ is a root and so $(x + 2)$ is a factor. Dividing this out leads to,

$$\begin{array}{r|l} & x^2 + 3x + 1 \\ \hline x + 2) & x^3 + 5x^2 + 7x + 2 \\ & -x^3 - 2x^2 \\ \hline & 3x^2 + 7x \\ & -3x^2 - 6x \\ \hline & x + 2 \\ & -x - 2 \\ \hline & 0 \end{array}$$

Note that $x^2 + 3x + 1$ has no real roots and so, unless we wish to work in terms of complex numbers, we obtain the factorization

$$x^3 + 5x^2 + 7x + 2 = (x + 2)(x^2 + 3x + 1)$$

EXAMPLE B.4

Simplify the rational function

$$\frac{6x^3 - 9x^2 - 9x + 6}{2x - 1}$$

Solution

We note that $x = 0.5$ is a root of the numerator and so $2x - 1$ is a factor. Proceeding as before, leads to

$$
\begin{array}{r|l}
 & 3x^2 \quad - 3x - 6 \\
\hline
2x - 1) & 6x^3 - 9x^2 \quad - 9x + 6 \\
 & -6x^3 + 3x^2 \\
\hline
 & \quad -6x^2 \quad - 9x \\
 & \quad 6x^2 \quad - 3x \\
\hline
 & \qquad -12x + 6 \\
 & \qquad 12x - 6 \\
\hline
 & \qquad\qquad 0
\end{array}
$$

That is,

$$\frac{6x^3 - 9x^2 - 9x + 6}{2x - 1} = 3x^2 - 3x - 6 = 3(x^2 - x - 2)$$

Furthermore, we note that $x^2 - x - 2 = (x + 1)(x - 2)$ and so obtain,

$$\frac{6x^3 - 9x^2 - 9x + 6}{2x - 1} = 3(x + 1)(x - 2)$$

EXAMPLE B.5

Simplify the rational function

$$\frac{2x^3 - 9x^2 + 15}{x - 1}$$

Solution

We note that $x = 1$ is *not* a root of the numerator and so $(x - 1)$ is *not* a factor of $2x^3 - 9x^2 + 15$. We therefore expect a remainder at the end of the division process. We proceed as follows,

$$
\begin{array}{r|l}
 & 2x^2 - 7x \quad - 7 \\
\hline
x - 1) & 2x^3 - 9x^2 \qquad + 15 \\
 & -2x^3 + 2x^2 \\
\hline
 & \quad -7x^2 \\
 & \quad 7x^2 - 7x \\
\hline
 & \qquad -7x + 15 \\
 & \qquad 7x \quad - 7 \\
\hline
 & \qquad\qquad 8
\end{array}
$$

Note that 8 cannot be divided by x and so the division ends with a remainder of 8. We therefore express the rational function as

$$\frac{2x^3 - 9x^2 + 15}{x - 1} = 2x^2 - 7x - 7 + \frac{8}{x - 1}$$

EXAMPLE B.6

Use long division to demonstrate that

$$\frac{x^3 + 3x^2 - 4x - 12}{x^2 + 5x + 6} = x - 2$$

Solution

This requires a minor extension of the process illustrated so far. In particular, we have more terms to subtract at each stage.

$$\begin{array}{r r}
 & x \quad - 2 \\ \cline{2-2}
x^2 + 5x + 6) & x^3 + 3x^2 \quad - 4x - 12 \\
 & -x^3 - 5x^2 \quad - 6x \quad\quad \\ \cline{2-2}
 & -2x^2 - 10x - 12 \\
 & 2x^2 + 10x + 12 \\ \cline{2-2}
 & 0
\end{array}$$

We therefore see that,

$$\frac{x^3 + 3x^2 - 4x - 12}{x^2 + 5x + 6} = x - 2$$

B.3 QUESTIONS

Question B.1. Factorize the quartic polynomial

$$x^4 - 18x^3 + 97x^2 - 180x + 100$$

Question B.2. Factorize the quintic polynomial

$$x^5 + 14x^4 + 67x^3 + 110x^2 - 32x - 160$$

Question B.3. Simplify the rational function

$$\frac{x^4 - 4x^3 + 3x^2 - 11x - 4}{x - 2}$$

Question B.4. Simplify the rational function

$$\frac{12x^3 - 3x^2 - 12x + 3}{4x - 1}$$

Question B.5. Perform the long division

$$\frac{x^4 - 1}{x - 1}$$

Question B.6. Perform the long division

$$\frac{x^4 + 2x^3 - 3x^2 - 8x - 4}{x^2 - x - 2}$$

Question B.7. Perform the long division

$$\frac{x^4 + 2x^3 - 3x^2 - 8x + 4}{x^2 - x - 2}$$

Question B.8. Perform the long division

$$\frac{x^3 + 3x^2 + 5x + 3}{x^2 + 2x + 1}$$

B.4 SOLUTIONS

Solution B.1. The first factor, $(x - 1)$, is obtained from the observation that $x = 1$ is a root of the quartic. Dividing this out leads to,

$$
\begin{array}{r|rrrrr}
 & & x^3 & -17x^2 & +80x & -100 \\
\hline
x - 1 & x^4 & -18x^3 & +97x^2 & -180x & +100 \\
 & -x^4 & +x^3 & & & \\
\cline{2-3}
 & & -17x^3 & +97x^2 & & \\
 & & 17x^3 & -17x^2 & & \\
\cline{3-4}
 & & & 80x^2 & -180x & \\
 & & & -80x^2 & +80x & \\
\cline{4-5}
 & & & & -100x & +100 \\
 & & & & 100x & -100 \\
\cline{5-6}
 & & & & & 0
\end{array}
$$

That is,

$$x^4 - 18x^3 + 97x^2 - 180x + 100 = (x-1)(x^3 - 17x^2 + 80x - 100)$$

Note that $x = 2$ is a root of $x^3 - 17x^2 + 80x - 100$ and so

$$\begin{array}{r|l}
 & x^2 - 15x + 50 \\
\hline
x-2 & x^3 - 17x^2 + 80x - 100 \\
 & -x^3 + 2x^2 \\
\hline
 & -15x^2 + 80x \\
 & 15x^2 - 30x \\
\hline
 & 50x - 100 \\
 & -50x + 100 \\
\hline
 & 0
\end{array}$$

The resulting quadratic can be factorized directly as $(x-10)(x-5)$ and so,

$$x^4 - 18x^3 + 97x^2 - 180x + 100 = (x-1)(x-2)(x-10)(x-5)$$

Solution B.2. It is clear that $x = 1$ is a root and so $(x-1)$ is a factor. Dividing this out leads to

$$\begin{array}{r|l}
 & x^4 + 15x^3 + 82x^2 + 192x + 160 \\
\hline
x-1 & x^5 + 14x^4 + 67x^3 + 110x^2 - 32x - 160 \\
 & -x^5 + x^4 \\
\hline
 & 15x^4 + 67x^3 \\
 & -15x^4 + 15x^3 \\
\hline
 & 82x^3 + 110x^2 \\
 & -82x^3 + 82x^2 \\
\hline
 & 192x^2 - 32x \\
 & -192x^2 + 192x \\
\hline
 & 160x - 160 \\
 & -160x + 160 \\
\hline
 & 0
\end{array}$$

Note that $x = -2$ is a root of the resulting quartic polynomial, and so

$$
\begin{array}{r l}
 & x^3 + 13x^2 \ + 56x \ + 80 \\
\cline{2-2}
x+2) & x^4 + 15x^3 + 82x^2 + 192x + 160 \\
 & -x^4 \ - 2x^3 \\
\cline{2-2}
 & 13x^3 + 82x^2 \\
 & -13x^3 - 26x^2 \\
\cline{2-2}
 & 56x^2 + 192x \\
 & -56x^2 - 112x \\
\cline{2-2}
 & 80x + 160 \\
 & -80x - 160 \\
\cline{2-2}
 & 0
\end{array}
$$

Note that $x = -4$ is a root of the resulting cubic polynomial, and so

$$
\begin{array}{r l}
 & x^2 \ + 9x + 20 \\
\cline{2-2}
x+4) & x^3 + 13x^2 + 56x + 80 \\
 & -x^3 \ - 4x^2 \\
\cline{2-2}
 & 9x^2 + 56x \\
 & -9x^2 - 36x \\
\cline{2-2}
 & 20x + 80 \\
 & -20x - 80 \\
\cline{2-2}
 & 0
\end{array}
$$

The resulting quadratic can be factorized directly as $(x + 5)(x + 4)$ and so,

$$x^5 + 14x^4 + 67x^3 + 110x^2 - 32x - 160 = (x - 1)(x + 2)(x + 5)(x + 4)^2$$

Solution B.3. Note that $x = 2$ is *not* a root of the numerator and so $(x - 2)$ is *not* a factor; we therefore expect a remainder from the division.

$$
\begin{array}{r|rrrrr}
\multicolumn{1}{r}{} & & x^3 & -2x^2 & -x & -13 \\ \cline{2-6}
x-2 & x^4 & -4x^3 & +3x^2 & -11x & -4 \\
\multicolumn{1}{r}{} & -x^4 & +2x^3 & & & \\ \cline{2-3}
\multicolumn{1}{r}{} & & -2x^3 & +3x^2 & & \\
\multicolumn{1}{r}{} & & 2x^3 & -4x^2 & & \\ \cline{3-4}
\multicolumn{1}{r}{} & & & -x^2 & -11x & \\
\multicolumn{1}{r}{} & & & x^2 & -2x & \\ \cline{4-5}
\multicolumn{1}{r}{} & & & & -13x & -4 \\
\multicolumn{1}{r}{} & & & & 13x & -26 \\ \cline{5-6}
\multicolumn{1}{r}{} & & & & & -30
\end{array}
$$

That is,

$$\frac{x^4 - 4x^3 + 3x^2 - 11x - 4}{x - 2} = x^3 - 2x^2 - x - 13 - \frac{30}{x - 2}$$

Solution B.4. We note that $x = 0.25$ is a root of the numerator and so $(4x - 1)$ is a factor.

$$
\begin{array}{r|rrrr}
\multicolumn{1}{r}{} & & 3x^2 & & -3 \\ \cline{2-5}
4x-1 & 12x^3 & -3x^2 & -12x & +3 \\
\multicolumn{1}{r}{} & -12x^3 & +3x^2 & & \\ \cline{2-3}
\multicolumn{1}{r}{} & & & -12x & +3 \\
\multicolumn{1}{r}{} & & & 12x & -3 \\ \cline{4-5}
\multicolumn{1}{r}{} & & & & 0
\end{array}
$$

That is,

$$\frac{12x^3 - 3x^2 - 12x + 3}{4x - 1} = 3(x^2 - 1)$$

Solution B.5. We consider the numerator as $x^4 + 0x^3 + 0x^2 + 0x - 1$ and proceed as

$$
\begin{array}{r|l}
 & x^3 + x^2 + x + 1 \\
\hline
x - 1 & x^4 \qquad\qquad\qquad - 1 \\
 & -x^4 + x^3 \\
\hline
 & \quad x^3 \\
 & \quad -x^3 + x^2 \\
\hline
 & \qquad x^2 \\
 & \qquad -x^2 + x \\
\hline
 & \qquad\quad x - 1 \\
 & \qquad\quad -x + 1 \\
\hline
 & \qquad\qquad 0
\end{array}
$$

That is,

$$\frac{x^4 - 1}{x - 1} = x^3 + x^2 + x + 1$$

Solution B.6. We have

$$
\begin{array}{r|l}
 & x^2 + 3x + 2 \\
\hline
x^2 - x - 2 & x^4 + 2x^3 - 3x^2 - 8x - 4 \\
 & -x^4 + x^3 + 2x^2 \\
\hline
 & \quad 3x^3 - x^2 - 8x \\
 & \quad -3x^3 + 3x^2 + 6x \\
\hline
 & \qquad 2x^2 - 2x - 4 \\
 & \qquad -2x^2 + 2x + 4 \\
\hline
 & \qquad\qquad 0
\end{array}
$$

That is,

$$\frac{x^4 + 2x^3 - 3x^2 - 8x - 4}{x^2 - x - 2} = x^2 + 3x + 2$$

Solution B.7. We have

$$
\begin{array}{r r}
 & x^2 + 3x + 2 \\ \cline{2-2}
x^2 - x - 2) & x^4 + 2x^3 - 3x^2 - 8x + 4 \\
 & - x^4 \;\; + x^3 + 2x^2 \qquad\qquad \\ \cline{2-2}
 & 3x^3 \;\; - x^2 - 8x \quad\;\; \\
 & - 3x^3 + 3x^2 + 6x \quad\;\; \\ \cline{2-2}
 & 2x^2 - 2x + 4 \\
 & - 2x^2 + 2x + 4 \\ \cline{2-2}
 & 8
\end{array}
$$

That is,

$$
\frac{x^4 + 2x^3 - 3x^2 - 8x + 4}{x^2 - x - 2} = x^2 + 3x + 2 + \frac{8}{x^2 - x - 2}
$$

Solution B.8. We have

$$
\begin{array}{r r}
 & x + 1 \\ \cline{2-2}
x^2 + 2x + 1) & x^3 + 3x^2 + 5x + 3 \\
 & - x^3 - 2x^2 \;\; - x \qquad \\ \cline{2-2}
 & x^2 + 4x + 3 \\
 & - x^2 - 2x - 1 \\ \cline{2-2}
 & 2x + 2
\end{array}
$$

That is,

$$
\frac{x^3 + 3x^2 + 5x + 3}{x^2 + 2x + 1} = x + 1 + \frac{2x + 2}{x^2 + 2x + 1}
$$

Bibliography

Although this book is intended to be self-contained, reading around a subject is always recommended. I provide a list of texts that I have found useful during my long study of mathematics and fundamental actuarial science since high school. This list should not be considered as exhaustive and the reader is encouraged to find alternative texts that suit their own style of learning.

Atkinson, K., 1993. Elementary Numerical Analysis. Wiley, New Delhi.

Binmore, K., Davies, J., 2002. Calculus: Concepts and Methods. Cambridge University Press, Cambridge.

Bostock, L., Chandler, S., 1994. The Core Course for A-Level. Stanley Thornes, Cheltenham.

Garrett, S.J., 2013. An Introduction to the Mathematics of Finance: A Deterministic Approach. Butterworth-Heinemann, Oxford.

Institute and Faculty of Actuaries, 2015. Subject CT4 Models Core Reading. IFoA, Oxford.

Kreyszig, E., 2006. Advanced Engineering Mathematics, ninth ed. Wiley, Hoboken, NJ.

Matthews, J.H., 1987. Numerical Methods for Mathematics, Science and Engineering. Prentice-Hall, Englewood Cliffs, NJ.

Nicholson, W.K., 1995. Linear Algebra with Applications. PWS Publishing Company, Boston.

Soper, J., 2003. Mathematics for Economics and Business. Blackwell, Oxford.

Spivak, M., 2006. Calculus. Cambridge University Press, Cambridge.

Steffensen, A.R., Johnson, L.M., 1991. Introductory Algebra. HarperCollins, New York, NY.

Ummer, E.K., 2012. Basic Mathematics for Economics, Business and Finance. Routledge, Abington.

Waldron, P., Harrison, M., 2011. Mathematics for Economics and Finance. Routledge, Abingdon.

INDEX

Note: Page numbers followed by *f* indicate figures and *t* indicate tables.

Q

R

S

Printed in the United States
By Bookmasters